Sensors and Applications in Measuring and Automation Control Systems

Sergey Y. Yurish

Editor

Sensors and Applications in Measuring and Automation Control Systems

Book Series: Advances in Sensors: Reviews, Vol. 4

International Frequency Sensor Association Publishing

Sergey Y. Yurish
Editor

Sensors and Applications in Measuring and Automation Control Systems
Advances in Sensors: Reviews, Vol. 4

Published by International Frequency Sensor Association (IFSA) Publishing, S. L., 2016
E-mail (for print book orders and customer service enquires): ifsa.books@sensorsportal.com

Visit our Home Page on http://www.sensorsportal.com

ISBN: 978-84-617-7596-5
e-ISBN: 978-84-617-7597-2
BN-20161230-XX
BIC: TJFC

Acknowledgments

As Editor I would like to express my undying gratitude to all authors, editorial staff, reviewers and others who actively participated in this book. We want also to express our gratitude to all their families, friends and colleagues for their help and understanding.

Contents

Chapter 11
Design, Implementation and Characterization of Time-to-Digital Converter on Low-Cost FPGA 205

Chapter 12
New Approaches to Extend Lifetime in Wireless Sensor Network Based on Optimal Placement of Sensor Nodes and Using Duty Cycle Technique...........231

Chapter 16
Multifunction Sensing System for Wireless Monitoring of Chronic Wounds in Healthcare...315

Chapter 17
Direction of Arrival Using Smart Antenna Array..329

Chapter 18
UAV Control Based on Optimized Neural Network 355

Chapter 21
Artificial Intelligence Based Medical Computer Vision Image Registration System and Algorithm..**405**

Chapter 22
SmartLab Magnetic: A Modern Student Laboratory on Magnetic Materials and Circuits..**425**

Chapter 23
Recent Advances in Characterization of Sol-gel Based Materials for Sensor Applications **441**

Chapter 24

Contributors

Murtadha Al-Mamury Centre for Electronic Systems Research, College of Engineering Design and Physical Science, Brunel University, United Kingdom
Collage of Agriculture, Kerbala University, Iraq

Hamed Al-Raweshidy Centre for Electronic Systems Research, College of Engineering Design and Physical Science, Brunel University, United Kingdom

Wamadeva Balachandran Centre for Electronic Systems Research, College of Engineering Design and Physical Science, Brunel University, United Kingdom

Marko Beko Universidade Lusófona de Humanidades e Tecnologias, Lisbon, Portugal
CTS, UNINOVA – Campus FCT/UNL, Caparica, Portugal

Larissa Brentano Capeletti Institute of Chemistry, Universidade Federal do Rio Grande do Sul, Av. Bento Gonçalves 9500 - CEP 91501-970, Porto Alegre, RS, Brazil

Bri Seddik Material and Instrumentation (MIN), Electrical Engineering Department, High school of technology ESTM, Moulay Ismail University, Morocco

Francisco Bulnes Research Department in Mathematics and Engineering, GI-TESCHA, México

Cinzia Caliendo Institute for Photonics and Nanotechnologies, IFN-CNR, Via Cineto Romano 42, 00156 Rome, Italy

S. Cardoso INESC Microsystems and Nanotechnologies and IN, Rua Alves Redol 9, 1000-029, Lisbon, Portugal

Luan Casagrande Universidade Federal de Santa Catarina (UFSC), Florianópolis, Brazil

Surya Venkatasekhar Cheemalapati Department of Chemical and Biomedical Engineering, University of South Florida, Tampa, FL, USA

Sudhir Cherukulappurath Physics Department, Goa University, Goa, India

João Costa Telecommunication and Computer Dept. ISEL, R. Conselheiro Emídio Navarro, 1959-007 Lisboa, Portugal
CTS-UNINOVA, Quinta da Torre, Monte da Caparica, 2829-516, Caparica, Portugal

Yuri Crotti Universidade Federal de Santa Catarina (UFSC), Florianópolis, Brazil

Paulo E. Cruvinel Embrapa Instrumentation (CNPDIA), Rua XV de Novembro 1452, 13560-970 São Carlos, SP, Brazil

Renan Cunha Universidade Federal de Santa Catarina (UFSC), Florianópolis, Brazil

Dadouche F. ICube, UMR 7357, Université de Strasbourg and CNRS, 23, Rue du Loess, 67037 Strasbourg, France

Viviane Dalmoro Institute of Chemistry, Universidade Federal do Rio Grande do Sul, Av. Bento Gonçalves 9500 - CEP 91501-970, Porto Alegre, RS, Brazil

Rui Dinis Instituto de Telecomunicações, Lisbon, Portugal
DEE/FCT/UNL, Caparica, Portugal

Glib Dorozinsky V. Lashkaryov Institute of Semiconductor Physics NAS of Ukraine, Kyiv, Ukraine

Alessandro Fantoni Telecommunication and Computer Dept. ISEL, R. Conselheiro Emídio Navarro, 1959-007 Lisboa, Portugal, CTS-UNINOVA, Quinta da Torre, Monte da Caparica, 2829-516, Caparica, Portugal

Kleber R. Felizardo Embrapa Instrumentation (CNPDIA), Rua XV de Novembro 1452, 13560-970 São Carlos, SP, Brazil, Department of Electrical and Computer Engineering, University of São Paulo (EESC/USP), Avenida Trabalhador São-Carlense, 400, 13566-590 São Carlos - SP, Brazil

Orlando Frazão INESC TEC - INstituto de Engenharia de Sistemas e Computadores - Tecnologia e Ciência, Rua do Campo Alegre, 687, 4169-007 Porto, Portugal

I. Giouroudi Institute of Sensor and Actuator Systems, Vienna University of Technology, Vienna, Austria, Biophysics Institute, Department of Nanobiotechnology, BOKU - University of Natural Resources and Life Sciences, Muthgasse 11/II, 1190 Vienna, Austria

Soumyajit Goswami IBM India Private Limited, Saltlake, Sector V, Kolkata-700091, WestBengal, India

Kunihiko Gotoh Tokyo Institute of Technology, Nagatsutacho 4259, Midori-ku, Kanagawa, 226–8503, Japan

Vilson Gruber Universidade Federal de Santa Catarina (UFSC), Brazil

Sarah de Rezende Guerra Universidade Federal de Santa Catarina (UFSC), Florianópolis, Brazil

Bamba Gueye University Cheikh Anta Diop of Dakar, Senegal

Muhammad Hamidullah Institute for Photonics and Nanotechnologies, IFN-CNR, Via Cineto Romano 42, 00156 Rome, Italy

Alex Hariz School of Engineering, University of South Australia, Adelaide, Australia

Ihedrane Mohammed Amine Material and Instrumentation (MIN), Electrical Engineering Department, High school of technology ESTM, Moulay Ismail University, Morocco

Noboru Ishihara Tokyo Institute of Technology, Nagatsutacho 4259, Midori-ku, Kanagawa, 226–8503, Japan

Hiroyuki Ito Tokyo Institute of Technology, Nagatsutacho 4259, Midori-ku, Kanagawa, 226–8503, Japan

G. Kokkinis Institute of Sensor and Actuator Systems, Vienna University of Technology, Vienna, Austria

Pekka Koskela VTT Technical Research Centre of Finland Ltd., Espoo, Finland

Koji Kurihara Network Systems Laboratory, Fujitsu Laboratories Ltd., Kamikodanaka 4–1–1, Kawasaki Nakahara-ku, Kanagawa, 211–8588, Japan

Jong-Ha Lee Keimyung University, School of Medicine, Dept. of Biomedical Engineering, Daegu, South Korea

Léonard J. ICube, UMR 7357, Université de Strasbourg and CNRS, 23, Rue du Loess, 67037 Strasbourg, France

Pascal Lorenz University of Mulhouse-Haute Alsace, France

Paula Louro Telecommunication and Computer Dept. ISEL, R. Conselheiro Emídio Navarro, 1959-007 Lisboa, Portugal, CTS-UNINOVA, Quinta da Torre, Monte da Caparica, 2829-516, Caparica, Portugal

Mikko Majanen VTT Technical Research Centre of Finland Ltd., Espoo, Finland

Malass I. ICube, UMR 7357, Université de Strasbourg and CNRS, 23, Rue du Loess, 67037 Strasbourg, France

Nadarajah Manivannan Centre for Electronic Systems Research, College of Engineering Design and Physical Science, Brunel University, United Kingdom

Roderval Marcelino Universidade Federal de Santa Catarina (UFSC), Brazil

Isaías Martínez Electronic Engineering Division, TESCHA, México

Javier Martinez-Roman Universitat Politècnica de Valencia, Instituto de Ingeniería Energética, C. Vera s/n 46022, Valencia, Spain

Volodymyr Maslov V. Lashkaryov Institute of Semiconductor Physics NAS of Ukraine, Kyiv, Ukraine

Kazuya Masu Tokyo Institute of Technology, Nagatsutacho 4259, Midori-ku, Kanagawa, 226–8503, Japan

Heitor V. Mercaldi Embrapa Instrumentation (CNPDIA), Rua XV de Novembro 1452, 13560-970 São Carlos, SP, Brazil, Department of Electrical and Computer Engineering, University of São Paulo (EESC/USP), Avenida Trabalhador São-Carlense, 400, 13566-590 São Carlos, SP, Brazil

Emmanuel Mukwevho North West University, 1, Albert Luthuli & Cnr University road, Private Bag X 2046, Mmabatho, 2735, South Africa

Bonex Mwakikunga DST/CSIR National Centre for Nano-Structured Materials, PO Box 395, Pretoria 0001, South Africa

Diery Ngom University Alioune Diop of Bambey, Senegal

Kohei Ohba Tokyo Institute of Technology, Nagatsutacho 4259, Midori-ku, Kanagawa, 226–8503, Japan

Vilma A. Oliveira Department of Electrical and Computer Engineering, University of São Paulo (EESC/USP), Avenida Trabalhador São-Carlense, 400, 13566-590 São Carlos, SP, Brazil

Victor Ovchinnikov Department of Aatlo Nanofab, School of Electrical Engineering, Aalto University, Espoo, Finland

Elmer A. G. Peñaloza Embrapa Instrumentation (CNPDIA), Rua XV de Novembro 1452, 13560-970 São Carlos, SP, Brazil, Department of Electrical and Computer Engineering, University of São Paulo (EESC/USP), Avenida Trabalhador São-Carlense, 400, 13566-590 São Carlos, SP, Brazil

Manuel Pineda-Sanchez Universitat Politècnica de Valencia, Instituto de Ingeniería Energética, C. Vera s/n 46022, Valencia, Spain

Ruben Puche-Panadero Universitat Politècnica de Valencia, Instituto de Ingeniería Energética, C. Vera s/n 46022, Valencia, Spain

Anna Pyayt Department of Chemical and Biomedical Engineering, University of South Florida,Tampa, FL, USA

Isabel Rodrigues Telecommunication and Computer Dept. ISEL, R. Conselheiro Emídio Navarro, 1959-007 Lisboa, Portugal

Valentine Saasa DST/CSIR National Centre for Nano-Structured Materials, PO Box 395, Pretoria 0001, South Africa

Angel Sapena-Baño Universitat Politècnica de Valencia, Instituto de Ingeniería Energética, C. Vera s/n 46022, Valencia Spain

Arghya Sarkar MCKV Institute of Engineering, 243, G. T. Road (N), Liluah, Howrah, 711204, India

Samarjit Sengupta Department of Applied physics, University of Calcutta, Kolkata, 700001, India

Susana Silva INESC TEC - INstituto de Engenharia de Sistemas e Computadores - Tecnologia e Ciência, Rua do Campo Alegre, 687, 4169-007 Porto, Portugal

Vitor Silva Telecommunication and Computer Dept. ISEL, R. Conselheiro Emídio Navarro, 1959-007 Lisboa, Portugal, CTS-UNINOVA, Quinta da Torre, Monte da Caparica, 2829-516, Caparica, Portugal

Skilitsi A. ICube, UMR 7357, Université de Strasbourg and CNRS, 23, Rue du Loess, 67037 Strasbourg, France

Takashi Suganuma Tokyo Institute of Technology, Nagatsutacho 4259, Midori-ku, Kanagawa, 226–8503, Japan

Slavisa Tomic ISR/IST, LARSyS, Lisbon, Portugal

Milan Tuba Faculty of Computer Science, John Naisbitt University, Belgrade, Serbia

Turko T. ICube, UMR 7357, Université de Strasbourg and CNRS, 23, Rue du Loess, 67037 Strasbourg, France

Uhring W. ICube, UMR 7357, Université de Strasbourg and CNRS, 23, Rue du Loess, 67037 Strasbourg, France

Yuriy Ushenin V. Lashkaryov Institute of Semiconductor Physics NAS of Ukraine, Kyiv, Ukraine

Mikko Valta VTT Technical Research Centre of Finland Ltd., Espoo, Finland

Manuel Augusto Vieira Telecommunication and Computer Dept. ISEL, R. Conselheiro Emídio Navarro, 1959-007 Lisboa, Portugal, CTS-UNINOVA, Quinta da Torre, Monte da Caparica, 2829-516, Caparica, Portugal

Manuela Vieira Telecommunication and Computer Dept. ISEL, R. Conselheiro Emídio Navarro, 1959-007 Lisboa, Portugal, CTS-UNINOVA, Quinta da Torre, Monte da Caparica, 2829-516, Caparica, Portugal
DEE-FCT-UNL, Quinta da Torre, Monte da Caparica, 2829-516, Caparica, Portugal

Koichiro Yamashita Network Systems Laboratory, Fujitsu Laboratories Ltd.Kamikodanaka 4–1–1, Kawasaki Nakahara-ku, Kanagawa, 211–8588, Japan

Yoshihiro Yoneda Tokyo Institute of Technology, Nagatsutacho 4259, Midori-ku, Kanagawa, 226–8503, Japan

Omar Zamudio Electronic Engineering Division, TESCHA, México

João Henrique Zimnoch dos Santos Institute of Chemistry, Universidade Federal do Rio Grande do Sul, Av. Bento Gonçalves 9500 - CEP 91501-970, Porto Alegre, RS, Brazil

Preface

It is my great pleasure to present the fourth volume from our popular Book Series 'Advances in Sensors: Reviews' started by the IFSA Publishing in 2012. The Vol. 4 of this book series is now published as an Open Access Book in order to significantly increase the reach and impact of this volume, which also published in two formats: electronic (pdf) with full-color illustrations and print (paperback).

The fourth volume titled 'Sensors and Applications in Measuring and Automation Control Systems' contains twenty four chapters with sensor related state-of-the-art reviews and descriptions of latest advances in sensor related area written by 81 authors from academia and industry from 5 continents and 20 countries: Australia, Austria, Brazil, Finland, France, Japan, India, Iraq, Italia, México, Morocco, Portugal, Senegal, Serbia, South Africa, South Korea, Spain, UK, Ukraine and USA. Many of contributors of this Book Series are members of the editorials board of different journals related to the field and some of them are International Frequency Sensor Association (IFSA) members.

Like the first three volumes of this Book Series, the fourth volume also has been organized by topics of high interest. In order to offer a fast and easy reading of each topic, every chapter in this book is independent and self-contained. All chapters have the same structure: first an introduction to specific topic under study; second particular field description including sensing or/and measuring applications. Each of chapter is ending by well selected list of references with books, journals, conference proceedings and web sites.

This book ensures that our readers will stay at the cutting edge of the field and get the right and effective start point and road map for the further researches and developments. By this way, they will be able to save more time for productive research activity and eliminate routine work.

Built upon the Book Series 'Advances in Sensors: Reviews' - a premier sensor review source, it presents an overview of highlights in the field and becomes. Coverage includes current developments in physical sensors and transducers, chemical sensors, biosensors, sensing materials, signal conditioning, energy harvesters and sensor networks. Sure, we would have liked to include even more topics, but it is difficult to cover everything due to reasonable practical restrictions. With this unique combination of information in each volume, the 'Advances in Sensors: Reviews' Book Series will be of value for scientists and engineers in industry and at universities, to sensors developers, distributors, and users.

I hope that readers enjoy this book and that can be a valuable tool for those who involved in research and development of various physical sensors, chemical sensors, biosensors and measuring systems.

I shall gratefully receive any advices, comments, suggestions and notes from readers to make the next volumes of Advances in Sensors: Reviews book Series very interesting and useful.

Dr. Sergey Y. Yurish

Editor
IFSA Publishing   *Barcelona, Spain*

Chapter 1
Irregular Arrays of Silver Nanoparticles on Multilayer Substrates and Their Optical Properties

Victor Ovchinnikov

1.1. Introduction

Plasmonic nanostructures are widely used in sensors, metamaterials, solar cells, photonics and spectroscopy [1-5]. Effective application of these structures is based on localized surface plasmon resonance (LSPR) demonstrated in ultraviolet, visible and infrared. The wavelength of LSPR depends on material and geometry of nanostructures, their interaction with each other and electromagnetic properties of environment, including substrate and capping layers. Despite on near field nature, LSPR can be observed in far field optical measurements due to variation in optical properties of the studied structures. Extinction is the most popular method of LSPR registration due to its simple implementation and straightforward interpretation, i.e., maximum position and width of extinction peak correspond LSPR wavelength and damping, respectively. However, extinction can be measured only for non-opaque structures, e.g., for nanostructures on transparent substrates or for plasmonic colloidal particles. At the same time, most of sensors are realized on opague substrates. Furthermore, extinction spectra are not effective for overlapped peaks, when spectral deconvolution is not obvious.

LSPRs on opaque substrates can be visualized by different kinds of reflection and scattering measurements. However, peaks and troughs of specular reflectance do not correspond to LSPRs and spectrum analysis becomes problematic. Scattering measurements require special arrangement of light illumination (dark field) to separate low scattering signal from strong reflection background. It limits range of measured samples by plasmonic nanostructures on substrate surface and, for example, plasmonic nanoparticles inside of dielectric matrix cannot be analyzed. In case of correlated scattering centers, i.e., when array of coupled nanostructures is analyzed, correlation between scattered peaks and LSPRs is broken and LSPRs should be observed in specular reflectance. Additionally, scattering results are obtained in arbitrary units and cannot be

Victor Ovchinnikov
Department of Aatlo Nanofab, School of Electrical Engineering, Aalto University, Espoo, Finland

used for comparison of different experiments. This happens, due to the problems with measurement of scattered reference spectrum for calibration procedure. In contrast, reflection reference spectrum can be easily obtained for any material. Combination of total reflectance and diffuse reflectance is especially useful for analyzing plasmon structures, because the last one provides measurement of scattering in absolute values.

In this chapter, we continue our study of reflection from irregular arrays of nanoparticles [6, 7] and propose to use total reflectance for identification of LSPR on opaque and transparent substrates. It is demonstrated that wavelength position of LSPR correlates with peak and trough of reflectance in a clear way. Furthermore, overlapped peaks manifest themselves separately in reflectance spectra and can be easily distinguished.

This chapter is organized in a following way. In the subsequent Section 1.2, the details of sample preparation and the measurement procedures are presented. In Section 1.3, the results of the work are demonstrated for quartz and silicon substrates by scanning electron microscope (SEM) images as well as reflection and extinction spectra of the fabricated samples. The effect of additional dielectric layers on reflectance of silver nanoparticles is discussed in Section 1.4. Specific features appearing in UV part of reflection spectra are considered in Section 1.5. In Section 1.6, the conclusions are drawn.

1.2. Experimental

Quartz or crystalline Si wafers (100 mm in diameter, 0.5-mm-thick) were used as substrates. An Al_2O_3 layer was grown on the substrate by atomic layer deposition (ALD) and a SiO_2 layer was created by thermal oxidation of the Si wafer. Silver layers with a thickness of 15 nm were deposited by electron-beam evaporation with deposition rate 0.5 nm/s. Nanoparticle arrays were fabricated by ion beam mixing (IBM) or annealing of silver films. In case of IBM, Ag films were irradiated by 400 keV Ar ions at normal incidence and at low (1×10^{16} Ar/cm^{-2}) or high (2×10^{16} Ar/cm^{-2}) ion fluences to produce the nanoparticles as reported elsewhere [8]. Some samples were processed by IBM with Xe ions at dose 6×10^{15} Xe/cm^{-2}. In the case of annealing, silver films were heated at 350 °C during 10 minutes. Annealing was done in a diffusion furnace in nitrogen ambient. A 13.56 MHz driven parallel plate reactor was used for dry etching. To cover the nanoparticles with a SiO_2 layer plasma enhanced chemical vapor deposition (PECVD) technique was applied. Metal deposition and ion irradiation were performed at room temperature. To avoid silver oxidation, ALD and PECVD processes were run at reduced temperatures 200 °C and 170 °C, respectively. Further details about samples and processing can be found elsewhere [9, 10]. To examine the nanoparticle formation in the structures created, the images of the samples were taken with a Zeiss Supra 40 field emission scanning electron microscope. Three such images, depicting effect of IBM dose and mixing ions are shown in Fig. 1.1 (a, b, c). One more image of the sample prepared by annealing of silver film is presented in Fig. 1.1 (d). It cost to note, that IBM particles are partly submerged in the substrate (see inserts in Fig. 1.1). This burrowing of nanoparticle is larger for small nanoparticles, for high dose (Fig. 1.1(b)) of IBM and for Xe mixed samples (Fig. 1.1 (c)) The details of nanoparticle size distribution and sample surface morphology can be found elsewhere [8, 11].

26

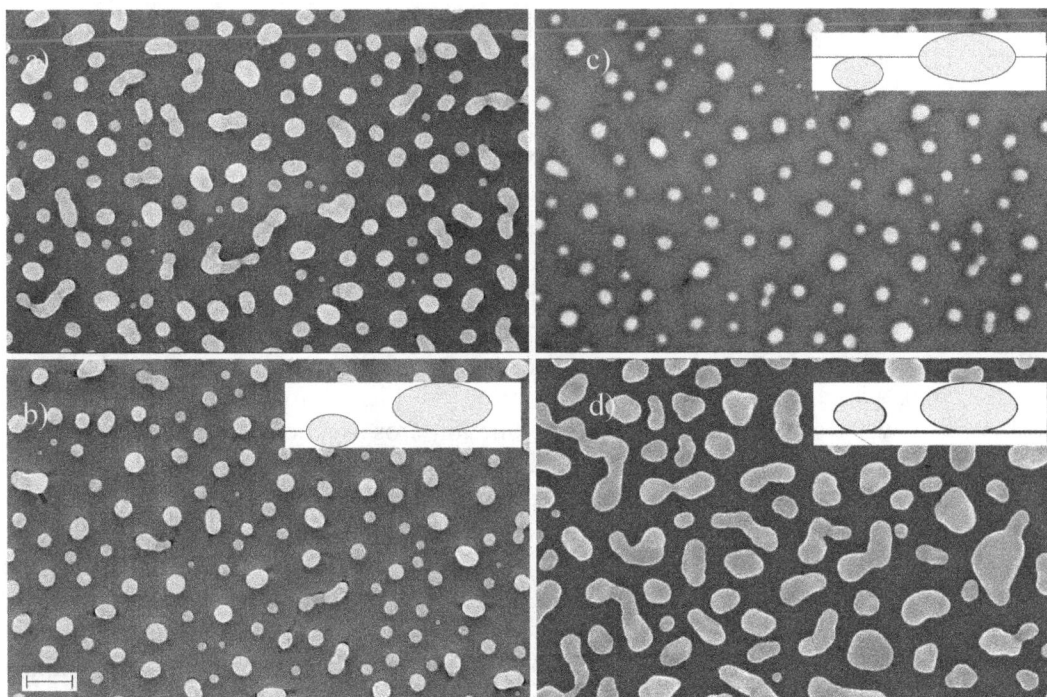

Fig. 1.1. Plan SEM images of (a) low, and (b) high dose Ar IBM samples, (c) Xe IBM sample, and (d) annealed sample. Scale bar is 200 nm. Inserts demonstrate arrangement of large and small nanoparticles on the sample surface.

Optical extinction and reflection spectra were measured with a Perkin Elmer Lambda 950 UV-VIS spectrometer in the range from 250 to 850 nm. Reflection spectra at the angle of light incidence 8° were obtained by using an integrating-sphere detector incorporated in the spectrometer. Either total reflectance or diffuse reflectance only can be measured by placing spectral on plate at the specular reflectance angle or removing it, respectively.

1.3. Ag Nanoparticles on a Dielectric Substrate

In this section, we study spectra of silver nanostructures on quartz and silicon substrates without additional sublayers. In subsections 1.3.1 and 1.3.2, Ag nanoparticles on quartz substrates are discussed, while the subsection 1.3.3 is devoted to Ag nanoparticles on silicon.

1.3.1. Spectra of Ag Nanoparticles on Quartz Substrate

Silver nanostructures on a transparent substrate without additional layers between nanostructures and the substrate are studied in this subsection. It simplifies spectrum analysis, because extinction spectrum can be also taken into consideration. In Fig. 1.2 (a, b) extinction, reflection and scattering of silver nanoislands on quartz substrate are demonstrated for high and low dose of IBM, respectively. The corresponding SEM images of the samples are given in Fig. 1.1 (a, b). In the spectra, there are distinctly visible two areas: right one (wavelength more than 400 nm) with broad and intense peak in visible (VIS) range and left one (wavelength is less than 400 nm) with weak peak in ultraviolet (UV) range. Further, we call these parts as VIS and UV, respectively. The high amplitude peak is usually attributed to dipolar LSPR, whereas the low amplitude one is connected with quadrupolar LSPR [3, 11, 12]. Theoretically, LSPR exhibits itself at the same wavelength in extinction and scattering [2]. However, it is valid only for isolated nanoparticles without size variation. In Fig. 1.2 (a, b) we observe difference in peak positions for extinction and diffuse reflection, while coinciding for extinction and total reflection peaks. Standard explanation of the observed difference is the size variation of plasmon nanoparticles. Scattering cross-section is higher for larger nanoparticles possessing lower frequency LSPR, while extinction cross-section is higher for smaller nanoparticles having LSPR at higher frequency. As a result, extinction and scattering peaks are separated. The same argument is used for explaining an increased full width at a half maximum (FWHM) of peaks in comparison with calculated ones [2]. Peak asymmetry is usually explained by shape deviation of nanoparticles from sphere to ellipsoid. It results in splitting of one LSPR in two separate resonances (redshifted and blueshifted), which can lead to observable shape of dipolar peak. In Fig. 1.2 (a, b) the low dose sample demonstrates redshift of scattering in comparison with the high dose sample, because with increasing of IBM dose particle shape variation diminishes and particle size distribution converges to smaller size.

UV resonance manifestation is usually attributed to valley near 360 nm in total reflection, as well as to peak at 350 nm in extinction [8, 11, 13] and is ascribed to quadrupolar resonance. There is also a peak at 330 nm clearly visible in total reflection of low dose sample in Fig. 1.2 (b). As a whole, UV features are more intense in low dose sample than in high dose one, but dipolar peak intensity is practically the same in both samples.

According to Fig. 1.2, intensity of diffuse reflection is 20 times less than intensity of total reflection. Therefore, we can consider total reflection as specular one and assume that samples scatter most of radiation in direction of specular reflection. It is only possible, if all radiating points, i.e., silver nanoparticles work in phase and have similar radiation patterns. If we consider our samples as diffracting gratings, then specular reflection is possible at the zero-order grating condition on the period Λ, which is expressed as [14].

$$\Lambda < \frac{\lambda}{n \sin \theta + n}, \qquad (1.1)$$

where λ is the wavelength, θ is the incident angle and n is the refractive index of ambient. For $\theta = 8°$ and $n = 1$ the inequality (1.1) is simplified to $\Lambda < \lambda$. This condition is fulfilled for all wavelengths in our experiments and leads to specular reflection of the arrays despite of scattering of any individual nanoparticle.

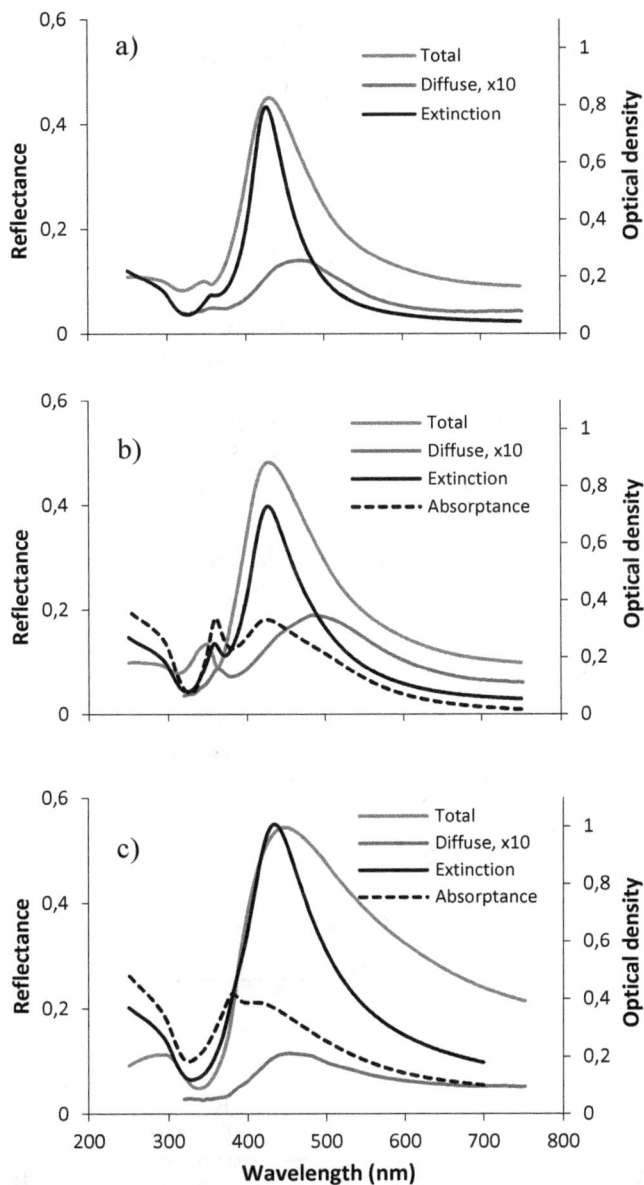

Fig. 1.2. Spectra of (a) high, and (b) low dose Ar mixed samples, and (c) SiO$_2$ capped sample.

To get additional information about sample optical properties, variation of dielectric environment was done. To increase or decrease ε, Ag nanoparticles were covered (capped) by oxide layer or elevated on pillars, respectively. Capping was realized by IBM of oxide covered silver film. A 12 nm thick film of PECVD SiO$_2$ was deposited on Ag film before IBM, what resulted in Ag nanoparticles embedded inside of SiO$_2$ matrix. Fig. 1.2 (c) shows spectra of Ag nanoparticles capped by SiO$_2$. Intensities of VIS extinction and total reflection became higher than in uncapped sample. It can be connected with higher silver amount in capped sample, because capping layer prevents Ag sputtering during IBM. Additionally, VIS peaks are much wider, due to increased extinction and reflection at long wavelengths. The sharp UV valley at 350 nm in total reflection disappeared. Additionally, in Fig. 1.2 (b, c) is also shown absorptance

$$A = 1 - T - R,$$
(1.2)

where T and R are the transmittance and reflectance, respectively. In comparison with uncapped samples, the quadrupolar peak of absorptance (380 nm) and dipolar peak of total reflectance were redshifted on 30 nm and 20 nm, respectively. It happened due to increasing of ambient ε (nanoparticles are embedded in SiO_2 matrix).

The high and low dose samples were additionally treated by reactive ion etching (RIE) during 5 and 15 minutes as reported elsewhere [5, 15, 16]. As a result, oxide between Ag nanoislands was removed and SiO_2 pillars with a height of 15 and 50 nm were fabricated. Ag nanoparticles were saved at the top of pillars. The purpose of experiment was to change the dielectric environment and to reduce possible coupling between nanoparticles. The obtained spectra of pillar samples are shown in Fig. 1.3.

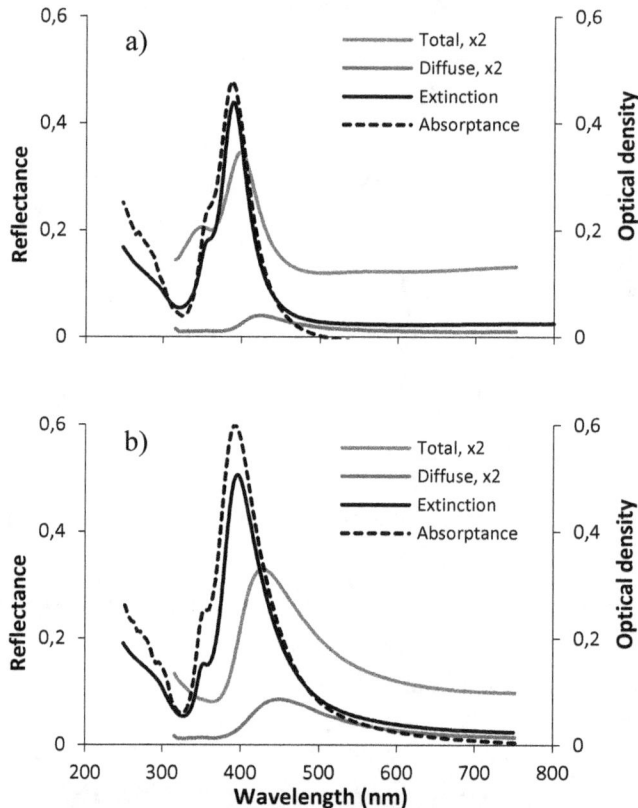

Fig. 1.3. Spectra of samples after RIE with different time: (a) high dose sample, 5 minutes, and (b) low dose sample, 15 minutes.

In both samples intensity of extinction and total reflection decreased 1.5 and 3 times, respectively, but intensity of absorptance and scattering increased 2 and 2.5 times, respectively. The dipolar extinction, absorptance and scattering peaks were blueshifted on 30 - 50 nm, due to replacing SiO_2 ($\varepsilon = 2.1$ at 633 nm) between nanoparticles by air ($\varepsilon = 1$). The UV peak of absorptance was replaced by corresponding shoulder. At the same time, the peak of total reflection was shifted only in high dose sample (Fig. 1.3 (a)) and saved its position in low dose sample. The UV peak/valley in reflectance converted to shoulder for high dose sample and to broad depression for low dose sample.

Comparison of spectra in Figs. 1.2, 1.3 leads to conclusion that peaks of extinction and diffuse reflection are connected with average positions of LSPRs. However, these positions are ascribed to LSPRs of different nanoparticles (small and large, respectively), what results in peak deviation exceeding 50 nm. At the same time, behavior of total reflection is still unclear and several moments should be considered in details. First of all, it concerns invariability of total reflection peak in Fig. 1.2 (b) and Fig. 1.3 (b), despite of changing of dielectric environment. According to LSPR theory [2] the peak must be blueshifted in RIE treated sample due to decreasing of effective dielectric constant ε. Then, it is unclear why dipolar and quadrupoalr LSPRs are observed in different way in total reflection, i.e., as peak and valley, respectively. And at last, it is required to consider the difference in peak positions of absorptance and total reflection of the same sample, which varies from zero in Fig. 1.2 (a, b) to 30 nm in Fig. 1.3 (b).

1.3.2. Analysis of Obtained Results

In disordered array of nanoparticles LSPR happens only in part of nanoparticles at any wavelength. Other particles reflect in nonresonant mode like a continuous film consisting from mixture of silver and air. This reflection is strong, creates interference and should be distinguished from LSPRs in reflection spectra. In case of absorption measurements, background from nonresonant particles is weak and does not affect on resonance peak position.

The identification of resonance features in total reflection, can be done by comparison of obtained spectrum with nonresonant one, e.g., spectrum of silver film. In Fig. 1.4 reflectance of low dose IBM sample is given together with reflectance of 8 nm thick silver film on glass. It can be supposed that dipolar resonance is responsible for increasing of reflectance in the range of 380 - 480 nm and decreasing of reflectance for wavelengths beyond 480 nm. In turn, quadrupolar resonance is connected with formation of UV valley at 360 nm. However, intensity of light reflection depends not only on LSPR of particles, but also on radiation pattern of particles and interference of emitted light. Therefore, possible reflection from quartz substrate and angular dependence of radiation intensity should be taken into account.

Interference is inseparably linked with phase change in the system. In case of plasmon particle near the interface, phase variation happens due to LSPR and reflection from interface. Nanoparticle radiates light with phase shift $\Delta\varphi_{pl}$ appearing between incident and emitted radiation [1]

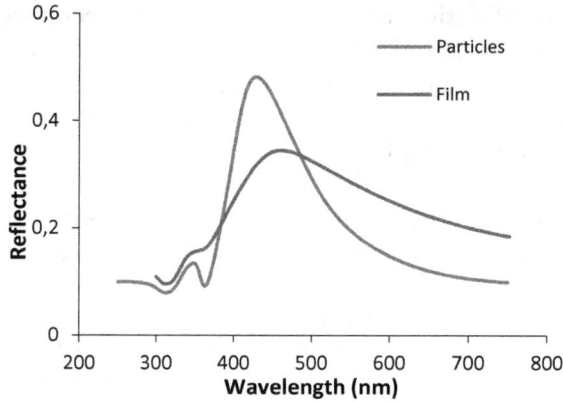

Fig. 1.4. Comparison of reflection from Ag nanoparticles and silver film.

$$\Delta\varphi = arctan\frac{2\beta\omega}{\omega_0^2-\omega^2},\qquad(1.3)$$

where β is the damping constant, ω_0 is the plasmon resonance frequency and ω is the frequency of incident wave. According to (1.3), $\Delta\varphi_{pl}$ is changed from 0° at low frequency to 180° at high frequency and is equal 90° at the resonance frequency. In case of interacting particles, the original LSPR is splitted in longitudinal and transverse resonances (Fig. 1.5) and the same takes place with phase shift. Longitudinal LSPR is redshifted and have $\Delta\varphi_{pl}$ in the range of 0° - 90°, transverse LSPR is blueshifted and have $\Delta\varphi_{pl}$ in the range of 90° - 180°.

(a)   (b)

Fig. 1.5. Interaction of nanoparticles: (a) longitudinal LSPR, (b) transverse LSPR.

Radiation emitted by nanoparticle is reflected by substrate with variable intensity and phase, which depend on nanoparticle and substrate properties. High reflectance and broad phase variation facilitate appearing of interference. Reflected light propagates back to nanoparticle and interact with its electromagnetic field. As a result, substrate and a nanoparticle create an optical cavity in which the nanoparticle is one of the mirrors. Length of the cavity is equal to nanoparticle height (around 45 nm for IBM samples [8]). The particle emits light during plasmon excitation with phase shift $\Delta\varphi_{pl}$ according to (1.3). After that total phase shift of optical cavity is $\Delta\varphi_{pl} + \Delta\varphi_{refl}$, where $\Delta\varphi_{refl}$ is the phase shift upon reflection at the substrate.

For example, at wavelength more than 500 nm (dipolar depression in Fig. 1.4) $\Delta\varphi_{pl}$ close to 0°, because it corresponds longitudinal resonance of two interacting particles. At the same time, reflectance phase at air/SiO$_2$ interface is 180°, what provides conditions for destructive interference. As a result, exciting electromagnetic field near plasmon nanoparticles is reduced, which leads to decreasing of nanoparticle radiation at these wavelength in comparison with Ag film. In similar way, the dipolar peak in Fig. 1.4 is formed at wavelength of transverse LSPR of interacting particles, due to $\Delta\varphi_{pl}$ close to 180° and constructive interference. In the sample with capping layer dipolar destructive interference is not possible, what leads to increasing of reflectance in the red part of the spectrum (Fig. 1.2 (c)).

Direction of light radiation is characterized by angular intensity distribution, i.e., by radiation pattern. In isotropic space, dipole and quadrupole have symmetric two-lobe and four-lobe radiation patterns. However, near the interface the radiation pattern is distorted and most of radiation energy is emitted in space with higher ε [18]. In case of quadrupole, additionally the direction of maximum radiation into substrate is changed and correlates with critical angle of total internal reflection of considered interface [19]. The largest part of light is emitted at angles more than critical ones, i.e., closer to interface than to surface normal. As a result, most of quadrupolar radiation is trapped in the substrate, what leads to trough in reflection and peak in extinction.

Importance of interface for formation of total reflection is illustrated by Figs. 1.2 (c), 1.3. In both cases, the sharp quadrupolar valley at 360 nm is not observed, because back reflection plane disappears, what leads to changing of radiation pattern. After RIE treatment common supporting plane of Ag nanoparticles is replaced by individual pillars, what results in diminishing of specular reflection and increasing of nonresonant scattering. In the sample with capping SiO_2 layer the reflection plane disappears at all, because substrate and capping layer have the same refractive index. In both cases, the quadrupolar resonance is observed only in absorptance as 350 nm shoulder (Fig. 1.3) and 380 nm peak (Fig. 1.2 (c)). Redshift of quadrupolar peak to 380 nm after capping by SiO_2 layer is explained by increasing of ε.

Submerging of nanoparticles after IBM is size sensitive and leads to result, when small nanoparticles are more burrowed, than large ones. This arrangement is saved during dry etching. Simplified profile of sample after RIE processing is shown in Fig. 1.6. Burrowing of small nanoparticles causes increased radiation in substrate direction, what decreases total reflection from small nanoparticles. Due to this, total reflection mainly formed by large size particles. On the other hand, extinction is defined by small size particles. As a result, peaks of total reflection and extinction have different positions. In comparison with high pillars (Fig. 1.3 (b)), the total reflection of low pillar sample is blushifted and have broad valley near 500 nm. These features are connected with nanoparticle interaction and reflection from quartz surface. Indeed, transversal interaction (Fig. 1.5 (b)) provides short-wavelength LSPR with $\Delta\varphi_{pl}$ close to 180° and constructive interference (dipolar peak) at 390 nm. At the same time, longitudinal interaction (Fig. 1.5 (a)) leads to long-wavelength LSPR with $\Delta\varphi_{pl}$ close to 0° and destructive interference at 500 nm. In case of high pillar sample, interaction between nanoparticles is broken, due to decreasing of ambient dielectric constant. Therefore, reflection is dominated by separate nanoparticle LSPR and save its peak position at 420 nm.

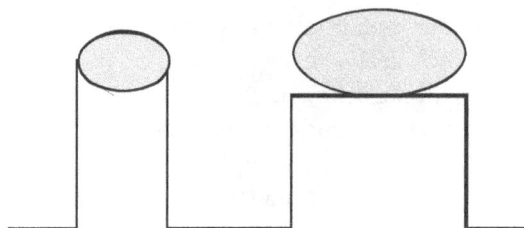

Fig. 1.6. Arrangement of small and large nanoparticles on top of pillar after RIE treatment.

The analysis of samples on quartz substrate demonstrates that total reflection should be carefully used for LSPR identification in disordered arrays of nanoparticles. Dipolar resonance corresponds $\Delta\varphi_{pl} = 90°$ and can be found between destructive and constructive interference areas, i.e., on the right slope of total reflection peak. In case of known scattering, the possible position of dipolar LSPR can be between peaks of total and diffuse reflection.

1.3.3. Reflection of Ag Nanoparticles on Si Substrate

Fig. 1.7 demonstrates reflection and scattering of silver nanoparticles on bare silicon substrate. This sample was prepared by annealing of silver film deposited on Si substrate with native oxide. The UV part of total reflection looks similar to reflection of IBM samples on quartz (Fig. 1.2 (a, b)), but with increased peak intensity at 330 nm and redshifted minimum of reflectance to 380 nm. The VIS part of total reflection is dissimilar to spectra of Fig. 1.2 and contents only one very broad peak at 720 nm.

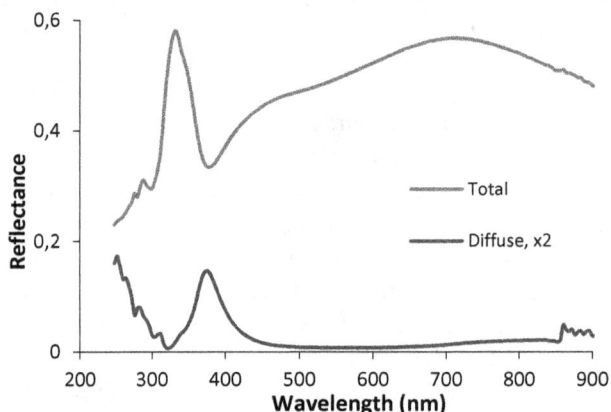

Fig. 1.7. Total and diffuse reflection of silver nanoparticles on silicon substrate.

Silicon has much higher dielectric constant ($\varepsilon = 15$ at 633 nm) than quartz. Due to this and higher surface density of nanoparticles (Fig. 1.1 (d)), dipolar coupling is very strong in this sample. It leads to splitting of original LSPR peak in blue and red ones. In this case, scattering nanoparticles are not coupled ones and diffusion reflection shows position of

34

original LSPR at 830 nm. Therefore, VIS spectrum is mainly formed by blue component of split LSPR. Phase shift of radiated light differs in coupled and isolated nanoparticles. Coupled particles oscillate with $\Delta\varphi_{pl} > 90°$ and $\Delta\varphi_{pl} < 90°$ at the frequencies of blue and red peaks, respectively. It means, that constructive interference happens for whole VIS part of total reflection in the optical cavity with nanoparticle and results in observed spectrum. Position of dipolar LSPR cannot be identified in reflection spectrum in case of Si substrate.

1.4. Ag Nanoparticles on Multilayer Substrates

In the previous section, we demonstrated that nanoparticle and substrate form an optical cavity. It, in turn, leads to intensity variation of electromagnetic field near the nanoparticle. Multilayer substrate provides additional reflection interfaces, what result in multiple optical cavities including the same nanoparticle. Fig. 1.8 shows spectra of Xe mixed Ag nanoparticles on 100 nm thick Al_2O_3 layer above Si substrate. Here is also given reflectance spectrum of 15 nm thick Ag film above Al_2O_3 before IBM. This spectrum demonstrates transparency of 15 nm thick silver film and high quality interference in Al_2O_3 optical cavity with Si and Ag mirrors. The minimum at 290 nm and maximum at 380 nm are close to calculated interference extrema, obtained with bulk silver optical constants.

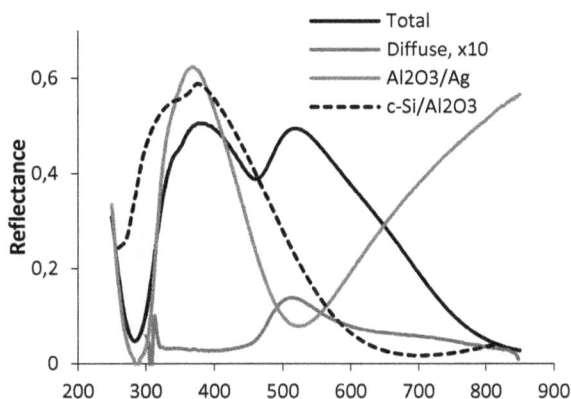

Fig. 1.8. Total and diffuse reflection of silver nanoparticles on 100 nm thick Al_2O_3 layer deposited on silicon substrate.

After formation of nanoislands, the minimum at 530 nm is replaced by maximum at 520 nm and minimum at 470 nm. In this sample, the new optical cavity is created between reflecting surface of Si substrate and Ag nanoparticle on Al_2O_3 layer. Therefore, presence of the additional optical cavity creates two local extrema visible in total reflection. When in the optical cavity one of the mirrors is replaced with plasmonic structure, the phase shift balance for extrema is [1]

35

$$2\Delta\varphi_{prop} + \Delta\varphi_{refl} + \Delta\varphi_{pl} = N\pi, \qquad (1.4)$$

where $\Delta\varphi_{prop}$ is the phase shift due to propagation of the wave through cavity, $\Delta\varphi_{refl}$ is the phase shift upon reflection at the cavity mirror and N is the integer. From diffuse reflection spectrum in Fig. 1.8 follows, that LSPR of Ag nanoparticles on Al_2O_3 is situated below 510 nm. Wavelength of local minimum in Fig. 1.8 is 460 nm, what corresponds $2\Delta\varphi_{prop} = 1.47\pi$ ($n_{Al2O3} = 1.69$). Additional phase shift $\Delta\varphi_{pl} = 0.53\pi$ provides condition for destructive interference ($\Delta\varphi_{refl} = \pi$) in the Al_2O_3 optical cavity at 460 nm according to (1.4). The dipolar resonance itself is redshifted on 0.03π from the local minimum, because $\Delta\varphi_{pl} = 0.53\pi$ and $\Delta\varphi = 0.5\pi$ at the resonance.

Fig. 1.9 (a) demonstrates total and diffuse reflection for Ar mixed Ag nanoparticles prepared on Si substrate covered by 20 nm of SiO_2.

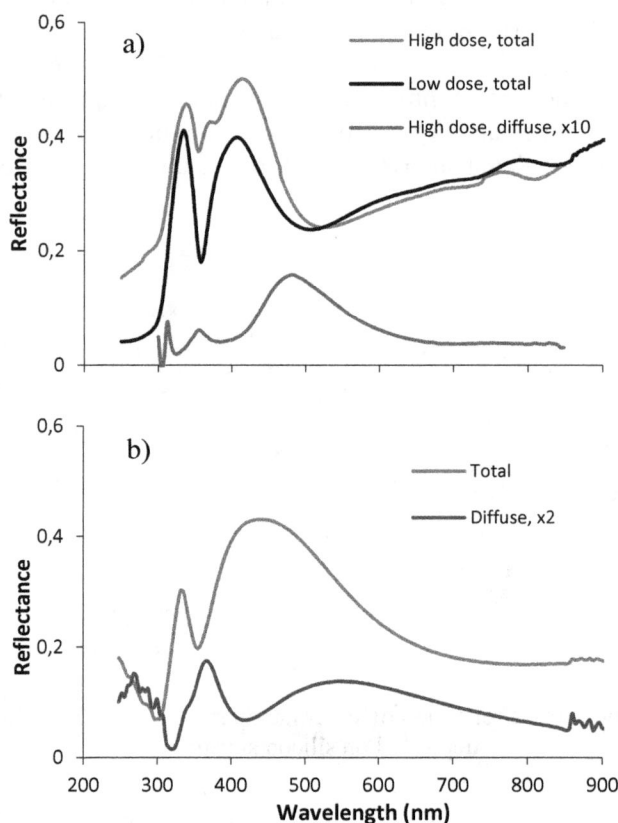

Fig. 1.9. Total and diffuse reflection of silver nanoparticles created by (a) IBM and by (b) Ag annealing on oxidized Si substrate.

The fabrication procedure is the same as for samples on quartz substrate in Fig. 1.2 (a, b). The obtained spectra look similar to spectra in Fig. 1.2 (b) in UV part. They include the same peak at 330 nm and the valley at 360 nm, however, extremum intensity is much higher than in Fig. 1.2 (b). The trough of quadrupolar LSPR is deeper, due to more effective trapping of light between Si and nanoparticles. Nevertheless, VIS parts of the spectra in Fig. 1.2 (b) and Fig. 1.9 (a) are different. The depression around 500 nm is ascribed to longitudinal LSPR and has interference origin ($\Delta\varphi_{pl}$ is close to zero, $\Delta\varphi_{refl} = \pi$ at SiO$_2$/Si interface). At wavelengths shorter than 510 nm transverse LSPR of Ag nanoparticles in optical cavities increases reflectance similarly to sample on quartz substrate. The reflection spectrum of high dose sample (Fig. 1.9 (a)) demonstrates one more quadrupolar resonance at 390 nm. It is attributed to Ag nanoparticles fully submerged in SiO$_2$ during IBM [11].

Spectra of Ag nanoparticles prepared by annealing on oxidized Si substrate (36 nm thick SiO$_2$) are given in Fig. 1.9 (b). They are similar to spectra in Fig. 1.6 (a) in UV part. The VIS part of Fig. 1.9 (b) has a broader dipolar peak and redshifted valley in comparison with similar features in Fig. 1.9 (a). Larger size of silver nanoparticles (Fig. 1.1 (d)) redshifts position of dipolar LSPR to 550 nm, how it is visible from scattering in Fig. 1.9 (b). Additionally, high surface density of nanoparticles (Fig. 1.1 (d)), results in splitting of original LSPR in the same way as for Si substrate. Part of optical cavities with coupled Ag nanoparticles acquires $\Delta\varphi_{pl} > 90°$ at blue peak wavelength and creates constructive interference in the range of 450-600 nm, what leads to increased FWHM of the VIS peak. It cost to note, that interference takes place upon reflection both from air/SiO$_2$, and SiO$_2$/Si interfaces. Therefore, conditions for constructive interference are fulfilled at different wavelengths, due to variation of $\Delta\varphi_{prop}$. It leads to broadening of dipolar peak at 430 nm. The same is valid for destructive interference and broadening of valley near 800 nm.

1.5. UV Features of the Spectra

All studied samples, exclude the 100 nm Al$_2$O$_3$/Ag and capped ones, demonstrate 330 nm peak in the UV part of reflection spectra. Moreover, in the RIE processed samples, this feature was observed in as prepared nanoparticle arrays and disappeared after RIE (Fig. 1.3). Therefore, existing of reflective surface below Ag nanoparticles is essential for obtained results. Wavelength of UV peak (330 nm) does not depend on nanostructure shape and substrate ε. The peak can be attributed to variation of reflection phase and intensity at Ag/air and Ag/substrate interfaces according with wavelength and coincides with maximum of silver refractive index n_{Ag}. Removing of reflective surface by pillars or capping changes reflection conditions and the mentioned UV feature disappears.

Position of UV peak in absorption (360 nm) strictly coincides with the peak of diffuse reflection (Fig. 1.2 (b), Fig. 1.7 and Fig. 1.9). It means that this feature is a quadrupole LSPR [12]. However, its position weakly depends on dielectric environment and for Si substrate with high ε the valley is shifted only to 375 nm (Fig. 1.7). At the same time, quadrupolar LSPR of submerged in oxide Ag particles is shifted to 390 nm (Fig. 1.9 (a)). Additionally, extinction coefficient of silver has minimum at the same wavelength 360 nm, what facilitates destructive interference. We believe that all these factors contribute in stable position of 360 nm valley.

1.6. Conclusions

It has been demonstrated that peaks and valleys in reflection spectrum of disordered nanoparticles on multilayer substrate do not correspond directly to plasmon resonances. The sample reflectance is modulated by constructive and destructive interferences in the optical cavity, containing the nanoparticle. Nevertheless, it is possible to identify position of LSPR with accuracy of FWHM of resonance band. This position is situated between two local extrema of reflectance spectrum, corresponding phase shift variation during LSPR. Additionally, LSPR positions can be unmasked by proper design of studied samples, e.g., by eliminating reflecting surfaces or by variation of propagation phase shift in the optical cavity. The spectrum features in UV range may be attributed either to quadrupolar resonance or to variation of material properties. In the first case, the valley position depends on dielectric environment and geometry of plasmonic structures. In the second case, the peak is fixed at 330 nm and is observed only in nanostructures having back reflecting surface. The obtained results can be used in analysis and design of plasmonic nanostructures on opaque substrates.

Acknowledgements

This research was undertaken at the Micronova Nanofabrication Centre, supported by Aalto University.

References

[1]. R. Ameling, L. Langguth, M. Hentschel, M. Mesch, P.V. Braun and H. Giessen, Cavity-enhanced localized plasmon resonance sensing, *Applied Physics Letters*, Vol. 97, Issue 25, 2010, pp. 253116-1-253116-3.

[2]. E. C. Le Ru and P. G. Etchegoin, Principles of surface-enhanced Raman spectroscopy and related plasmonic effects, *Elsevier*, 2009.

[3]. S. Pillai, K. R. Catchpole, T. Trupke, and M. A. Green, Surface plasmon enhanced silicon solar cells, *Journal of Applied Physics*, Vol. 101, Issue 9, 2007, pp. 093105-1-093105-8.

[4]. P. Biagioni, J.-S. Huang, and B. Hecht, Nanoantennas for visible and infrared radiation, *Reports on Progress in Physics*, Vol. 75, Issue 2, 2012, pp. 024402-1-024402-40.

[5]. A. Shevchenko, V. Ovchinnikov, and A. Shevchenko, Large-area nanostructured substrates for surface enhanced Raman spectroscopy, *Applied Physics Letters*, Vol. 100, Issue 17, 2012, pp. 171913-1-171913-4.

[6]. V. Ovchinnikov, Reflection from irregular array of silver nanoparticles on multilayer substrate, in *Proceedings of the 9th International Conference on Quantum, Nano/Bio, and Micro Technologies (ICQNM 2015)*, Venice, Italy, 23-28 August 2015, pp. 16-21.

[7]. V. Ovchinnikov, Reflection from Disordered Silver Nanoparticles on Multilayer Substrate. *Sensors & Transducers*, Vol. 193, Issue 10, October 2015, pp. 170-178.

[8]. V. Ovchinnikov and A. Priimagi, Anisotropic plasmon resonance of surface metallic nanostructures prepared by ion beam mixing, in *Proceedings of the 1st International Conference on Quantum, Nano and Micro Technologies (ICQNM' 07)*, Guadeloupe, French Caribbean, 2-6 January 2007, pp. 3-8.

[9]. V. Ovchinnikov, Analysis of furnace operational parameters for controllable annealing of thin films, in *Proceedings of the 8th International Conference on Quantum, Nano/Bio, and Micro Technologies (ICQNM'14)*, Lisbon, Portugal, 16-20 November 2014, pp. 32-37.

[10]. V. Ovchinnikov, Effect of thermal radiation during annealing on self-organization of thin silver films, in *Proceedings of the 7th International Conference on Quantum, Nano and Micro Technologies (ICQNM'13)*, Barcelona, Spain, 25-31 August 2013, pp. 1-6.

[11]. V. Ovchinnikov, Formation and characterization of surface metal nanostructures with tunable optical properties, *Microelectronics Journal*, Vol. 39, Issue 3-4, 2008, pp. 664-668.

[12]. E. Thouti, N. Chander, V. Dutta, and V. K. Komarala, Optical properties of Ag nanoparticle layers deposited on silicon substrates, *Journal of Optics*, Vol. 15, Issue 3, 2013, pp. 035005-1-035005-7.

[13]. V. Ovchinnikov and A. Shevchenko, Surface plasmon resonances in diffusive reflection spectra of multilayered silver nanocomposite films, in *Proceedings of the 2nd International Conference on Quantum, Nano and Micro Technologies (ICQNM'08)*, Sainte Luce, Martinique, 10-15 February 2008, pp. 40-44.

[14]. T. Sang, L. Wang, S. Ji, Y. Ji, H. Chen, and Z. Wang, Systematic study of the mirror effect in a poly-Si subwavelength periodic membrane, *Journal of the Optical Society of America A*, Vol. 26 Issue 3, 2009, pp. 559-565.

[15]. V Ovchinnikov, A Malinin, S Novikov, and C Tuovinen, Silicon nanopillars formed by reactive ion etching using a self-organized gold mask, *Physica Scripta*, Vol. T79, 1999, pp. 263-265.

[16]. A. Shevchenko and V. Ovchinnikov, Magnetic excitations in silver nanocrescents at visible and ultraviolet frequencies, *Plasmonics*, Vol. 4, Issue 2, 2009, pp. 121-126.

[17]. H. W. Edwards, Interference in thin metallic films, *Physical Review*, Vol. 38, 1931, pp. 166-173.

[18]. L. Luan, P. R. Sievert and J. B. Ketterson, Near-field and far-field electric dipole radiation in the vicinity of a planar dielectric half space, *New Journal of Physics*, Vol. 8, 2006, p. 264.

[19]. H. F. Arnoldus, J. T. Foley, Highly directed transmission of multipole radiation by an interface, *Optics Communications*, Vol. 246, 2005, pp. 45–56.

Chapter 2
Structural, Optical and Light Sensing Properties of Carbon-ZnO Films Prepared by Pulsed Laser Deposition

Valentine Saasa, Emmanuel Mukwevho, Bonex Mwakikunga

2.1. Introduction

There has been tremendous interest in ZnO [1-4] as well as carbon materials [5-7]; these are already in various technological devices. Work on the combination of ZnO and carbon materials is gaining interest [8-10] since these composites benefit from the good properties of both popular materials. For instance, carbon modified ZnO materials have been presented for solar cells [11-13] and sensors [14- 16, 21, 22] to name a few.

In light sensing, single carbon nanotubes mixed with ZnO have been employed [17-18]. The current voltage characteristics obtained on the SWNTs –ZnO devices were found to be non-rectifying and Ohmiclike in the dark and when exposed to optical radiation. While the conductivity of SWNTs thin films was observed to increase by ~ 7 % under UV illumination, similarly tested SWNT-ZnO samples consistently exhibited a decrease in conductivity upon exposure to UV radiation yielding a net negative change of ~ 10 % associated with the formation of ZnO-SWNT interface. This effect as well as that of strong optical gating was explained within the model of interface mediated charging/discharging effects. In the present study, the aim was to push the UV and visible light sensing efficiency of the multi-wall carbon- nanotube-modified ZnO (MWCNT-ZnO) to higher values than 10 % by developing MWCNT-ZnO materials at varying processing temperatures from 300 K to 1173 K. This ensures that for the same starting MWCNT-ZnO composite, the proportions of MWCNT/ZnO are tuned by the process temperatures. In this work, the structural, morphological and optical properties of MWCNT-ZnO processed at varying temperatures are reported and how the process temperature impinges on structure, surface roughness and optical band gap energy of the MWCNT-ZnO composites calculated from reflectivity as well as the UV and visible light sensing of these

Valentine Saasa

DST/CSIR National Centre for Nano-Structured Materials, PO Box 395, Pretoria 0001, South Africa composites is also discussed. It is demonstrated the light sensing efficiency can be pushed to above 50 %.

2.2. Experiments

Many methods have been employed in combining carbon nanotubes with other metal oxide. Methods like in-situ; whereby the growth of CNT are achieved within the same process and ex-situ, where the decoration is performed in a separate step following the synthesis of CNT [19]. The Pulsed laser deposition technique has never been reported on the preparation of carbon nanotubes with Zinc oxide. In this study, PLD technique has been employed in the synthesis of carbon modified ZnO.

Puled Laser Deposition (PLD) is a physical vapour deposition process carried out in a vacuum system that shares some process characteristics common with molecular beam epitaxy and some with sputter deposition. In PLD, as shown in Fig. 2.1, a pulsed laser is focused onto a target of the material to be deposited. For satisfactorily high laser energy density, each laser pulse vaporizes or ablates a small amount of the material creating a plasma plume. The ablated material is ejected from the target in a highly forward-directed plume. The ablation plume provides the material flux for film growth [20].

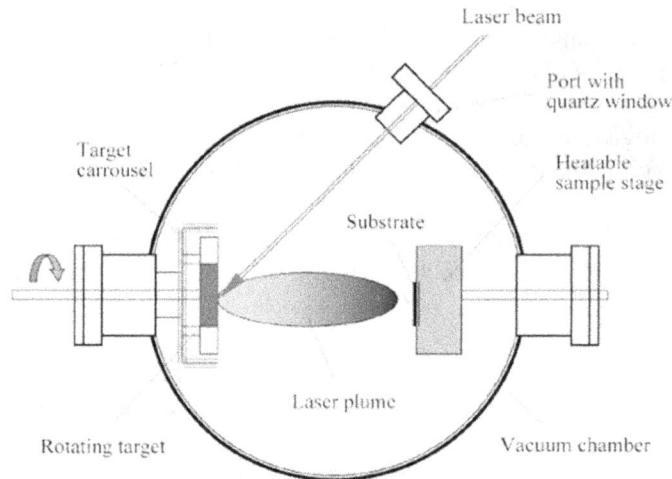

Fig. 2.1. Schematic view of the PLD system [21].

Several features make PLD attractive for complex material film growth. These include stoichiometric transfer of material from the target, generation of energetic species, hyperthermal reaction between the ablated cations and the background gas in the ablation plasma, and compatibility with background pressures ranging from ultrahigh vacuum (UHV) to 1 Torr [21].

The conditions of the preparation of carbon modified ZnO films with pulsed laser deposition (PLD) are as follows; a 3-cm radius pellet of pre-mixed ZnO and CNT powder was first prepared in a press of maximum 10 tones equivalent pressure. The pellet was sintered at 473 K for 5 hrs in order to avoid fracturing during laser deposition. PLD was accomplished by employing a KrF excimer laser emitting a pulse of wavelength 268 nm, a pulse repetition rate of 20 pulses per second. Each deposition was carried out on Si substrate and varying the substrate temperatures from 300 K, 373 K, 473 K, 573 K in that order up to 1173 K.

The films were characterized for their structural, morphological and optical properties by respectively employing the following instruments: the XRD using CuKα radiation of wavelength 1.5418 Å recorded in the 2θ range from 5 to 90, SEM (Auriga Cobra FIB FESEM) at a working distance of 100 nm and a Varian Cary UV-Vis-IR Spectrophotometer with a diffuse reflectance spectrometer (DRS) attachment respectively. AFM topography examinations were carried out with the Multimode AFM NanoScope Version (R) IV (VEECO instrument) using a $0.5 - 2$ Ω-cm phosphorous (n) doped Si tip with a radius of curvature of 10 nm and the aspect ratio of 1:1. Imaging was done in tapping mode (TM) and varying the scan rates and magnifications. Rutherford Backscattering Spectrometry (RBS) and Proton Induced X-ray Emission (PIXE) measurements were carried out using a collimated 2 MeV He+ beam at GNS Science. For RBS measurements, the backscattered particles were collected using a surface barrier detector placed at 165 with a target current ~15 nA. PIXE measurements were simultaneously carried out, and the X-rays were detected using liquid nitrogen cooled Canberra detector. More details on how all the instrumentation mentioned in the study is explained below. This includes their working principles, and the information they can provide about the material.

2.2.1. X-Ray Diffraction (XRD)

X-Ray diffraction is mostly used analytical technique in material science for characterization of crystalline materials. It provides information about elemental analyses such as structures, crystal orientation and phases [22].

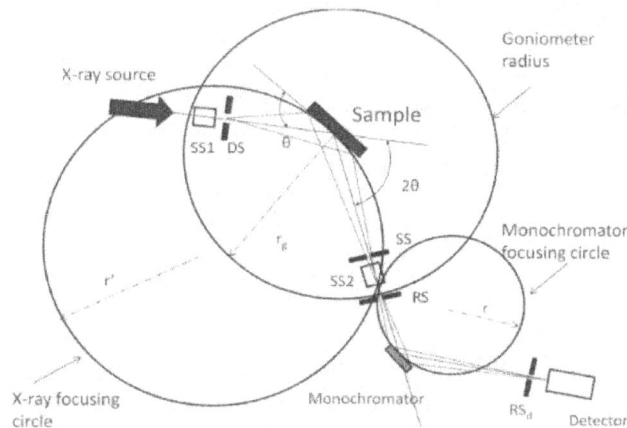

Fig. 2.2. Schematic principle of operation of the XRD [23].

When a monochromatic X-ray beam with a well-defined wavelength (λ) that is, a wavelength shorter than the spacing (d_{hkl}) between atoms in a crystal, is projected onto a material at an angle (θ), X-ray diffraction peaks are produced by constructive interference of the scattered (reflection) beam from each set of lattice h k l planes at a specific angle. Constructive interference gives the diffraction peaks according to the Bragg's condition by varying the angle (θ). The Bragg's law condition in Fig. 2.3 and Eq. (2.1) is satisfied by different d spacing in the polycrystalline materials [24].

$$2d\mathrm{Sin}\theta_B = n\lambda \qquad (2.1)$$

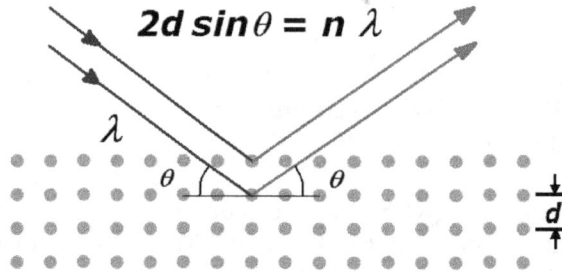

Fig. 2.3. Bragg' law describing the diffraction of X-ray in a crystal. Inter-planer spacing d sets the difference in path length for the ray scattered from the top plane and the ray scattered from the bottom plane gives the different in path as 2d sin θ_B [21].

The profile angular position and intensities of the resultant diffracted peaks of radiation produces a pattern, which is the characteristic of the material. The particle size of the crystal (L) can be calculated from the peak broadening B_L (full width at half maximum (FWHM) using Deby-Scherre model (2.2), where KI a constant with a value close to unity but often taken as 0.94 [24-25].

$$BL = \frac{K\lambda}{L\cos\theta} \qquad (2.2)$$

From Debye-Scherrer model, B_L and L are reciprocally related; the greater the broadening the smaller the particle size and vice-versa. A model for particle size and stain was also proposed by G.K. Williamson and H.W. Hall, their principle relies on the assumption that the size broadening B_L and stain broadening B_e, vary quite differently with respect to Bragg's angle θ as expressed as follows [24-25].

$$B_{total} = BL + Be \; \frac{K\lambda}{L\cos\theta} + {}^{\eta}\mathcal{E}\tan\theta \qquad (2.3)$$

$$B_{total}\cos\theta = \frac{K\lambda}{L} + {}^{\eta}\mathcal{E}\sin\theta \qquad (2.4)$$

The quantity (ε) gives the strain value which is calculated from the intercept of (2.3), the particle size is from the slope while η is a constant value often taken as 4 or 5. The size contribution varies as $1/\cos\theta$ and strain contribution varies as $\tan\theta$, hence both contributions present the integral of the product of the broadening function [25].

2.2.2. Scanning Electron Microscope (SEM)

Scanning electron microscope uses electron beam as a probing source to obtain image. Unlike TEM which is based on transmission mode, SEM relies upon finely focused electron beam that is scanned across the sample [26]. The scanning mode operation of SEM makes it usable at a very small cathode to anode's potential difference4 (1 to 30 kV). A typical SEM schematic diagram is shown in Fig. 2.4. Its principle is similar to that of TEM, when beam of primary electron from anode collides with sample; it transfers part of its energy to the sample, which causes electrons from the sample to be dislodged. The dislodged electrons are secondary electrons, and are attracted and collected by a positively biased secondary electron detector, and converted to image [26-27].

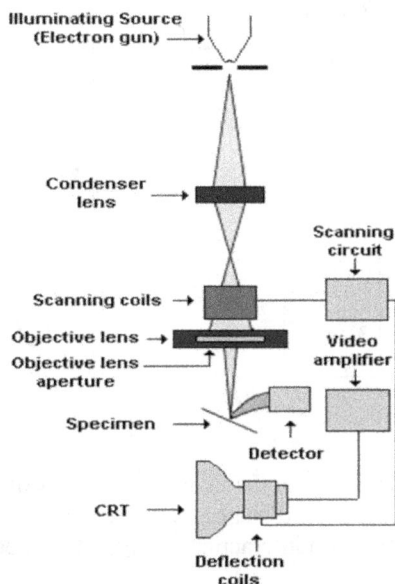

Fig. 2.4. Schematic diagram of a Scanning electron microscope (26).

2.2.3. Atomic Force Microscope (AFM)

Atomic Force Microscope Spectroscopy uses a principle of attaching the tip of the cantilever to one end of the biomolecules to be analyzed, the other end being held by the substrate. The cantilever is then moved using a piezocontroller to exert a tensile load on the molecule [28].

In AFM, the sample surface is scanned with a sharp probe or tip, situated at the apex of a flexible cantilever that is often a diving board or V-shaped, made of silicon. The AFM utilizes a piezoelectric scanner that moves the sample in 3 dimensions by a subnanometer amount when a voltage is applied (Fig. 2.5 (a)) [29]. To form an image, the tip is brought close to the sample and raster-scanned over the surface, causing the cantilever to be deflected due to probe-sample interactions. A line-by-line image of the sample is formed

45

as a result of this deflection, which is itself detected using laser light reflected off the back surface of the cantilever onto a position-sensitive to a photodiode detector. Forces acting between the sharp probes (tip) placed in close contact with the sample result in a measurable deformation of the cantilever to which the probe is attached. The cantilever bends vertically upwards or downwards because of repulsive or attractive interactions, respectively. The forces acting on the tip vary, depending on the sample nature, imaging mode and conditions used in the measurements [30, 31].

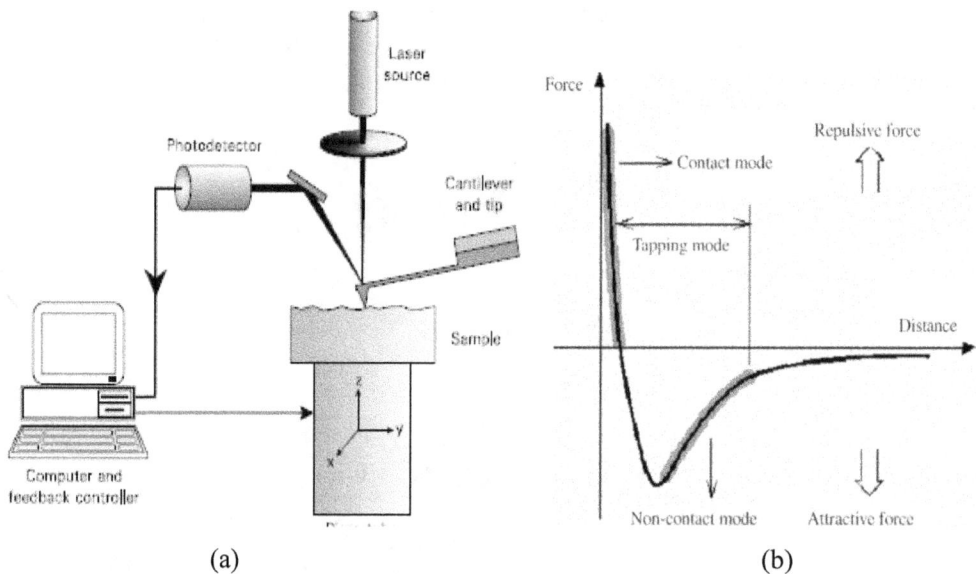

(a)   (b)

Fig. 2.5. (a) Schematic representation of the main components of an atomic force microscope (AFM) and (b) illustrative force vs. distance curve between the scanning tip and sample that reflects the main interaction during AFM measurements.

2.2.4. Ultraviolet/Visible/Infrared (UV/Vis/IR) Spectrophotometer

Ultraviolet/Visible/Infrared (UV/Vis/IR) spectroscopy is a technique used to quantify the light that is absorbed and scattered by a sample. In (UV/Vis/IR) a sample is placed between a light source and a photodetector, the intensity of a beam of light is measured before and after passing through the sample (Fig. 2.6) [32].

Nanoparticles have optical properties that are sensitive to size, shape, concentration, agglomeration state, and refractive index near the nanoparticle surface, which makes UV/Vis/IR spectroscopy a valuable tool for identifying, characterizing, and studying these materials.

The standard UV-Vis analysis is performed with an Agilent 8453 single beam diode array spectrometer, which collects spectra from 200-1100 nm using a slit width of 1 nm. Deuterium and tungsten lamps are used to provide illumination across the ultraviolet, visible, and near-infrared electromagnetic spectrum. Spectra are typically collected from

1 mL of sample dispersion, but we can test volumes as small as 100 μL using a microcell with a path length of 1 cm [32].

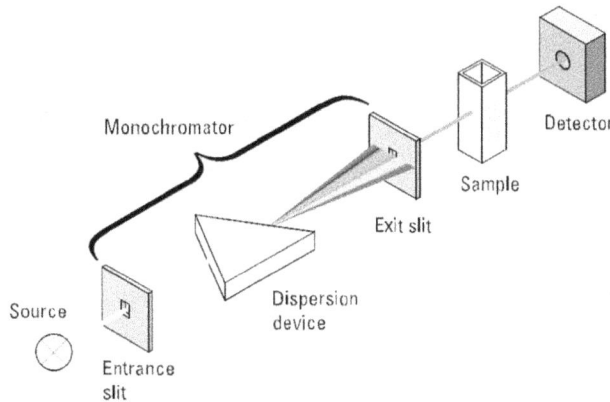

Fig. 2.6. Representative of a simple UV-visible spectrometer.

2.2.5. Rutherford Backscattering Spectrometry RBS

Rutherford Backscattering Spectrometry (RBS) is a widely used nuclear method for the near surface layer analysis of solids. A target is bombarded with ions at an energy in the MeV-range (typically 0.5–4 MeV), and the energy of the backscattered projectiles is recorded with an energy sensitive detector, typically a solid state detector allows quantitative determination of the composition of a material and depth profiling of individual elements. RBS quantifies without the need for reference samples, has a good depth resolution of the order of several nm, and a very good sensitivity for heavy elements of the order of parts-per-million (ppm). The analyzed depth is usually about 2 μm for incident He-ions and about 20 μm for incident protons [33].

RBS includes all types of elastic ion scattering with incident ion energies in the range 500 keV – several MeV. Usually protons, ^{4}He, and sometimes lithium ions are used as projectiles at backscattering angles of typically 150– 170.

The most often used scattering geometries are shown in Fig. 2.7. In IBM geometry incident beam, exit beam and surface normal of the sample are in the same plane, with

$$\alpha + \beta + \theta = 180° \tag{2.5}$$

In Cornell geometry, incident beam, exit beam and the rotation axis of the sample are in the same plane and

$$\cos(\beta) = -\cos(\alpha)\cos(\theta). \tag{2.6}$$

Cornell geometry has the advantage of combining a large scattering angle, which is wishful for optimized mass resolution, and grazing incident and exit angles, which optimizes depth resolution.

47

IBM   Cornell

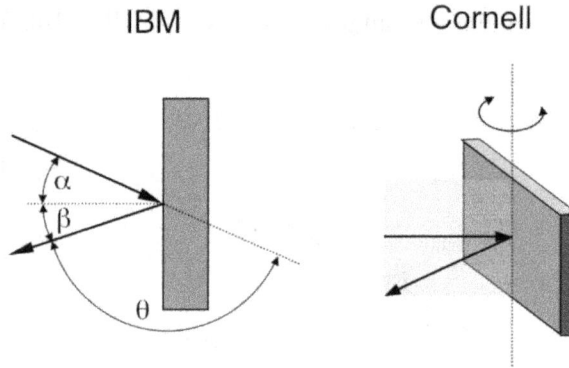

Fig. 2.7. Left: IBM geometry; Right: Cornell geometry. Incident angle α, exit angle β, and scattering angle θ.

The energy E_1 of a backscattered projectile with incident energy E_0 and mass M_1 after scattering is given in the laboratory system by

$$E_1 = KE_0, \qquad (2.7)$$

where the kinematic factor K is given by

$$K = \frac{M_1{}^2}{(M_1 + M_2)^2} \left\{ \cos\theta \pm \int_{-\infty}^{\infty} e^{-x^2} dx \left[\left(\frac{M_2}{M_1}\right)^2 \right]^{1/2} [\sin 2\,\theta]^{1/2} \right\}^2 , \qquad (2.8)$$

where θ is the scattering angle and M_2 is the mass of the target nucleus initially at rest. For $M_1 < M_2$ only the plus sign in eq. 2.8 applies. If $M_1 > M_2$ then eq. 2.8 has two solutions, and the maximum possible scattering angle θ_{max} is given by

$$\theta_{max} = \sin^{-1}\left(\frac{M_2}{M_1}\right) \qquad (2.9)$$

The kinematic factor K, as a function of target mass M_2, is shown in Fig. 2.8 for incident protons, ^{4}He, and ^{7}Li ions at a scattering angle of 165°. If two different target elements with mass difference ΔM_2 are present, then the energy separation $\Delta E1$ of particles backscattered from the two masses is given by

$$\Delta E_1 = E_0 \frac{dk}{dkm2} \Delta M_2 \qquad (2.10)$$

As can be seen from Fig. 2.8, best energy separation and mass resolution are obtained for light target elements where the derivative dK/dM_2 is steep, while for heavy elements the mass resolution gets small. The mass resolution for heavier elements can be improved by using higher incident energies or heavier incident projectiles. However, the deteriorated energy resolution of solid state detectors for heavier ions generally compensates the increased energy separation, and a gain in mass resolution by using heavier incident ions is only obtained with magnetic or electrostatic spectrometers. With solid state detectors the optimum mass resolution is obtained for projectiles with mass M_1 in the rage 4-7 [33].

48

$$\theta = 165°$$

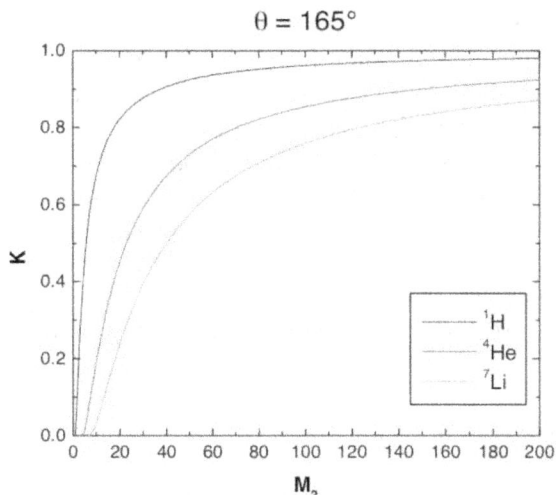

Fig. 2.8. Kinematic factor K at a scattering angle $\theta = 165°$ as a function of target mass M_2 for incident protons, 4He, and 7Li.

2.2.6. Proton Induced X-ray Emission (PIXE)

Proton Induced X-ray Emission (PIXE) is an X-ray spectrographic technique, which can be used for the non-destructive, simultaneous elemental analysis of solid, liquid or aerosol filter samples. The X-ray spectrum is initiated by energetic protons exciting the inner shell electrons in the target atoms. The expulsion of these inner shell electrons results in the production of X-rays. The energies of the X-rays, which are emitted when the created vacancies are filled again, are uniquely characteristic of the elements from which they originate and the number of X-rays emitted is proportional to the mass of that corresponding element in the sample being analyzed.

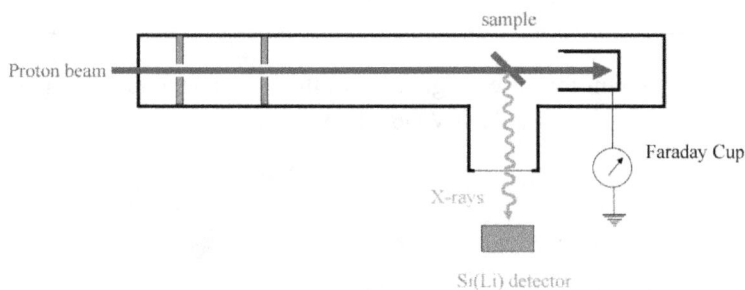

Fig. 2.9. PIXE arrangements.

The generation of X-rays in a sample is very strongly influenced by the bombarding proton. The probability of X-ray production depends upon both the total number of incident protons and the proton energy (measured in million electron volts (MeV)). The total number of incident protons can be expressed as proton current (measured in micro amps). The greater the proton current the greater is the probability for X-ray production. As proton energy changes, so does the probability for X-ray production. If quantitative analysis is to be assured, then both of these factors must be accurately known. Protons, while interacting with the sample matrix to produce X-rays, can be envisioned as undergoing collisions, much as a billiard ball on a pool table. Each successive collision will cause a transfer of kinetic energy from the projectile (proton) to the stationary object (target atom). While energy transfer per collision is small, a large number of collisions will eventually reduce proton energy significantly until it comes to rest. As the proton's energy decreases, so does its ability to produce X-rays. Since instrument calibration is performed at a specific proton energy, knowledge of the proton energy loss is essential for quantitative analysis.

The thin film materials were exposed to the ultraviolet light (254 nm) for 10 sec responds and 10 second recovery as well as visible light and the resistance of the films were simultaneously collected by the Keithley Semiconductor Characterization System (SCS) model 4200. The sample was placed in a chamber having a window with an aperture that could be opened or closed by sliding especially for the case of visible light sensing. For UV light sensing, a UV source was placed on top of the open aperture and the source could be turned ON or OFF while the resistance of the film is being monitored in time. The schematic of the light sensing system is given in Fig. 2.10.

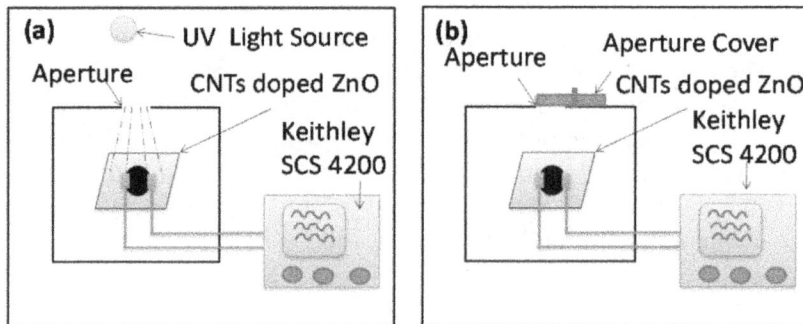

Fig. 2.10. Schematic diagram for the configuration of the experiment during (a) UV light sensing, and (b) Visible light sensing.

2.3. Results

2.3.1. Morphological Properties

The surface morphology of MWCNT-ZnO thin films deposited on a silicon substrate shows an evenand smooth coverage of the ZnO on the silicon substrate. In Fig. 2.11 are shown scanning electron microscope images of (a) carbon nanotubes mixed with ZnO starting material (pellet surface) showing that the carbon nanotubes and the ZnO powder are well mixed (b-d) typical SEM images of deposited substrates showing the carbon nanotubes are not as evenly distributed as expected. The carbon nanotube doped zinc oxide visualised at 100 nm shows the distributed rod shaped particles (Fig. 2.11 (b-d)). The carbon nanotubes are observed to become more straightened after deposition than before. This suggests that the flimsy carbon nanotubes are coated with stiffer material of ZnO.

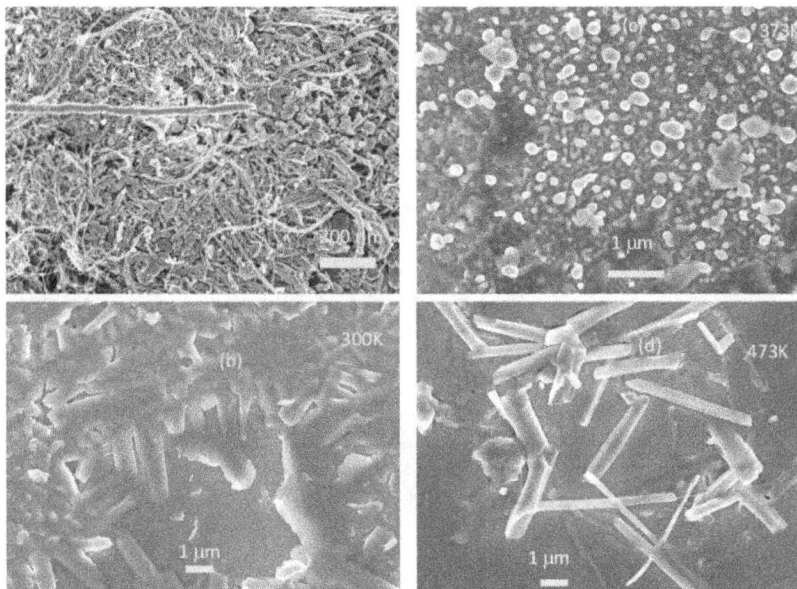

Fig. 2.11. (a) SEM image of carbon modified zinc oxide target before pulsed laser deposition, (b) deposited at 300 K, (c) deposited at 373 K, and (d) deposited at 473 K.

2.3.2. Atomic Force Microscopy

Surface topology of the MWCNT-ZnO thin films is presented in Fig. 2.12. The surface sections observed by AFM show no obvious evidence of carbon nanotubes distributed evenly throughout the surface. This observation is supported by SEM above where carbon nanotubes are seen in isolated islands on the surface. Apart from the different morphologies seen in the materials as the substrate temperatures are increased, there seems to be a pattern in the increment and decrement of the surface roughness of the samples. The values of RMS roughness are obtained via the following equation:

$$R_q = \sqrt{\frac{1}{N}\sum_{i=1}^{N}\left(z_i - \widetilde{z}\right)^2} \, , \tag{2.11}$$

where N is the number of data points, z_i and z ~ are the ith position and the average surface level respectively. The values of roughness are summarized in Table 2.1 along with crystallite sizes from XRD, band gaps from reflectance and other measurements whose results are given in the forthcoming sections. It will be noted that the pattern of surface roughness changes point to the bulk structural changes in the materials which will be discussed.

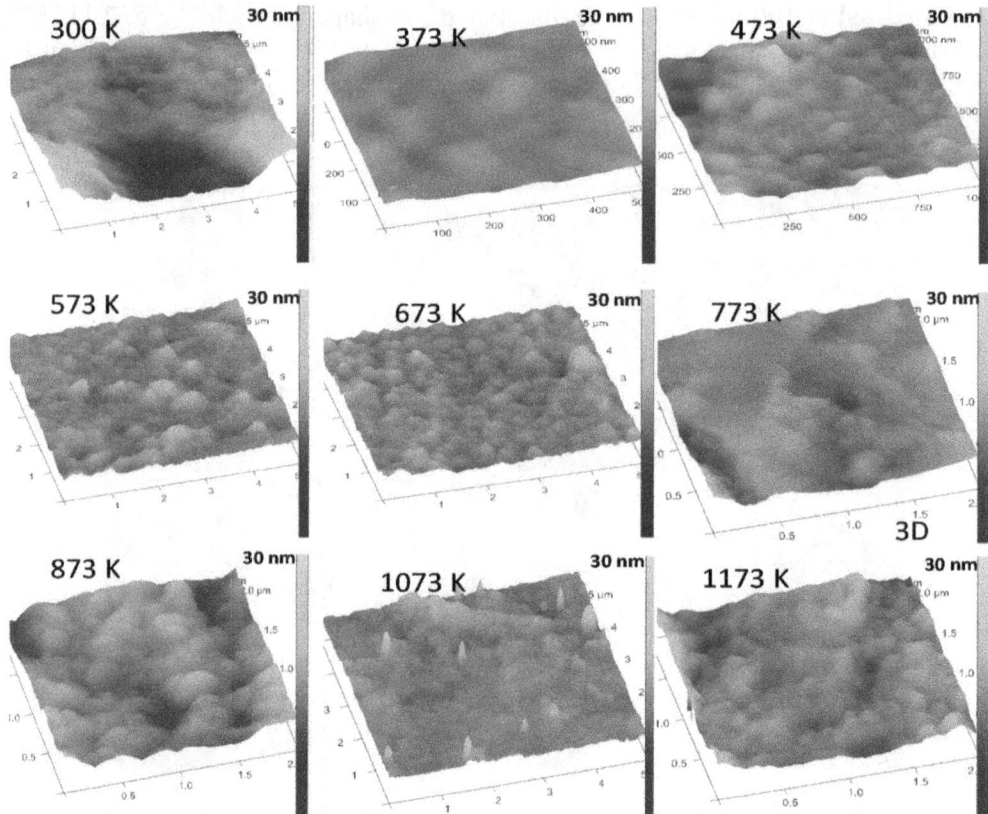

Fig. 2.12. The surface topology of the MWCNT-ZnO materials show a variety of morphologies that indicates that the carbon nanotubes are not uniformly distributed at the surface.

Table 2.1. Summary of roughness, crystallite size, calculated optical band gap energies, UV and Vis light response results for carbon modified ZnO samples synthesized on Si substrates at various temperatures.

Substrate Temp	Roughness, (nm)	ZnO (002) (nm)	Band Gap (eV)	Response (UV)	Response (Vis)	τ_{90} (UV)	τ_{90} (Vis)
301	4.1	19	2.9	0.05	0.2	0.1	1.3
373	1.65	43	3.4	0.02	0.04	0.07	1.1
473	10	22	3.1	0.2	0.01	0.06	1.0
573	3.47	39	3.1				
673	4.75	61	3.1	0.05	0.05	0.05	0.9
773	39.9	57	3.6				
873	13.3	33	4.3	0.5		0.05	1.0
1073	1.86	20	4.3		0.5		0.8
1173	12.9	19	2.7	0.05	0	0.04	

2.3.3. Structural Properties

Pure ZnO wurzite hexagonal structure show XRD strong peaks at 2θ values (and Miller (hkl) indices) of 31.7° (100), 34.422° (002) and 36.253° (101) as referenced from the International Crystal and Crystallography Data (ICCD) powder diffraction file (PDF) number 36-1451. For pure cubic ZnO, the PDF file 65-0523 shows three major peaks: 36.415° (111), 42.298° (200) and 61.360° (220) and PDF 13-0311 presents 31.927 (111), 37.024 (200) and 63.355 (311). Fig. 2.13 shows the XRD diffracto-grams of nanostructured carbon nanotubes modified zinc oxide.

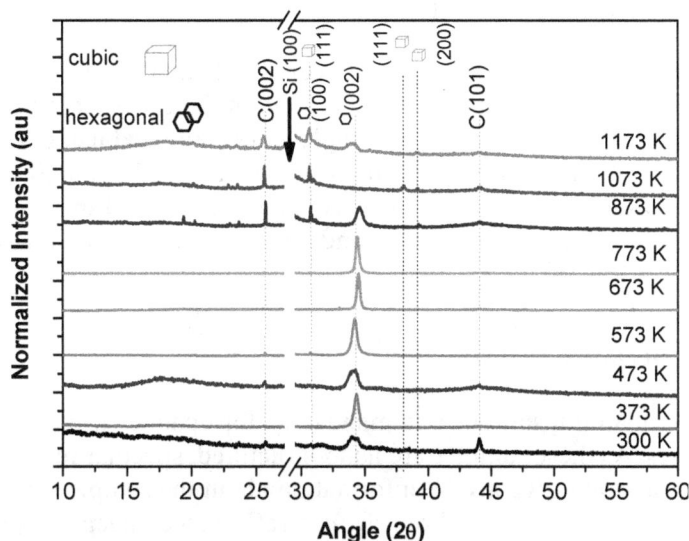

Fig. 2.13. X-ray diffraction spectra for carbon-modified ZnO thin films prepared at various substrate temperatures show a first class: from 300 K to 773 K assigned to the wurzite ZnO structure and new peaks when the films are prepared beyond 873 K which are assigned to cubic ZnO structure with carbon segregated to the surface.

It is clear that the samples prepared at substrate temperatures from 300 K to 773 K present a wurzite hexagonal structure. There are major shifts in the MWCNT-ZnO spectra from the pure ZnO: for instance in the pure ZnO the strongest alignment of the crystallite is in the (102) plane direction whereas in the carbon modified samples [300-773 K] the preferential alignment is in the {002} direction. This preferential orientation starts to disappear when the MWCNT-ZnO are prepared above the temperature of 873 K and this peak completely disappears at a temperature 1073 K. A new preferential orientation emerges at this temperature which matched the cubic ZnO structures with PDF numbers 65-0523 and 13-0311. It should be noted that the intensity peak at 2θ = 28° corresponds to (111) crystal plane of the silicon substrate which is observed in all the materials except 673 K. Also the intensity peak of carbon at 35° corresponds to (311). Fig. 2.13 shows the XRD diffracto-grams of nanostructured carbon nanotubes modified zinc oxide. It is clear that the samples prepared at substrate temperatures from 300 K to 773 K present a wurzite hexagonal structure. There are major shifts in the MWCNT-ZnO spectra from the pure

ZnO: for instance in the pure ZnO the strongest alignment of the crystallite is in the (102) plane direction whereas in the carbon modified samples [300-773K] the preferential alignment is in the {002} direction. This preferential orientation starts to disappear when the MWCNT-ZnO are prepared above the temperature of 873 K and this peak completely disappears at a temperature 1073 K. A new preferential orientation emerges at this temperature which matched the cubic ZnO structures with PDF numbers 65-0523 and 13-0311.

It should be noted that the intensity peak at $2\theta = 28°$ corresponds to (111) crystal plane of the silicon substrate which is observed in all the materials except 673 K. Also the intensity peak of carbon at 35° corresponds to (311).

XRD points to the substrate temperature between 873 K and 1073 K at which there is a transformation from hexagonal to cubic orientation in these carbon modified ZnO materials. The crystallite sizes for all the sample were estimated using Scherrer equation given by $D = (K\lambda)/(\beta\cos\theta)$ where D is the volume weighted crystallite diameter, K is constant which is taken to be equal to 0.9; λ is the wavelength of the X-ray photons, β is the full width at half maximum of the peak and θ is the Bragg angle in radian. The results are summarized in Table 2.1.

2.3.4. Diffuse Reflectance Spectroscopy

Reflectance rather than transmittance is mandatory for opaque surfaces as is the case in the present samples. As these samples are nano-structured, specular reflectance, suited for smooth surfaces, does not give all the information about the sample especially when the surface is rough. Rather, we carried out diffuse reflectance spectroscopy (DRS) which captures reflectance which would otherwise have been lost had we employed specular configuration. In DRS, a hemispherical concave mirror is used to focus both the specular and diffuse reflectance onto the detector. Fig. 2.14. PIXE spectra of as-deposited and MWCNT-ZnO films synthesized at varying substrate temperatures on Si (100). The reflectance data obtained can be used to acquire some information on the proportion of incident photons that are absorbed and hence the band gap energy of the material in the sample. The Kubelka- Munk equation [34] which relates the reflectance to absorption coefficient and scattering coefficient was use and is given here as

$$\frac{\alpha}{S} = \frac{(1-R)^2}{2R} \tag{2.12}$$

where α is the absorption coefficient, S is the scattering coefficient and R is the reflectance. In order to elucidate band gaps we employ the well know Tauc equation [35] given as

$$(\alpha h v)^n = E_g - h v \tag{2.13}$$

where h is the Planck's constant, E_g is the optical band gap of the material in the thin film and n can be with equal to 2 (for direct band gap material) or equal to ½ (for an indirect band gap material). Combining Equations 2.12 and 2.13, one can get

$$\left(S\frac{(1-R)^{\gamma}}{2R}h\nu \right)^{n} = E_{g} - h\nu \tag{2.14}$$

Note that the scattering coefficient, S, can be wavelength dependent. However it is known that for particle size greater than 5 μm, S can be wavelength independent [34]. In this analysis we also assumed that S is wavelength independent since the beam size impinges on larger than 1 mm in the far field and therefore the detector in the DRS observes a wavelength independent scattering from the nano-structured surface.

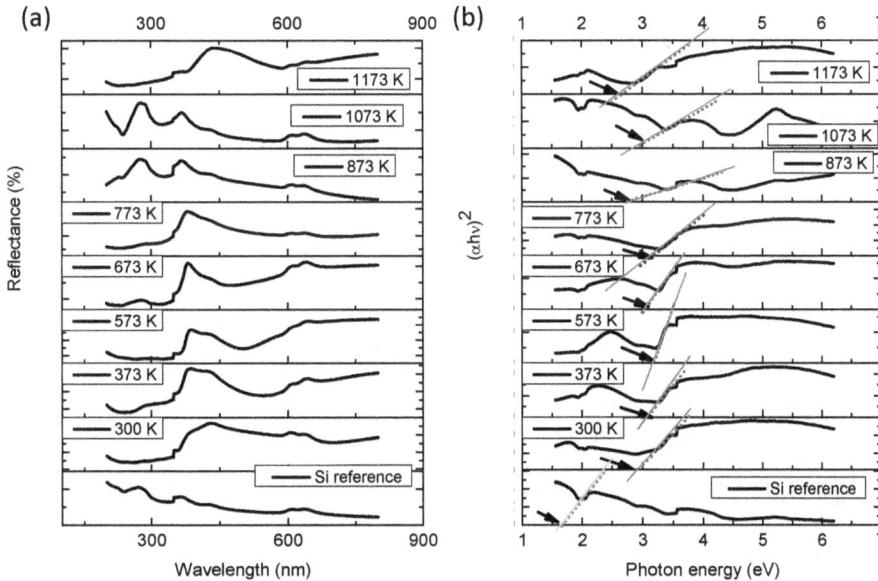

Fig. 2.14. (a) Reflectance vs wavelength for all samples compared to the Si reflectance as the substrate. The arrows show the main absorption edge; (b) The Tauc plots for all samples including the Si reference calculated from the Kubelka Munk function. The arrows indicate where on the photon energy axis the linear fit the optical band is extracted from for each sample.

The reflectance-versus-wavelength data are plotted for all substrate temperatures in Fig. 2.14 (a). In comparison the Si substrate, all MWCNT-ZnO samples have a higher reflectance in all visible and infrared wavelengths. However in this range, the Si substrate influences the reflectance spectrum as there are two humps between 600 nm and 700 nm in all spectra emanating from Si. For MWCNT -ZnO surfaces, there is a downward step around 320 nm indicating an absorption edge.

In order to pursue the absorption edge study and calculation of the optical band gap energies of each sample, a Tauc plot based on Equation 2.14 is displayed in Fig. 2.14 (b). Coloured lines are inserted in each spectrum in order to guide the reader's eye; these lines are the linear plots of Equation 2.14 when n = 2 as it well known that ZnO is a direct band gap material assuming that the allotrope of carbon in ZnO is mostly graphitic or simply carbon nano-tube which are also known to be direct band gap materials.

55

2.3.5. Ultra Violet and Visible Sensing

The functionalized MWCNT-ZnO sensors synthesized at substrate temperatures of 300, 373, 473 up to 1173 K were tested using UV light (254 nm) and visible light as a stimuli for the interval of 10 second during exposure and 10 second during recovery (Fig. 2.15). The response, S, has been defined as S = [R-R_0]/R_0 [19, 20]. It was found that all the materials respond to both the UV light and visible light except for the sample synthesized at a substrate temperature of 773 K. In addition, for the MWCNT-ZnO material synthesized at 1173 K, response to visible light is negligible; the streaks in the almost flat time profile could be due to changes in pressure as the light chopper is opened and closed.

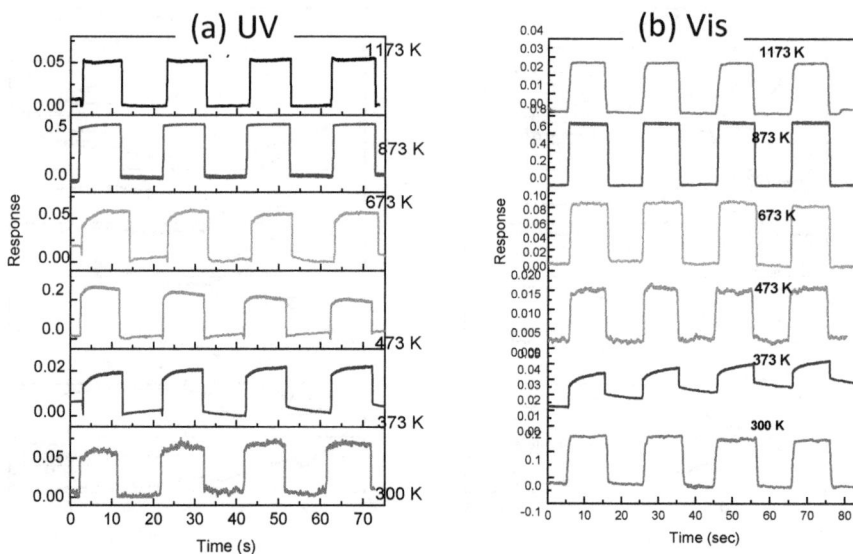

Fig. 2.15. Temporal response to (a) UV light and (b) Visible light of the MWCNT-ZnO samples synthesized at different temperatures show the variation of response when the light is ON and OFF. The response and recovery time of MWCNT-ZnO sensor films towards UV and visible light were extracted from the fitting of the following exponential association curve in Origin™ software:

$$S(t) = A\left[1 - \exp\left(-\frac{\tau}{t}\right)\right], \tag{2.15}$$

where S(t) is the response (or recovery) of the sensor to (or from) the exposure as a function of time, A is the amplitude of the response and τ is the response (or recovery) lifetime of the response when the response equals $A(1-e^{-1}) \sim 0.65$ A.

The response times are traditionally defined as time taken by the sensor output to reach 90 % of its saturation after applying or switching off the target analyte and it is called τ_{90}. With the above fitting, τ_{90} values are obtained through calculated τ from Equation 2.15 as follows:

$$\tau_{90} = \frac{-\tau}{\ln\left(1 - \dfrac{S_{90}}{A}\right)} \qquad (2.16)$$

The response time of the materials calculated from Equation 2.16 are plotted Fig. 2.16 from the results the material that is synthesised at 600 °C has the longest response time which means slowest of them all.

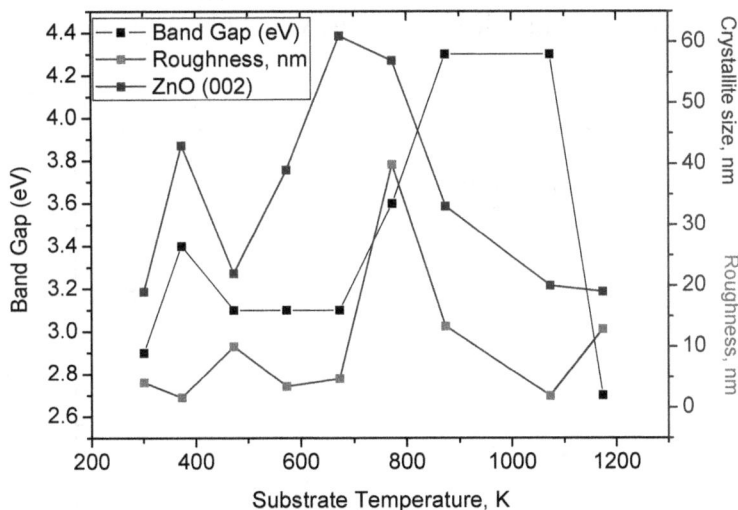

Fig. 2.16. A chart that shows a plot of band gap energy, roughness and crystallite size for carbon modified ZnO materials on Si substrate as functions of the substrate temperature.

The extracted responses and response time values (τ_{90}) are summarized in Table 2.1, alongside calculated band gap energies from reflectivity [Eq. 3.4, 3.5] and from Fig. 2.14 as well as crystallite size from XRD and roughness from AFM.

When band gap energies, crystallite size and roughness of the MWCNT-ZnO materials are plotted against substrate temperature, it is found that band gaps are in the range from 3.1 eV to 3.4 eV which is acceptable for wurzite hexagonal structure of ZnO up to about 600 K. Above this temperature (800 K – 1073 K), the optical band gap energies drastically increase to above 4 eV only to subside into the 2.6 eV range at 1173 K. This may suggest that at around 800 K there is a transition from one structure to another type of structure. This has been confirmed by XRD which suggests a transition from hexagonal to cubic structure of ZnO between 773 K and 873 K. Similar trends are seen in this Fig. 2.10 for crystallite size as well roughness although the transition temperature are slightly shifted to lower temperatures.

In addition, plots of response versus substrate temperature in Fig. 2.17 reveal that there is low sensitivity in the MWCNT-ZnO synthesized at lower than 673 K, which is the range of temperatures that gave rise to the wurzite hexagonal structure. However there is an abrupt rise in response in the MWCNT-ZnO materials synthesized at substrate temperatures in the range 800 to 1073 K in which range the cubic structure of ZnO dominates. The response in the range reaches the maximum value of 50 % which much higher than the reported values for MWCNT-ZnO materials [13, 14]. There is another abrupt and sharp decrease in response as the substrate temperature is increased from 1073 K to 1173 K. This may indicate that there are some changes on the surface of the sample synthesized at 1173 K and the present RBS results confirmed a drastically reduced thickness in the deposition of film at this temperature. This suggests that either the sample loses its integrity as MWCNT-ZnO and becomes zinc carbides or the ZnO actually start to evaporate from the coating leaving behind very little material for good conduction and hence meaningful sensing.

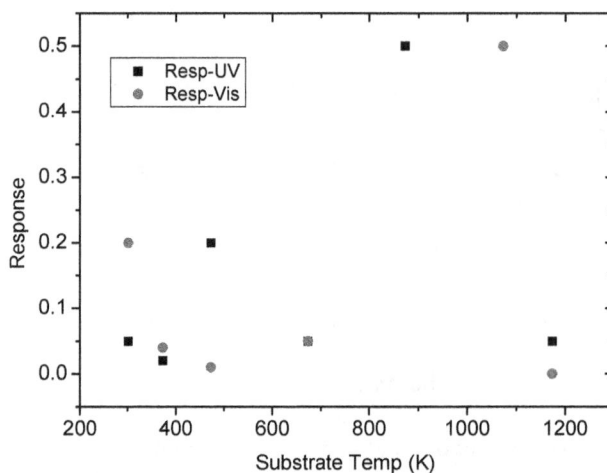

Fig. 2.17. Response time versus substrate temperature at which the MWCNT-ZnO were synthesised showing the response to both UV and visible light is lower when the samples adopt the hexagonal structure but higher when MWCNT-ZnO is cubic.

There is an interesting trend in the plots of response times of the carbon modified ZnO against their substrate temperatures. It can be seen in Fig. 2.18 that the response to UV is generally faster than the response to visible light as the response times to UV are consistently smaller than those to visible light. In addition, one observes decreasing trends of the response times in both UV response and visible light response as the substrate temperatures are increased from 300 K (1.3 s in Vis and 0.1 s in UV) to 1173 K (0.8 s for Vis and 0.03 s for UV). This indicates that the MWCNT-ZnO materials become faster in light response as the substrate temperatures are increased from 300 K up to 600 K. Above 600 K the response times are more or less constant (0.8 s for Vis and 0.03 s for UV). In the constant regime, there is a small hump at 873 K which confirms the transition temperature from hexagonal to cubic structures of ZnO as already shown through XRD, optical band gap energies as well as morphological properties of roughness.

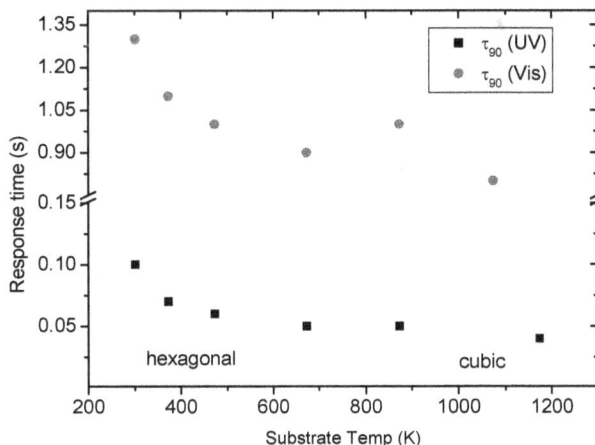

Fig. 2.18. A plot of response time as function of the substrate temperature of MWCNT-ZnO nanomaterials showing longer response times (or slower response) when the MWCNT-ZnO is hexagonal than when the MWCNT-ZnO transforms into cubic form.

2.4. Conclusions

The present report has shown that MWCNT-ZnO thin films on Si have been prepared by pulsed laser deposition of a ZnO/carbon-nano-tube pellet. The films were obtained at varying substrate temperature from 300 K to 1173 K in increments of 100K. XRD has shown that the wurzite hexagonal structure of ZnO transforms into cubic structure in the temperature range of between 774 K and 873 K. This temperature range has been found to be display drastic changes in optical band gaps (3.1 eV → 4.0 eV), surface roughness (10 nm → 40 nm) as well as crystallite sizes (20 nm → 60 nm). Similar changes in UV and visible light sensing properties such as response (10 % →50 %) and response times (1.2 s → 0.8 s for visible light and 0.1 s → 0.03 s for UV sensing) are witnessed at these temperatures. The chapter has shown that it is possible to enhance the light sensing efficiency in MWCNT-ZnO materials from the current value of 10 % to a new maximum of 50 % by varying the substrate temperatures. And these maxima are obtained in materials that were synthesized at substrate temperatures in the ranges between 873 K and 1073 K.

Acknowledgements

The financial support from the HGER27S project of the DST/CSIR is acknowledged. Partial support was also obtained from the India-Brazil-South Africa (IBSA) project number HGE24X and the South Africa- Taiwan project number HGER31X.

References

[1]. M. Willander, K. Khun, Z. H. Ibupoto, ZnO based potentiometric and amperometric nanosensors, *J. Nanosci. & Nanotechnol.*, 14, 9, 2014, pp. 6497-6508.

[2]. B. W. Mwakikunga, Progress in ultrasonic spray pyrolysis for condensed matter sciences developed from ultrasonic nebulization theories since Michael Faraday, *Crit. Rev. in Solid State and Mater. Sci.,* 39, 2014, pp. 46-80.

[3]. G. H. Mhlongo, D. E. Motaung, S. S. Nkosi, H. C. Swart, G. F. Malgas, K. T. Hillie, Mwakikunga, B. W., Temperature-dependence on the structural, optical, and paramagnetic properties of ZnO nanostructures, *Appl. Surf. Sci.,* 293, 2014, pp. 62-70.

[4]. D. E. Motaung, G. H. Mhlongo, I. Kortidis, S. S. Nkosi, G. F. Malgas, B. W. Mwakikunga, S. S. Ray, G. Kiriakidis, Structural and optical properties of ZnO nanostructures grown by aerosol spray pyrolysis: Candidates for room temperature methane and hydrogen gas sensing, *Appl. Surf. Sci.,* 279, 2013, pp. 142-149.

[5]. A. Boyd, I, Dube, G. Fedorov, and M. Paranjape, Gas sensing mechanism of carbon nanotubes: From single tubes to high-density networks, *Carbon,* 69, 2014, pp. 417-423.

[6]. J Goldberger, D. J. Sirbuly, M. Law, P. Yang, ZnO nanowire transistors, *Phy Chem B Lett,* 109, 2005, pp. 9-14.

[7]. Z-Y. Fan, J. G. Lu, Chemical sensors and electronic noses based on 1-D metal oxide nanostructures, *IEEE Transactions on Nanotechnology,* 5, 2006, pp. 393-397.

[8]. S. S. Nkosi, B. Yalisi, D. E. Motaung, J. Keartland, E. Sideras-Haddad, A. Forbes, B. W. Mwakikunga, Optical constants correlated electrons-spin of micro doughnuts of Mn-doped ZnO films, *Appl. Surf. Sci.,* 265, 2013, pp. 860–864.

[9]. K. T. Roro, B. Mwakikunga, N. Tile, B. Yalisi, A. Forbes, Effect of accelerated thermal ageing on the selective solar thermal harvesting properties of multiwall carbon nanotube/nickel oxide nanocomposite coatings, *Int. J. Photoenergy,* 2012, 2012, Article ID 678394.

[10]. R. Taziwa, E. L. Meyer, E. Sideras-Haddad, R. M. Erasmus, E. Manikandan and B. W. Mwakikunga, Effect of Carbon Modification on the Electrical, Structural, and Optical Properties of TiO_2 Electrodes and Their Performance in Labscale Dye-Sensitized Solar Cells, *Int J. Photoenergy,* Vol. 2012, Article ID 904323, 9 pages.

[11]. B. W. Mwakikunga, A. E. Mudau, N. Brink, C. J. Willers, Flame temperature trends in reacting vanadium and tungsten ethoxide fluid sprays during CO_2-laser pyrolysis, *Appl. Phys. B.,* 105, 2011, pp. 451-462.

[12]. K. T. Roro, N. Tile, B. Mwakikunga, B. Yalisi, A. Forbes, Solar absorption and thermal emission properties of multiwall carbon nanotube/nickel oxide nanocomposite thin films synthesized by sol–gel process, *Mater. Sci. & Eng. B: Solid-State Materials for Advanced Technology,* 177, 2012, pp. 581-587.

[13]. N. Kouklin, M. Omari, Optical-gating and carrier modulation effects in single-walled carbon nanotube-oxide interfaces for opto-electronic applications, *ECS Transactions,* 25, 21, 2010, pp. 27-31.

[14]. S. Sen, D. Chowdhary, N. A. Kouklin, Negative photoconduction of planar heterogeneous random network of ZnO-carbon nanotubes, *Appl. Phys. Lett,* 91, 9, 2007, pp. 093125.

[15]. F. Fang, J. Futter, A. Markwitz, and J. Kennedy, UV and humidity sensing properties of ZnO nanorods prepared by the arc discharge method, *Nanotechnol,* 20, 2009, pp. 245502.

[16]. V. J. Kennedy, A. Markwitz, U. D. Lanke, A. McIvor, H. J. Trodahl, and A. Bittar, Ion beam analysis of ion-assisted deposited amorphous GaN, *Nucl. Instrum. Method Phys. Res. B,* 190, 2002, pp. 620-624.

[17]. T. J. McCarthy, S.-P. Ngeyi, J.-H. Liao, D. C. DeGroot, T. Hogan, C. R. Kannewurf, M. G. Kanatzidis, Use of molten alkali-metal polythiophosphate fluxes for synthesis at intermediate temperatures. Isolation and structural characterization of ABiP2S7 (A = K, Rb), *Chem. Mater.,* 5, 1993, pp. 331-340.

[18]. A. Simo, B. W. Mwakikunga, B. T. Sone, B. Julies, R. Madjoe, M. Maaza, VO_2 nanostructures based chemiresistors for low power energy consumption hydrogen sensing, *Int. J. Hydrogen Energy,* 39, 2014, pp. 8147–8157.

[19]. S. Mallakpour, E. Khadem, Carbon nanotube-metal oxide nanocomposites: Fabrication, properties and applications, *Chemical Engineering J.,* 302, 2016, pp. 344-367.

[20]. R Eason, Pulsed Laser Deposition of thin film, *John Wiley & Sons,* 2006.

[21]. B. Fults, J. Howe, Transmission Electron Microscopy and Diffractometry of Materials, Springer, 2003.

[22]. B. D. Cullit, Element of X-Ray diffraction, Addison Wesley, 1956.

[23]. A. G. Marangoni, M. F. Peyronel, X-Ray Powder Diffraction: Analytical methods, procedures & theory for physical characterization of fats, *Lipids Library,*
http://lipidlibrary.aocs.org/Biochemistry/content.cfm?ItemNumber=40299

[24]. B. R. Rehani, P. B. Joshi, K. N. Lad, A. Pratap, Crystallite size Estimation of elemental and composite silver nano-powders using XRD principles, *Indian J. Pure & App. Phys.*, 44, 2006, pp. 157-161.

[25]. http://pd.cheem.ucl.ac.uk/pdnn/peaks/size.htm

[26]. R. F. Egerton, Physical Principles of Electron Microscopy. An introduction to TEM, SEM and AEM, *Springler,* 2005.

[27]. M. Govender, Synthesis of Tungsten Oxide Nanostructure by Laser Pyrolysis, MSc. Dissertation, *University of Witwatersrand,* 2011.

[28]. R. Agrawal, K. Lee and D. Ponnavolu, AFM Spectroscopy, in Micro/Nano Science and Engineering, *Northwestern University,* Evanston, USA, 2004.

[29]. F. L. Leite, L. H. C. Mattoso, O. N. Oliveira Jr, P. S. P. Herrmann Jr. The Atomic Force Spectroscopy as a Tool to Investigate Surface Forces: Basic Principles and Applications, in Modern Research and Educational Topics in Microscopy, A. Méndez-Vilas and J. Díaz (Eds.), *Formatex,* 2007.

[30]. G. Meyer, and N. M. Amer, Novel optical approach to atomic force microscopy, *Appl. Phys. Lett.*, 53, 1045, 1988.

[31]. S. Alexander, L. Hellemans, O. Marti, J. Schneir, V. Elings, P. K. Hansma, M. Lonhmire and J. Gurley, An atomic-resolution atomic-force microscope implemented using an optical lever, *J. Appl. Phys.* 65, 164, 1989.

[32]. Nanocomposix, UV/VIS/IR Spectroscopy Analysis of Nanoparticles, nanocomposix.com, 2012.

[33]. M. Mayer, Rutherford Backscattering Spectrometry, *Workshop on Nuclear Data for Science and Technology,* 2003.

[34]. B. W. Mwakikunga, S. Motshekga, L. Sikhwivhilu, M. Moodley, M. Scriba, G. Malgas, A. Simo, B. Sone, M. Maaza, S. S. Ray, A classification and ranking system on the H 2 gas sensing capabilities of nanomaterials based on proposed coefficients of sensor performance and sensor efficiency equations, *Sensors and Actuators B: Chemical,* 184, 2013, pp. 170–178.

[35]. B. W. Mwakikunga, M. Maaza, K. T. Hillie, C. J. Arendse, T. Malwela, E. Sideras-Haddad, From phonon confinement to phonon splitting in flat single nanostructures: a case of VO 2@ V 2 O 5 core–shell nano-ribbons, *Vibrational Spectroscopy,* 61, 2012, pp. 105–111.

[36]. T.-J. Huang, C.-C. Ma, Characterization of response of ZnO/LiNbO3-based surface acoustic wave delay line photodetector, *Jap J Appl. Phys.,* 47, 2008, pp. 6507-6512.

[37]. N. M. Kiasari, S. Soltanian, B. Gholamkhass, P. Servati, Environmental gas and light sensing using ZnO nanowires, *IEEE Transactions on Nanotechnology*, 13, 2014, pp. 368- 374.

Chapter 3
Fiber Optic Sensing Based on Multimode Interference

Susana Silva and Orlando Frazão

3.1. Introduction

The appearance of the multimode fiber, in the early 1970's [1], was a major breakthrough for the achievement of high-speed telecommunications and quickly gave rise to the development of several optical fiber sensors. The multimode interference (MMI) effects in multimode fibers rapidly earned the attention of researchers due to the high potential in integrated optics and, in fact, led to the production of new optical devices [2].

In the past, multimode fiber-based structures were extensively studied for the development of microbend sensors since the intensity modulation induced by microbend loss in such fibers could provide a simple transduction mechanism to detect environmental changes such as pressure, temperature, acceleration, and electromagnetic fields [3]. In 1997, Donlagic *et al.* [4] presented the first microbend sensor configuration based on a multimode fiber section with gradient index profile (GI-MMF) spliced between two singlemode fibers (SMFs) thus forming a SMF–GI-MMF–SMF, that was defined by the authors as an SMS structure. In this case, the GI-MMF provided no MMI effects to form the microbend sensor. In a different perspective, multimode fibers with step-index profile (SI-MMF) make use of interference between high order modes along its length [5]. Only in recent years, the research on this topic became more attractive due to the possibility of creating new MMI-based fiber devices.

Nowadays, the typical MMI-based fiber device relies on a SI-MMF (or, for sake of simplicity, MMF), which is the basis of the SMS structure; it can also be the combination of an MMF section spliced at the end of an SMF (SMF-MMF tip) in order to act as a wavelength tunable fiber lens [5] or a fiber tip structure [6]. Regardless the type of configuration used, the MMI-based fiber structure can act as a bandpass or band-rejection filter depending on the length of the MMF used [7-9]. Bandpass filters have unique

Susana Silva
INESC TEC - INstituto de Engenharia de Sistemas e Computadores - Tecnologia e Ciência, Porto, Portugal

spectral characteristics that make them highly attractive for applications such as wavelength-tunable filtering [10, 11] and high-power lasers and amplifiers [12, 13]; while band-rejection filters are often used for the measurement of physical parameters such as temperature, strain and refractive index (RI) [14].

In overall, MMI-based fiber devices are simple structures that have been increasingly used for sensing applications due to the performances possible to be achieved, namely, in the measurement of RI. This Chapter addresses this topic, starting by the theoretical concepts of light propagation in a MMI-based fiber device, in particular the typical SMS configuration. Afterwards, an historical overview of fiber optic sensors based on MMI concept is presented with focus on the measurement of RI. The last section finishes with concluding remarks.

3.2. Operating Principle of MMI-based Fiber Device

The typical MMI-based fiber device relies on the SMS configuration and the theoretical model of light propagation along such fiber structure has been extensively studied [5, 7, 8]. In this section, the operating principle of such fiber device is described as follows. Consider that the SMS fiber structure is formed by a section of MMF with length L spliced between two SMFs and interrogated in transmission. In this case, the SMFs and MMF are assumed to be aligned along the same axis; they also have circular cross-sections and step-index profiles. The schematic of the SMS fiber configuration is presented in Fig. 3.1.

To better understand the behavior of the SMS structure, the linearly polarized mode approximation is used. Consider that the light propagating along the input SMF enters the MMF with an approximate Gaussian-shaped field distribution $E(r)$ that serves as the input field for the MMF section.

$$E(r) = e^{-(r/W)^2} e^{-j\beta_0 z},$$

(3.1)

where β_0 is the longitudinal propagation constant for the SMF guided mode LP_{01}, r and z are the radial and longitudinal axis, and W is the half-width at half-maximum spot size of the Gaussian beam [5]. Note that the time dependence $e^{j\omega t}$ is implicit in the field profile.

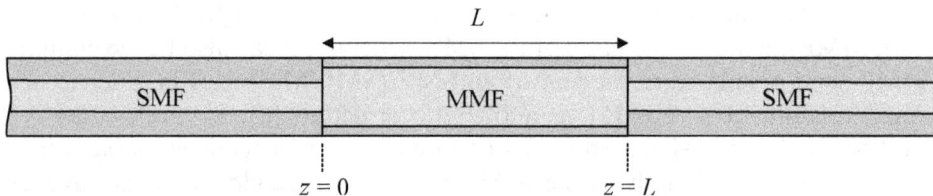

Fig. 3.1. Generic scheme of the SMS fiber structure.

Due to the circular cross-section of the SMF and cylindrical symmetry characteristic of the fundamental mode LP_{01}, the input light at $z = 0$ (see Fig. 3.1) is assumed to have a field distribution $E(r) = E(r,0)$.

When the light launches the MMF, at $z = 0$, the input field can be decomposed by the eigenmodes LP_{lm} of the MMF. Considering the ideal alignment of the fibers, as mentioned above, the radiating modes are neglected. Moreover, the cylindrical symmetry condition of the input field allows only pure radial modes LP_{0m} to be excited (i.e., there is no contribution from azimuthal modes) [5]. Therefore, the input field profile at $z = 0$ (see Fig. 3.1) can be represented by a finite summation over the guided modes only, as follows [7]:

$$E(r,0) = \sum_{m=1}^{M} c_m \varphi_m(r),$$ (3.2)

where c_m is the field expansion coefficient of the mth mode of the MMF, M is the total number of radial modes, and $\varphi_m(r)$ is the field profile of LP_{0m}.

The input field excites a specific number of guided modes LP_{0m} inside the MMF. As the light propagates along the MMF section with length L, the field distribution at a distance z can be written as [7]:

$$E(r,z) = \sum_{m=1}^{M} c_m \varphi_m(r) e^{j\beta_m z},$$ (3.3)

where β_m is the longitudinal propagation constant of the mth mode of the MMF. The field expansion coefficient c_m can be calculated by the following cross-correlation expression [8]:

$$c_m = \frac{\int_0^\infty E(r,0) \varphi_m(r) dr}{P_m},$$ (3.4)

where P_m is the power of each guided mode and is given by:

$$P_m = \int_0^\infty |\varphi_m(r)|^2 r dr,$$ (3.5)

The power coupling efficiency η determines the amount of the input field power that couples to each specific mode in the MMF. The field excitation coefficient c_m is related to the power coupling efficiency η through $c_m^* = \sqrt{\eta}$, where the modified field excitation coefficient c_m^* is defined as [8]:

$$c_m^* = c_m \sqrt{P_m / P_S} , \tag{3.6}$$

where P_s is the power of the source and is given by:

$$P_S = \int_0^\infty |E(r,0)|^2 r dr , \tag{3.7}$$

Therefore, power coupling efficiency η can be determined by the overlap integral between $E(r,0)$ and $\varphi_m(r)$ as follows [8]:

$$\eta = \frac{\left| \int_0^\infty E(r,0)\varphi_m(r) r dr \right|^2}{\int_0^\infty |E(r,0)|^2 r dr \int_0^\infty |\varphi_m(r)|^2 r dr} , \tag{3.8}$$

The coupling efficiency will reach a peak when the length of the MMF section is selected so that the field distribution at the end of said fiber is an image of the input field. That length is referred to as the *self-imaging* distance [2] and the deviation from this condition causes the coupling efficiency to drop. This parameter will depend not only on the MMF dimensions but it is also strongly wavelength dependent.

The principle of self-imaging in multimode waveguides was stated by Soldano *et al.* [2] as follows: "*Self-imaging is a property of multimode waveguides by which an input field profile is reproduced in single or multiple images at periodic intervals along the propagation direction of the guide.*"

When the light field propagating along the input SMF enters the MMF section, several modes of MMF are excited with different propagation constants β_m. Hence, interference between different modes occurs, while the light beam propagates along the remainder of the MMF section.

At the exit end of the MMF section, the phase difference between two consecutive modes, LP_{0m} and LP_{0n}, is determined by the MMF length and the difference between the longitudinal propagation constants of the two modes, $\beta_m - \beta_n$, which can be expressed as [5]:

$$\beta_m - \beta_n = \frac{u_m^2 - u_n^2}{2ka^2 n_{co}} \qquad m > n , \tag{3.9}$$

where $u_m = \pi(m - 1/4)$ and $u_n = \pi(n - 1/4)$ are the roots of the Bessel function of zero order, a is the MMF core radius, $k = 2\pi / \lambda$ is the free space wave number, and n_{co} is the refractive index of the fiber core.

66

The condition for constructive interference between LP_{0m} and LP_{0n} is [5]:

$$\left(\beta_m - \beta_n\right)z = 2\pi N, \qquad (3.10)$$

where N is an integer. This means that are several locations z_N, corresponding to local maxima along the MMF, where field condensation occurs. The self-image condition should be a specific case of constructive interference, i.e. a specific integer N multiple of the location z_N characteristic of this effect. From equation (3.10), one can obtain z_N as follows:

$$z_N = \frac{8ka^2 n_{co}}{\pi(m-n)[2(m+n)-1]}N. \qquad (3.11)$$

Since two consecutive modes are considered, $n = m - 1$ and equation (3.11) is simplified according to:

$$z_N = \frac{8ka^2 n_{co}}{\pi(4m-3)}N. \qquad (3.12)$$

Moreover, there is a specific location z_p where the phase difference between the two excited modes equals 2π (i.e., $N = 1$). This is the so called beat length and is given by:

$$z_p = \frac{8ka^2 n_{co}}{\pi(4m-3)}. \qquad (3.13)$$

One can conclude that $z_N = Nz_p$. From this expression it is possible to obtain the self-imaging distance z_i as follows [5]:

$$z_i = \frac{16a^2 n_{co}}{\lambda}, \qquad (3.14)$$

where λ is the operational wavelength. Thus, the self-imaging distance z_i pertains to where the light field at the input of the MMF section is replicated, in both amplitude and phase, on the output of the MMF for a specific wavelength. In this specific case, only the first self-image (and multiples) will exhibit minimum losses – the self-imaging effect provides an optical spectrum in the form of a bandpass filter [8]. In the case of the fractional planes where the coupling coefficient exhibit minimum losses, one can obtain band-rejection filtering [9]. The optical spectrum presents loss peaks that are produced by destructive interference between a set of specific higher-order modes at the exit end of the MMF. In overall, MMI-based structures rely on the characteristics of the MMF used, namely, its diameter and length, but also depend on coupling efficiency between MMF and SMF. The following section provides an historic overview of MMI-based fiber structures where its ability to be used for sensing strongly relies on the aforementioned aspected that characterize a fiber structure based on multimode interference.

3.3. Historic Overview of MMI-based Fiber Sensing

The first sensor based on MMI concept that used a step-index MMF was studied in 2003 by Mehta *et al.* [15], which developed the theoretical model and experimental setup of a displacement sensor. The structure consisted of an SMF-MMF tip combined with a planar mirror, as depicted in Fig. 3.2. This structure makes use of the self-imaging concept; the planar mirror reflects the light in the self-imaging point back to the MMF and thus forming the displacement sensor. In 2006, Li *et al.* [6] developed a temperature sensor by means of the fiber tip concept, as shown in Fig. 3.3. The structure consisted of an SMF-MMF tip which is interrogated in reflection. This sensor relied on the interference between different modes at the MMF end facet, which are wavelength sensitive to temperature variation. The same research group also studied the effect of certain packaging materials such as ceramic to perform temperature compensation [16] and also aluminum to acquire high sensitivity to temperature [17].

Later, Wang *et al.* [18] presented a numerical simulation for an optical fibre refractometer based on self-imaging concept – the SMS structure relied on an MMF core section where the surrounding liquid sample worked as the cladding medium, as shown in Fig. 3.4.

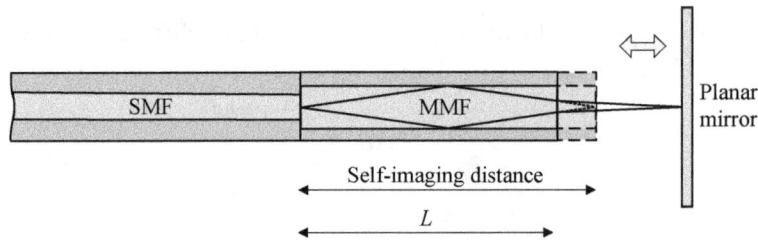

Fig. 3.2. Schematic of the SMF-MMF tip + mirror configuration [15].

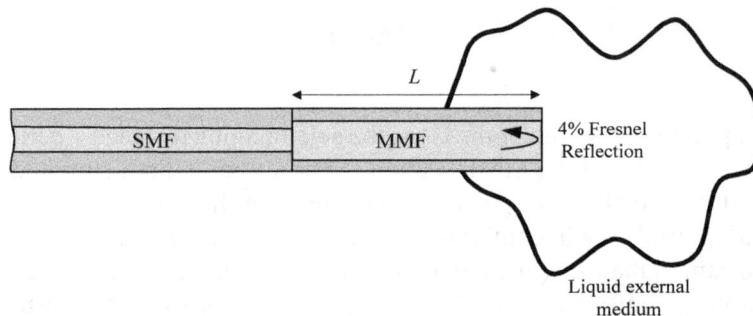

Fig. 3.3. Schematic of the SMF-MMF tip sensor [6].

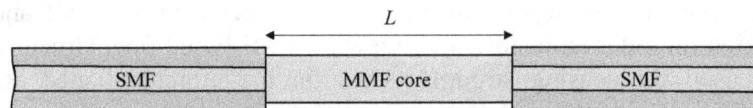

Fig. 3.4. Schematic of the SMS structure based on an MMF core section [18].

Interference between different modes occurs while the light propagates along the MMF core section and it will be wavelength-sensitive to different external RI. The RI resolution was estimated to be 5.4×10^{-5} RIU within the range from 1.33 to 1.45. A few years later, this from research group extended the study to taper-based SMS (STMS) structures [19, 20]. From numerical analysis, an STMS could offer a maximum resolution of 2.59×10^{-5} RIU within the dynamic range 1.346 to 1.416 [19]; while experiments with this type of structure resulted in an average sensitivity of 487 nm/RIU over an RI range of 1.33-1.44 [20]. A maximum sensitivity of 1913 nm/RIU was achieved for RI from 1.43 to 1.44 with a corresponding resolution of 5.23×10^{-6} RIU.

In 2010, Gao *et al.* [21] developed a temperature sensor based on a SMS structure with an etched MMF. In a different perspective, Hatta *et al.* [22] proposed a ratiometric power measurement scheme based on two SMS fiber structures for the measurement of strain with temperature compensation. The research group also demonstrated that attaching a piezoelectric to the SMS structure a voltage sensor could be attained by induced strain variations [23]. Zhang *et al.* [24] reported instead the feasibility of a SMS fibre structure as a refractometer sensor where an MMF section with different HF-corroded cladding diameters was used. Results showed that the resolution could be improved by using an MMF core with full HF-corroded cladding. The RI resolution achieved for this sensor was 7.9×10^{-5} RIU for RI range from 1.33 to 1.38.

In 2011, an SMF-MMF tip sensor for liquid level measurement was reported by Antonio-Lopez *et al.* [25]. The sensing structure was based on a pure silica rod (coreless-MMF) spliced to an SMF to form a bandpass filter. The MMF tip was coated with a thin layer of gold to act as mirror, so that the self-image was back-reflected and coupled into the SMF. Therefore, when the coreless-MMF was immersed in a liquid and its level was changed, a correlated shift of the wavelength peak was found. In a different perspective, Gong *et al.* [26] developed an SMS fiber structure for curvature sensing. The transmitted optical spectrum had several loss bands that presented different sensitivities to curvature. Using that ability, Silva *et al.* [27] proposed a temperature and strain independent curvature sensor by means of a similar SMS configuration. Wu *et al.* [28] produced, in turn, a displacement sensor based on a bent SMS structure. Later, the research group used this concept to create a single-side bent SMS structure to measure both displacement and temperature simultaneously and independently [29]. A simple scheme based on two SMS fiber structures in series was also developed by Wu *et al.* [30] for simultaneous measurement of strain and temperature. In a different line of research, Silva *et al.* [31] characterized a SMS structure based on a large-core, air-clad photonic crystal fiber (PCF) as a refractometer sensor – the schematic and detail of the PCFs used are shown in Fig. 3.5.

Using two distinct large-core air-clad PCF geometries, one for RI measurement and other for temperature compensation, it was possible to implement a sensing head that was sensitive to RI changes induced by temperature variations in water. The system presented a resolution of 3.4×10^{-5} RIU for a variation of 1 °C and maximum sensitivity of 800 nm/RIU was attained in the RI range from 1.3246 to 1.3266. On the other hand, Coelho *et al.* [14] used the outer cladding of a large-core air-clad PCF as an optical waveguide, which in turn acts as an MMF section in the SMS structure. The fiber device allowed simultaneous measurement of temperature and strain. A maximum sensitivity of 322.08 nm/RIU was also attained for the measurement of RI in the range 1.338-1.403. Wu *et al.* [32] investigated the influence of different etched MMF core diameters and lengths on the sensitivity of an SMS fiber based refractometer. Numerical analysis has shown that the sensitivity to external RI increased when the etched MMF core diameter decreased. Experimental results have shown that using an MMF core with 80 μm-diameter, a maximum sensitivity of 1815 nm/RIU for a RI range from 1.342 to 1.437 could be achieved.

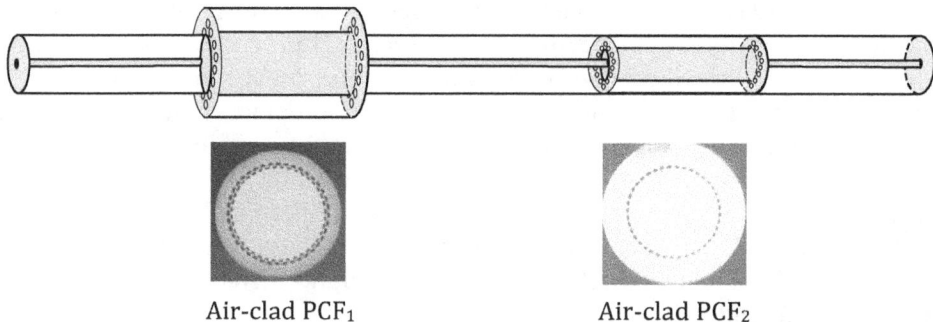

Air-clad PCF₁   Air-clad PCF₂

Fig. 3.5. Schematic of the SMS structure based on a large-core air-clad PCF section [31].

In 2012, Gao *et al.* [33] developed an SMS structure based on an etched MMF for the measurement of RI via intensity variations of the bandpass spectral filter. A resolution of 4.6×10^{-6} RIU in the RI range from 1.34 to 1.43 was attained. An SMF-MMF tip configuration was reported by Silva *et al.* [34] for the measurement of different RI liquids by means of the measurand-induced intensity variation of the reflected band-rejection spectral filter. A maximum resolution of 3.8×10^{-4} RIU in the RI range from 1.30 to 1.38 was achieved. Later, Silva *et al.* [35] improved the SMS configuration proposed by Wu *et al.* [32] by using a coreless-MMF with 55 μm-diameter, which could allow obtaining a maximum sensitivity of 2800 nm/RIU in the RI range from 1.42-1.43. This sensor proved to have an ultrahigh sensitivity to temperature variations in the liquid RI of 1.43 − a maximum sensitivity of −1880 pm/°C was attained. In a different line of research, the research group proposed a MMI sensor for curvature and temperature discrimination based on the combination of SMS bandpass and band-rejection filters [36]. On the other hand, Biazoli *et al.* [37] presented instead the real-time monitoring of the fabrication process of tapering down a SMS structure based on a coreless-MMF.

The bandpass filter was used for the measurement of RI variations of the external liquid medium − a maximum sensitivity of 2946 nm/RIU in the RI range of 1.42-1.43 was obtained. Xue *et al.* [38] studied the sensitivity enhancement of an SMS fiber structure in the measurement of surrounding RI, by coating the MMF core section with a high RI overlay of 1.578. Numerical analysis shown that, for RI ranges from 1.31 to 1.35 and 1.0 to 1.03, average sensitivities of 900 and 206 nm/RIU could be achieved with overlay thicknesses of 340 and 470 nm, respectively. Later, a micro-displacement sensor based on a SMS bandpass filter was developed by Antonio-Lopez *et al.* [39]. The fiber structure was placed inside a ferrule with index matching liquid in order to increase the effective length of the coreless-MMF. A wavelength variation could be achieved with the change of the MMF length.

In recent years, researchers have been used the potential of the SMS structure based on a coreless-MMF for RI sensing. Liu *et al.* [40] proposed in 2014 a refractometric system relying on this type of configuration placed inside a fiber ring cavity laser (104 μm-diameter coreless-MMF). The SMS device was used not only as a band-rejection filter but also as sensing head for the measurement RI in aqueous solution. A sensitivity of ~ 131.64 nm/RIU in the RI range from 1.333 to 1.3707 was experimentally achieved. In 2015, Bai *et al.* [41] studied numerically and experimentally the influence of the MMF length on the sensitivity of the SMS fiber structure to RI sensing. A coreless-MMF with 80 μm-diameter was used with different lengths in order to obtain a bandpass filter in distinct operation wavelengths. In practice, a maximum sensitivity of 1923 nm/RIU was achieved when the RI of the surrounding liquid ranged from 1.334 to 1.434. In this case, the coreless-MMF had 20.2 mm in length and operated at 1220 nm. Therefore, maximum RI sensitivity was achieved for shorter operating wavelengths. In a different approach, Zhou *et al.* [42] used a coreless-MMF as a fiber tip sensor for reflective measurements of different liquid RI. The MMF had a similar diameter with the SMF and its tip was coated with a gold thin film in order to be wavelength sensitive to RI. A sensitivity of 141 nm/RIU and a resolution of 2.8×10^{-5} were obtained in the RI range from 1.33 to 1.38.

Most recently, researchers have been used cascaded SMS structures for RI sensing. Liu *et al.* [43] proposed a configuration based on two coreless-MMFs with different lengths, spliced between SMFs, in order to for the two SMS fiber structures. In the RI range of 1.3288-1.3666, the corresponding sensitivities were 148.60 nm/RIU and 119.27 nm/RIU for the sensors with 25 and 30 mm-length MMFs, respectively. Chen *et al.* [44] used instead a coreless-MMF-based SMS followed by a standard MMF-based SMS structure. Simultaneous measurement of temperature and RI was performed by means of the wavelength shift of the band-rejection peaks. Experimental results showed RI and temperature sensitivities of 113.66 nm/RIU (RI range from 1.333 to 1.381) and 9.2 pm/°C, respectively.

The MMI-based device has often been used as an optical power splitter either, where the purpose is to combine that ability with other fiber devices such as fiber Bragg gratings (FBGs) or even to create new optical devices such as Mach–Zehnder interferometers (MZIs). In 2007, Li *et al.* [45] proposed the combination of a SMS structure with a FBG in order to perform temperature and strain discrimination. Frazão *et al.* [46] demonstrated an all fiber MZI based on a SMS structure combined with a long-period grating (LPG) for the measurement of curvature. In this line of research, Nguyen *et al.* [47] combined two SMS structures in line to form a MZI. This sensor has shown to be highly sensitive to temperature variations. Later, Zhou *et al.* [48] presented a FBG followed by an SMS structure for simultaneous measurement of temperature and strain. In the same line of research, Zhang *et al.* [49] studied a sensing scheme based on the combination of an SMF-MMF tip with a Fizeau etalon to measure temperature and strain simultaneously. In

2010, Jin *et al.* [50] developed a fiber tip-based refractometer that relied on a SMS bandpass filter followed by a tilted FBG. This configuration presented a sensitivity of 28.5 µW/RIU in the RI range from 1.37 to 1.43. Recently, Wu *et al.* [51] reported instead a refractometer sensor based on the combination of a SMS bandpass filter followed by an FBG. Experimentally, a maximum sensitivity of 7.33 nm/RIU in the RI range from 1.324 to 1.439 was achieved.

The MMI device has also been used as a tool for the development of interrogation techniques. Wu *et al.* [52] proposed an SMS band-rejection filter as the interrogating system for dynamic temperature compensation of an FBG strain sensor. On the other hand, Frazão *et al.* [53] combined an SMF-MMF tip with a simple interrogation method that used two FBGs as discrete optical sources, in order to measure RI variations of liquids. In this case, the measurand information was encoded in the relative intensity variation of the reflected signals. A resolution of 1.75×10^{-3} RIU was achieved with this technique. Following, Table 3.1 summarizes the various types of MMI-based fiber optic structures that already been developed for RI sensing.

3.4. Conclusions

Fiber-optic sensors have been proposed as a promising and alternative solution with respect to conventional sensors. In this chapter, an overview of fiber-optic sensors based on the MMI concept was presented, for the measurement of physical parameters and with focus on the measurement of RI. It has been shown that the SMS structure is a flexible configuration where distinct MMFs have been used, ranging from standard MMF, coreless-MMF and microstructured fibers. The coreless-MMF has been extensively used due to its high sensitivity to the external medium. The SMS structure is also flexible in terms of spectral selection according with the length and characteristics of the MMF used. Bandpass and band-rejection filters are well suited for the discrimination of physical parameters such as temperature and curvature, or temperature and strain. This ability to discriminate parameters is an important feature when the purpose is to adapt the sensing head to real-case scenarios where, for example, curvature or strain of the fiber sensor may not be avoided or, in the case of biological environments, where cross-sensitivity to temperature is a drawback in the monitoring process of such systems. Sensing liquid refractive index is of utmost importance in a wide range of applications, either for quality control of chemical products, food analysis or even monitoring the environment. For instance, monitoring salinity and temperature allows the detection of water pollution, contributing for sea life control and preservation. The feasibility of MMI-based fiber sensors has permitted the development of potential solutions for practical applications due to higher performance in resolution or sensitivity achievable with this kind of sensors. Therefore, the great contribution of MMI-based fiber structures has been in the measurement of liquid RI. In the future, it is expected to implement fiber-optic sensors for RI sensing in real-case scenarios and an attractive technology to apply in industry.

Table 3.1. MMI-based fiber refractometric sensors.

RI range	Sensing head	Sensitivity (nm/RIU)/ Resolution (RIU)	Ref
1.33 – 1.45	SMF- MMF-SMF (100 μm-diameter MMF)	5.4×10^{-5} RIU	[18]
1.346 – 1.416	SMF-tapered MMF-SMF (40 μm-diameter MMF)	2.59×10^{-5} RIU	[19]
1.33 – 1.44	SMF-tapered MMF-SMF (30 μm-diameter MMF)	487 nm/RIU	[20]
1.43 – 1.44		1913 nm/RIU, 5.23×10^{-6} RIU	
1.33 – 1.38	SMF- etched MMF-SMF (105 μm-diameter MMF)	7.9×10^{-5} RIU	[24]
1.3246 – 1.3266	SMF- Large Core Air Clad PCF-SMF	800 nm/RIU, 3.4×10^{-5} RIU	[31]
1.338 – 1.403	SMF- outer cladding, large core air clad PCF-SMF	322.08 nm/RIU, 7.2×10^{-4} RIU	[14]
1.342 – 1.437	SMF- etched MMF-SMF (80 μm-diameter MMF)	1815 nm/RIU	[32]
1.34 – 1.43	SMF- etched MMF-SMF (81.5 μm-diameter MMF)	4.6×10^{-6} RIU	[33]
1.30 – 1.38	SMF-MMF tip	3.8×10^{-4} RIU	[34]
1.42 – 1.43	SMF- coreless-MMF-SMF (55 μm-diameter MMF)	2800 nm/RIU	[35]
1.42 – 1.43	SMF- tapered coreless-MMF-SMF (55 μm-diameter MMF)	2946 nm/RIU	[37]
1.31 – 1.35	SMF- MMF with high RI overlay coating -SMF	900 nm/RIU	[38]
1.0 – 1.03		206 nm/RIU	
1.333 – 1.3707	SMF- coreless-MMF-SMF (104 μm-diameter MMF)	131.64 nm/RIU	[40]
1.334 – 1.434	SMF- coreless-MMF-SMF (80 μm-diameter MMF)	1923 nm/RIU	[41]
1.33 – 1.38	SMF- coreless-MMF tip with gold mirror	141 nm/RIU, 2.8×10^{-5} RIU	[42]
1.3288 – 1.3666	Two cascaded SMF- coreless-MMF-SMF SMS_1: 25 mm-length MMF, SMS_2: 30 mm-length MMF	SMS_1: 148.60 nm/RIU SMS_2: 119.27 nm/RIU	[43]
1.333 – 1.381	Cascaded SMS: SMF- coreless-MMF-SMF + SMF-MMF-SMF	113.66 nm/RIU	[44]
1.37 – 1.43	SMS-tilted FBG tip	28.5 μW/RIU	[50]
1.324 – 1.439	SMS-FBG	7.33 nm/RIU	[51]
1.3 – 1.38	SMF-MMF tip + FBGs	-5.87/RIU, 1.75×10^{-3} RIU	[53]

Acknowledgements

This work was supported by Project "CORAL – Sustainable Ocean Exploitation: Tools and Sensors, NORTE-01-0145-FEDER-000036, financed by the North Portugal Regional Operational Programme (NORTE 2020), under the PORTUGAL 2020 Partnership Agreement, and through the European Regional Development Fund (ERDF). S.S. received a Pos-Doc fellowship (ref. SFRH/BPD/92418/2013) also funded by FCT – Portuguese national funding agency for science, research and technology.

References

[1]. D. Gloge, Bending loss in multimode fibers with graded and ungraded core index, *Appl. Opt.,* Vol. 11, No. 11, 1972, pp. 2506-2518.

[2]. L. B. Soldano, and E. C. M. Pennings, Optical multi-mode interference devices based on self-imaging: principles and applications, *J. Lightwave Technol.,* Vol. 13, No. 4, 1995, pp. 615–627.

[3]. N. Lagakos, J. H. Cole, and J. A. Bucaro, Microbend fiber-optic sensor, *Appl. Opt.,* Vol. 26, No. 11, 1987, pp. 2171-2180.

[4]. D. Donlagic, and M. Zavrsnik, Fiber-optic microbend sensor structure, *Opt. Lett.,* Vol. 22, No. 11, 1997, pp. 837-839.

[5]. W. S. Mohammed, A. Mehta, and E. G. Johnson, Wavelength tunable fiber lens based on multimode interference, *J. Lightwave Technol.,* Vol. 22, No. 2, 2004, pp. 469-477.

[6]. E. Li, X. Wang, and C. Zhang, Fiber-optic temperature sensor based on interference of selective higher-order modes, *Appl. Phys. Lett.,* Vol. 89, 2006, 091119.

[7]. Q. Wang, G. Farrell, and W. Yan, Investigation on single mode-multimode-single mode fiber structure, *J. Lightwave Technol.,* Vol. 26, No. 5, 2008, pp. 512-519.

[8]. W. S. Mohammed, P. W. E. Smith, and X. Gu, All-fiber multimode interference bandpass filter, *Opt. Lett.,* Vol. 31, No. 17, 2006, pp. 2547-2549.

[9]. Q. Wang, and G. Farrell, Multimode fiber based edge filter for optical wavelength measurement: application and its design, *Microw. Opt. Technol. Lett.,* Vol. 48, No. 5, pp. 900-902, 2006.

[10]. A. Castillo-Guzman, J. E. Antonio-Lopez, R. Selvas-Aguilar, D. A. May-Arrioja, J. Estudillo-Ayala, and P. LiKamWa, Widely tunable erbium-doped fiber laser based on multimode interference effect, *Opt. Express,* Vol. 18, No. 2, 2010, pp. 591-597.

[11]. J. E. Antonio-Lopez, A. Castillo-Guzman, D. A. May-Arrioja, R. Selvas-Aguilar, and P. LiKamWa, Tunable multimode-interference bandpass fiber filter, *Opt. Lett.,* Vol. 35, No. 3, 2010, pp. 324-326.

[12]. X. Zhu, A. Schulzgen, H. Li, L. Li, V. L. Temyanko, J. V. Moloney, and N. Peyghambarian, High-power fiber lasers and amplifiers based on multimode interference, *IEEE J. Sel. Top. Quant. Electr.,* Vol. 15, No. 1, 2009, pp. 71-78.

[13]. L. Man, Y. Qi, Z. Kang, Y. Bai, and S. Jian, Tunable fiber laser based on the refractive index characteristic of MMI effects, *Opt. Laser Technol.,* Vol. 57, 2014, pp. 96-99.

[14]. L. Coelho, J. Kobelke, K. Schuster, and O. Frazão, Multimode interference in outer cladding large-core air-clad photonic crystal fiber, *Microw. Opt. Technol. Lett.,* Vol. 54, No. 4, 2012, pp. 1009-1011.

[15]. A. Mehta, W. S. Mohammed, and E. G. Johnson, Multimode interference-based fiber-optic displacement sensor, *J. Lightwave Technol.,* Vol. 22, No. 8, 2003, pp. 1129-1131.

[16]. E. Li, Temperature compensation of multimode-interference-based fiber devices, *Opt. Lett.,* Vol. 32, No. 14, 2007, pp. 2064-2066.

[17]. E. Li, and G.-D. Peng, Wavelength encoded fiber optic temperature sensor with ultra high sensitivity, *Opt. Commun.,* Vol. 281, 2008, pp. 5768-5770.

[18]. Q. Wang, and G. Farrell, All-fiber multimode-interference based refractometer sensor: proposal and design, *Opt. Lett.,* Vol. 31, No. 3, 2006, pp. 317-319.

[19]. P. Wang, G. Brambilla, M. Ding, Y. Semenova, Q. Wu, and G. Farrell, Investigation of single-mode-multimode-single-mode and single-mode-tapered-multimode-single-mode fiber structures and their application for refractive index sensing, *J. Opt. Soc. Am. B,* Vol. 28, No. 5, 2011, pp. 1180-1186.

[20]. P. Wang, G. Brambilla, M. Ding, Y. Semenova, Q. Wu, and G. Farrell, High-sensitivity, evanescent field refractometric sensor based on a tapered, multimode fiber interference, *Opt. Lett.,* Vol. 36, No. 12, 2011, pp. 2233-2235.

[21]. R. X. Gao, Q. Wang, F. Zhao, B. Meng, and S. L. Qu, Optimal design and fabrication of SMS fiber temperature sensor for liquid, *Opt. Commun.,* Vol. 283, 2010, pp. 3149-3152.

[22]. A. M. Hatta, Y. Semenova, Q. Wu, and G. Farrell, Strain sensor based on a pair of single-mode-multimode-single-mode fiber structures in a ratiometric power measurement scheme, *Appl. Opt.,* Vol. 49, No. 3, 2010, pp. 536-541.

[23]. M. Hatta, Y. Semenova, G. Rajan, and G. Farrell, A voltage sensor based on a singlemode-multimode-singlemode fiber structure, *Microw. Opt. Technol. Lett.,* Vol. 52, No. 8, 2010, pp. 1887-1890.

[24]. J. Zhang, and S. Peng, A Compact SMS refractometer based on HF corrosion scheme, in *Proceedings of the Symposium on Photonics and Optoelectronic (SOPO),* 2010, pp. 1-4.

[25]. J. E. Antonio-Lopez, J. J. Sanchez-Mondragon, P. L. Wa, and D. A. May-Arrioja, Fiber-optic sensor for liquid level measurement, *Opt. Lett.,* Vol. 36, No. 17, 2011, pp. 3425-3427.

[26]. Y. Gong, T. Zhao, Y.-J. Rao, and Y. Wu, All-fiber curvature sensor based on multimode Interference, *IEEE Photon. Technol. Lett.,* Vol. 23, No. 11, 2011, pp. 679-681.

[27]. S. Silva, O. Frazão, J. Viegas, L. A. Ferreira, F. M. Araújo, F. X. Malcata, and J. L. Santos, Temperature and strain-independent curvature sensor based on a singlemode/multimode fiber optic structure, *Meas. Sci. Technol.,* Vol. 22, 2011, 085201.

[28]. Q. Wu, Y. Semenova, P. Wang, A. M. Hatta, and G. Farrell, Experimental demonstration of a simple displacement sensor based on a bent single mode-multimode-single mode fiber structure, *Meas. Sci. Technol.,* Vol. 22, 2011, 025203.

[29]. Q. Wu, A. M. Hatta, P. Wang, Y. Semenova, and G. Farrell, Use of a bent SMS structure for simultaneous measurement of displacement and temperature sensing, *IEEE Photon. Technol. Lett.,* Vol. 23, No. 2, 2011, pp. 130-132.

[30]. Q. Wu, Y. Semenova, A. M. Hatta, P. Wang, and G. Farrell, Singlemode-multimode-singlemode fiber structures for simultaneous measurement of strain and temperature, *Microw. Opt. Technol. Lett.,* Vol. 53, No. 9, 2011, pp. 2181-2185.

[31]. S. Silva, J. L. Santos, F. X. Malcata, J. Kobelke, K. Schuster, and O. Frazão, Optical refractometer based on large-core air-clad photonic crystal fibers, *Opt. Lett.,* Vol. 36, No. 6, 2011, pp. 852-854.

[32]. Q. Wu, Y. Semenova, P. Wang, and G. Farrell, High sensitivity SMS fiber structure based refractometer – analysis and experiment, *Opt. Express,* Vol. 19, No. 9, 2011, pp. 7937-7944.

[33]. R. X. Gao, W. J. Liu, Y. Y. Wang, Q. Wang, F. Zhao, and S. L. Qu, Design and fabrication of SMS fiber refractometer for liquid, *Sensor. Actuat. A-Phys.,* Vol. 179, 2012, pp. 5-9.

[34]. S. Silva, O. Frazão, J. L. Santos, and F. X. Malcata, A reflective optical fiber refractometer based on multimode interference, *Sensor. Actuat. B-Chem.,* Vol. 161, 2012, pp. 88-92.

[35]. S. Silva, E. G. P. Pachon, M. A. R. Franco, J. G. Hayashi, F. X. Malcata, O. Frazão, P. Jorge, and C. M. B. Cordeiro, Ultrahigh-sensitivity temperature fiber sensor based on multimode interference, *Appl. Opt.,* Vol. 51, No. 16, 2012, pp. 3236-3242.

[36]. S. Silva, E. G. P. Pachon, M. A. R. Franco, P. Jorge, J. L. Santos, F. X. Malcata, C. M. B. Cordeiro, and O. Frazão, Curvature and temperature discrimination using multimode

interference fiber optic structures – A proof of concept, J. Lightwave Technol., Vol. 30, No. 23, 2012, pp. 3569-3575.

[37]. R. Biazoli, S. Silva, M. A. R. Franco, O. Frazão, and C. M. B. Cordeiro, Multimode interference tapered fiber refractive index sensors, Appl. Opt., Vol. 51, No. 24, 2012, pp. 5941-5945.

[38]. L.-L. Xue, and L. Yang, Sensitivity enhancement of RI sensor based on SMS fiber structure with high refractive index overlay, J. Lightwave Technol., Vol. 30, No. 10, 2012, pp. 1463-1469.

[39]. J. E. Antonio-Lopez, P. LiKamWa, J. J. Sanchez-Mondragon, and D. A. May-Arrioja, All-fiber multimode interference micro-displacement sensor, Meas. Sci. Technol., Vol. 24, 2013, 055104.

[40]. Z. Liu, Z. Tan, B. Yin, Y. Bai, and S. Jian, Refractive index sensing characterization of a singlemode-claddingless-singlemode fiber structure based fiber ring cavity laser, Opt. Express, Vol. 22, No. 5, 2014, pp. 5037-5042.

[41]. X. Bai, H. Wang, S. Wang, S. Pu, and X. Zeng, Refractive index sensing characteristic of single-modemultimode-single-mode fiber structure based on self-imaging effect, Opt. Eng., Vol. 54, No. 10, 2015, 106103.

[42]. X. Zhou, K. Chen, X. Mao, P. Wei, and Q. Yu, A reflective fiber-optic refractive index sensor based on multimode interference in a coreless silica fiber, Opt. Commun., Vol. 340, 2015, pp. 50-55.

[43]. X. Liu, X. Zhang, Y. Liu, Z. Liu, and W. Peng, Multi-point fiber-optic refractive index sensor by using coreless fibers, Opt. Commun., Vol. 365, 2016, pp. 168-172.

[44]. Y. Chen, Y. Wang, R. Chen, W. Yang, H. Liu, T. Liu, and Q. Han, A hybrid multimode interference structure based refractive index and temperature fiber sensor, IEEE Sensors J., Vol. 16, No. 2, 2016, pp. 331-335.

[45]. E. Li, Sensitivity-enhanced fiber-optic strain sensor based on interference of higher order modes in circular fibers, IEEE Photon. Technol. Lett., Vol. 19, No. 16, 2007, pp. 1266-1268.

[46]. O. Frazão, J. Viegas, P. Caldas, J. L. Santos, Araújo F. M., L. A. Ferreira, and F. Farahi, All-fiber Mach-Zehnder curvature sensor based on multimode interference combined with a long-period grating, Opt. Lett, Vol. 32, No. 21, 2007, pp. 3074-3076.

[47]. L. V. Nguyen, D. Hwang, S. Moon, D. S. Moon, and Y. Chung, High temperature fiber sensor with high sensitivity based on core diameter mismatch, Opt. Express, Vol. 16, No. 15, 2008, pp. 11369-11375.

[48]. D.-P. Zhou, L. Wei, W.-K Liu, Y. Liu, and J. W. Y. Lit, Simultaneous measurement for strain and temperature using fiber Bragg gratings and multimode fibers, Appl. Opt., Vol. 47, No. 10, 2008, pp. 1668-1672.

[49]. J. Zhang, Y. Zhang, W. Sun, and L. Yuan, Multiplexing multimode fiber and Fizeau etalon a simultaneous measurement scheme of temperature and strain, Meas. Sci. Technol., Vol. 20, 2009, 065206.

[50]. Y. Jin, X. Dong, H. Gong, and C. Shen, Refractive Index Sensor Based On Tilted Fiber Bragg Gratting Interacting With Multimode Fiber, Microw. Opt. Technol. Lett., Vol. 52, No. 6, 2010, pp. 1375-1377.

[51]. Q. Wu., S., B. Yan, Y. Ma, P. Wang, C. Yu, and G. Farrell, Fiber refractometer based on a fiber Bragg grating and single-mode–multimode–single-mode fiber structure, Opt. Lett., Vol. 36, No. 12, 2011, pp. 2197-2199.

[52]. Q. Wu, A. M. Hatta, Y. Semenova, and G. Farrell, Use of a single-multiple-single-mode fiber filter for interrogating fiber Bragg grating strain sensors with dynamic temperature compensation, Appl. Opt., Vol. 48, No. 29, 2009, pp. 5451-5458.

[53]. O. Frazão, S. O. Silva, J. Viegas, L. A. Ferreira, F. M. Araújo, and J. L. Santos, Optical fiber refractometry based on multimode interference, Appl. Opt., Vol. 50, No. 25, 2011, pp. E184-188.

Chapter 4
Enlarged Spectral Sensitivity Outside the Visible Spectrum in Tandem a-SiC:H pi'n/pin Photodiodes

Manuela Vieira, Paula Louro, Manuel Augusto Vieira, Isabel Rodrigues, Vitor Silva, Alessandro Fantoni, João Costa

4.1. Introduction

The Light Emitting Diodes (LED) are an effective lighting technology due to its high brightness, long life, energy efficiency, durability, affordable cost, optical spectrum and its colour range for decorative purposes. Their application as communication devices with a photodiode as receptor, has been used for many years in hand held devices, to control televisions and other media equipment, and to transfer data at higher rates between computational devices [1]. This communication path has been employed in the near infrared (NIR) range, but due to the increasing LED lighting in homes and offices, the idea to use them for visible light communications (VLC) has come up recently. Newly developed technologies, for infrared telecommunication systems, allow increase of capacity, distance, and functionality, leading to the design of new reconfigurable active filters [2-4], that enhance the transmission capacity and the application flexibility of optical communication. Efforts have to be considered, namely the Wavelength Division Multiplexer (WDM) based on a-SiC:H light controlled filters, when different visible signals are encoded in the same optical transmission path [5]. They can be used to achieve different filtering purposes, such as: amplification, switching, and wavelength conversion.

In this chapter, it is demonstrated that the same a-SiC:H device under front and back controlled near ultraviolet optical bias acts as a reconfigurable active filter in the visible and near infrared ranges, taking advantages of the visible spectrum for wireless communications. In consequence, bridging the visible spectrum to the telecom gap offers

Manuela Vieira
Telecommunication and Computer Dept. ISEL, R. Conselheiro Emídio Navarro, 1959-007 Lisboa, Portugal

the opportunity to provide alternative and additional low cost services to improve operative production processes in office, home and automotive networks.

The Section 4.1, gives the introduction and in Section 4.2, some experimental results are presented. In Section 4.3, the bias controlled selector is analyzed and in Section 4.4, the Wavelength Division Multiplexed (WDM) based on SiC technology is described. In Section 4.5, the optoelectronic model gives insight the physics of the device, the decoding algorithm is presented in Section 4.6, In Section 4.7 some applications are reported and finally, in Section 4.8, the conclusions are presented.

4.2. Experimental Details

4.2.1. Device Configuration

The light tunable filter is built using a double pi'n/pin a-SiC:H photodetector produced by Plasma Enhanced Chemical Vapor Deposition (PECVD).

The device has Transparent Conductive Oxide (TCO) front and back biased optical gating elements as depicted in Fig. 4.1.

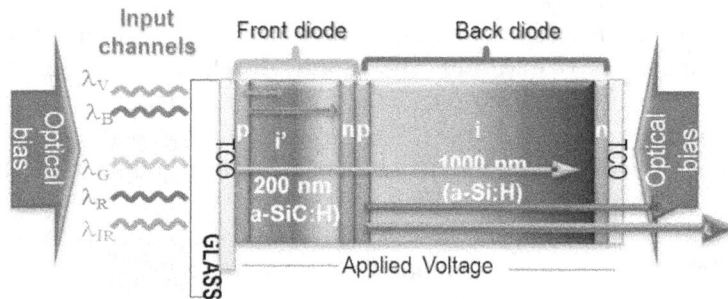

Fig. 4.1. Device configuration and operation.

The active device consists of a p-i'(a-SiC:H)-n/p-i(a-Si:H)-n heterostructure with low conductivity doped layers. The deposition conditions and optoelectronic characterization of the single layers were described elsewhere [6].

The thicknesses and optical gap of the front i'- (200 nm; 2.1 eV) and back i- (1000 nm; 1.8 eV) layers are optimized for light absorption in the blue and red ranges, respectively [7].

4.2.2. Device Operation

Monochromatic (infrared, red, green, blue and violet; $\lambda_{IR,R,G,B,V}$) pulsed communication channels (input channels) are combined together, each one with a specific bit sequence, impinge on the device and are absorbed according to their wavelengths (see arrow magnitudes in Fig. 4.1).

The combined optical signal (multiplexed signal; MUX) is analyzed by reading out the generated photocurrent under negative applied voltage (-8 V), with and without near ultraviolet background ($\lambda_{Background}$=390 nm) and different intensities, applied either from the front (λ_F) or the back (λ_B) sides. The device operates within the visible range using as input color channels the square wave modulated low power light supplied by near-infrared/visible (VIS/NIR) LEDs. In Fig. 4.2 a, the 524 nm input channel is displayed under front, back and without UV irradiation. The arrows indicate the enhancement (solid line) or quenching (dot line) of the dark signal, respectively under front and back irradiation. In Fig. 4.2 b, the polychromatic mixture of four different input channels (400 nm, 524 nm, 697 nm and 850 nm) under front and back 2800 μWcm^{-2} irradiation, is displayed. At the top, the input channels wavelength and their bit sequences are displayed to guide the eyes.

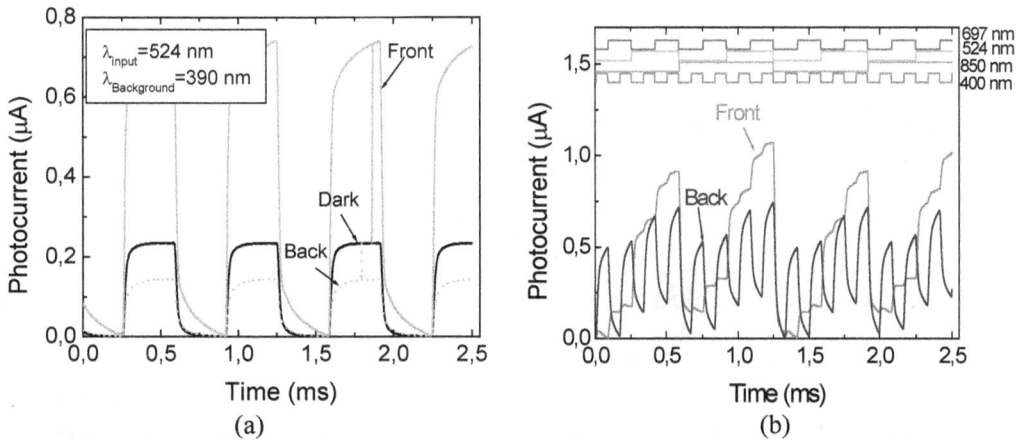

Fig. 4.2. (a) 524 nm input channel under front, back and without (dark) background irradiation; (b) MUX signals and under front and back λ=390 nm irradiation and different bit sequences.

4.3. Bias Controlled Selector

4.3.1. Optical Bias Controlled Filter

The spectral sensitivity was tested through spectral response measurements [8] without background and under 390 nm front and back backgrounds of variable intensities. The spectral gain (α), defined as the ratio between the signal with and without irradiation was inferred.

In Fig. 4.3, the spectral gain (α) is displayed under steady state irradiations. In Fig. 4.3 a, the light was applied from the front (λ_F) and in Fig. 4.3 b, the irradiation occurs from the back side (λ_B). The background intensity (ϕ) was increased from 5 μWcm^{-2} to 3800 μWcm^{-2}.

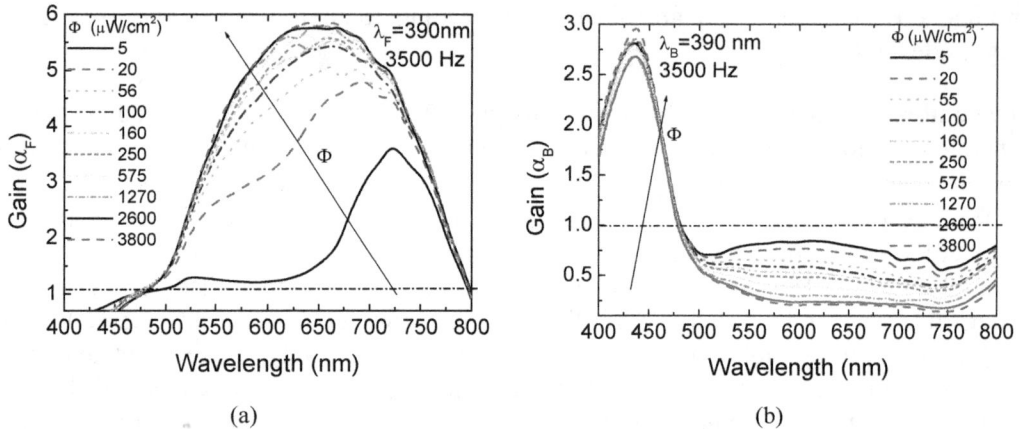

(a)   (b)

Fig. 4.3. (a) Front (λ_F), and (b) back (λ_B) spectral gains ($\alpha_{F,B}$) under λ=390 nm irradiations.

Results show that, the optical gains have opposite behaviors. Under front irradiation (Fig. 4.3a) and low flux, the gain is high in the infrared region, presents a well-defined peak at 725 nm and strongly quenches in the visible range. As the power intensity increases, the peak shifts to the visible range and can be deconvoluted into two peaks, one in the red range that slightly increases with the power intensity of the background and another in the green range that strongly increases with the intensity of the ultraviolet (UV) radiation. In the blue range, the gain is much lower. This shows the controlled high-pass filtering properties of the device under different background intensities. Under back bias (Fig. 4.3 b) the gain in the blue/violet range has a maximum near 420 nm that quickly increases with the intensity. Moreover, it strongly decreases for wavelengths higher than 450 nm, acting as a short-pass filter. Thus, back irradiation, tunes the violet/blue region of the visible spectrum whatever the flux intensity, while front irradiation, depending on the background intensity, selects the infrared or the visible spectral ranges. Here, low fluxes select the near infrared region and cut the visible one, the red part of the spectrum is selected at medium fluxes, and high fluxes tune the red/green ranges with different gains.

4.3.2. Nonlinear Spectral Gain

To analyze the effect of the background intensity in the input channels, several monochromatic pulsed lights separately (850 nm, 697 nm, 626 nm, 524 nm, 470 nm, 400 nm; input channels) or combined (MUX signal) illuminated the device at 12000 bps [9].

Steady state optical bias with different intensities was superimposed separately from the front and back sides and the photocurrent measured. For each individual channel the photocurrent gain under irradiation was determined. In Fig. 4.4, these gains are displayed as a function of the background lighting under front (Fig. 4.4 a) and back (Fig. 4.4 b) irradiation.

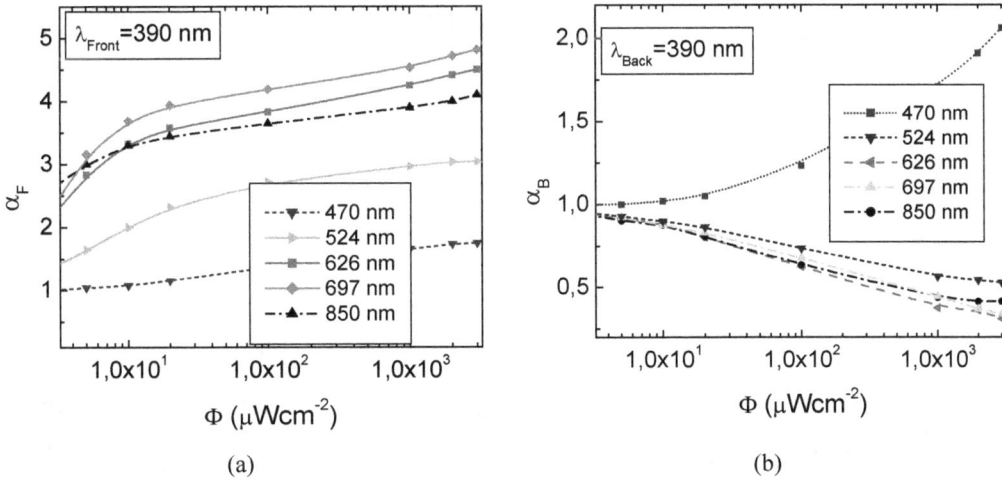

(a)

(b)

Fig. 4.4. Front (a) and back (b) optical gains as a function of the background intensity for different input wavelengths in the VIS/NIR range.

Results show that, even under transient conditions and using commercial visible and NIR LEDs, the background side and intensity changes the signal magnitude of the input channels.

The gain depends mainly on the channel wavelength and to some extent on the lighting intensity. Even across narrow bandwidths, the photocurrent gains are quite different. This nonlinearity allows identification of the different input channels in the visible/infrared ranges.

4.4. Wavelength Division Multiplexer

4.4.1. Input Channels

Four monochromatic (400 nm, 470 nm, 697 nm and 850 nm) pulsed lights with different intensities, separately (input channels) or combined (MUX signal) illuminated the device at 12000 bps.

Steady state 390 nm front and back optical bias with 2800 μWcm^{-2} intensity was superimposed separately and the photocurrent was measured. In Fig. 4.5 a, the blue and violet transient signals are presented under front and back irradiations while in Fig. 4.5 b, the red and infrared signals are displayed.

In Table 4.1, the measured optical gains for five different input channels are displayed.

Back irradiation enhances, differently, the input signals in the short wavelength range (Fig. 4.5 a) while front irradiation increases them otherwise in the long wavelength range (Fig. 4.5 b). This side dependent effect is used to enhance or to quench the input signals allowing their recognition and providing the possibility for selective tuning of the visible and IR input channels.

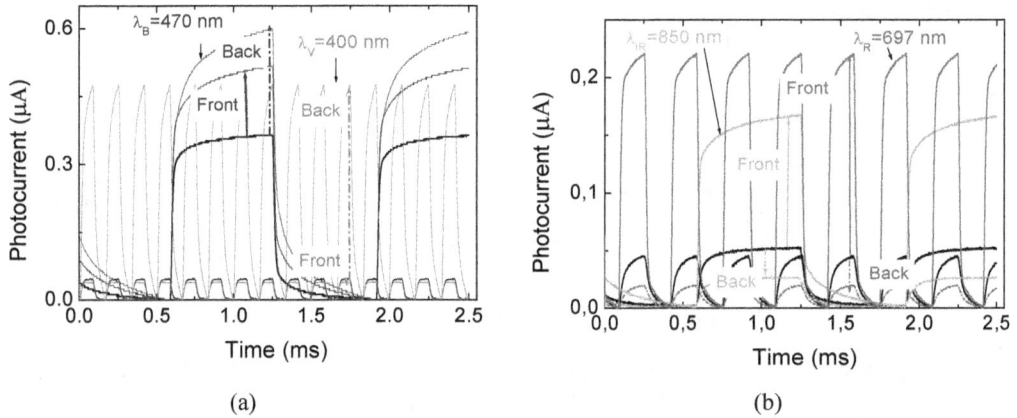

Fig. 4.5. Input signals under front and back 390 nm background irradiation: (a) violet and blue channels; (b) red and infrared channels.

Table 4.1. Optical gains under 390 nm front (αFront) and back (α_{Back}) irradiations.

	$\lambda=400$ nm	$\lambda=470$ nm	$\lambda=524$ nm	$\lambda=626$ nm	$\lambda=697$ nm	$\lambda=850$ nm
α_{Back}	11.6	1.8	0.6	0.4	0.4	0.4
α_{Front}	0.9	1.5	3.2	4.7	4.3	3.5

4.4.2. MUX Signal

In Fig. 4.6, two MUX signals due to the input signals of Fig. 4.2 a and Fig. 4.5 are displayed without (dark) and under front and back irradiation. On top, the signals used to drive the input channels are shown to guide the eyes into the *on/off* channel states. Results show that, the background side affects the form of the MUX signal, enhancing or quenching different spectral ranges. In Fig. 4.6 a all the *on/off* states are possible so, without optical bias, 2^4 ordered levels are detected and correspond to all the possible combinations of the *on/off* states. Under, either front or back irradiation, each of those four channels, by turn, are enhanced or quenched differently (Fig. 4.6, Table 4.1) resulting in an increase magnitude of red/green under front irradiation or of the blue/violet one, under back lighting. Since the gain of the four input channels is different ($\alpha_{F,B}$; Table 4.1) this nonlinearity allows identifying the different input channels in a large visible/infrared range. In Fig. 4.6 b, both 400 nm and 697 nm channels have the same bit sequence which corresponds to only 2^3 ordered levels, however since the optical gains of both channels are quite different under front and back irradiation (Table 4.1) it is possible to identify them. Under back irradiation the MUX signal receive its main contribution from the 400 nm channel while under front irradiation it is mainly weighed by the long wavelength channels. By comparing front and back irradiation is possible to decode the transmitted information.

Under front irradiation, near-UV radiation is absorbed at the beginning of the front diode and, due to the self-bias effect, increases the electric field at the back diode where the red/infrared incoming photons (see Fig. 4.1) are absorbed according to their wavelengths (see Fig. 4.3) resulting in an increased collection. Under back irradiation the electric field decreases mainly at the back i-n interface enhancing the electric field at the front diode quenching it at the back one. This leads to an increased collection of the violet/blue input signals.

So, by switching between front to back irradiation the photonic function is modified from a long- to a short-pass filter allowing, alternately selecting the red/infrared channels or the blue and violet ones, thus, making the bridge between the visible and the infrared regions.

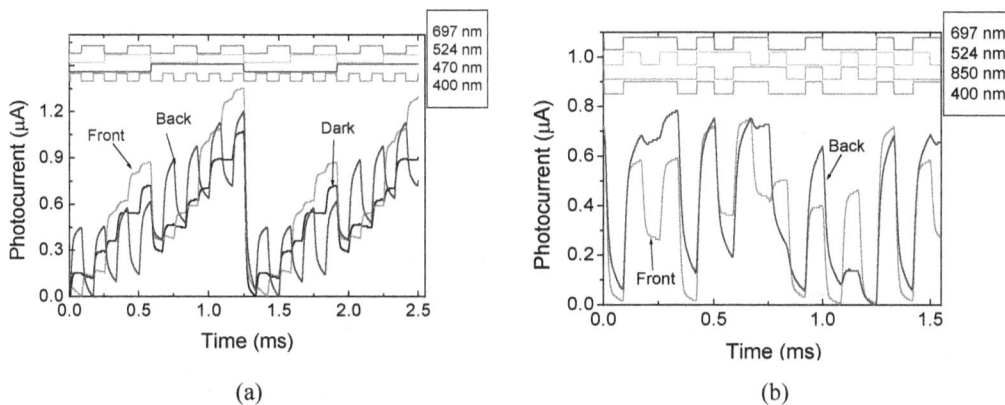

(a)   (b)

Fig. 4.6. MUX signals: (a) without background and under front and back λ=390 nm irradiation and different bit sequences; (b) Front and back irradiation and two channels (400 nm and 697 nm) with the same bit sequence.

4.5. Optoelectronic Model

Based on the experimental results and device configuration a two connected phototransistor model (Fig. 4.7 a), made out of a short- and a long-pass filter was developed [5] and upgraded to include several input channels. The *ac* circuit representation is displayed in Fig. 4.7 b and is supported by the complete dynamical large signal Ebers-Moll model with series resistances and capacitors. The charge stored in the space-charge layers is modelled by the capacitor C_1 and C_2. R_1 and R_2 model the dynamical resistances of the internal and back junctions under different *dc* bias conditions. The operation is based upon the following strategic principle: the flow of current through the resistor connecting the two transistor bases is proportional to the difference in the voltages across both capacitors (charge storage buckets). The modified electrical model developed is the key of this strategic operation principle. Two optical gate connections ascribed to the different light penetration depths across the front (Q_1) and back (Q_2) phototransistors were considered to allow independent blue (I_1), red/infrared (I_2) and green (I_3, I_4) channels transmission. Four square-wave current sources with different intensities are used; two of them, I_1 and I_2, with different frequencies to simulate the input blue and red channels and the other two, I_3 and I_4, with the same frequency but different intensities, to simulate the green channel due to its asymmetrical absorption across both front and back phototransistors.

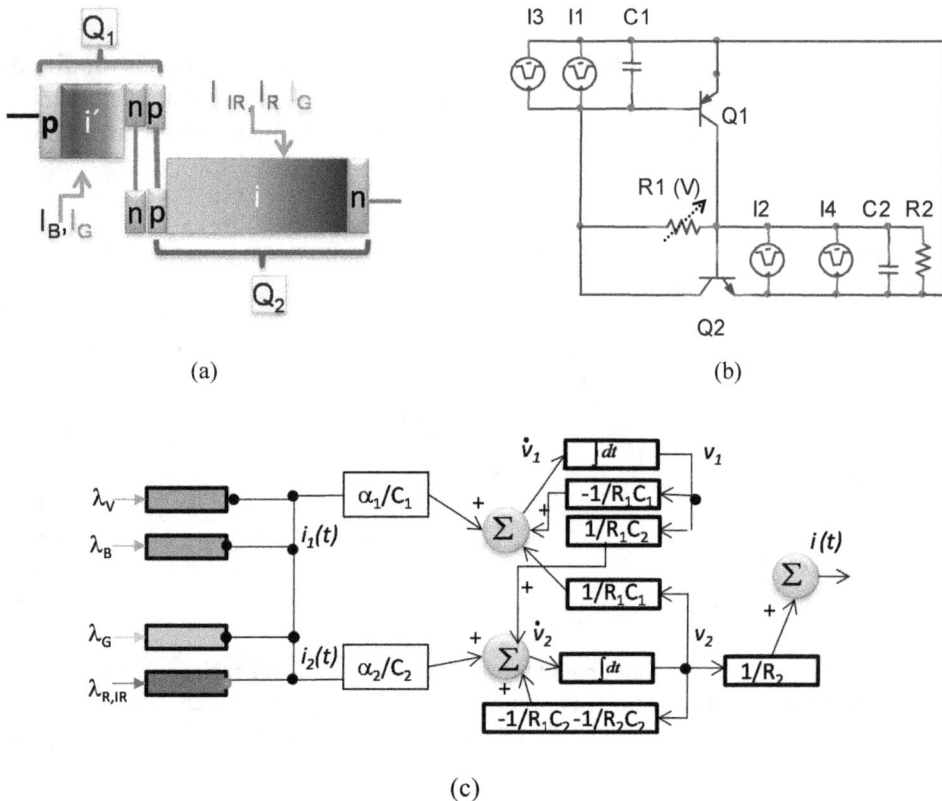

(a)

(b)

(c)

Fig. 4.7. (a) Two connected transistor model, (b) equivalent electric circuit, and (c) block diagram of the optoelectronic state model.

In Fig. 4.7 c, the block diagram of the optoelectronic state model is displayed. The resistors (R_1, R_2) and capacitors (C_1, C_2) synthesize the desired filter characteristics. The input signals, $\lambda_{IR,R,G,B,V}$ model the input channels and $i(t)$ the output signal. The amplifying elements, α_1 and α_2 are linear combinations of the optical gains of each impinging channel, respectively into the front and back phototransistors and account for the enhancement or quenching of the channels (Fig. 4.3 and Fig. 4.4) due to the steady state irradiation. Under front irradiation we have: $\alpha_2 >> \alpha_1$ and under back irradiation $\alpha_1 >> \alpha_2$. This affects the reverse photo capacitances, $(\alpha_{1,2}/C_{1,2})$ that determine the influence of the system input on the state change.

A graphics user interface computer program was designed and programmed based on the MATLAB® programming language, to ease the task of numerical simulation. This interface allows selecting model parameters, along with the plotting of bit signals and compare simulated and experimental photocurrent results. To simulate the input channels we have used the individual magnitude of each input channel without background lighting (Fig. 4.2 a and Fig. 4.5), and the corresponding gain at the simulated background intensity (Table 4.1). Fig. 4.8, presents results of a numerical simulation with 3000 $\mu W/cm^2$ front and back λ=390 nm irradiation and the experimental outputs of Fig. 4.2 b and Fig. 4.6 b, respectively.

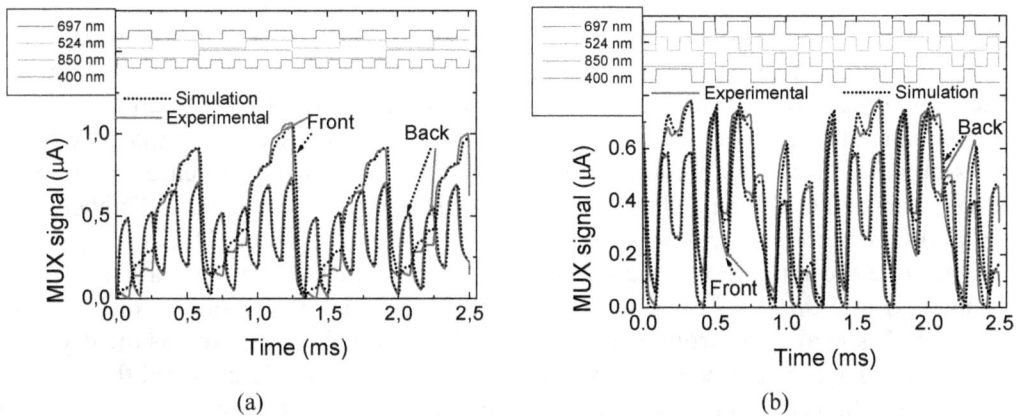

(a)   (b)

Fig. 4.8. Numerical simulation with front and back λ=390 nm irradiation, and different channel wavelength combinations and bit sequences.

Values of R_1=10 kΩ, R_2=1 kΩ, C_1=1 nF, C_2=20 nF (Fig. 4.7c) were used during the simulation process. On top of the figures, the drive input LED signals guide the eyes into the different *on/off* states and correspondent wavelengths

A good fitting between experimental and simulated results was achieved. The plots show the ability of the presented model to simulate the sensitivity behavior of the proposed system in the visible/infrared spectral ranges. The optoelectronic model with light biasing control has proven to be a good tool to design optical filters. Furthermore, this model allows for extracting theoretical parameters by fitting the model to the measured data (internal resistors and capacitors). Under back irradiation higher values of C_2 were obtained confirming the capacitive effect of the near-UV radiation on the device that increases the charge stored in the space charge layers of the back optical gate of Q_2 modelled by C_2 [10].

4.6. Decoding Algorithm

Results show that the background side changes the shape of the MUX signal, enhancing or quenching different spectral ranges. In Fig. 4.8 a all the *on/off* states are possible so, 2^4 ordered levels are detected and correspond to all possible combinations of the *on/off* states. Under, either front or back irradiation, each of those four channels, by turn, are enhanced or quenched differently (Fig. 4.5, Table 4.1) resulting in an increased magnitude of red/green under front irradiation or of the blue/violet one, under back lighting. Since the gain of the input channels is different ($\alpha_{F,B}$; Table 4.1) this nonlinearity allows identifying the different input channels in a large visible/infrared range. Under front irradiation the MUX signal presents sixteen separate levels each one ascribed to one of the of the 2^4 possible combinations of the *on/off* states and weighted by their optical gains. So, by assigning each output level to a four digit binary code weighted by the optical gain of the each channel, the signal can be decoded. A transmission rate of 15 kbps was achieved.

The decoding algorithm is based on a proximity search [11]. Each time slot is translated to a vector in multidimensional space. The vector components' are computed as a function of the sampled currents I_1 and I_2, where I_1 and I_2 are the currents measured under front and back optical bias in the respective time slot. The result is then compared with all vectors obtained from a calibration sequence. The color bits of the nearest calibration point are assigned to the time slot. Euclidian metric is applied to measure distances. We have used this simple algorithm to perform 1 -to-16 demultiplexer (DEMUX) function and to decode the multiplex signals. As proof of concept the decoding algorithm was implemented in *Matlab* [12] and tested using different binary sequences. In Fig. 4.9 a random MUX signal under front and back irradiation is displayed as well as the decoding results. A good agreement between the signals used to drive the LED's and the decoded sequences is achieved. In all tested sequences the RGBV signals were correctly decoded.

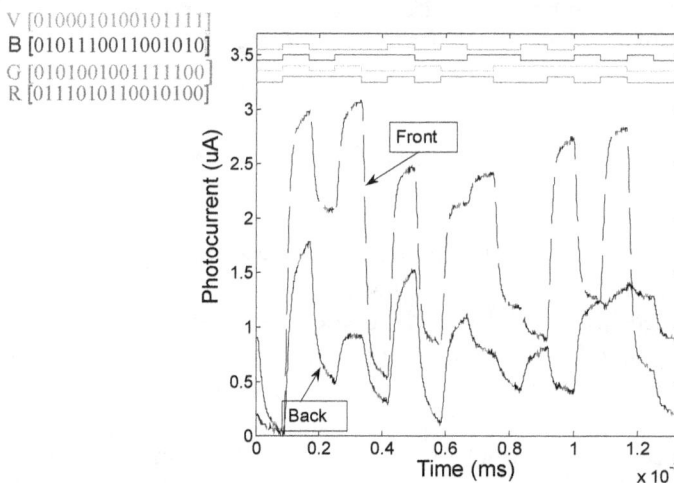

Fig. 4.9. DEMUX signals and decoded RGBV binary bit sequences.

The DEMUX sends the input logic signal to one of its 2^n (n is the number of color channels) outputs, according to the optoelectronic demux algorithm. So, by means of optical control applied to the front or back diodes, the photonic function is modified, respectively from a long-pass filter to pick the red/infrared channels to a short-pass filter to select the violet channel, giving a step reconfiguration of the device. The green and blue channels are selected by combining both active long- and short-pass filters into a band-pass filter. In practice, the decoding applications far outnumber those of demultiplexing. Multilayer SiC/Si optical technology can provide a smart solution to communication problems by providing a possibility of optical bypass for the transit traffic by dropping the fractional traffic that is needed at a particular point.

4.7. Applications

4.7.1. VIS/NIR Wavelength Selector based on a Multilayer pi'n/pin a-SiC:H Optical Filter

In Fig. 4.10, the polychromatic mixture of three (RGB MUX), four (RGBV MUX) or five (RGBVI MUX) input channels (V (400 nm), B (470 nm), G (524 nm), R (626 nm) and I (700 nm)) under front (Fig. 4.10a) and back (Fig. 4.10b) lighting are displayed. At the top, the input channels wavelengths and their bit sequences guide the eyes.

Results show that under front irradiation, if only three channels are involved, the MUX signal presents 2^3 different ordered levels ascribed each one to the eight *on/off* possible states. If four channels are considered, the 2^4 possible combinations are shown and, finally, if five channels are simultaneously impinging on the device, the number of visible levels increases to 2^5.

The levels are ordered accordingly the optical gains (Fig. 4.5, Table 4.1). Under front irradiation, in the higher levels the 624 nm and 700 nm channels are always in the *on* state and the presence of the violet channel is the less weighted. Under back irradiation due to the different enhancement of the violet and blue optical gains, if only three channels are present the blue channel is selected otherwise the violet will be tuned. This nonlinearity provides the possibility for selective tuning the different wavelengths allowing their recognition [13, 14].

In Fig. 4.11, both 400 nm and 700 nm channels have the same bit sequence which corresponds to only 2^3 ordered levels, however since the optical gains of both channels are quite different under front and back irradiation (Table 4.1) it is possible to identify them. Under back irradiation the MUX signal receives its main contribution from the 400 nm channel, while under front irradiation it is mainly weighed by the long wavelength channels. By comparing front and back irradiation is possible to decode the transmitted information [12].

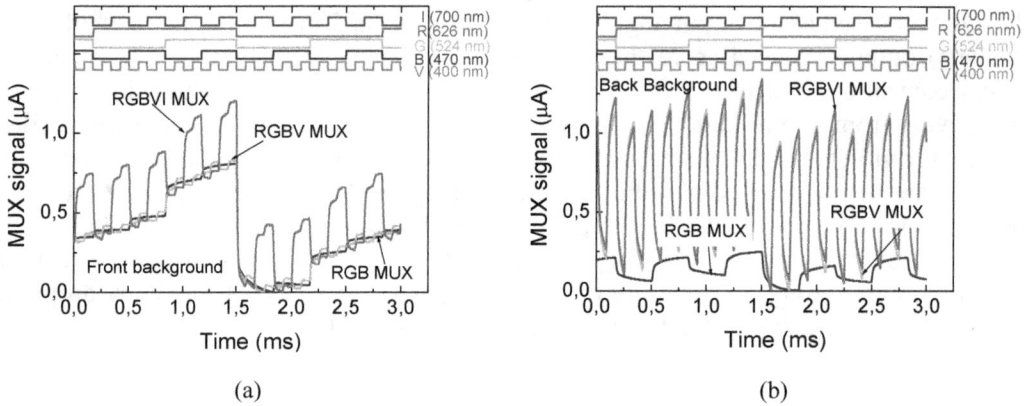

Fig. 4.10. Polychromatic mixture of three, four and five input channels under front (a), and back (b) irradiation.

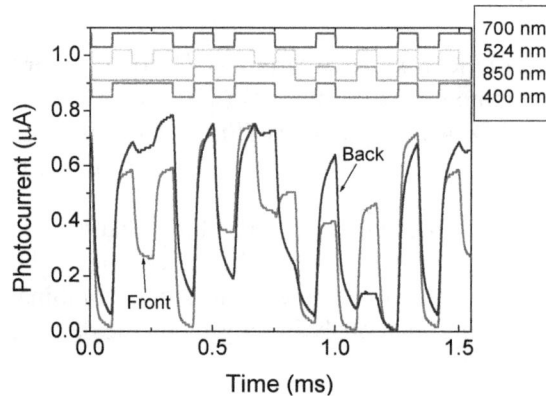

Fig. 4.11. MUX signals under front and back irradiation and two channels (400 nm and 700 nm) with the same bit sequence.

Taking into account the device configuration (Fig. 4.1), the front background irradiation is absorbed at the beginning of the front diode and, due to the self-bias effect [5], increases the electric field at the back diode where the red/infrared incoming photons (see Fig. 4.1) are absorbed accordingly to their wavelengths resulting in an increased collection. Under back irradiation the electric field decreases mainly at the i-n back interface quenching the long wavelength input signals in different ways. This effect may be due to the increased absorption under back irradiation that increases the number of carriers generated by the long wavelength photons. So, by switching between front and back irradiation the photonic function is modified from a long- to a band-pass filter allowing, alternately selecting the long and the short wavelength channels.

Fig. 4.12, the MUX signals due to two possible combinations of 700 nm, 626 nm 524 nm 470 nm and 400 nm input channels with different bit sequences are displayed, under front and back 2800 µWcm^{-2} irradiation. The driving signals are displayed on the top of the figures.

Results confirm that the magnitude and shape of the combined signal depends mainly on the channel wavelength through its own gain. Under front and back irradiation, the gains are quite different, so each of those five channels is enhanced or quenched differently. This side dependent effect is used to filter the input signals allowing their recognition and providing the possibility for selective tuning each input channel. Under front irradiation the MUX signal presents 2^5 separate levels each one ascribed to one of the of the thirty-two possible combinations of the *on/off* states. So, by assigning each output level to a five digit binary code weighted by the optical gain of the each channel, the signal can be decoded. A transmission rate of 30 Kbps was achieved.

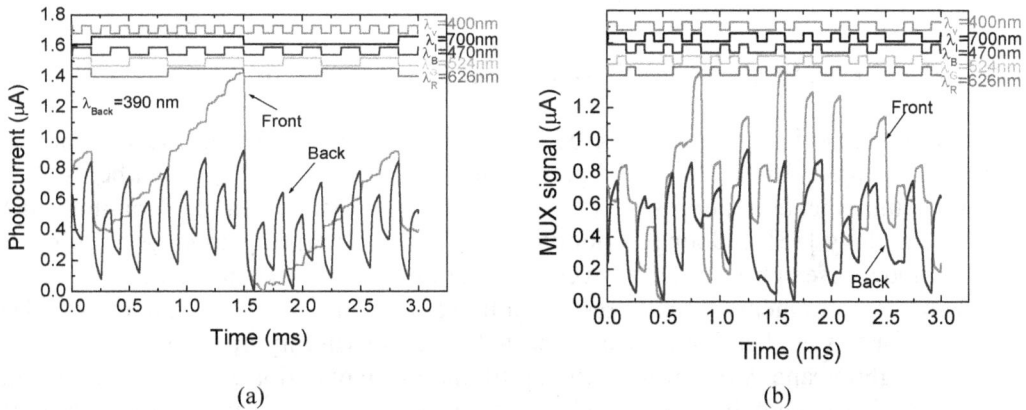

Fig. 4.12. MUX signals and under front and back λ=390 nm irradiation and different bit sequences: (a) standard sequence, and (b) random sequence.

Fig. 4.13, presents results of a numerical simulation with 2800 µW/cm^2 front and back λ=390 nm irradiation. Fig. 4.13 a uses the MUX signal of Fig. 4.12 a, whereas in Fig. 4.13 b a random sequence based on Fig. 4.12 b is used. Values of R_1=10 kΩ, R_2=1 kΩ, C_1=1 nF, C_2=20 nF were used during the simulation process. On top of the figures the drive input LED signals guide the eyes into the different *on/off* states and correspondent wavelengths.

A good fitting between experimental and simulated results was achieved. Furthermore, this model allows for extracting theoretical parameters by fitting the model to the measured data (internal resistors and capacitors). Under back irradiation higher values C_2 were obtained confirming the capacitive effect of the near-UV radiation on the device that increases the charge stored in the space charge layers of the back optical gate of Q_2 modelled by C_2.

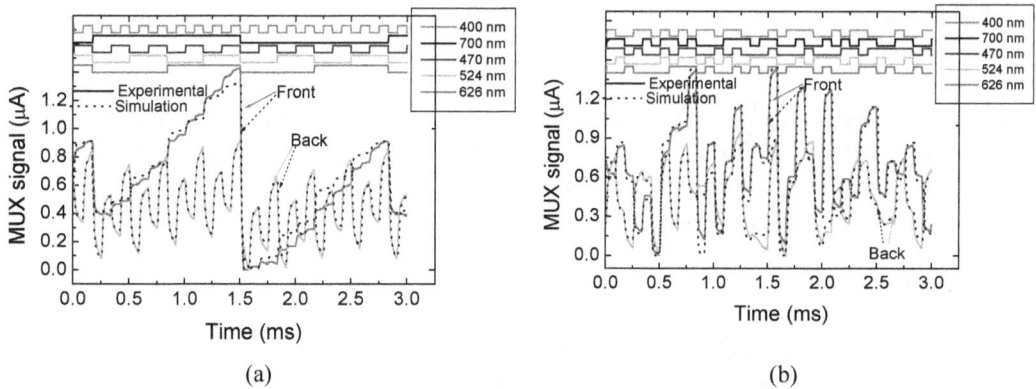

Fig. 4.13. Numerical simulation with front and back λ=390 nm irradiation, and different bit combinations.

4.7.2. Indoors Localisation Using Visible Light Communication

The lighting market has recently suffered tremendous changes due to the development of energy saving white LEDs. Besides lighting purposes the same devices can be used for wireless communication purposes when integrated in Visible Light Communication (VLC) technology [15]. Indoor positioning for navigation purposes in large buildings is currently under research in order to overcome the difficulties associated with the use of GPS in such environments, usually restricted by attenuation and shadowing effects. The motivation for this application is also supported by the possibility of taking advantage of an existing lighting and WiFi infrastructure [16], Indoors navigation can be used in several applications that extend from the guidance of users inside large buildings (museums, shopping malls, convention centers, airports, etc.) or provide location-based services and advertisements to the users when available in consumer electronic products or even for the automation of some inventory management processes through the location detection of products inside large warehouses [17].

There are different positioning algorithms for indoors positioning using VLC that can be broadly divided into three categories: triangulation, scene analysis, and proximity. In this chapter we use a novel approach to determine localization using a communication system that operates in the visible range with four white RGB LEDs and a photodetector device based on two stacked multilayered a-SiC:H/a-Si:H structures [18, 19]. The internal red and blue chips of the white LEDs are modulated at different frequencies and the device photocurrent is measured under different biasing conditions. The decoding strategy takes advantage of the external adjustment of the device sensitivity that makes possible the identification of the transmitted wavelength and of the output photocurrent Fourier analysis [20]. A navigation path is analyzed and decoded using the proposed positioning algorithm.

The proposed indoors navigation system involves wireless communication, computer based algorithms and smart sensor and optical sources network, which constitutes a multidisciplinary approach framed in cyber-physical systems.

4.7.2.1. Positioning System Design

The proposed localization system is based on the use of four white tri-chromatic LEDs placed on the ceiling and a pinpin photodetector located at the ground level (Fig. 4.14). At a fixed distance the photodetector device was centered inside this square.

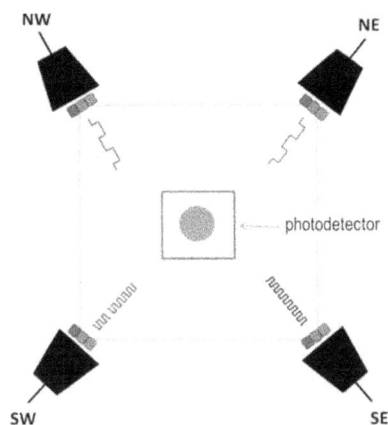

Fig. 4.14. Experimental assembly of LEDs and photodetector positioning system.

The red chips of the left white LEDs were modulated at different frequencies while the same procedure was done with the blue chips of the right LEDs. It was assumed that each white LED cone of light would overlap in the central region of the square. In the lateral and corner parts this intersection would be partial due to the radiation patterns superposition of the closest two or three LEDs. The identification of the different signals will be the key for the identification of the photodetector position and thus for the indoor navigation based on the lighting infra-structure.

It was assumed that each LED semiconductor chip emitted light only perpendicular to the semiconductor's surface, and a few degrees to the side, which results in a light cone pattern. When the LEDs are adjacent there is a superposition of the light pattern of each LED, giving rise to spatial regions that can be assigned to different light conditions (Fig. 4.15 a).

As the white LEDs are placed in a square geometry, the superposition of the four LEDs results in nine different regions (Fig. 4.15 b), where the presence of two, three or four optical signals coexist.

Fig. 4.16, shows the simplified cross-section structure of the multilayer heterostructure used to detect the transmitted information and the biasing steady state illumination supplied by monochromatic light of fixed wavelength.

Background steady state light is supplied by violet LEDs (390 nm) that illuminates the device by the back or the front side. The white light produced by the RGB LEDs is directed from the front side and in each LED the red of the blue chips were modulated with a specific bit sequence. The device was reverse biased at − 8 V and the photocurrent was measured between the front and back electrical contacts.

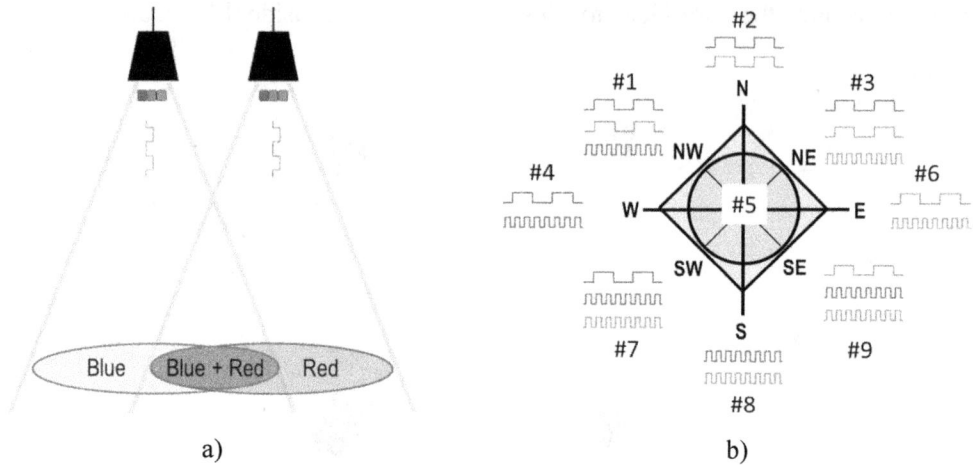

a)   b)

Fig. 4.15. (a) Light cone pattern superposition from adjacent LEDs; (b) Cardinal directions assigned to the superposition of the emission cones pattern of adjacent LEDs.

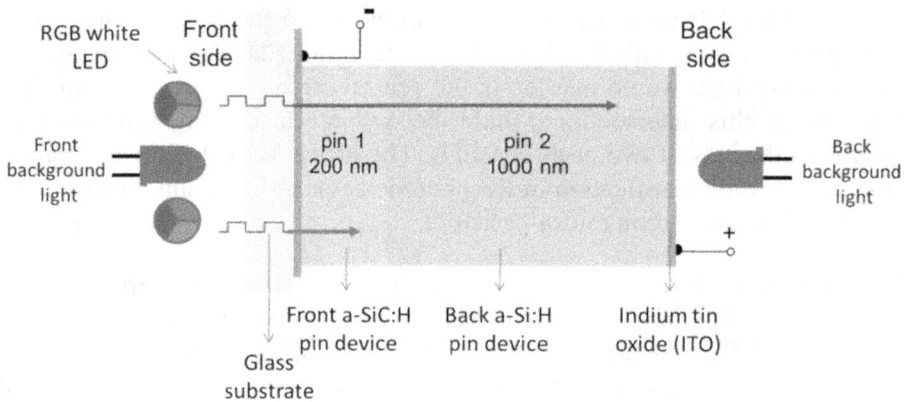

Fig. 4.16. Simplified cross-section view of the photodetector.

4.7.2.2. Optoelectronic Characteristics

The characterization of the optical sources was done through the measurement of the output spectra of each biased chip junction of the RGB white LED with the driving current. In Fig. 4.17 it is plotted the normalized output spectra of the red and blue chips of the RGB white LEDs used in this experiment and the dependence of peak intensity with the driving current.

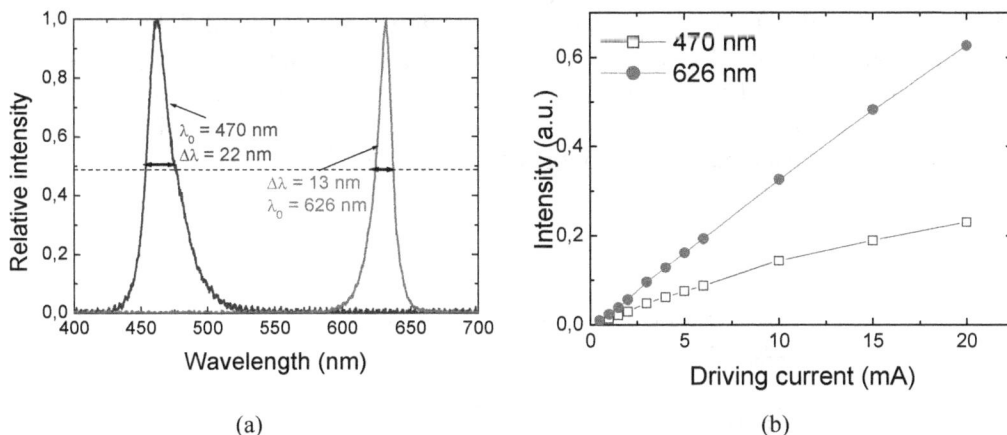

(a)   (b)

Fig. 4.17. Optoelectronic characteristics of the red and blue chips: (a) output spectrum;
(b) output intensity dependence on driving current.

The output spectra cover the wavelengths assigned to the blue, green and red regions, with
wavelengths centered, respectively at 470 nm and 626 nm. The full width half height
(FWHH) is 22 nm for the blue chip and 13 nm for the red chip, which is in agreement
with the usual design of these chips adjusted for the white color perception. As it can be
seen from Fig. 4.17 a) the trend is similar for both wavelengths, as the optical intensity is
enhanced with the current. When the current increases, the minority carrier concentration
increases also, and thus the rate of recombination is enhanced, which results in an
increased output of light intensity (Fig. 4.17 b). However, the increase in the output light
power is not linear with the LED current. At high current levels a strong injection of
minority carriers occurs, which leads to the recombination time to be dependent on this
concentration, and hence on the current itself. This leads to a nonlinear recombination rate
with the current. The optical characteristics of the white tri-chromatic LEDs are
summarized in Table 4.2.

Table 4.2. Optical characteristics of the white LEDs at 25 °C.

	Red	Green	Blue
Dominant wavelength, (nm)	619 - 624	520 - 540	460 - 480
Luminous intensity, (mcd)	355 - 900	560 - 1400	180 - 505
Spectral bandwidth @ 20 mA	24	38	28

The output characteristics of the photodetector in transient mode are shown in Fig. 4.18
without and under background light from both front and back sides. The experiment was
done using square waveform driving currents to modulate the red and blue chips of the
white LEDs, and the output transient signal was measured under pulsed illumination of
the red and the blue chips individually. On top of each graph the correspondent optical
signal is displayed.

Results show that the use of steady state illumination as background light changes the device spectral sensitivity. For long wavelengths (red) it is observed an amplification of the photocurrent under front optical bias while under back optical bias the signal is reduced. For shorter wavelengths (blue) the opposite trend is observed with a small amplification under back bias and a minor reduction under front bias.

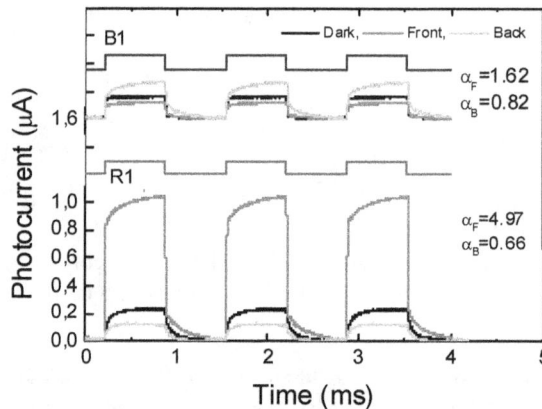

Fig. 4.18. Transient photocurrent measured under pulsed illumination of the (a) red, and (b) blue chips without optical bias and under front and back optical bias.

This means that the modulated signal of the red chip will be enhanced under front light and shortened under back illumination, while the blue signal will be amplified under back light and slightly reduced under front light. The quantification of the signal amplification under front and back bias is determined by the optical gain (α_F and α_B for the front and back gains, respectively), defined at each wavelength (λ) as the ratio between the signal magnitudes measured with and without optical bias. The gains observed for the signals used in this experiment are listed in Table 4.3.

Table 4.3. Optical gains of each individual signal.

	α_F	α_B
R_1	4.97	0.66
R_3	5.00	0.67
B_1	0.82	1.62
B_3	0.78	1.34

The analysis of the optical gains shows that red signals exhibit an amplification factor around 5 when measured under front bias, while under back bias they are reduced 30 %. On the other hand, the blue signals suffer amplification near to 1.5 under back illumination and a reduction of 20 % under front light background.

4.7.2.3. Analysis of Transient Signals from RGB White LEDs

In order to analyse the photocurrent signal when the red and blue chips of the tri-chromatic white LED are transmitting a different signal, the internal LEDs were pulsed using different time dependent biasing currents. The location identification is based on the analysis of the device photocurrent, which results from the optical excitation induced by the optical signals. Thus it is important for the system to be able to detect the combination of two, three of four optical signals. The output photocurrent signals measured under different optical bias conditions (with and without front and back optical violet bias) are displayed in Fig. 4.19. The condition assigned when all optical signals are off corresponds to the reference level.

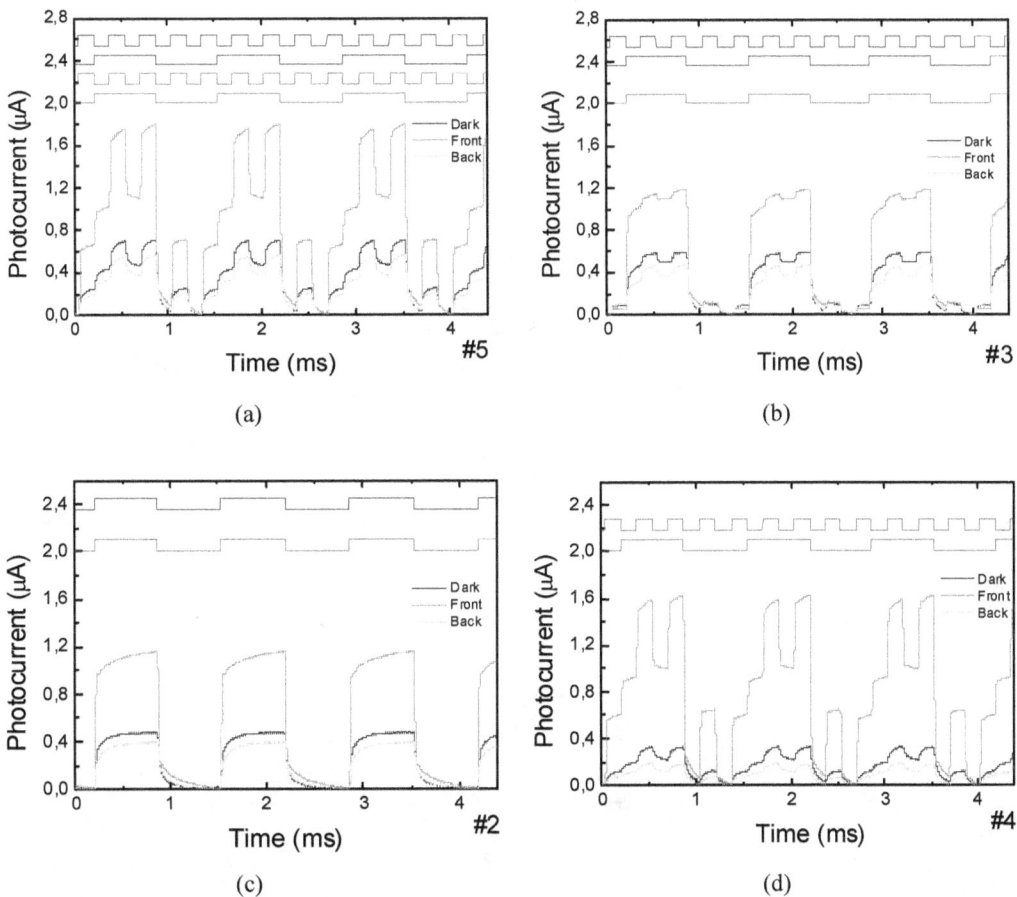

Fig. 4.19. Photocurrent measured without and under front and back optical bias when the device at: (a) central position, #5; (b) corner position, #3, (c) position #2, and (d) position # 4. At the top it is shown the input modulated optical signals from red and blue chips.

The single optical signals are displayed at the top of the figure to help the reader with the different on-off optical states. Fig. 4.19 a displays the transient signal using different optical bias conditions during the signal acquisition process when the photodetector is located at the central region (receiving 4 optical signals at position #5), in Fig. 4.19 b at a corner region (3 optical signals at position #3) and in Figs. 4.19 c and 4.19 d at side regions (2 optical signals at positions #2 and #4). In each region each optical signal will induce an optical excitation in the photodetector and thus contribute differently to the photocurrent. In positions #5, #3 and #2 the detector receives optical signals from the red and blue signals modulated at different frequencies. In these cases the signal measured under back illumination is similar to the signal measured without background illumination, which is due to the presence of both red and blue wavelengths that exhibit opposite behaviors under back illumination. The red light quenches the signal and the blue light amplifies it. On the other hand, the photocurrent under front illumination results in an amplified signal compared with the signal without optical background bias, due to the high amplification factor of the red light. From the evaluation of the amplification factors it is observed the values of $\alpha_F=2.53$, $\alpha_B=0.80$ for the signals of position #5 (Fig. 4.19a), $\alpha_F=2.02$, $\alpha_B=0.81$ for position #3 and $\alpha_F=2.42$, $\alpha_B=0.83$ for position #2 (Fig. 4.19c), which are the key elements for the detection of blue and red channels. In position #4 there is only the presence of two red channels of different frequencies and thus the signal under front optical background is strongly amplified ($\alpha_F=4.86$) and cut around the half under back bias ($\alpha_B=0.58$).

In Fig. 4.20 is displayed the dependence observed between the amplification factors of the mixed signals measured in positions #1 to #9. As this quantity is obtained from the ratio between the signal magnitudes measured with and without optical bias, it results from the ratio of maximum photocurrent values obtained when all optical signals are ON.

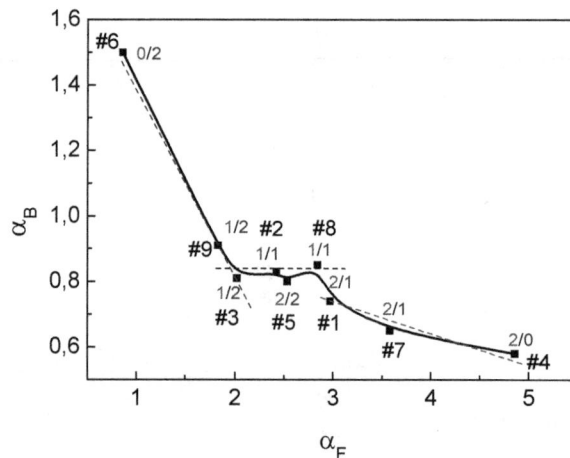

Fig. 4.20. Dependence observed between the amplification factors of the mixed signals measured in positions #1 to #9. The colored dashed lines illustrate different trends assigned to balanced (black dashed line) or unbalanced contribution the channels (red and blue dashed lines).

The trend of the plot shown in Fig. 4.20 shows three distinct regions, that correspond to the strong presence of blue light ($\alpha_F<2$ and $\alpha_B>0.8$), to the balanced contribution of both red and blue channels ($2.42 < \alpha_F < 2.84$ and $\alpha_B \cong 0.80$) and to the strong contribution of the red light ($\alpha_F > 3$). These trends allow thus the detection of the signals received by the photodetector/DEMUX device.

4.7.2.4. Decoding Strategy

As shown in Fig. 4.15b each region is characterized by different light signals. As proof of concept we have developed a computer algorithm to detect the sensor's relative position based on the measured photocurrents. The program uses two simple steps. In the first step, taking advantage of the wavelength filtering properties of the photodiode, front biasing is used to detect the red wavelength and back biasing to detect the blue wavelength. Because signals of two different modulating frequencies may share the same light wavelength an additional step is necessary to detect which frequencies are present. The modulus of the complex Fourier coefficient of the photocurrent is calculated for each of the relevant frequencies and compared with a predefined threshold value. Only when the calculated coefficient is superior to the threshold the respective signal is considered present. Using this approach it was possible to correctly identify the regions illustrated in Fig. 4.15 b based only on the measured photocurrent. As an example, Fig. 4.21 shows the absolute value of the Fourier coefficients at the center position. In this case for both the back and front photocurrent we can see that there are two strong peaks at 750 Hz and 3 kHz. This suggests that the two modulating frequencies are present in both wavelengths which is in agreement with the center position being illuminated by all four LEDs.

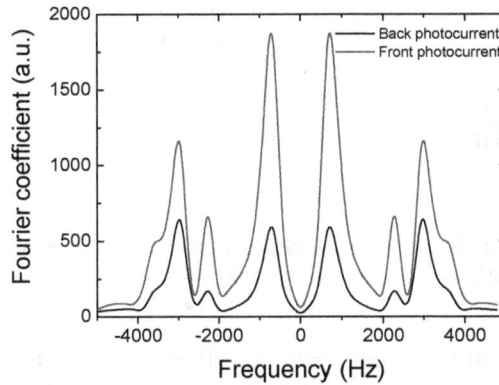

Fig. 4.21. Frequency spectra of the front and back photocurrent signals measured at the central position.

4.7.2.5. Navigation Performance

In order to test the navigation performance of the system, the photocurrent signal was measured placing the detector at the center of the geometrical assembly defined by the four LEDs. Then the detector was moved towards different directions defining 3 paths to NW, NE and SW (Fig. 4.22 a). In each of these directions the signal was measured and compared to the signal obtained under the assumption that in that region only two or three of the LEDs were ON (Figs. 4.22 b, c and d).

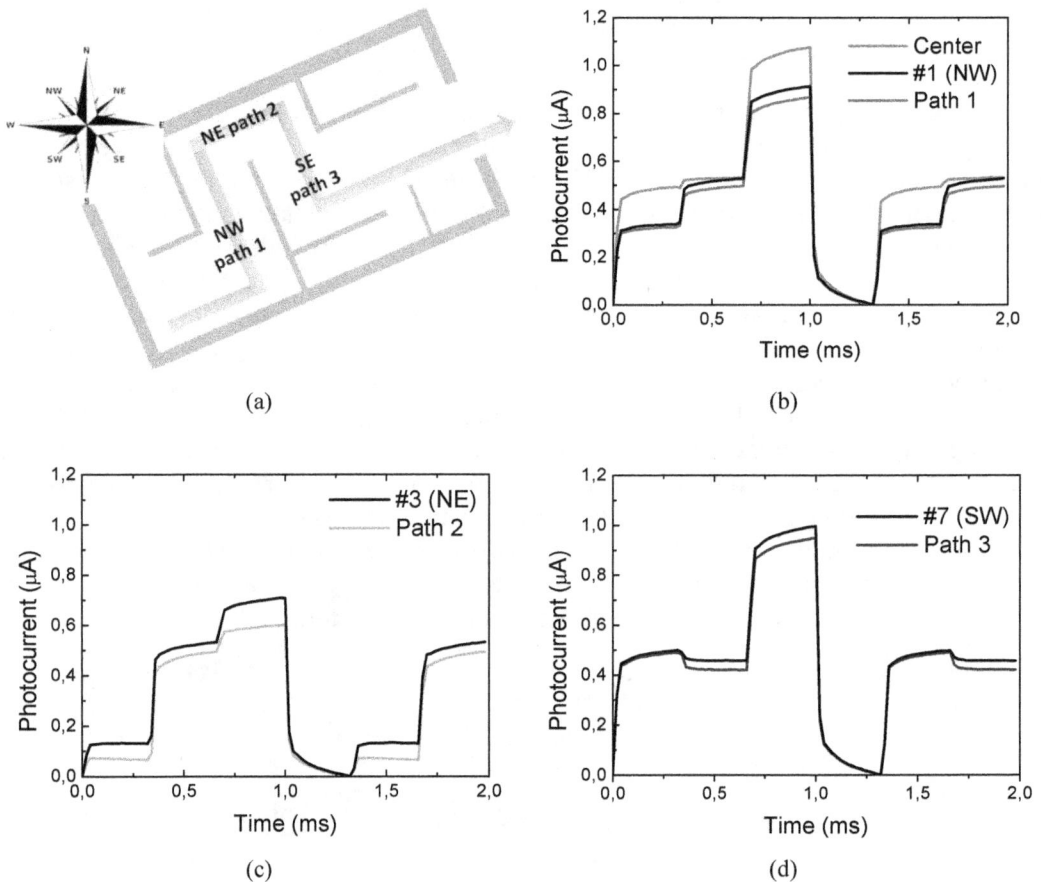

(a)

(b)

(c)

(d)

Fig. 4.22. (a) Paths of the detector; Front photocurrent signal measured at different detector positions: (b) Center and NW, (c) NE and, (d) SW.

Results show that the output signal develops differently depending on the impinging optical signals, which are related to the detector position. In plots of Figs. 4.22 b c and d the dark line corresponds to the front photocurrent signal measured under the LEDs pulsed illumination that corresponds to that cardinal direction (Fig. 4.15 b). The red, green and blue lines correspond to the photocurrent signals measured with the detector following the paths shown in Fig. 4.22 a. The observed signals agreement is the key for the adequate localization.

4.7.2.6. Summary

An indoors positioning system based on the use of VLC has been presented. Results showed that the use of a pinpin double photodiode based on a a-SiC:H heterostucture as photodetector and modulated RBG white LEDs as optical sources enable navigation capabilities to the proposed system, as the position of the moving detector can be inferred along time. The decoding strategy used takes advantage of the filtering properties of the device to infer the modulated wavelength and the Fourier transform for frequency determination. Future work involves intensity studies and the analysis of the position accuracy of the photodetector.

4.8. Conclusions

We experimentally and theoretically demonstrate the use of near-ultraviolet steady state illumination to increase the spectral sensitivity of a double a-SiC/Si pi'n/pin photodiode beyond the visible spectrum (400 nm-880 nm). The concept is extended to implement a 1 by 4 wavelength division multiplexer with channel separation in the visible/near infrared ranges.

Results show that, the pi'n/pin multilayered structure becomes reconfigurable under front and back irradiation, acting as data selector in the VIS/NIR ranges. The device performs WDM optoelectronic logic functions providing photonic functions such as signal amplification, filtering and switching. The opto-electrical model with light biasing control has proven to be a good tool to design optical filters in the VIS/NIR. An optoelectronic model was presented and proven to be a good tool to design optical filters in the VIS/NIR range. A decoding algorithm to decode the information was presented. Some applications using the same device in the VLC domain were reported.

Acknowledgements

This work was supported by FCT (CTS multi annual funding) through the PIDDAC Program funds (UID/EEA/00066/2013).

References

[1]. T. Komiyama, K. Kobayashi, K. Watanabe, T. Ohkubo, and Y. Kurihara, Study of visible light communication system using RGB LED lights, in *Proceedings of the IEEE SICE Annual Conference,* 2011, pp. 1926–1928.

[2]. S. S. Djordjevic et al., Fully Reconfigurable Silicon Photonic Lattice Filters With Four Cascaded Unit Cells, *IEEE Photonics Technology Letters,* 23, No. 1, 2011, pp. 41-44.

[3]. P. P. Yupapin and P. Chunpang, An Experimental Investigation of the Optical Switching Characteristics Using Optical Sagnac Interferometer Incorporating One and Two Resonators, *Optics & Laser Technology,* Vol. 40, No. 2, 2008, pp. 273-277.

[4]. S. Ibrahim et al., Fully Reconfigurable Silicon Photonic Lattice Filters with Four Cascaded Unit Cells, in *Proceedings of the Optical Fibre Communications Conference, OSA/OFC/NFOEC,* San Diego, 21 Mar 2010, paper OWJ5.

[5]. M. Vieira, P. Louro, M. Fernandes, M. A. Vieira, A. Fantoni, and J. Costa, Three Transducers Embedded into One Single SiC Photodetector: LSP Direct Image Sensor, Optical Amplifier and Demux Device, Advances in Photodiodes, *InTech,* Chap. 19, 2011, pp. 403-425.

[6]. M. Vieira, P. Louro, M. Fernandes, M. A. Vieira, A. Fantoni, and J. Costa Advances in Photodiodes, *InTech,* Chap. 19, 2011, pp. 403-425.

[7]. M. A. Vieira, P. Louro, M. Vieira, A. Fantoni, and A. Steiger-Garção, , Light-activated amplification in Si-C tandem devices: A capacitive active filter model, *IEEE Sensor Jornal,* 12, No. 6, 2012, pp. 1755-1762.

[8]. M. A. Vieira, M. Vieira, P. Louro, V. Silva, and A. S. Garção, Photodetector with integrated optical thin film filters, *Journal of Physics: Conference Series,* 421, March 2013, 012011.

[9]. M. Vieira, M. A. Vieira, I. Rodrigues, V. Silva, and P. Louro, Tuning optical a-SiC/a-Si active filters by UV bias light in the visible and infrared spectral ranges, *Phys. Status Solidi,* C, 2014, pp. 1-4.

[10]. M. Vieira, M. A. Vieira, I. Rodrigues, V. Silva, P. Louro, A. Fantoni, UV Irradiation to Increase the Spectral Sensitivity of a-SiC:H pi'n/pin Photodiode Beyond the Visible Spectrum Light, in *Proceedings of the 6th International Conference on Sensor Devices, Technologies and Applications (SENSORDEVICES' 15),* Venize, Italy, 2015, pp. 44-50.

[11]. M. A. Vieira, M. Vieira, P. Louro, V. Silva, J. Costa, A. Fantoni, SiC Multilayer Structures as Light Controlled Photonic Active Filters, *Plasmonics,* 8, 1, 2013, pp. 63-70.

[12]. M. A. Vieira, M. Vieira, J. Costa, P. Louro, M. Fernandes, A. Fantoni, Double pin Photodiodes with two Optical Gate Connections for Light Triggering: A capacitive two-phototransistor model, *Sensors & Transducers,* Vol. 10, Special Issue, February 2011, pp. 96-120.

[13]. M. Vieira, M. A. Vieira, P. Louro, A. Fantoni, V. Silva, SiC pinpin Photonic Filters for Linking the Visible Spectrum to the Telecom Gap, *Microelectronic Engineering*, Vol. 126, 25 August 2014, pp. 179–183.

[14]. M. Vieira, M. A. Vieira, V. Silva, I Rodrigues, P. Louro, A. Fantoni, Wide Spectral Sensitivity of Monolithic a-SiC:H pi'n/pin Photodiode Outside the Visible Spectrum, *Sensors & Transducers,* Vol. 193, Issue 10, October 2015, pp. 33-40.

[15]. K. Panta and J. Armstrong, Indoor localisation using white LEDs, *Electron. Lett.*, 48, 4, 2012, pp. 228–230.

[16]. W. Zhang, M. I. S. Chowdhury, and M. Kavehrad, Asynchronous indoor positioning system based on visible light communications, *Optical Eng.,* 53, 4, 2014, 045105.

[17]. T. Tanaka and S. Haruyama, New position detection method using image sensor and visible light leds, in *Proceedings of the 2nd IEEE Int. Conf. on Machine Vision (ICMV'09),* 2009, pp. 150–153.

[18]. P. Louro V. Silva, M. A. Vieira, M. Vieira, Viability of the use of an a-SiC:H multilayer device in a domestic VLC application, *Phys. Stat. Sol.*, C11, 2014, pp. 1703–1706.

[19]. P. Louro, V. Silva, J. Costa, M. A. Vieira and M. Vieira, Transmission of Signals Using White LEDs for VLC Application, *MRS Advances,* 2016, in press.

[20]. P. Louro, V. Silva, J. Costa, M. A. Vieira and M. Vieira, Added transmission capacity in VLC systems using white RGB based LEDs and WDM devices, *Proc. SPIE 9899, Optical Sensing and Detection IV*, 98990F, 2016.

Chapter 5
Trends in Improving the Accuracy of SPR-Devices

Volodymyr Maslov, Yuriy Ushenin, Glib Dorozinsky

5.1. Introduction

It is known that optical methods possess a high operation speed and enable to reach high accuracy and sensitivity in measurements. One of the promising optical methods for analysis of various compounds and micro-objects as well as processes at the molecular level is the refractometric method based on surface plasmon resonance (SPR) phenomenon. As compared with traditional measuring methods, the SPR-method provides possibility to study processes of molecular interaction in micrometer-thickness layers in the real-time scale; low value of the sample volume required for measurements (less than 10 μl); the method does not require any markers or fluorescent labels for studying the analyte [1].

5.2. Principle of Operation of Devices Based on Surface Plasmon Resonance and Main Areas of their Application

The principle of operation of SPR-devices lies in determination of changes in the analyte refraction index (RI) by observing the shift of the analyte reflection curve R(θ) minimum. The preferential majority of SPR-devices are designed using the Kretschmann geometry (Fig. 5.1a) that consists of a laser (1), prism for total internal reflection (TIR) (2), sensitive element (3) and photodetector (4) [2].

The most widely spread sources for exciting surface plasmons are lasers. The reflection characteristic R(θ) is the dependence of the intensity of laser light on the angle of its incidence onto the surface of sensitive element (SE) within the range of angles higher than the TIR angle at the boundary SE – analyte (Fig. 5.1b). The analyte RI value is related

Volodymyr Maslov
V. Lashkaryov Institute of Semiconductor Physics NAS of Ukraine, Kyiv, Ukraine

with the value of the minimum inherent to the reflection curve $\Delta\theta0$ via the following parameters of device optical scheme elements: laser light wavelength, refraction indexes of SE, TIR prism and analyte. As a plasmon-carrying layer, there mainly used is gold or silver.

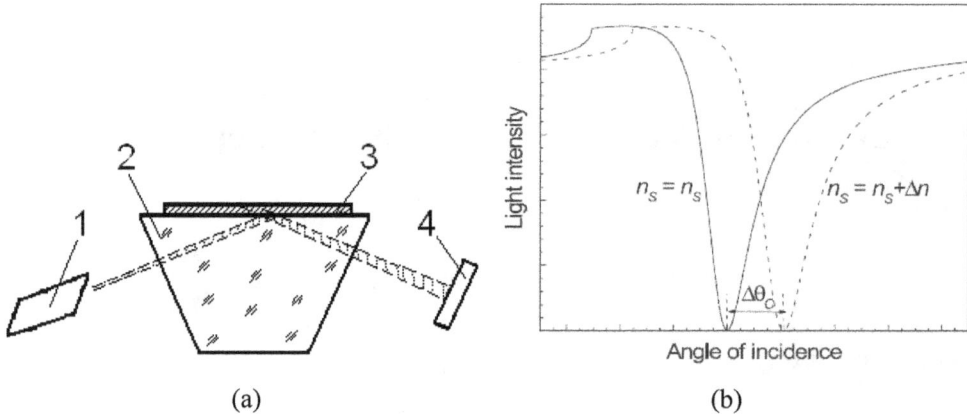

(a) (b)

Fig. 5.1. Optical scheme of the SPR-device based on the Kretschmann geometry (a) and angular dependences for the intensity of light reflected from the boundary SE-analyte before (-) and after (---) changes in the analyte RI by the value Δn (b) [3].

If the SE metal layer is sufficiently thin (< 200 nm), then a considerable part of electromagnetic wave decaying in metal can reach the opposite surface of this layer. Then SPR becomes sensitive to properties of the medium contacting with metal. A position of the minimum in the reflection curve depends on electric polarization (dielectric permittivity) of this medium.

When properties of the metal layer or the refraction index of medium being above this layer are changed, the reflection minimum is essentially shifted. Therefore, originally measurements of reflection characteristics under SPR conditions were only considered as a very sensitive method for studying the optical properties and states of metal surfaces [4]. In what follows, it became used as the most exact method of refractometry for determining the refraction indexes of liquids and gases.

Improvement of SPR-devices and creation of highly stable receptors [5] enabled to perform an efficient monitoring of drugs both in the process of producing them and in the process of their application [6-10]. SPR-devices are widely used for revealing bacteria, parasites and viruses, namely: Staphylococcus aureus [11, 12], Pichia pastoris [13], Leishmania [14], Mycobacterium tuberculosis [15], human adenovirus [16], Epstein-Barr virus [17], avian influenza H5N1 [18], Dengue virus [19], hepatitis B [20] and HIV [21]. Besides, the SPR method is used for prevention and manifestation of embolism of arteries at early stages [22], of early birth [23], sclerosis [24], Alzheimer's desease [25, 26], for investigation of liver [27] and kidneys [28]. This method is also used for determination of nerve activity [29], glucose level [30] and blood group [31]. It is worth to note application of the SPR method for monitoring and curing oncological deseases. Due to its high sensitivity [32-36], this technology can be efficiently used to reveal cancer at its early stages, which is very important for timely curing. SPR-devices are also used for curing the definite oncological deseases both to develop medical drugs [37 - 40] and for direct cure. The most widely spread oncological deseases that can be diagnosed at the early stages by using the SPR method are as follows: cerebral glioma [44], skin melanoma [45] and ovarian carcinoma [46].

Enhancement of accuracy and sensitivity of measurements enables to widen the area for applying the SPR-devices in instrument making industry for detection of ecologically and fire dangerous volatile substances as vapors of methanol, acetone, benzenein air of paint zone or section. These devices are used for quantitative and qualitative analyses of the following air components: CO, H_2, NO_2 as well as vapors of organic substances [47, 48]. Also, they can be used for integrated controlling water quality [49], determination of metal ions content in water and food, in particular, iron [50], silver [51], cadmium [52], chromium [53], determination of availability and amount of pesticides as well as herbicides in water and food [54, 55], outcrop of parasites and creation of efficient means to throw off them [56-59]. SPR phenomenon is also used in analytical equipment for determining poison and explosive substances. In particular, it enables to detect such poison substances as phenols [60], neurotoxins [61], ochrotoxins [62], enterotoxins [63], palitoxins [64] and cyanides [65]. SPR can be applied for revealing such explosive substances as trinitrotoluol [66], for checking motor oils [67], detecting fume and methanol [69] in air.

To further widen the areas for application of SPR-devices and to reduce the detection limit of concentrations of studied substances (analytes), it is necessary to enhance sensitivity and accuracy of these devices.

This review is devoted to the analysis of main modern tendencies in increasing the accuracy of devices based on SPR phenomenon.

Our analysis of literature data for the last 25 years enabled to ascertain the tendency of development of SPR-devices with regard to enhancing their sensitivity and lowering their detection limit. Starting from 1990, the detection limit for the analyte concentration in buffer was lowered by 160 times from 8 ng/ml down to 50 pg/ml (Fig. 5.2 a) [68]. Also increased is the number of publications (Fig. 5.2 b).

With account of the growth in the number of publications devoted to improvement and application of SPR-devices (Fig. 5.2 b), one can draw a conclusion that this direction of scientific researches is topical and promising. Being based on the above analysis, it can be expected that the detection limit will reach the range 3...10 pg/ml in 2020 - 2025. In future, decreasing the detection limit will be limited by availability of thermal noises and noises caused by friction of liquid flow in a cuvette, as well as technological possibilities in preparation of analytes. In the most cases, enhancement of the sensitivity is related with increasing the electromagnetic field at the surface of a sensitive element. The detection limit value can be considerably reduced by using the systems with a high value of the ratio signal-to-noise, namely: interferometry, ellipsometry or polarimetry [3-5].

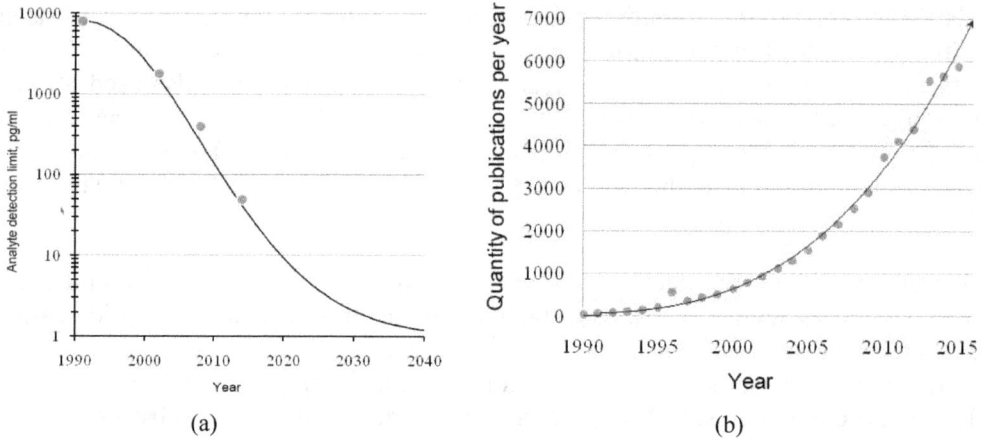

(a)   (b)

Fig. 5.2. Advancement in the analyte detection limit (a), and growth of the quantity of publications in the field of SPR sensors (b) for the last 25 years [2].

In recent decades, one can observe development of sensitive facilities based on SPR [6, 7] for optical detection of small biological and chemical objects in gases and liquids [8]. SPR-devices are applied in various fields of analytical activity including molecular recognition, identification of immune deseases, etc. [9-15]. The SPR-devices are widely used in scientific researches, in medicine, pharmacology and ecological monitoring [16-18], therefore, enhancement of accuracy inherent to these devices seems to be a topical task. Therefore, leading firms producing SPR-devices as well as scientific community improve both technology and design of SPR-devices to increase their sensitivity and accuracy in measurements.

In refractometric SPR sensors with direct conversion, the analyte refraction index directly changes characteristics of light wave, namely: the angle and wavelength of SPR excitation, intensity, phase and polarization of radiation (Fig. 5.3). And the main performances of these devices are sensitivity, accuracy and dynamic range. Just accuracy is the most important parameter. The sensor response Y to the given amplitude of the measured value X can be represented using the transfer function $Y = F(X)$ that can be defined by the theoretical model of this sensor or by the results of its calibration. However, the magnitude of the measured value X_{meas} differs from the true value X_0 by the amplitude of error of measurements. This error is a consequence of many reasons accompanying the process of measurements.

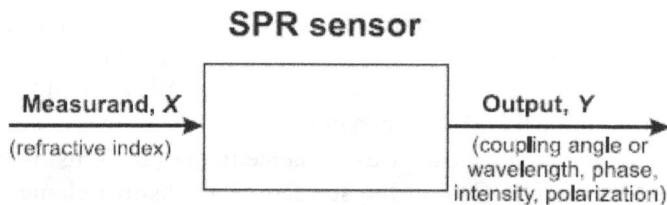

Fig. 5.3. Scheme of measurements for direct conversion of the input signal [1].

The analysis of construction and operation regimes of devices with angular scanning has shown that the main structural-and-technological factors influencing the nature and amplitude of basic measurement errors are as follows: construction of the sensor and technology of its manufacturing, temperature regime of device operation, excitation wavelength, value of analyte flow through the measuring cuvette, parameters of the device kinematic scheme, and the process of mathematical processing the measurement results. Some additional measurement errors are caused by stochastic processes, namely: interphase fluctuations related with random processes of adsorption-desorption at the boundary SE – analyte, effects of microscopic air bubbles at the SE surface, supply line interferences, electromagnetic interferences, effects of friction – slipping in the kinematic scheme, and so on.

The influence of these additional errors is usually minimized by device construction and application of complementary equipment in the course of measurements. For example, in devices of the "Plasmon" series [87], pumping the analyte through the measuring cuvette is performed using the 8-channel peristaltic pump developed at V. Ye. Lashkaryov Institute of Semiconductor Physics, NAS of Ukraine. The speed of pumping was chosen within the range (5...20) ± 1 μl/min and is kept constant, which provides minimal effect on analyte RI and changes in its pressure during pumping through the measuring cuvette. Mechanical vibrations are reduced due to special dampers between elements of the kinematic scheme and device case.

In practice, SPR registration is realized by measuring energetic relations between amplitudes of incident and reflected light. Changing the angle of light incidence or its wavelength, one can plot the dependence of light reflection on these parameters. The narrower this dependence, the higher accuracy of measurements of the minimum position can be reached [88].

Let us consider in more detail the following principal ways to increase the accuracy of SPR measurements for analytes: increasing the wavelength of radiation exciting SPR, improvement of construction and technology for deposition of the SE metal layer, reducing the influence of the temperature factor on measurement results as well as results of numeric methods for processing data of measurements with applying a special software.

5.3. The Way to Increase the Wavelength of Radiation Exciting SPR

Changes in the wavelength of exciting radiation cause changes in the refraction index of analyte and changes of optical scheme elements in the measuring facility along with SE. As a result, the shape of reflection characteristic is also changed (Fig. 5.4 a), which leads to increasing the absolute error in measurements of minimum angular position and, as a consequence, to growing the absolute error in RI of analyte. Besides, the increase in the wavelength results in narrowing the dynamic range for measurements of analyte RI in the regime Slope [2], makes the list of studied substances and chemical reactions more narrow, too (Fig. 5.4 b).

(a)

(b)

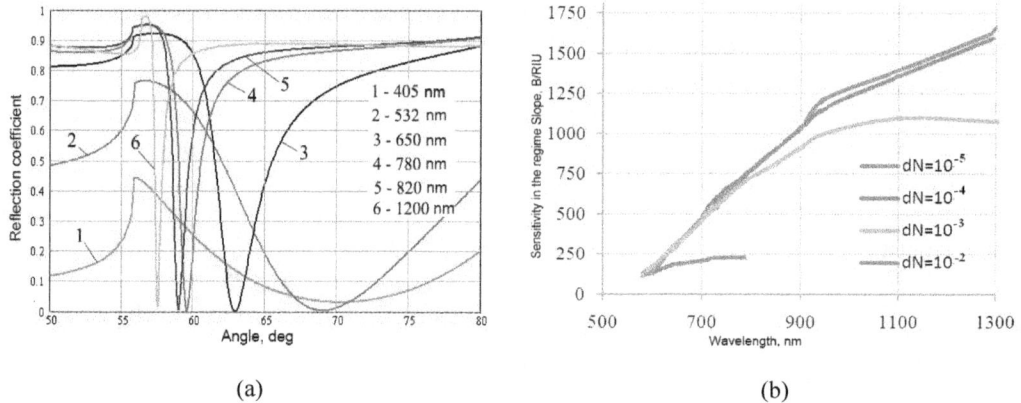

Fig. 5.4. Results of the numeric analysis for the calculated dependence R(θ) for various wavelengths from the optical range (a) and dependences for the device sensitivity in the Slope regime on the excitation wavelength under changing the analyte refraction index within the interval 10-5 to 10-2 (b) [2].

To reach the maximum in sensitivity, it is recommended to use the sources for SPR excitation with a wavelength taken from the range λ = 700…1000 nm [89]. But transfer to the infrared spectral range increases temperature effect on analyte. The optimal wavelength for excitation within this range was chosen as 850 nm from the viewpoint of minimal errors in determining the analyte RI as well as minimal action of electromagnetic radiation on this analyte [90, 91]. Experiments showed an increasing response of SPR-sensor and widening the measurement range (Fig. 5.5).

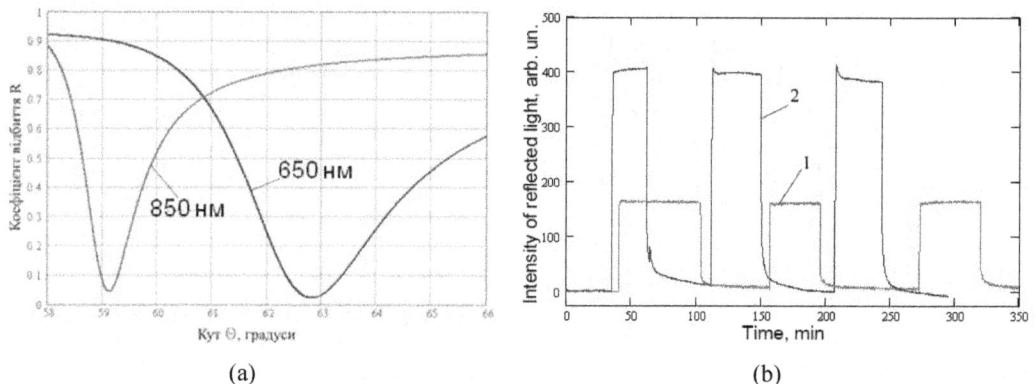

(a)

(b)

Fig. 5.5. Experimental dependences R(θ) (a), kinetics of distilled water substitution with sodium chloride solution (b) for the laser wavelengths 650 (1) and 850 nm (2) [2].

In the course of experiments, distilled water in the measuring cuvette was substituted with 9 % solution of sodium chloride. At first, the reflection characteristics for analytes R(θ) were measured, then the kinetics in the regime Slope was done. It was experimentally ascertained that growth in the device laser wavelength from 650 up to 850 nm provided reduction of the absolute error in measurements of analyte RI by 5.5 times: from $\pm 6.2 \times 10^{-5}$ down to $\pm 1.1 \times 10^{-5}$ [92]. Data of the numeric analysis showed that further increase in the wavelength from 850 up to 1200 nm has no effect on the value of absolute error in measurements of analyte RI and can be only used to enhance the device sensitivity in the Slope operation regime [70].

5.4. Improvement of Fabrication and Construction of the Device Sensitive Element

When creating the analytical devices, it is important to analyze the influence of SE material and technology for its making on the shape of reflection characteristics R(θ) as well as accuracy in determination of its minimum position. In practice, gold and silver are preferentially used as SE material. Copper and aluminum are not practically used: copper – through its high capability to be oxidized, and aluminum – through its very high value of the imaginary part of dielectric permittivity εi, which essentially widens the reflection characteristic. Analyzed in the works [93-95] is the problem of optimal choosing the metal and exciting light wavelength from the viewpoint of reaching the maximal sensitivity and chemical inertness of SE operation surface. It is known that usage of silver layers enables to obtain rather narrow minimum in the R(θ) characteristic as compared with that of other metals. However, the gold operation surface of SE is more stable and chemically inert. Therefore, just the gold layer is most widely used as carrier of surface plasmons.

An important factor here is the influence of surface relief inherent to the gold layer on light absorption, since just the surface is characterized by availability of a strong electric field. Therefore, surface roughness of metal layer defines an essential effect on propagation of surface plasmons, which, as a result of energy dissipation, leads to early decay of plasmons and reduction of their phase velocity [96-99]. In this case, the shape of the dispersion curve is changed, and the resonance frequency of surface plasmons is shifted. If the wavelength of incident light is fixed, with growing the SE surface roughness the minimum position of the reflection characteristics is shifted to the side of higher angles, the reflection amplitude in the resonance minimum (R_{min}) increases, the reflection characteristic is widen, and, as a consequence, the error of determining the minimum position grows.

Let us consider the influence of technology for preparation of SE metallic layer. The principal factors influencing the structure and properties of metallic layers made of thermal evaporation in vacuum are the speed of deposition and temperature of substrate [100, 101]. In the case of gold metallic layer of SE, the deposition speed 4 to 5 nm/s provides a layer with a maximal density as well as reproducible optical parameters and smooth homogeneous surface. At the same time, low speeds of deposition results in fine-dispersed, rough and friable structure of deposited layers, while high deposition speeds lead to the coarse-grained structure of surface [102]. To reduce the influence of substrate relief on the surface roughness of deposited metal layers, the substrate surface is prepared using traditional optical technology that is usually used when making optical parts [103].

An alternative way to act on the structure and properties of SE metal layers is the temperature annealing [104]. For the layers crystallizing under conditions of considerable overcooling at room temperature, the temperature annealing is an efficient stabilizing factor [105]. This thermal treatment decreases the concentration of defects in the crystalline lattice, the structure of these layers transfers to more stable thermodynamic state, which is accompanied by more stable optical properties [106]. In polycrystalline layers of gold and silver, the most essential changes in the structure with increasing of grain sizes take place for the first 5 – 10 min of annealing at relatively low temperatures (not exceeding 300 °C) [107]. To provide a minimal roughness, it is recommended to anneal at the temperature 120 °C [108].

One of the promising technological ways to enhance the accuracy of measurements is to narrow the reflection characteristic $R(\theta)$, which can be realized by decreasing the roughness of the SE metal layer due to changing the geometry of mutual arrangement of the substrate and evaporator. It was ascertained experimentally that when the substrate is placed at the angle 45° between its normal and direction to the evaporator, and the SE metal layer is deposited multiply, the surface roughness of this layer is decreased by 2.5 times: from 2 down to 0.8 nm. It resulted in narrowing the reflection characteristic and increasing the sensor response by 1.5 times, when analyzing liquid substances, and by 2 times for gases, in the Slope regime of measurements. Due to narrowing the SPR curve, the absolute error of measuring the analyte RI was 5-fold decreased: from $\pm\, 7 \times 10^{-6}$ down to $\pm\, 1.2 \times 10^{-6}$. The obtained results were confirmed by the authors of the work [109]. Implementation of this new technology for preparation of the SE metal layer not only decreased the absolute error of measuring the analyte RI but, in addition, increased the sensitivity due to growing the steepness of SPR curve slopes and ordering the structure of SE surface [110, 111].

5.5. Decreasing the Influence of Temperature Factor on Results of Measuring the Analyte Refraction Index

The influence of temperature on the accuracy of SPR-devices can be expressed via the absolute error of measurement results δNT caused by temperature changes of RI inherent to elements of the device optical scheme and analyte. The value of this error depends on the range of temperature changes in the course of measurements as well as values of RI temperature coefficients of optical elements and analyte.

The principal elements of the SPR optical scheme, which have the main effect on the value of this error are the half-pentaprism and SE metal layer [112]. Besides, the result of analyte RI measurements is influenced by the temperature dependence of laser wavelength. The temperature coefficient for the laser wavelength is approximately (0.12...0.15) nm/K [113]. Temperature changes of RI n and extinction coefficient k inherent to SE metal layer are, first of all, related with decreasing the density of charge carriers (electrons) in it, which is conditioned by thermal volume expansion of the metal layer and thermal vibrations of the metal crystalline lattice. The density of electrons in metal layer is in inverse proportion to temperature and the coefficient of thermal volume expansion [114].

Temperature changes of the half-pentaprism RI are determined by the value of the temperature coefficient inherent to material from which it is made of. In the devices of Plasmon series, half-pentaprisms are made of glasses of K8 (nP = 1.5145) and Φ1 (nP = 1.6154) types for measuring gas-like and liquid analytes, respectively. The temperature coefficients of refraction indexes for these types of glasses are as follows: 1.2×10^{-6} K^{-1} for type K8 and 3×10^{-6} K^{-1} for Φ1 [115]. For the distilled water (nA = 1.3314), as a liquid analyte, the temperature coefficient of RI is equal 1×10^{-4} K^{-1}, while that of dried air (nA = 1.00028) is close to -2.5×10^{-5} K^{-1} [116]. Being based on the known values of temperature coefficients for the elements of SPR-device optical scheme, one can draw a conclusion that the most essential influence on changes in results of measurements can be rendered by temperature changes in the analyte RI.

To improve the device construction, it is insufficient to have information about temperature coefficients for components of its optical scheme, but it is also necessary to know main sources for heat release. With this aim, it was offered and experimentally checked the way to determine these sources by using the thermographic method [117, 118]. Being based on the results of investigations, the authors developed the way and made a special facility (thermal box) to stabilize temperature, which enabled to keep its value with the absolute error no higher than ± 0.5° C in all the volume of operation chamber where the device and reservoirs with analytes were placed [119]. The results of experiments allowed ascertaining that temperature stabilization of measuring equipment reduces the absolute temperature error for the analyte RI: in the case of gas-like analytes – by 18 times (from 7×10^{-6} down to 3.8×10^{-7}), while for the liquid ones – by 3.2 times (from 2.5×10^{-5} down to $7.6 \times 10_{-6}$). This decrease of the temperature error due to stabilization of temperature for the device and analyte is explained by reduction of the temperature drift observed for the reflection characteristic minimum when measuring the kinetics.

5.6. Numeric Methods for Enhancing the Measurement Accuracy of SPR-Devices

The absolute error in measurements of the analyte RI when changing the excitation wavelength depends on the shape of reflection curve and is related with the way of approximation of measurement results, since this approximation is necessary to compensate the long step of angular scanning by SPR-facility. The absolute approximation error contains two components: methodical error and that caused by changing the width of reflection characteristic $R(\theta)$. The methodical error depends on symmetry of the reflection curve $R(\theta)$, angular range of approximation and amplitude range of approximation related with it. In the course of modeling, decreasing the methodical approximation error is possible if performing the following two conditions: i) approximation is only made in a limited bottom part of the reflection curve $R(\theta)$ near the minimum; ii) reflection curve has the shape close to the symmetric one within this part [5].

Approximation of the part should be performed around the minimum at the level below 25 % relatively to the maximum of the reflection characteristic $R(\theta)$. In the case, when considering this characteristic with clearly expressed asymmetry, the better results are provided by choosing the same non-symmetric positions for the origin and end of the processed curve part. In the work {120], the authors studied the influence of the power of approximating polynomial on the error in determination of the minimum position when approximating at the wavelength 650 nm. They noted that the polynomial of the 2-nd power is the most simple and reliable, but when the SPR curve is highly asymmetric the satisfactory results can be obtained only under considerable limitation of the level chosen for this approximation part (down to 5 %). Within the range of wavelengths chosen for modelling, the reflection characteristic $R(\theta)$ essentially changes its width, therefore the angular range for approximation was selected to be multiple to the reflection curve width at the amplitude level 20.

In addition, it was chosen the non-symmetric positions of the origin and end for the processed curve part. Thus, the authors realized conditions providing the decrease of methodic error due to insignificant dispersion of the amplitude range for approximation within the limits 10 - 23 %. When the wavelength grows, the absolute error slows down in accord with the expression $\delta n\,(\lambda) = 0.121 \times e^{-0.01\lambda}$, and within the range of wavelengths 850...1200 nm is confined within the limits $(2...3.5) \times 10^{-5}$ RIU, which is explained by higher symmetry and narrowness of the reflection curve around the minimum (at the level 20 %) [2]. Thus, usage of approximation enhances the measurement accuracy. But it leads to complication in calculations of the minimum position in the reflection curve. This method possesses one deficiency more: the error of approximation depends on the wavelength when operating within the visible range of wavelengths, which makes this method to be non-universal.

To decrease the error in measurements of RI, which is caused by heating the device or by growth of the ambient temperature, it seems purposeful to use compensation of this temperature effect [121]. To compensate the temperature drift of the reflection curve minimum, we used mathematical processing the measurement results (Fig. 5.6). Compensation was provided by taking into account the temperature coefficient of the analyte (distilled water) refraction index that is equal to $TKn\,(H_2O) = -1.15 \times 10^{-4}\ K^{-1}$ as well as temperature changes of water during measurements, which were determined by using the thermistor built-in into the case of flow-through cuvette in the SPR device.

Shown in Fig. 5.6 is the kinetics of the operation point without any compensation of temperature changes observed in the studied substance (1) and with compensation (2). The sharp increase in the operation point position at the beginning of the plot (2) is related with inertness of heat transfer processes from the liquid to thermistor. The SPR response to temperature changes was faster than that of thermistor, which resulted in over-compensation. Elimination of this undesirable phenomenon was provided using the second (reference) channel of device by determination of the difference signal between measuring channels as well as by additional averaging (with account of 10 sequential experimental data). This way allows not only eliminating this over-compensation but additional 12-fold reducing the amplitude of the noise track from $\delta N_L = 3.14 \times 10^{-5}$ down to $\delta N_L = 2.7 \times 10^{-6}$. In this case, the measurement error without compensation was 2.4×10^{-4} and with it $- 8.7 \cdot 10^{-5}$, while with compensation and additional averaging by 10 sequential data of measurements $- 2.1 \times 10^{-5}$.

110

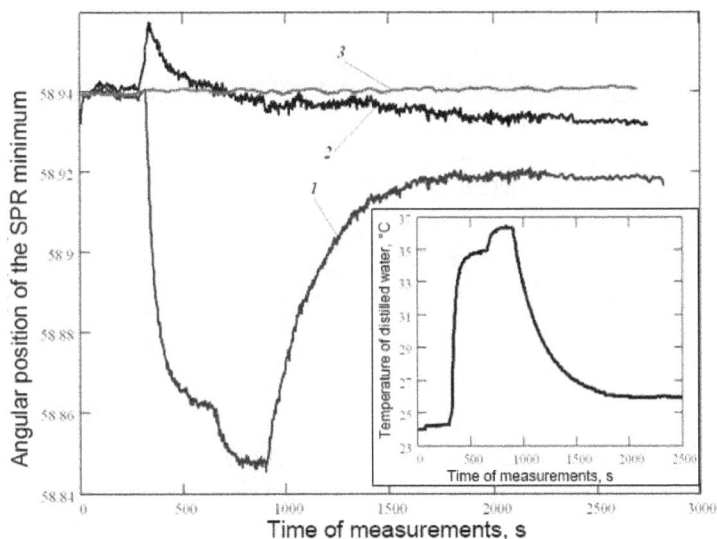

Fig. 5.6. Kinetics of the SPR minimum shift for distilled water when changing its temperature from 24 °C up to 36.6 °C without thermal compensation (1), with it (2) and with additional averaging by the data of 10 sequential measurements (3) [121].

Thus thermal compensation decreases the measurement error for the analyte refraction index at least by 3 times, which enables to essentially increase the device accuracy.

One of the promising numeric methods to reduce the errors in determination of the position of reflection characteristic minimum is the method of a mean line that is not related with any additional processing the experimental data (for instance, approximation or averaging). The method deals with direct data and uses simple arithmetic operations, namely: determination of the middle of a segment and the intersection point for two straight lines. Despite its simplicity, it provides reducing the measurement error when determining the angular position of the reflection characteristic minimum by more than 60 times (from 2.4×10^{-4} down to 4×10^{-6}).

From all the considered numeric methods for enhancing the accuracy of measuring the analyte RI, the most efficient is the method of a mean line. It provides the 5-fold decreased measurement error of RI (4×10^{-6} against 2×10^{-5}) than the methods based on approximation of the reflection minimum position or compensation of the temperature effect.

5.7. Conclusions

Necessity of widening the application ranges for SPR-devices and reducing the measurable concentrations of the studied substances (analytes) requires not only an enhanced sensitivity but improved measurement accuracy of these devices, too. The improved accuracy allowed increasing reliability in measurements of chemical and biological reactions in medicine as well as widening the application range for SPR-devices in other fields of science and technique.

The most widely spread are the devices with angular scanning the intensity of reflected light (reflection characteristics), and they, *per se*, serve as refractometers based on SPR phenomenon. The main measured parameters are the angular position of the reflection characteristic minimum and its shift in time, the value of which allows calculation of changes in the analyte refraction index and its concentration. The analysis of literature data has shown that from 1990 to 2015 improvement of SPR-devices resulted in decreasing the detection limit for the analyte concentration from 8 ng/ml down to 50 pg/ml. It enables to expect that the detection limit value in 2020 – 2025 will reach the range 3…10 pg/ml.

The most effective directions for improving the accuracy of these devices are as follows: increasing the wavelength of radiation exciting surface plasmons, modernization of technology for preparation of metal layer in the sensitive element and stabilization of temperature of the device and analyte during measurements. The investigations performed by the authors have shown that increasing the excitation wavelength from 650 up to 1200 nm results in decreasing the error when determining the analyte refraction index by 5.6 times: from $\pm 6.2 \times 10^{-5}$ down to $\pm 1.1 \times 10^{-5}$. As it follows from experiments, roughness of the metal layer surface in sensitive element can be reduced by 2.5 times (from 2 down to 0.8 nm) due to updating technology of its thermal deposition in vacuum on a glass substrate, which reduces the error by 5.8 times (from $\pm 7 \times 10^{-6}$ down to $\pm 1.2 \times 10^{-6}$). In both cases, this decrease in errors is reached due to narrowing the minimum of the reflection characteristics, which, in turn, was caused by increasing the excitation wavelength and decreasing the roughness of the SE metal layer. The decreasing in the temperature error during measurements of the analyte refraction index have been reached using stabilization of the temperature inherent to the device and reservoirs with analyte. In the case of gas-like analytes, the error has been 18-fold decreased (from 7×10^{-6} down to 3.8×10^{-7}), while for the liquid ones – by 3.2 times (from 2.5×10^{-5} down to 7.6×10^{-6}).

The considered here additional methods for increasing the measurement accuracy are numeric methods of data processing based on calculations of the exact position of the reflection characteristic minimum for a finite value of the scanning step, when the true minimum value is located between adjacent steps. Our comparison of the most widely used methods has shown that the most exact method is that offered by the author (mean line method). It provides decreasing the measurement error by more than 60 times (from $2.4 \cdot 10^{-4}$ down to $4 \cdot 10^{-6}$) as compared with measurements without any numeric processing the experimental data.

References

[1]. J. Homola, Surface Plasmon Resonance Based Sensors, Springer Series on Chemical Sensors and Biosensors, Series Editor: O. S.Wolfbeis, *Springer-Verlag,* Berlin Heidelberg, 2006.
[2]. E. Kretschmann, H. Reather, Radiative decay of nonradiative surface plasmon excited by light, *Z. Naturf.,* 23A, 1968, pp.2135-2136.

[3]. A. Shalabney, I. Abdulhalim, Electromagnetic field distribution in multilayer thin film structures and the origin of sensitivity enhancement in surface plasmon resonance sensors, *Sensors and Actuators A.,* Vol. 159, 2010, pp. 24-32.

[4]. E. Kretschmann, Die Bestimmung optischer Konstanten von Metallen durch Anregung von Oberflachenplasmaschwingugnen, *Z Phys,* 241, 1971, pp. 313-324.

[5]. M. Canovi, J. Lucchetti, M. Stravalaci, F. Re, D. Moscatelli, P. Bigini, M. Salmona, M. Gobbi, Applications of surface plasmon resonance (SPR) for the characterization of nanoparticles developed for biomedical purposes, *Sensors.* Vol. 12, 2012, pp. 16420-16432.

[6]. Y. B. Kang, P. R. Mallikarjuna, D. A. Fabian, A. Gorojana, C. L. Lim, E. L. Tan, Bioactive molecules: current trends in discovery, synthesis, delivery and testing, *IejSME*, Vol. 7, 2013, pp. 32-46.

[7]. A. Falsone, V. Wabitsch, E. Geretti, H. Potzinger, T. Gerlza, J. Robinson, T. Adage, M.M. Teixeira, A.J. Kungl, Designing CXCL8-based decoy proteins with strong anti-inflammatory activity in vivo, *Bioscience Reports,* Vol. 33, 2013, pp. 743-754.

[8]. Yi-Pin Chang, Yen-Ho Chu, Mixture-Based Combinatorial Libraries from Small Individual Peptide Libraries: A Case Study on α1-Antitrypsin Deficiency, *Molecules,* Vol.19, 2014, pp. 6330-6348.

[9]. P. Roach, D. J. McGarvey, M. R. Lees, C. Hoskins, Remotely Triggered Scaffolds for Controlled Release of Pharmaceuticals, *International Journal of Molecular Sciences*, Vol. 14, 2013, pp. 8585-8602.

[10]. T. Ktari, H. Baccar, M. B. Mejri, and A. Abdelghani, Calibration of Surface Plasmon Resonance Imager for Biochemical Detection, *International Journal of Electrochemistry.* Vol. 2012, 2012, pp. 1-5.

[11]. R. Kylväjä, M. Kankainen, L. Holm, B. Westerlund-Wikström, Adhesive polypeptides of Staphylococcus aureus identified using a novel secretion library technique in Escherichia coli, *BMC Microbiology*, Vol. 11, 2011, pp. 117-130.

[12]. L. Mocan, I. Ilie, C. Matea, F. Tabaran, E. Kalman, C. Iancu, T. Mocan, Surface plasmon resonance-induced photoactivation of gold nanoparticles as bactericidal agents against methicillin-resistant Staphylococcus aureus, *International Journal of Nanomedicine*, Vol. 9, 2014, pp. 1453-1461.

[13]. T. Shibui, K. Bando, S. Misawa High-level secretory expression, purification, and characterization of an anti-human Her II monoclonal antibody, trastuzumab, in the methylotrophic yeast Pichia pastoris, *Advances in Bioscience and Biotechnology*, Vol. 4, 2013, pp. 640-646.

[14]. P. Rodriguez, H. Rojas, M. Medina, J. Arrivillaga, Y. Francisco, F. Dager, V. Piscitelli, M. Caetano, A. Ferdinandez, J. Castillo, Study of Functionalized Gold Nanoparticles with Anti-gp63 IgG Antibody for the Detection of Glycoprotein gp63 in Membrane Surface of Leishmania Genus Parasites, *American Journal of Analytical Chemistry,* Vol. 4, 2013, pp. 100-108.

[15]. S.-H. Hsu, Y.-Y. Lin, S.-H. Lu, I-F. Tsai, Y.-T. Lu, H.-T. Ho, Mycobacterium tuberculosis DNA Detection Using Surface Plasmon Resonance Modulated by Telecommunication Wavelength, *Sensors*, Vol. 14, 2014, pp. 458-467.

[16]. N.V. Nesterov, L. M. Nosach, O. Yu. Povnytsya, S. D. Zahorodnya, H. V. Baranova, A. V. Holovan', Yu. V. Ushenin, R.V. Khrystosenko, Imunosensory test system for detection in blood serum of antibodies against human adenovirus, *Ukraine Patent No. 46973*, Claimed 27.07.2009; No. u200907930. Publ. 11.01.2010, No. 1.

[17]. R. V. Khrystosenko, N. V. Nesterova, E. V. Kostyukevych, S. D. Zahorodnyaya, H. V. Baranova, A .V. Holovan', Yu. V. Ushenyn, A. V. Samoylov, S. A. Kostyukevych, Imunosensor based on surface plasmon resonance to detect antibodies against the Epstein-Barr virus, *Optoelectronics and Semiconductor Equipment.* Vol.46, 2011, pp. 92-99.

113

[18]. H. Bai, R. Wang, B. Hargis, H. Lu, Y. Li, A SPR Aptasensor for Detection of Avian Influenza Virus H5N1, *Sensors*, Vol. 12, 2012, pp. 12506-12518.

[19]. D. Hu, S.R. Fry, J.X. Huang and other, Comparison of Surface Plasmon Resonance, Resonant Waveguide Grating Biosensing and Enzyme Linked Immunosorbent Assay (ELISA) in the Evaluation of a Dengue Virus Immunoassay, *Biosensors*, Vol. 3, 2013, pp. 297-311.

[20]. N. S. Heo, S. Zheng, M. H. Yang and other, Label-Free Electrochemical Diagnosis of Viral Antigens with Genetically Engineered Fusion Protein, *Sensors,* Vol. 12, 2012, pp. 10197-10108.

[21]. T. Christopeit, K. Overbo, U. H. Danielson, I. W. Nilsen, Efficient Screening of Marine Extracts for Protease Inhibitors by Combining FRET Based Activity Assays and Surface Plasmon Resonance Spectroscopy Based Binding Assays, *Marine Drugs*, Vol. 11, 2013, pp. 4279-4293.

[22]. G. Li, X. Li, M. Yang, M.-M. Chen, L.-C. Chen, X.-L. Xiong, A Gold Nanoparticles Enhanced Surface Plasmon Resonance Immunosensor for Highly Sensitive Detection of Ischemia-Modified Albumin, *Sensors*, Vol. 13, 2013, pp. 12794-12803.

[23]. C.-Y. Chen, C.-C. Chang, C. Yu, C.-W. Lin, Clinical Application of Surface Plasmon Resonance-Based Biosensors for Fetal Fibronectin Detection, *Sensors*, Vol. 12, 2012, pp. 3879-3890.

[24]. F. Real-Fernandez, I. Passalacqua, E. Peroni and other, Glycopeptide-Based Antibody Detection in Multiple Sclerosis by Surface Plasmon Resonance, *Sensors,* Vol. 12, 2012, pp. 5596-5607.

[25]. F. Yao, R. Zhang, H. Tian, X. Li, Studies on the Interactions of Copper and Zinc Ions with β-Amyloid Peptides by a Surface Plasmon Resonance Biosensor, *International Journal of Molecular Sciences,* Vol. 13, 2012, pp. 11832-11843.

[26]. E. Salvati, F. Re, S. Sesana and other, Liposomes functionalized to overcome the blood–brain barrier and to target amyloid-β peptide: the chemical design affects the permeability across an in vitro model, *International Journal of Nanomedicine*, Vol. 8, 2013, pp. 1749-1758.

[27]. X. Sheng, X. Zhu, Y. Zhang, G. Cui and other, Rhein Protects against Obesity and Related Metabolic Disorders through Liver X Receptor-Mediated Uncoupling Protein 1 Upregulation in Brown Adipose Tissue, *International Journal of Biological Sciences*, Vol. 8, 2012, pp. 1375-1384.

[28]. E. Gorodkiewicz, J. Breczko, A. Sankiewicz, Surface Plasmon Resonance Imaging biosensor for cystatin determination based on the application of bromelain, ficin and chymopapain, *Folia Histochemica et Cytobiologica*, Vol. 50, Issue 1, 2012, pp. 130-136.

[29]. H. Li, L. Zhang, X. Chen, J. Sun, D. Cui, SPR imaging combined with cyclic voltammetry for the detection of neural activity, *AIP Advances,* Vol. 4, 2014, pp. 031342-1 — 4.

[30]. J. C. Claussen, Nanoparticle biosensor for noninvasive glucose sensing, *SPIE Newsroom*, 2013, pp. 1-2.

[31]. N. Houngkamhang, A. Vongsakulyanon, P. Peungthum and other, ABO Blood-Typing Using an Antibody Array Technique Based on Surface Plasmon Resonance Imaging, *Sensors,* Vol.13, 2013, pp. 11913-11922.

[32]. S. F. Elliott, G. Allen, D. J. Timson, Biochemical analysis of the interactions of IQGAP1 C-terminal domain with CDC42, *World Journal of Biological Chemistry*, Vol. 3, Issue 3, 2012, pp. 53-60.

[33]. W. Bhati, A. Vishwa, Nanotechnology Method Comparison for Early Detection of Cancer, *I. J. Intelligent Systems and Applications,* Vol. 3, 2013 pp. 58-65.

[34]. S. Holler, S. Arnold, V. Dantham, A nanoplasmonic sensor detects cancer proteins at the single-molecule level, *SPIE Newsroom*, 2013, pp.1-2.

[35]. N. Ya. Hrydina, V. V. Biloshyts'kyy, A. M. Morozova, V. D. Rozumenko, N. H. Drahuntsova, O. M. Velychko, O. I. Veselova, V. P. Maslov, V. Yu. Ushenin, The use of verapamil and

ketamine in experiments on rats grafted with glioma 101.8, *Ukrainian Medical Journal,* T.11, No. 1-4, 2014, p. 8083.

[36]. Q. Zhao, R. Duan, J. Yuan, Y. Quan, H. Yang, M. Xi, A reusable localized surface plasmon resonance biosensor for quantitative detection of serum squamous cell carcinoma antigen in cervical cancer patients based on silver nanoparticles array, *International Journal of Nanomedicine*, Vol. 9, 2014, pp. 1097-1104.

[37]. Y. Yanase, T. Hiragun, K. Ishii, T. Kawaguchi and other, Surface Plasmon Resonance for Cell-Based Clinical Diagnosis, *Sensors*, Vol. 14, 2014, pp. 4948-4959.

[38]. M. Takami, Y. Takakusagi, K. Kuramochi and other, A Screening of a Library of T7 Phage-Displayed Peptide Identifies E2F-4 as an Etoposide-Binding Protein, *Molecules*, Vol. 16, 2011, pp. 4278-4294.

[39]. Chong Li, Yixin Wang, Xiaolin Zhang and other, Tumor-targeted liposomal drug delivery mediated by a diseleno bond-stabilized cyclic peptide, *International Journal of Nanomedicine*, Vol. 8, 2013, pp. 1051-1062.

[40]. T. Gerlza, B. Hecher, D. Jeremic and other, A Combinatorial Approach to Biophysically Characterise Chemokine-Glycan Binding Affinities for Drug Development, *Molecules,* Vol. 19, 2014, pp. 10618-10634.

[41]. Q. Huo, J. Colon, A. Coldero, J. Bogdanovic and other, A Facile Nanoparticle Immunoassay for Cancer Biomarker Discovery, *Journal of Nanobiotechnology*, Vol. 9, 2011, pp. 20-32.

[42]. T. F. Cabada, C. S. L. de Pablo, A. M. Serrano and other, Induction of cell death in a glioblastoma line by hyperthermic therapy based on gold nanorods, *International Journal of Nanomedicine,* Vol. 7, 2012, pp. 1511-1523.

[43]. J. Han, J. Li, W. Jia, L. Yao and other, Photothermal therapy of cancer cells using novel hollow gold nanoflowers, *International Journal of Nanomedicine*, Vol. 9, 2014, pp. 517-526.

[44]. N. Ya. Hrydyna, Y. V. Boltyna, Yu. V. Ushenyn, A. V. Lapyna, A. P. Kolycnychenko, O. N. Velychko, Age-related aspects of the research of the relationship between the aggregation of blood cells and chromosomal aberrations in peripheral blood lymphocytes in patients with brain gliomas, *Ageing and Longevity,* Vol. 17, №3, 2008, pp. 328-337.

[45]. M. Fardilha, J. Figueiredo, M. Espona-Fiedler and other, Phosphoprotein Phosphatase 1 Isoforms Alpha and Gamma Respond Differently to Prodigiosin Treatment and Present Alternative Kinase Targets in Melanoma Cells, *Journal of Biophysical Chemistry,* Vol. 5, 2014, pp. 67-77.

[46]. J. Yuan, R. Duan, H. Yang, X. Luo, M. Xi, Detection of serum human epididymis secretory protein 4 in patients with ovarian cancer using a label-free biosensor based on localized surface plasmon resonance, *International Journal of Nanomedicine,* Vol. 7, 2012, pp. 2921-2928.

[47]. G. Dharmalingam, N. A. Joy, B. Grisafe, M. A. Carpenter, Plasmonics-based detection of H_2 and CO: discrimination between reducing gases facilitated by material control, *Beilstein Journal of Nanotechnology,* Vol. 3, 2012, pp. 712-721.

[48]. K. V. Kostyukevych, R. V. Khristosenko, Yu. M. Shirshov, S. A. Kostyukevych, A. V. Samoylov, V. I. Kalchenko, Multi-element gas sensor based on surface plasmon resonance: recognition of alcohols by using calixarene films, *Semiconductor Physics, Quantum Electronics and Optoelectronics*, Vol. 14, No. 3, 2011, pp. 313-320.

[49]. L. N. Maslyuk, A. V. Samoylov, Yu. V. Ushenyn, R. V. Khrystosenko, Using plasmon resonance spectrometer for the study of water structuring process, *Actual Problems of Transport Medicine,* 4, 14, 2008, pp. 90-95.

[50]. N. Cennamo, G. Alberti, M. Pesavento and other, A Simple Small Size and Low Cost Sensor Based on Surface Plasmon Resonance for Selective Detection of Fe(III), *Sensors*, Vol. 14, 2014, pp. 4657-4671.

[51]. G. Sener, E. Ozgur, E. Yilmaz and other, Preparation of Ion Imprinted SPR Sensor for Real-Time Detection of Silver(I) Ion from Aqueous Solution, *Proceedings of the International Conference Nanomaterials: Applications and Properties,* Vol. 1, No. 2, 2012, pp. 1-2.

[52]. P. Mulpur, S. Patnaik, A. Chunduri and other, Detection of Cd2+ Ions Using Surface Plasmon Coupled Emission on Engineered Silver-α Nano Alumina Thin Film Hybrids, *Soft Nanoscience Letters*, Vol. 3, 2013, pp. 27-31.

[53]. Y. Wang, M. Huang, X. Guan, Z. Cao and other, Determination of trace chromium (VI) using a hollow-core metal-cladding optical waveguide sensor, *OSA Optics Express*, Vol. 21, No. 23, 2013, pp. 31130-31136.

[54]. M. Minunni, M. Mascini, Detection of pesticide in drinking water using real-time biospecific interaction analysis, *Anal. Lett,* Vol. 26, 1993, pp. 1441-1460.

[55]. C. Mouvet, R. D. Harris, C. Maciag and other, Determination of simazine in water samples by waveguide surface plasmon resonance, *Anal. Chim. Acta,* Vol. 338, 1997, pp. 109-117.

[56]. J. Svitel, A. Dzgoev, K. Ramanathan, B. Danielsson, Surface plasmon resonance based pesticide assay on a renewable biosensing surface using the reversible concanavalin A monosaccharide interaction, *Biosensors and Bioelectronics*, Vol. 15, 2000, pp. 411-415.

[57]. P. M. Boltovets, V. R. Boyko, B. A. Snopok, Surface capturing of virion-antibody complexes: Kinetic study, *Mat. -wiss. u. Werkstofftech*, Vol. 44, No. 2-3, 2013, pp. 112-118.

[58]. E. Rouah-Martin, J. Mehya, B. van Dorst and other, Aptamer-Based Molecular Recognition of Lysergamine, Metergoline and Small Ergot Alkaloids, *International Journal of Molecular Sciences*, Vol. 13, 2012, pp. 17138-17159.

[59]. H. Ali Mondal, A. Roy, S. Gupta, S. Das, Exploring the Insecticidal Potentiality of Amorphophallus paeonifolius Tuber Agglutinin in Hemipteran Pest Management, *American Journal of Plant Sciences*, Vol. 3, 2012, pp. 780-790.

[60]. C. Pirvu, C. C. Manole, Electrochemical surface plasmon resonance for in situ investigation of antifouling effect of ultra thin hybrid polypyrrole/PSS films, *Electrochimica Acta,* Vol. 89, 2013, pp. 63-71.

[61]. J. Halliwell, C. Gwenin, A Label Free Colorimetric Assay for the Detection of Active Botulinum Neurotoxin Type A by SNAP-25 Conjugated Colloidal Gold, *Toxins.* Vol. 5, 2013, pp. 1381-1391.

[62]. M. Heurich, Z. Altintas, I. E. Tothill, Computational Design of Peptide Ligands for Ochratoxin A, *Toxins,* Vol. 5, 2013, pp. 1202-1218.

[63]. K. B. Turner, D. Zabetakis, P. Legler and other, Isolation and Epitope Mapping of Staphylococcal Enterotoxin B Single-Domain Antibodies, *Sensors*, Vol. 14, 2014, pp. 10846-10863.

[64]. A. Alfonso, M.-J. Pazos, A. Fernandez-Araujo and other, Surface Plasmon Resonance Biosensor Method for Palytoxin Detection Based on Na+,K+-ATPase Affinity, *Toxins*, Vol. 6, 2014, pp. 96-107.

[65]. R. Nanjunda, E. A. Owens, L. Mickelson and other, Selective G-Quadruplex DNA Recognition by a New Class of Designed Cyanines, *Molecules,* Vol. 18, 2013, pp. 13588-13607.

[66]. R. Yutabe, T. Onodera, K. Toko, Fabrication of an SPR Sensor Surface with Antifouling Properties for Highly Sensitive Detection of 2,4,6-Trinitrotoluene Using Surface-Initiated Atom Transfer Polymerization, *Sensors*, Vol. 13, 2013, pp. 9294-9304.

[67]. G. V. Dorozinsky, A. I. Liptuga, V. I. Gordienko, V. P. Maslov, V. V. Pidgornyi, Diagnostics of motor oil quality by using the device based on surface plasmon resonance phenomenon, *Scholars Journal of Engineering and Technology (SJET)*, Vol. 3, 2015, pp. 372-374.

[68]. G. V. Dorozinsky, V. P. Maslov, N. V. Kachur, R. L. Filonchuk, The multichannel smoke detector, *Ukraine Patent № 91922,* Claimed 13.01.2014, No. u201400277, Publ. 25.07.2014, Bul. No. 14.

[69]. G. V. Dorozinsky, M. V. Lobanov, V. P. Maslov, Detection of methanol vapor by surface plasmon resonance, *East European Journal of Advanced Technologies*, Vol. 4, 76, 2015, pp. 4-7.

[70]. G. V. Dorozinsky, V. P. Maslov, Yu. V. Ushenin, Sensor devices based on surface plasmon resonance, *NTUU 'KPI'*, Kyiv, 2016. Online available: http://ela.kpi.ua/handle/123456789/15312.

[71]. W. Yuan, H. P. Hoa, S. Y. Wua, Y. K. Suen, and S. K. Kong, Sensitivity-enhancement methods for surface plasmon sensors, *Sens. Actuators A*, 151, 2009, pp. 23–28.

[72]. M. Meunier, P. N. Prasad, A. V. Kabashin, and S. Patskovsky, Self-noise-filtering phase-sensitive surface plasmon resonance biosensing, *Opt. Express*, Vol. 18, 2010, pp. 14353–14358.

[73]. H. V. Beketov, O. S. Klymov, I. Ye. Matyash, B. K. Serdeha at all, Physical principles of polarimetry high informative ability: monograph, *NTUU 'KPI'*, Kyiv, 2013.

[74]. A. V. Zayats, I. I. Smolyaninov, and A. A. Maradudin, Nano-optics of surface plasmon polaritons, *Phys. Rep.*, Vol. 408, 2005, pp. 131–314.

[75]. J. M. Pitarke, V. M. Silkin, E. V. Chulkov, and P. M. Echenique, Theory of surface plasmons and surface-plasmon polaritons, *Rep. Prog. Phys.*, Vol. 70, 2007, pp. 1–87.

[76]. I. Abdulhalim, M. Zourob, and A. Lakhtakia, Surface plasmon resonance for biosensing: a mini-review, *Electromagnetics (UK)*, Vol. 28, 2008, pp. 214–242.

[77]. H. Raether, Surface Plasmons on Smooth and Rough Surfaces and on Gratings, *Springer Tracts in Modern Physics*, Vol. 111, 1988.

[78]. B. Liedberg, C. Nylander, and I. Lunstrom, Surface plasmon resonance for gas detection and biosensing, *Sens. Actuators*, Vol. 4, 1983, pp. 299–304.

[79]. C. E. H. Berger, T. A. M. Beumer, R. P. H. Kooyman, and J. Greve, Surface plasmon resonance multisensing, *Anal. Chem.*, Vol. 70, 1998, pp. 703–706.

[80]. K. A. Peterlinz, and R. M. Georgiadis, T. M. Herne, and M. J. Tarlov, Characterization of DNA probes immobilized on gold surfaces, *J. Am. Chem. Soc.*, Vol. 119, 1997, pp. 3401–3402.

[81]. V. Owen, Real-time optical immunosensors—A commercial reality, *Biosens. Bioelectron. (UK)*, Vol. 12, 1, 1997, pp. i–ii.

[82]. J. Homola, S. S. Yee, and G. Gauglitz, Surface plasmon resonance sensors: review, *Sensors and Actuators B: Chemical*, 54, Vol. 1, 1999, pp. 3–15.

[83]. J. Homola, Present and future of surface plasmon resonance biosensors, *Analytical and Bioanalytical Chemistry*, Vol. 377, 2003, pp. 528–539.

[84]. I. D. Voitovych, S. G. Korsunskyi, Sensors based on plasmon resonance: principles, technologies, applications, *Stal'*, Kyiv, 2011.

[85]. N. Gridina, V. Maslov, Yu. Ushenin, Tumor-associated inflammation and cerebral gliomas, *Saarbrucken: Lambert Academic Publishing*, 2013.

[86]. Nuno Miguel Matos Pires, Tao Dong, Ulrik Hanke, Nils Hoivik, Recent developments in Optical detection technologies in Lab-on-a-Chip devices for biosensing applications, *Sensors*, Vol. 14, 2014, pp. 15458-15479.

[87]. Ye. F. Vyenher, S. A. Zyn'o, Ye. P. Matsas, A. V. Samoylov, Yu. V. Ushenin, at al., Spectrometer poverhnevogo plasmon resonance Plasmon-6, in *Proceedings of the Scientific Conference SENSOR'07*, Odessa, Ukraine, 5 June 2007, p. 111.

[88]. K. Kurihara, K. Nakamura, K. Suzuki, Asymmetric SPR sensor response curve-fitting equation for the accurate determination of SPR resonance angle, *Sensors and Actuators B: Chemical*, Vol. 86, 2002, pp. 49-57.

[89]. H. E. De Bruijn, R. P. H. Kooyman, J. Greve Choice of metal and wavelength for surface-plasmon resonance sensors: some considerations, *Applied Optics*, Vol. 31, Issue 4, 1992, pp. 440-442.

117

[90]. A. V. Samoylov, Yu. V. Ushenin, R. V. Khrystosenko, The sensor for the analysis of biochemical environments, *Patent Ukraine No. 58775*, Claimed. 27.09.2010, No. u201009975, Publ. 26. 04. 2011, Bul. No. 8.

[91]. Yu. V. Ushenin, A. V. Samoylov, R. V. Khrystosenko, Increased sensitivity of the sensors changes the refractive index of the medium based on surface plasmon resonance, *Technology and Design of Electronic Equipment*, No. 1, 2011, pp. 12-14.

[92]. G. V. Dorozinsky, Evaluation of major errors of measurement of the refractive index of the analyte device based on the phenomenon of surface plasmon resonance, *Scientific Bulletin of Kharkiv Polytechnic Institute*, No. 16, 2015, pp. 39-46.

[93]. H. E. De Bruijn, R. P. H. Kooyman, J. Greve, Choice of metal and wavelength for surface-plasmon resonance sensors: some considerations, *Applied Optics*, Vol. 31, Issue 4, 1992, pp. 440-442.

[94]. J. Davies, Surface plasmon resonance - the technique and its applications to biomaterial processes, *Nanobiology*, Vol. 3, 1994, pp. 5-16.

[95]. Z. Salamon, H. A. Macieod, G. Tollin, Surface plasmon resonance spectroscopy as a tool for investigating the biochemical and biophysical properties of membrane protein systems. II: Applications to biological systems, *Biochemical et Biophysica Acta*, 1331, 1997, pp. 131-152.

[96]. D. L. Mylls, Surface polaritons, *Science*, 1985.

[97]. V. A. Kosobukyn, The effect of the gain of an external electric field near the surface of the metal and its manifestation in spectroscopy, *Surface. Physics, Chemistry, Mechanics*, 12, 1983, pp. 5-21.

[98]. A. J. Braundmeier, E. T. Arakawa, Effect of surface roughness on surface plasmon resonance adsorption, *Journal Physics Chemistry Solids*, Vol. 35, 1974, pp. 517-520.

[99]. W. H. Weber Modulated surface-plasmon resonance for in situ metal-film surface studies, *Physical Review Letters*, Vol. 39, 1977, pp. 153-156.

[100]. L. Kholland Deposition of thin films in vacuum, *State Energy Publishing*, 1963.

[101]. V. P. Severdenko, E. Y. Tochytskyy, The structure of the thin metal films, *Science and Technology*, Minsk, 1968.

[102]. *R.* Hlanh, L. Mayssel, Thin Films Technology, Vol. 1, *Soviet Radio, Moscow, 1977*.

[103]. V. P. Maslov, Physical and technological problems of precision parts connection of opto-electronic devices: a monograph, *NTUU 'KPI'*, Kyiv, 2012.

[104]. R. A. Tun, The structure of thin films, *Thin Films Physics*, Vol. 1, 1967, pp. 224-274.

[105]. E. I. Tochytskyy, Crystallization and thermal processing of thin films, *Science and Technology*, Minsk, 1976.

[106]. Ya. D. Vyshnyakov, Packing defects in the crystalline structure, *Science*, Moscow, 1970.

[107]. V. P. Kostyuk, Y. N. Shklyarevskyy, Influence of preparation conditions silver and copper layers on their optical properties, *Optics and Spectroscopy*, Vol. 29, 1, 1970, pp. 195-197.

[108]. B. A. Snopok, E. V. Kostyukevich, S. I. Lysenko, P. M. Lytvyn, O. S. Lytvyn, S. V. Mamykin, S. A. Zynio, S. A. Kostyukevich, P. E. Shepeliavii, Yu. M. Shirshov, E. F. Venger, Optical biosensors based on the surface plasmon resonance phenomenon: optimization of the metal layer parameters, *Semiconductor Physics, Quantum Electronics and Optoelectronics*, Vol. 4, № 1, 2001, pp. 56-69.

[109]. N. -H. Kim, M. Choi, J. W. Leem, J. S. Yu, T. W. Kim, T. -S. Kim, K. M. Byun, Improved biomolecular detection based on a plasmonic nanoporous gold film fabricated by oblique angle deposition, *Optics Express*, Vol. 23, Issue 14, 2015, pp. 18777-18785.

[110]. G. Dorozinsky, T. Doroshenko, V. Maslov Influence of technological factors on sensitivity of analytical devices based on surface plasmon resonance, *Journal of Sensor Technology*, Vol. 5, 2015, pp. 54-61.

[111]. G. V. Dorozinsky, V. P. Maslov, The device for analyzing liquids and gases, *Patent of Ukraine No. 108149*, Claimed 09.09.2013; No. a201310852, Publ. 25.03.2015, Bul. No. 6.

[112]. G. V. Dorozinsky, Analysis of the main errors of measurement of the refractive index of the analyte device 'Plasmon-6', *East European Scientific Journal*, Vol. 5, 2016, pp. 54-61.

[113]. Yu. V. Bayborodyn, Fundamentals of laser technics, 2-nd. ed., *Vyshcha Shkola*, Kyiv, 1988.

[114]. S. K. Özdemir, G. Turhan-Sayan, Temperature Effects on Surface Plasmon Resonance: Design Considerations for an Optical Temperature Sensor, *Journal of Light Wave Technology*, V. 21, Issue 3, 2003, pp. 805-815.

[115]. S. P. Hlaholyev, Quartz glass. Property, production, appliance, *GHTI*, Moscow, 1934.

[116]. A. A. Revel', A. M. Ponomar'ova, Short Guide physico-chemical values. 8th Ed, *Chemistry*, Leningrad, 1983.

[117]. G. Dorozinsky, V. Dunaevsky, V. Maslov, Thermal-vision method of investigations and control of device based on surface plasmon resonance, *Universal Journal of Control and Automation*, Vol. 2, 2013, pp. 34-39.

[118]. G. V. Dorozinsky, V. I. Dunayevskiy, V. P. Maslov Method of increasing the accuracy of the device based on the phenomenon of surface plasmon resonance, *Patent of Ukraine No. 84770.* Claimed. 10. 06. 2013, No. u201307356, Publ. 25. 10. 2013, Bul. No. 20.

[119]. G. V. Dorozinsky, Yu. V. Ushenin, A. V. Samoylov, V. P. Maslov Method study biomolecular and biochemical reactions in liquid and gaseous environments using surface plasmon resonance phenomenon, *Patent of Ukraine No. 77080,* Claimed. 25. 07. 2012, No. u201209152, Publ. 25. 01. 2013, Bul. No. 2.

[120]. Yu. M. Shyrshov, A. V. Samoylov, R. V. Khrystosenko, Yu. V. Ushenyn, V. M. Myrskyy, Analysis and numerical simulation of PPR spectrometers with manual scan on the corner: an algorithm for determining the angular position of the minimum, *Registration, Storage and Processing*, Vol. 6, No. 3, 2004, pp. 3-18.

[121]. Yu. V. Ushenin, V. P. Maslov, G. V. Dorozinsky, T. A. Turu, N. V. Kachur, Application of temperature sensors for improving the device based on the phenomenon of surface Plasmon, *Sensor Electronics and Microsystem Technologies*, Vol. 13., № 1, 2016, pp. 33-40.

Chapter 6
GaPO$_4$-based SHAPM Sensors for Liquid Environments

Cinzia Caliendo, Muhammad Hamidullah

6.1. Introduction

Sensors based on the propagation of acoustic waves use a detection mechanism based on the perturbation of the acoustic waves characteristics. Any acoustic wave device is potentially a sensor: when the acoustic wave propagates on the surface of the device material, any perturbation that affects the propagating medium (i.e. temperature, pressure, relative humidity, mass loading, electric loading, viscosity loading, and so on) result in a change of the acoustic wave velocity and/or attenuation.

Acoustic wave devices are described by the characteristics of the wave propagation, i.e. the wave velocity and the particle displacement components. Many combinations of polarization and velocity are possible, depending on the acoustic waveguide material types, its crystallographic orientation, the wave propagation direction, and the boundary conditions. Longitudinal waves are polarized parallel to the propagation direction, while shear horizontal and shear vertical waves are polarized parallel and normal to the propagating surface. All the acoustic wave sensors are able to work in gaseous environment, regardless of their polarization: only a subset of them can be used for the design of biosensors working in liquid environment. Thus biosensors based on electroacoustic devices require a careful design to be able to work in liquid environment and to show high sensitivity to the liquid properties (such as viscosity and conductivity). Sensors based on the propagation of in-plane polarized waves do not radiate appreciable energy into the liquids contacting the sensor surface: Love waves, surface transverse waves (STW), shear horizontal acoustic plate modes (SHAPM) and shear horizontal surface acoustic waves (SHSAW) are examples of acoustic waves whose shear horizontal polarization ensures no coupling between the liquid and the elastic propagating medium. On the contrary, electroacoustic devices based on the propagation of waves with a shear vertical displacement component are not suitable for liquid application, since they radiates compressional waves into the liquid, thus causing excessive damping. An exception to

Cinzia Caliendo

Institute for Photonics and Nanotechnologies, IFN-CNR, Via Cineto Romano 42, 00156 Rome, Italy this rule occurs for devices based on the propagation of the elliptically polarized fundamental anti-symmetric Lamb mode, A0, when it propagates at a velocity lower than the sound velocity in the liquid. Among the piezoelectic materials, the most commonly used are quartz (SiO_2), lithium tantalate ($LiTaO_3$), and lithium niobate ($LiNbO_3$). Each has specific advantages and disadvantages, which include temperature coefficient of delay (TCD), electromechanical coupling efficiency $K2$, and propagation velocity. Quartz shows specific cut angle and wave propagation direction that are suitable to obtain a low or high first order temperature dependence of the wave velocity. The latter condition is suitable when an acoustic wave temperature sensor has to be designed. On the contrary, $LiNbO_3$ and $LiTaO_3$ don't show any temperature stable cut but exhibit larger electroacoustic coupling efficiency than quartz. Gallium orthophosphate, $GaPO_4$, is a relatively new and still poorly explored piezoelectric material that has the unique advantage to be able to withstand temperature as high as 900°C without losing its piezoelectric properties. Its chemical inertness makes it suitable for the implementation of sensors able to work in harsh environment and in extreme conditions. Recently this material has been studied for biosensing applications: results in the development of a $GaPO_4$ micro balance for thermo gravimetric analysis and for affinity sensor applications in aqueous liquids, such as biosensors, are described in reference [1], while in reference [2] the creation and arrangement of biomolecule-binding sensors based on TiO_2 nanofibrous scaffold grown on $GaPO_4$ crystal microbalances are described.

In the present chapter we theoretically investigated the propagation of the shear horizontal acoustic plate modes (SHAPMs) along y-rotated $GaPO_4$ single crystal piezoelectric substrates that are suitable for liquid sensing applications.

The organization of the present chapter begins with the calculation of the phase velocity and coupling efficiency dispersion curves of the SHAPMs in vacuum. Then the sensitivity dispersion curve of the fundamental SHAPM (SH_0) sensor is calculated for gravimetric detection in vacuum. Finally, a viscous Newtonian liquid is introduced that contacts one plate surface, and the wave velocity and attenuation changes are calculated for different liquid mass density-viscosity products by applying the perturbative approach. The effect of the plate thickness on the sensitivity of SH_0 mode sensor is investigated and compared with that of SH_0 mode sensor implemented on more conventional piezoelectric plate materials.

6.2. Theoretical Investigation of the SHAPMs Propagation in y-rot GaPO4

The SHAPMs are acoustic waves that propagate in finite thickness plates and employ input and output interdigital transducers (IDTs) to excite and detect the acoustic modes, as shown in Fig. 6.1. The piezoelectric plate acts as an acoustic waveguide that, unlike the surface acoustic wave (SAW) devices, confine the acoustic energy between the two plate sides as the wave propagates. The waveguide crystal can be employed as a physical barrier between the IDTs and the liquid medium that contacts the opposite plate side where the IDTs are located [3].

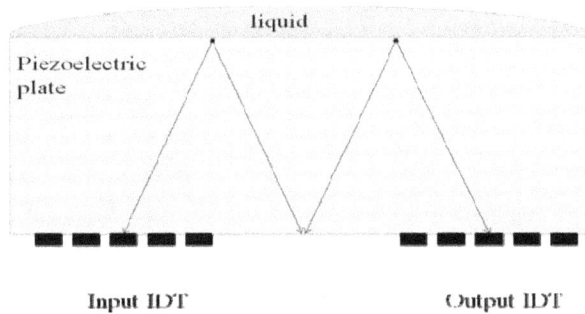

Fig. 6.1. The SHAPM sensor configuration.

The acoustic modes are in-plane (shear horizontal) polarized normal to the acoustic wave-vector. Since the wave interacts with both plate sides, either of them can be used as the sensing surface, with the advantage to isolate the sensing surface from the metal electrodes. SHAPMs exist on rotated Y-cuts of trigonal class 32 group crystals, which include the $GaPO_4$ and the quartz crystals, as an example.

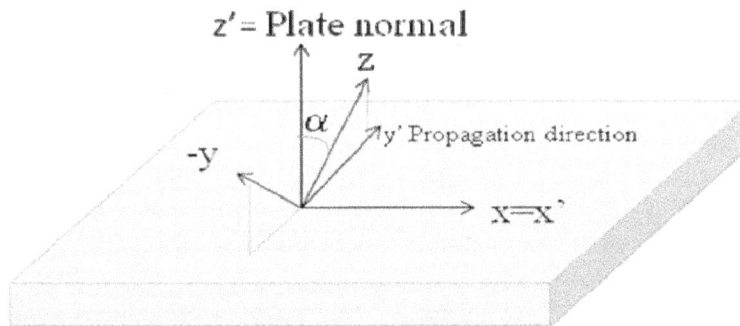

Fig. 6.2. The crystal cut and propagation direction of SHAPMs; the laboratory coordinates system x' y' z' and the $GaPO_4$ crystallographic coordinate system x y z.

The number of the SHAPMs that propagate in the piezoelectric waveguide structure depends on the substrate thickness, and each mode has the corresponding acoustic energy distributed throughout the bulk of the substrate. For each mode the maximum displacements occur on the top and bottom surfaces of the plate, but, for the fundamental mode, it occurs at the plate surfaces as well as in the whole plate depth. The field profile of the first four SHAPMs are shown in Fig. 6.3. The mode is shear horizontally polarized (the longitudinal and shear vertical particle displacement components, U_1 and U_3, are equal to zero, and only the shear horizontal component, U_2, is non null) in the plane of the plate so it can be used for liquid-based applications without suffering substantial loss. The acoustic field profile calculation was performed by using a Matlab program under the hypothesis of lossless material.

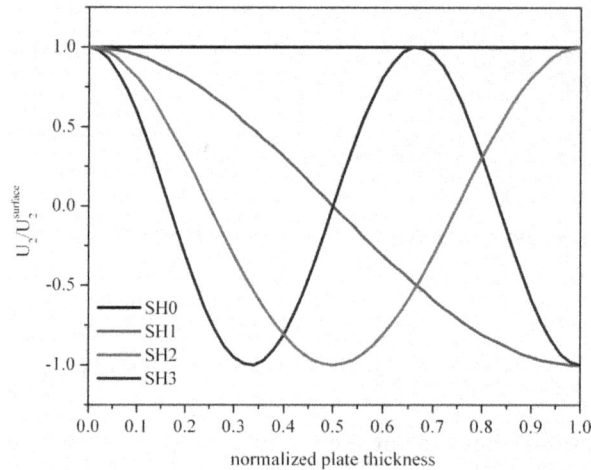

Fig. 6.3. The field profile of the first four SHAPMs.

6.3. Phase Velocity and Electromechanical Coupling Dispersion Curves

The phase velocity dispersion curves of the SHAPMs were calculated by using a Matlab routine in the approximation of lossless $GaPO_4$ in vacuum: the material data (mass density, elastic, piezoelectric, and dielectric constants) were extracted from Wallnofer et al. [4] and C. Reiter et. al. [5]. In performing the theoretical calculations, the $GaPO_4$ substrate was modeled as an anisotropic piezoelectric medium of finite thickness that extends indefinitely in the directions corresponding to the plate width and length. The phase velocity of the modes were theoretically studied for different z-axis tilt angle and electrical boundary conditions.

For plate thickness-to-wavelength ratio $h/\lambda \ll 1$, the phase velocity dispersion curves are easily distinguishable between the different modes, while with $h/\lambda \gg 1$ it is difficult to isolate a single acoustic mode since their velocity values are very close. As an example, Fig. 6.4 shows the phase velocity dispersion curves of the first six SHAPMs travelling along a 91°-y 90°-X $GaPO_4$ plate with normalized plate thickness up to $h/\lambda=1.4$.

As it can be seen, the fundamental mode, SH_0, is low-dispersive and exhibits no cut off thickness, while the higher order modes are highly dispersive and have a cut off thickness before which they do not propagate. Higher order modes can reach very high velocity: near the cut off the slope of the dispersion curves is near to be infinite.

In designing a SHAPM device, an important feature to be obtained is a low insertion loss, which can be achieved by selecting a crystallographic orientation with a large electromechanical coupling coefficient, K^2. The value of K^2 is directly related to the IDTs electrical-to-mechanical energy conversion efficiency. Two different coupling configurations, with the IDTs placed on one of the $GaPO_4$ plate surfaces, with or without a floating electrode on the opposite surface. The configuration called substrate/transducer (ST) refers to a coupling structure with the IDTs positioned on the $GaPO_4$ upper surface, opposite to the surface that contact the environment to be tested; when a floating metallic plane (M, metal) is placed at the GaPO$_4$ surface opposite to the one where the IDTs are located, the configuration is called metal/substrate/transducer (MST), as shown in Fig. 6.5.

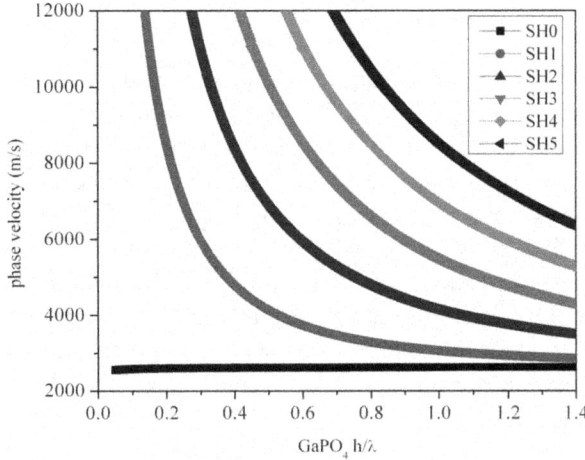

Fig. 6.4. The phase velocity dispersion curves of 91°Y 90°-X GaPO$_4$ plate up to h/λ=1.4.

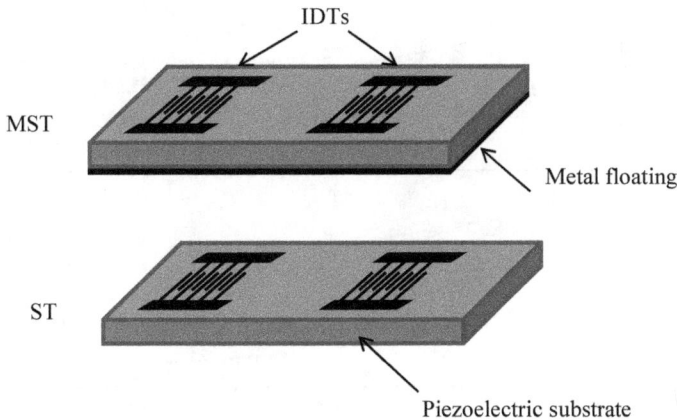

Fig. 6.5. The two coupling configurations, ST and MST.

The K^2 can be obtained by calculating the perturbation of the wave velocity when the tangential electric field component is shorted out at the plate surface: $K^2 = 2 \times (v_{free} - v_{met})/v_{free}$, where v$_{free}$ and v$_{met}$ are the phase velocities at the electrically opened and shorted surfaces of the plate. The v$_{met}$ is obtained by the insertion of a perfectly conductive and infinitesimally thin film at the interfaces where the IDTs and the floating plane are located in each of the two coupling structures. By denoting as v$_{ij}$ (for i, j = m, f) the wave velocity referred to the electrical boundary conditions at the lower (first index, i) and upper (second index, j) plate surface, the following approximated formulas were used to calculate the coupling constant of the two structures:

$$K_{ST}^2 = \frac{2 \cdot (v_{ff} - v_{fm})}{v_{ff}} \qquad (6.1)$$

$$K_{MST}^2 = 2 \cdot \frac{(v_{fm} - v_{mm})}{v_{fm}} \qquad (6.2)$$

The mechanical effect of the IDTs and floating electrode was ignored as they were assumed to be infinitely thin. The coupling efficiency of the SH_0 mode was calculated for different GaPO$_4$ plate thicknesses and y axis tilt angle: the K^2 dispersion curves are shown in Fig. 6.6 where the y axis tilt angle α is the running parameter. It can be observed that K^2 reaches the highest values (approx. 7 %) at h/λ=0.11 for 95°Y GaPO$_4$; all the K^2 dispersion curves reach a maximum and then they decreases with increasing the plate normalized thickness, asymptotically reaching the K^2 of the shear horizontal bulk acoustic wave (SHBAW). The ST configuration, as well as the MST, is quite efficient, but, unlike the MST, it offers the further advantage to be sensitive to the electrical properties of the liquid environment: this feature is particularly useful to test the electrical conductivity of the liquid phase.

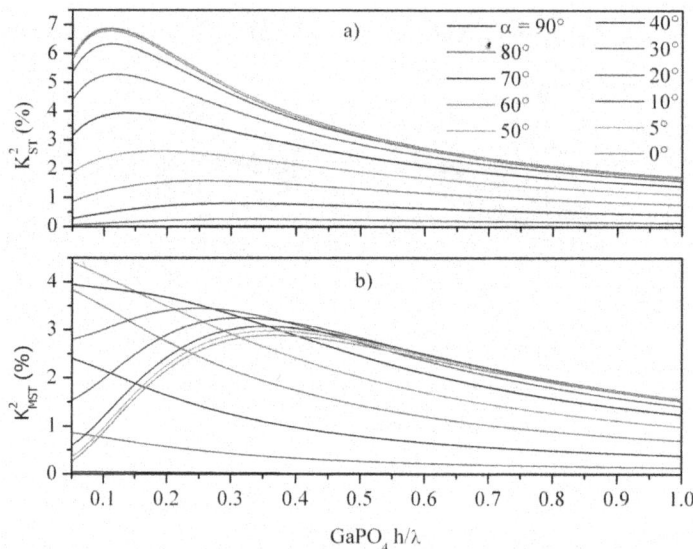

Fig. 6.6. The coupling efficiency vs. the GaPO$_4$ plate normalized thicknesses for different y axis tilt angle for the ST configuration, and b) for the MST configuration.

6.4. The Temperature Coefficient of Delay

The velocity of the acoustic waves is a function of the material constants (mass density, elastic, piezoelectric and dielectric constants) of the propagating medium, and these constants are affected by the temperature changes: as a result the wave velocity is sensitive to the temperature variations of the surrounding medium. The SH_0 mode phase velocity at different temperatures was theoretically estimated by modifying the mass density ρ, the elastic and piezoelectric constants, c_{ij} and e_{ij}, of GaPO4 according to their temperature coefficients available from the literature [5, 8]. The ρ was evaluated at different temperatures T as follows: $(T) = \rho(25°C) \cdot [1 - (\alpha_{11} + \alpha_{22} + \alpha_{33}) \cdot \Delta T])$ where $\rho(T)$ is the mass density at a certain temperature T, $\Delta T = (T - T_0) = T - 25°C$, ρ (25 °C) is the mass density at 25 °C and α_{ii} is the linear thermal expansion coefficient. The α_{ii}, c_{ij} and e_{ij} were calculated at different temperatures as follows:

$$\alpha_{ij}(T) = \alpha_{ij}(T_0) \cdot \left[1 + T_{\alpha ij}^1 \Delta T + T_{\alpha ij}^2 \Delta T^2 + T_{\alpha ij}^3 \Delta T^3 + T_{\alpha ij}^4 \Delta T^4\right] \qquad (6.3)$$

$$c_{ij}(T) = c_{ij}(T_0) \cdot \left[1 + T_{cij}^1 \Delta T + T_{cij}^2 \Delta T^2 + T_{cij}^3 \Delta T^3\right] \qquad (6.4)$$

$$e_{ij}(T) = e_{ij}(T_0) \cdot \left[1 + T_{eij}\Delta T\right] \qquad (6.5)$$

The $T_{\alpha ij}^n$ T_{cij}^n and T_{eij}^n are the n-order temperature coefficient of the α_{ij}, c_{ij} and e_{ij}, respectively. The temperature coefficient of velocity, TCV, was calculated as the relative velocity changes with respect to T= 300°C, as follow:

$$TCV = \frac{v_{T°C} - v_{300°C}/v_{300°C}}{T°C - 300°C}$$

The TCV of the 95°Y 90°-X GaPO4 plate with normalized thickness $h/\lambda = 0.11$ was calculated and found equal to 13 ppm/°C; this value decreases to 12.7 for h/λ equal to 0.23.

6.5. Viscous Liquid

When a viscous fluid contacts the lower surface of the SHAPM device, the shear wave penetrates into the adjacent fluid up to a depth $\delta = (2\eta/\omega\rho)^{1/2}$ that is determined by the operating frequency $f = v/\lambda$, and the liquid viscosity and density. Thus a thin layer of the order of micro meters moves synchronously with the vibrating surface while the bulk of the liquid is unaffected by the acoustic signal. The viscous coupling of the liquid to the plate mode causes a change in the SH_0 mode velocity and loss that were calculated by applying the perturbation formulas [6]:

$$\frac{\Delta v}{v_0} = -\frac{c_f \eta_l}{2\omega} \text{Im}\left(\frac{\zeta}{1+j\omega\tau}\right) \qquad (6.6)$$

$$\zeta^2 = \left(k^2 - \frac{\omega^2 \rho_l}{\mu_l}\right) + j\frac{\omega\rho_l}{\eta_l} \qquad (6.7)$$

$$\alpha = \frac{c_f \eta_l}{2v_0} \text{Re}\left(\frac{\zeta}{1+j\omega\tau}\right), \qquad (6.8)$$

where c_f is the mass sensitivity of the plate; ρ_l, μ_l and η_l are the liquid density, shear modulus and viscosity, and $\tau = \eta_l/\mu_l$ is the relaxation time; v_0 and v are the velocities of the mode travelling along the bare plate and the liquid loaded plate. The c_f coefficient of the SH_0 mode was calculated as the phase velocity relative change per unit added mass,

$$c_f = \frac{v'-v_0}{v_0}/(\rho \cdot h), \qquad (6.9)$$

being v_0 and v' the velocity of the mode travelling along the bare plate and the plate loaded by an added mass m=$\rho \cdot$h, being ρ and h the added mass density and thickness. c_f was calculated for different plate thicknesses and crystallographic orientations. The c_f calculations were performed under the hypothesis that the added mass consists of an ideal thin elastic film that moves synchronously with the oscillating surface.

As an example, Fig. 6.7 shows the mass sensitivity (to be divided by λ) for the 95°-Y 90°-X plate: as it can be seen, c_f is negative since the mode velocity decreases due to the effect of the mass loading, and it decreases with increasing the plate thickness.

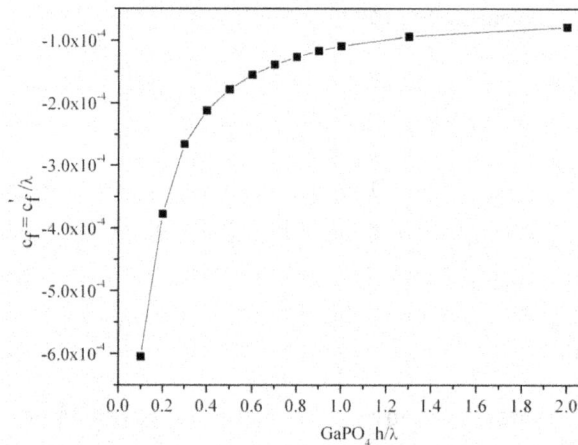

Fig. 6.7. The mass sensitivity coefficient (to be divided by λ) vs. the GaPO$_4$ plate normalized thickness for the 95°Y 90°-X orientation.

The GaPO$_4$ based gravimetric sensor can also be used to advantage in the characterization of film properties such as film thickness and surface area. It is also useful to monitor processes such as thin film deposition or removal, and materials modifications.

A typical SHAPM device implemented on AT 90°X 0.5 mm thick, with IDT period of 30 μm, shows a SH_0 mass sensitivity of -0.78 ppm-mm^2/ng at 169 MHz. if the plate thickness is reduced to 100 μm, the mass sensitivity is equal to -3.9 ppm mm^2/ng. Its major drawbacks is that the piezoelectric coupling is extremely low: K^2 is 0.012 % compared to 0.12 % for the SAW on the temperature compensated ST-quartz, and 4 % for the APM on ZX lithium niobate. However, as the plate thickness is decreased, the coupling improves and is 0.058 % at 0.1 mm (half the value of the SAW on ST-quartz) [8]. For a GaPO$_4$ plate 150 μm thick, for λ=75 μm (h/λ=2), the mass sensitivity is equal to -0.92 ppm mm^2/ng, and K^2=1 %, for the fundamental mode; if the plate thickness is reduced to 100 μm and λ=50 μm (h/λ=2), then Sm=-1.588 ppm mm^2/ng. Thus, with respect to the SHAPM based on conventional piezoelectric quartz plate, the GaPO$_4$ seems to be preferred.

When a liquid is loaded in the acoustic path of a SHAPM device, the mode velocity and attenuation change. The SH$_0$ mode velocity and attenuation shift, α and Δv/v$_0$, arising from viscous liquid entrainment, were calculated for different glycerol/water volume percentage, from 0 to 100 %. The glycerol/water mixture mass density and viscosity values were calculated by following the parameterization shown in [9]. Figs. 6.8 show the SH$_0$ velocity change and attenuation vs. $\sqrt{\rho\eta}$ for a 95°Y 90°-X GaPO$_4$ plate: the running parameter is the plate normalized thickness h/λ that varies from 0.1 to 2, being the plate thickness *h* fixed (150 μm). Since the mode velocity is dispersive, both the corresponding frequency f = v/λ and K^2 vary from 1.7 MHz to 35.4 MHz, and from 6.9 % to 1.0 %. Figs. 6.8 clearly show that the response of both amplitude and phase velocity depends on the plate thicknesses.

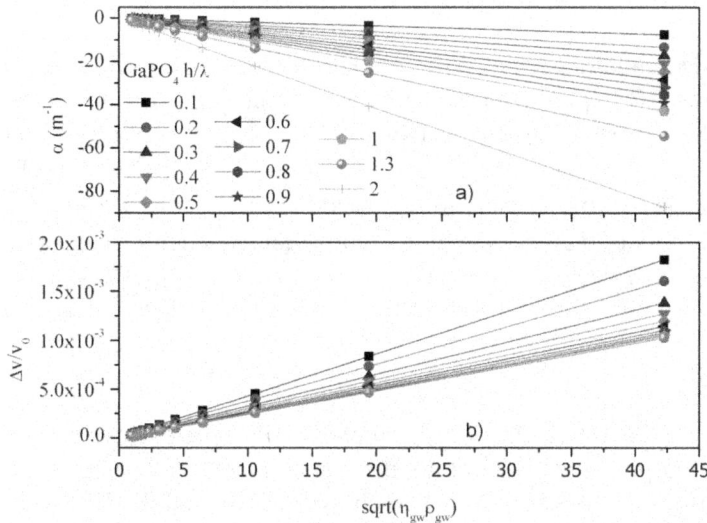

Fig. 6.8. The SH$_0$ Δα/k (a) and Δv/v$_0$ (b) vs the $\sqrt{\rho\eta}$: the mode travels along a GaPO$_4$ (0° 5° 90°) plate 150 μm thick. The plate normalized thickness is the running parameter.

According to equations 6 and 8, both the velocity and loss changes are affected by the c_f coefficient, the plate thickness and orientation: thus, the sensitivity of a mass and viscous loading sensor can be improved by choosing the proper plate thickness and orientation that ensure high c_f value. The slope of the linear fit of α and Δv/v$_0$ vs. $\sqrt{\rho\eta}$, i.e., the attenuation and velocity sensitivity to the viscosity-density product, are plotted vs the plate normalized thickness and shown in Fig. 6.9. The data are referred to a 95°Y90°-X GaPO$_4$ with normalized thickness h/λ that varies from 0.1 to 2, being the plate thickness *h* fixed (150 μm). Thus quite good K^2 values and sensitivities can be combined for the 95°Y crystallographic orientation in the 0.05 to 0.5 normalized plate thickness range.

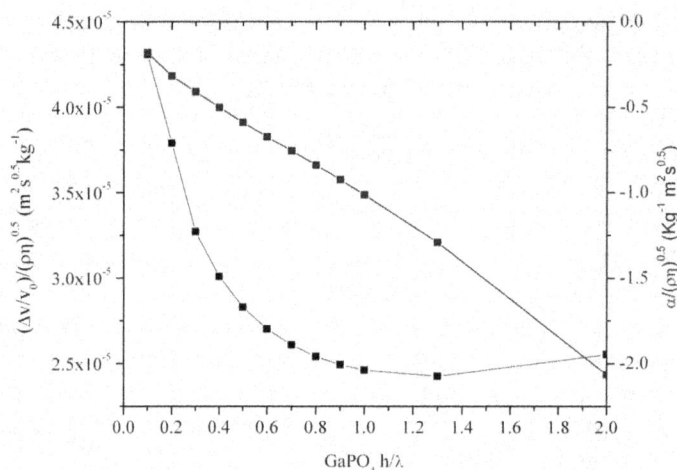

Fig. 6.9. The SH_0 α and $\Delta v/v_0$ per unit square root of $(\rho_{gw} \cdot \eta_{gw})$ vs the plate normalized thickness: the mode travels along a $GaPO_4$ (0° 5° 90°) plate 150μm thick.

The velocity and attenuation sensitivities of $GaPO_4$-based SHAPM sensor are comparable to those of quartz [10-12] and $LiTaO_3$ [13]. For ST-cut quartz with operating frequency of 158 MHz, the reported velocity sensitivity is -14 ppm $Kg^{-1}m^2s^{1/2}$ and attenuation sensitivities are 3.3 Np $Kg^{-1}m^2s^{1/2}$ and 5.7 Np $Kg^{-1}m^2s^{1/2}$ [11, 12]. For BT-cut quartz, the reported attenuation sensitivity is 4 Np $Kg^{-1}m^2s^{1/2}$. For $LiTaO_3$, T Sato et. al. reported the attenuation sensitivity of 0.55 Np $Kg^{-1}m^2s^{1/2}$ and velocity sensitivity of -22 ppm $Kg^{-1}m^2s^{1/2}$ [13].

6.6. Conclusion

The propagation of the SHAPMs along y-rotated $GaPO_4$ plates have been investigated by theoretical calculations with respect to the plate thickness, y axis rotation, electrical boundary conditions and temperature. The phase velocity and the K^2 dispersion curves of two coupling configurations have been theoretically studied specifically addressing the design of enhanced-coupling electroacoustic devices. The gravimetric sensitivity in vacuum, the attenuation and velocity sensitivities to the liquid mass density-viscosity product of sensors based on $GaPO_4$ plates have been theoretically studied with respect to the plate thickness. The $GaPO_4$-based mass sensors are proven to achieve quite good performances (high sensitivity, low TCV and enhanced coupling efficiency) that are important prerequisite for the design of sensing devices, to be used in the context of chemical, biological and physical quantities detection.

Acknowledgement

This publication is part of a project that has received funding from the European Union's Horizon 2020 research and innovation programme under grant agreement No 642688.

References

[1]. Krispel F., Reiter C., Schauperl R., Mannelli I. and Wallnofer W., Microbalance sensor applications using the piezoelectric crystal material GaPO$_4$, in *Proceedings of the 18th European Frequency and Time Forum (EFTF'04)*, Guildford, 2004, pp. 565-568.

[2]. Macagnano A., et al., Biomolecule binding sensor based on TiO$_2$ nanofibres grown on GaPO$_4$ crystal microbalance, in *Proceedings of the 2nd International Conference of Bio-Sensing Technology*, Amsterdam, The Netherland, 10-12 October 2011.

[3]. Caliendo, C., et al., K+ detection using shear horizontal acoustic modes., in *Proceedings of the IEEE Ultrasonics Symposium,* 1990.

[4]. Wallnöfer, W., P. W. Krempl, and A. Asenbaum, Determination of the elastic and photoelastic constants of quartz-type GaPO$_4$ by Brillouin scattering, *Physical Review B*, 49, 15, 1994, pp. 10075.

[6]. Reiter, C., Krempl, P. W., Wallnofer, W., Leuprecth, G., GaPO$_4$: A Critical Review of Material Data, *in Proceedings of the 9th European Frequency and Time Forum*, Besancon, France, 8–10 March 1995.

[6]. Auld, Bertram Alexander, Acoustic fields and waves in solids, *Wiley*, 1973.

[7]. D. Palmier, Ph. D. Thesis, *UMII Montpellier II*, Nov. 1996.

[8]. McAllister, Douglas J., An Acoustic Plate Mode Sensor for Biowarfare Toxins, Phase II. No. 96-02-FY1, *Biode Inc.,* Bangor Me, 1997.

[9]. Cheng, Nian-Sheng, Formula for the viscosity of a glycerol-water mixture., *Industrial & Engineering Chemistry Research*, 47, 9, 2008, pp. 3285-3288.

[10]. Martin, S. J., et al., Characterization of SH acoustic plate mode liquid sensors, *Sensors and Actuators*, 20, 3, 1989, pp. 253-268.

[11]. Ricco, A. J., and S. J. Martin, Acoustic wave viscosity sensor, *Applied Physics Letters*, 50, 21, 1987, pp. 1474-1476.

[12]. Soluch, Waldemar, and Magdalena Lysakowska, Properties of shear horizontal acoustic plate modes in BT-cut quartz, *IEEE Transactions on Ultrasonics, Ferroelectrics, and Frequency Control*, 58, 10, 2011, pp. 2239-2243.

[13]. Sato, Toshio, et al., Shear horizontal acoustic plate mode viscosity sensor, *Japanese Journal of Applied Physics*, 32, 5S, 1993, pp. 2392.

Chapter 7
Biosensors based on Magnetic Nanoparticles

I. Giouroudi, S. Cardoso, G. Kokkinis

7.1. Introduction

Microfluidics provide a rapidly growing platform for developing new systems and technologies for an ever-growing list of applications in biotechnology, life sciences, public health, pharmaceuticals and agriculture. Many chemical, biological, and biophysical processes and experiments take place in liquid environments. The chips used during these processes are called microfluidic devices [1-8]. Microfluidics is also often defined as the technology that moves nanoscale fluid volumes inside channels of micrometer size. Microfluidic systems for biomedical analysis usually consist of a set of units, which guarantees the manipulation, detection and recognition of bioanalytes (e.g. pathogens) in a reliable and flexible manner. On the other hand, a biosensor is a device used for the detection of bioanalyte consisting of a biological recognition component which interacts or reacts with the bioanalyte under investigation, and a transducer which converts this recognition component into a measurable electric output signal.

While optical and electrochemical techniques have long been used for biomedical diagnosis, they are not favorable for on-chip applications due to several technical challenges e.g. the size and the cost of the required instrumentation (laser for the excitation of fluorescent labels and detection optics) as well as photostability issues with time, narrow excitation range and broad emission spectra of the fluorescent labels [9-12]. A combination of magnetic microsensors with magnetic nanoparticles has provided a promising alternative that can fulfill the increasing requirements of such a portable robust devices [12-19]. These methods involve the labeling of the bioanalyte with magnetic nanoparticles and the detection of their stray field using integrated magnetoresistive (MR)/ magnetoimpedance (MI) sensors [12-23]. Such sensors are compatible with standard silicon IC technology, and thus suitable for integration into hand held, portable on-chip biosensing systems. They possess high sensitivity (having values of saturation field from

I. Giouroudi
Institute of Sensor and Actuator Systems, Vienna University of Technology, Vienna, Austria;
Institute for Biophysics, Department of Nanobiotechnology, BOKU - University of Natural Resources and Life Sciences, Vienna, Austria

0.1 to 10 kA/m and noise level in the range of a few nT/Hz1/2). Compared to the superconducting quantum interference device (SQUID)-based ultrasensitive magnetic detection the MR based sensors are operated at room temperature and have low power consumption in the range of 10 mW. That said, giant magnetoresistance (GMR) based biosensors [12-13] have emerged as excellent pathogen detection techniques at room temperature and as quantification methods of biological entities due to their high sensitivity, less complex instrumentation, compact size, and integration flexibility. Current efforts are to integrate these sensors within microfluidic devices to develop cost-effective, sensitive, and portable devices for rapid diagnosis of diseases [19, 24].

In the case of in-vitro cancer cell trapping and single cell analysis microscale methods such as on-chip magnetic microfluidic platforms show again great promise and superiority. Most of the reported methods utilizing microfluidics for the confinement of specific cancer cells are based on coating the microfluidic surfaces or microstructures, fabricated inside the microfluidic channels, with antibodies against epithelial cell markers or tumor-specific antigens such as EpCAM or PSMA [25-26]. The biggest disadvantage of this type of capturing method, which depends on a biological functionalization layer (e.g. antibodies, antigens), is that it is prone to time dependent changes of this functionalization layer such as aging and contamination. Long-term system stability is therefore an issue. This can be effectively overcome by utilizing the method reviewed in this chapter since no functionalization layer on top of the microfluidic surface is required.

7.2. Magnetic Nanoparticles

Throughout the last decades, magnetic nanoparticles (MPs) have facilitated laboratory diagnostics, cell sorting and analysis, DNA sequencing, medical drug targeting and have served as contrast agents for magnetic resonance imaging (MRI), as biomarkers for tumor therapy or cardiovascular disease. Their controllable size, which ranges from a few nanometers up to hundreds of nanometers, is one of their greatest advantages. It makes them very attractive for coupling with biological entities (e.g. viruses, bacteria, cells, genes etc.), of comparable sizes. Once their surface is functionalized with the appropriate bioligands they can bind and interact with biological entities thus providing a key method of labeling. Another major advantage is their magnetic nature; they can be controlled and manipulated by an external magnetic field gradient. The impressive developments in nano biotechnology enabled the modulation and tailoring of their composition, size, surface functionalization and magnetic properties.

In order for MPs to be applied in biomedicine they should be suspended in an appropriate carrier liquid, forming magnetic liquids, which are called ferrofluids. The most commonly used magnetic materials are iron oxides (e.g., γ-Fe_2O_3 or Fe_3O_4) and the carrier liquids are water or various oils [35-36]. Even though MPs suspended in liquids usually do not agglomerate due to magnetic dipole interaction, they still need to be protected against agglomeration due to van der Waals interactions [37]. This can be achieved by appropriate surface modification, which also acts as a functionalization layer for further conjugation with bioactive molecules. For example, MPs can be functionalized either with organic materials (e.g., polymers) [37-38], inorganic metallic materials such as gold or oxide materials (e.g., silica) [39-40]. Moreover, the biomedical applications of MPs impose strict requirements on their physical and chemical properties e.g., chemical composition, crystal structure, magnetic behavior, surface structure, adsorption properties, solubility and low toxicity [33]. In order to fulfill these requirements several fabrication methods have been developed worldwide and some details and extended reviews on MPs can be found in [34-39, 43-46].Finally, the detection of magnetic particles in static fluids or under fluidic flow requires sensors with high sensitivity to magnetic fields in the range of 1 mT and below [27, 40-42, 47-50].

7.3. Biosensors

7.3.1. GMR Microfluidic Biosensors

Novel, multiplex, portable microfluidic biosensors have been reported by the authors in [15-23]. These biosensors can be used for monitoring the presence of potentially hazardous pathogens (e.g. bacteria and viruses suspended in a static fluid) in recreational and drinking water supplies such as rivers, lakes and springs as well as in liquid food products such as milk in order to prevent human and animal infection. These biosensors can also be used for the quantification of biomolecules, such as proteins (or other biomarkers), antibodies or DNA strands conjugated with Fe_3O_4 nanoparticles. This way, the monitoring of pathogens and the quantification of biomolecules is simplified and accelerated due to real-time detection. Quantification hands-on time is reduced, and sample throughput can be increased using automation and efficient data evaluation with the appropriate software.

The reported biosensors use magnetic markers (magnetic particles – MPs) that are functionalized with bioligands directed against the pathogens to be detected and mixed with the liquid sample under investigation. If pathogens are present, compounds are being formed after mixing consisting of the functionalized MPs and the attached pathogens (LMPs).

The innovative aspect of this detection method is that the induced velocity on reference MPs (plain MPs – not functionalized with bioligands) and on MPs bound to pathogens (LMPs), while imposed to the same magnetic field gradient in a static fluid, is inversely proportional to their overall, non-magnetic volume [17-18, 21-22]. This is due to the enhanced Stokes drag force exerted on the LMPs resulting from their greater volume and altered hydrodynamic shape (due to the attached pathogens) as reported in [18]. In the case of biotinylated antibodies, it is the increased friction force at the interface between the modified MP and the biosensor's surface as reported in [22]. Detected differences in velocity between the LMPs and the reference MPs indicate the presence of pathogens in the static liquid sample. With this biosensing method the time needed for the MPs and the LMPs to travel a certain distance, when accelerated by an externally applied magnetic field, is measured utilizing spin valve (GMR) sensor pairs. The induced velocity is then automatically calculated from these measurements. This way, a compact and integrated solution with high sensitivity and reliability is offered. A method for signal acquisition and demodulation was also reported by the authors in [16-17]; expensive function generators, data acquisition devices and Lock-in Amplifiers are substituted by a generic PC sound card and an algorithm combining the Fast Fourier Transform (FFT) of the signal with a peak detection routine. This way, costs are drastically reduced and portability is enabled.

7.3.2. Working Principle

The biosensing platform developed by the authors consists of multiple microfluidic channels to enable multiplex detection. Different pathogens can be detected in different pairs of microfluidic channels.

In detail, each pair consists of two identical microfluidic channels; a reference channel and a detection channel. The reference channel is used in order to avoid "false positive" results; the values from the detection channel are simultaneously and automatically compared to the ones taken from the reference channel.

Plain (not functionalized) magnetic particles (MPs) are mixed with the liquid sample (e.g. water). After mixing, the resulting sample (consisting of the plain MPs suspended in the liquid) is introduced to the reference channel with a pipette. There is no flow thus the liquid inside the reference channel is static. At the same time, functionalized MPs are mixed with the same liquid sample. After mixing, the resulting sample (consisting of the LMPs suspended in the liquid) is introduced to the detection channel with a pipette. The LMPs consist of functionalized MPs (of same magnetic volume as the ones in the reference channel) with attached pathogens. The overall, non-magnetic volume of the LMPs is greater than that of the plain MPs. There is no flow thus the liquid inside the detection channel is static.

Current carrying microconductors are fabricated underneath the channels, using evaporation deposition technique and photolithography, in order to impose a magnetic field gradient to the MPs and LMPs and move them from the inlet to the outlet of the channels. The microconductors are automatically and sequentially switched on and off by a programmable microprocessor. There is no additional external flow in both channels.

Underneath the first (inlet) and the last (outlet) microconductors of each channel, GMR, spin valve sensors are fabricated. By measuring the time interval between the resistance change of the inlet GMR sensors and the outlet GMR sensors, the mean velocity of the MPs and LMPs during their acceleration is calculated.

Fig. 7.1 shows a schematic of a pair of microfluidic channels (reference and detection). An array of such pairs is fabricated on the biosensing platform. In each pair of channels different pathogens can be detected; this is achieved by using MPs functionalized with different bioligands (each ligand directed against a different pathogen) in each detection channel of each pair.

All four GMR sensors (two at the reference channel and two at the detection channel) have zero output before the liquid sample with the magnetically labeled pathogens (LMPs) and the reference suspension of MPs are introduced into the channels using a pipette. Once the liquids are introduced, a change of signal at the inlet sensors of both channels indicates the presence of MPs on top of them, thus the initialization of measurements. Afterwards, the microconductors are sequentially switched on and off accelerating the particles towards the outlet. The actuation time is optimized for continuous motion of plain MPs while LMPs will fall behind due to their smaller mobility. When the output of the outlet sensor of the reference channel changes from zero the measurement ends. If the output of the outlet GMR sensor of the detection channel remains zero, it is proven that the reference, plain MPs travelled the same distance faster than the LMPs inside the detection channel. This demonstrates that the non-magnetic volume of the functionalized MPs increased through the binding of pathogens. If we are testing for the presence of biotinylated antibodies it is the increased friction force between LMPs and the surface of the channel that caused this decrease in their velocity compared to the plain MPs inside the reference channel [22]. Once the last microconductor is switched off the MPs and LMPs are removed from the microfluidic channels by rinsing with distilled water. Moreover, the time the particles need to accumulate on the sensor's surface and the magnitude of the sensor's output determines their concentration. Additionally, the time until the sensor's signal is saturated is used to define the size difference between the LMPs and the plain MPs.

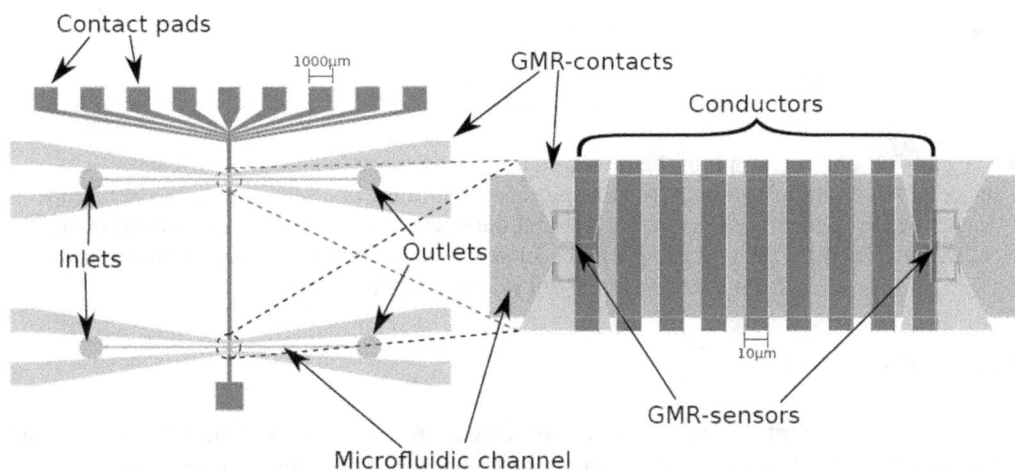

Fig. 7.1. Schematic of one pair of microfluidic channels with the current carrying microstructure, four GMR sensors (two in each channel), GMR-contact-pads and reference and detection channels with inlets and outlets. Several identical pairs are fabricated on the biosensing platform to enable multiplex pathogen detection.

Fig. 7.2 illustrates the microfluidic biosensor, developed by the authors, with the integrated GMR sensors and current carrying microconductors.

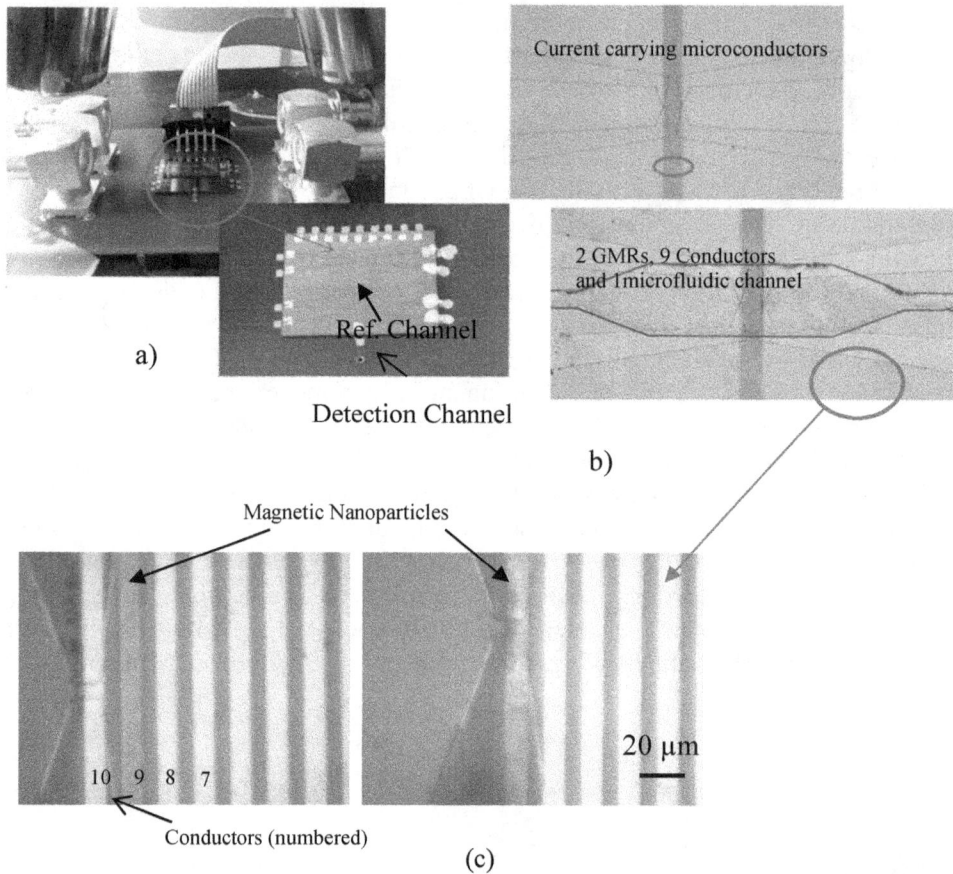

Fig. 7.2. a) Photograph of the reviewed biosensor consisting of the GMR sensors, the conducting microstructures and the measurement microfluidic channel; b) Microscope image of the conducting microstructures and the microfluidic channel; c) Detailed image of the microconductors and the GMR sensor with magnetic nanoparticles as they moved from the right to the left microconductor [17, 20].

7.3.3. Results

Successful experimental results were reported in [17] with magnetically labeled Escherichia Coli K12 "wild type" gram negative bacteria in water samples. Moreover, the authors presented in [15, 18] the systematic study on detection and quantification of commercially available Nanomag-D MPs using the above mentioned GMR microfluidic biosensor. The biosensor proved to be capable of detecting concentrations as low as 500 pg/µl of commercially available Nanomag-D MPs and quantifying them in a linear scale over a wide particle concentration range (1 ng/µl - 500 ng/µl). As several biological identities can be labeled by these magnetic markers, the developed sensor has potential for a rapid, portable, and reliable diagnosis of diseases.

7.3.4. Magnetic Microfluidic Counter

The authors of [50] recently reported on the development of a magnetic counter that identifies the presence of Streptococcus agalactiae (a Group B Streptococci) in milk samples. An integrated microfluidic platform is used for the detection, where 50 nm magnetic particles attached to Streptococcus agalactiae were dynamically detected by of magnetoresistive (MR) sensors. This device allows the analysis of raw milk without bridging the microfluidic channels, making this integrated platform very attractive for fast bacteriological contamination screening.

7.3.5. Working Principle

This portable device is composed of MR sensors, namely spin-valve (SV) sensors, integrated with a microfluidic platform and connected to an amplification and acquisition setup. The sensors are sensitive to the magnetic field created by the magnetically labeled bioanalyte flowing in microchannels above the sensors. This dynamic detection is based on immunoassay techniques since these antibodies anti-GB Streptococci (probes) recognize immunogenic proteins on bacteria cell walls (targets). The SV sensor detects the fringe field of the magnetic markers bound around the target analyte through the specific probe. This platform is described in detail in [51], and a proof of concept was demonstrated for milk samples. In [50] the authors successfully reduced the functionalized nanoparticles quantity to the limits where one can distinguish magnetic signal amplitude between milk control samples and milk samples with known bacterial concentrations. Fig. 7.3 shows a photo of the developed device.

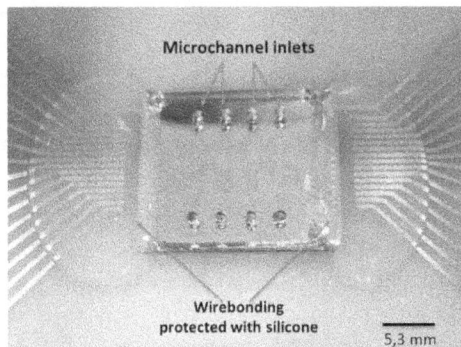

Fig. 7.3. Final device with the magnetoresistive chip bonded to the PDMS microchannels. The sensor´s wirebonding are protected with silicone.

7.3.6. Results

For the experiments the authors used milk samples mixed with a solution combining specific antibodies and magnetic nanoparticles, and analyzed. From the sensor readout, magnetic signature of the labeled cells showed no linear correlation between the detected peak number and the bacterial concentration. The milk samples with the anti-GB

Streptococci antibody revealed the most exuberant signal with *S. uberis* when compared with the other two bacteria-antibody pairs. Only the *Streptococcus agalactiae/* pAb anti-GB Streptococci pair evidenced no peaks higher than 200 μV. Despite that, these MR sensors could detect *Streptococcus agalactiae* and *Streptococcus uberis* in milk samples and assess its concentration from 0.1 cfu/μl (100 cfu/ml).

Comparison with PCR results showed sensitivities of 73 % and 41 %, specificity values of 25 % and 57 %, and PPV values of 35 % and 54 % for magnetic identification of streptococci species with an anti-*S. agalactiae* antibody and an anti-GB Streptococci antibody, respectively.

Magnetic detection of milk samples showed some microbiological and immunological constraints. Since bacterial cells have high variability on the number of immunogenic proteins per cell, the number of labeled sited through the antibody is also not well defined. This affects the quantification of the magnetic method. As a consequence, it was not possible to quantify the peaks profile (number, shape) for each bacterial concentration. The method, however, allow to determine their presence, and quantification may be done within lower/upper threshold limits. Simulations of the sensor output as a function of the nanoparticle distribution over the cells (using colonies/clusters configurations compatible with the experimentally observed in microscope) can provide indication on minimum and maximum numbers. Further work would be done towards a more accurate quantification based on simulations.

In any case, even a qualitative analysis is of great importance to dairy farms as they need to rapidly identify bovine causing bacteria and consequently target the antimicrobial therapy to be used, aiming at a more efficient cow treatment and disease control.

7.4. Cancer Cell Trapping

As mentioned in Section 7.1, developing patient specific therapies can be enabled by analyzing the cancer cells' metastasis-driving capabilities on the single cell level. A portable and cost effective microfluidic platform for trapping and studying the mechanical properties of single cancer cells in suspension is reported in [52]. The innovative aspect of that trapping method is that it uses MPs to label the cancer cells and then trap those using microchambers with integrated current carrying microconductors. These intact single cells can then be used for studying their mechanical properties and for additional testing and patient specific drug screening.

7.4.1. Working Principle

The magnetic microfluidic platform reported in [52] consists of a microfluidic channel made of PDMS and several trapping microchambers (microtraps) fabricated by photolithography and by patterning a dry photoresist thin film (Ordyl®). Fig. 7.4 shows the developed trapping platform.

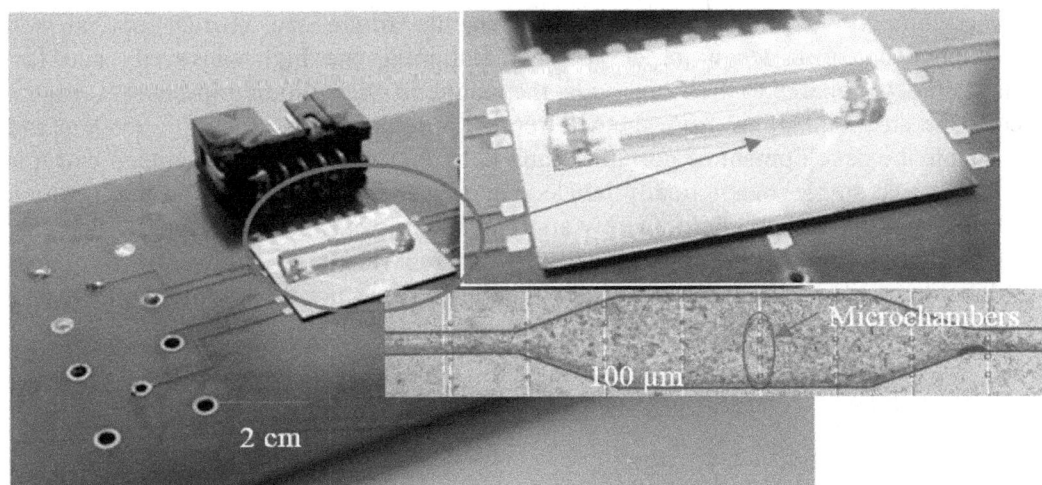

Fig. 7.4. Fully equipped PCB with the cell trapping chip in the middle.

The cancer cells in the liquid sample are labeled with commercially available functionalized MPs which have selective affinity to these cells. Current carrying gold (Au) microstructures (microconductors) are fabricated underneath the microtraps using evaporation deposition technique and photolithography; this way, the trapping of the magnetically labeled cells is achieved.

7.4.2. Results

Preliminary trapping experiments were successfully reported in [52] with Jurkat cells labeled with commercially available "M-450 Dynabeads". The cells were introduced to the channel through the inlet with an estimated velocity of about 50 μm/s. It was observed that the microchambers were not influencing the flow of the cells when the microconductors were switched off. In order to begin with the trapping procedure 150 mA were sequentially applied to the microconductors. Fig. 7.5 shows the trapping of a single magnetically labeled Jurkat cell. All pictures were taken with a 60-fold lens and above microconductor Nr. 4 (17×17 μm). After the microconductor was switched off it was clearly observed that the cell remains trapped.

7.5. Conclusions

In this chapter, we reviewed some of the most recent developments of the authors towards microfluidic biosensing using MR sensors and magnetic nanoparticles as well as towards cancer cell trapping. The presented biosensing systems are very promising candidates for lab-on-a-chip devices as compact, ultra-sensitive and inexpensive solutions for clinical diagnostics and biomedical applications in general. Through our study we conclude that several novel manipulation, separation and detection mechanisms based on magnetic methods are continuously emerging, proving that magnetic biosensing has the potential to

141

become competitive and probably replace in the future the current optical and fluorescence detection technologies, while maintaining the high sensitivity and fast readout time. Compared to these methods, the magnetic microfluidic biosensors measure directly the electrical signal from the MR sensor, provide a fully electronic readout and thus enable the development of portable, hand-held devices. We also concluded that it is possible to trap single, magnetically labeled cancer cells and hold them in that position for further analysis without permanently applying electric current and magnetic field. By using the configuration of the cell trapping platform reported by the authors the fluid flow over the microtraps is neither influenced by the trapped cells nor by the empty microtraps.

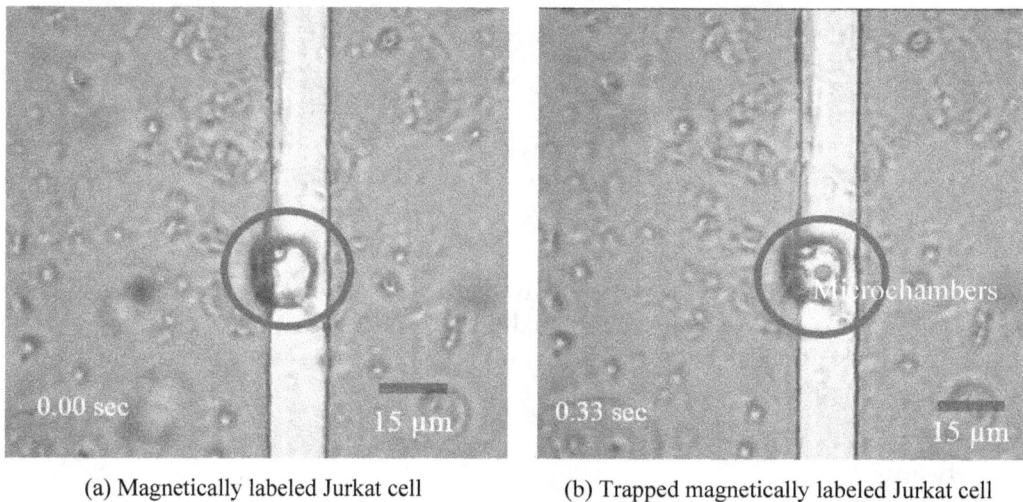

(a) Magnetically labeled Jurkat cell   (b) Trapped magnetically labeled Jurkat cell

Fig. 7.5. (a) A labelled Jurkat cell approaching the microtrap on the surface of the current carrying microconductor, appearing blur due to shallow depth of field and focusing on the microconductor's plane. (b) The labelled cell after it has been trapped.

Acknowledgements

The authors would like to acknowledge the financial support of the Austrian Science Fund (FWF) with Project Numbers.: P 24372-N19 and P28544-N30. Part of the reported experimental work was supported by the Fundação para a Ciência e a Tecnologia (FCT) Project under Grant EXCL/CTM-NAN/0441/2012 and the Institute of Nanoscience and Nanotechnology Associated Laboratory, Pest-OE/CTM/LA0024/2011.

References

[1]. J. Mairhofer, K. Roppert, and P. Ertl, Microfluidic Systems for Pathogen Sensing: A Review, *Sensors,* 9, 2009, p. 4804.
[2]. J. Loureiro, C. Fermon, M. Pannetier-Lecoeur, G. Arrias, R. Ferreira, S. Cardoso, and P. P. Freitas, Magnetoresistive detection of magnetic beads flowing at high speed in microfluidic channels *IEEE Trans. Magn.,* 45, 2009, p. 4873.

[3]. A. K. Balasubramanian, A. Beskok, and S. D. Pillai, In situ analysis of bacterial capture in a microfluidic channel, *J. Micromechanics Microengineering*, 17, 2007, p. 1467.

[4]. J. Chen, D. Chen, Y. Xie, T. Yuan, and X. Chen, A microfluidic chip for direct and rapid trapping of white blood cells from whole blood, *Nano-Micro Lett.*, 2, 2013, p. 66.

[5]. K.-S. Yun, D. Lee, H.-S. Kim, and E. Yoon, A microfluidic chip for measurement of biomolecules using a microbead-based quantum dot fluorescence assay, *Meas. Sci. Technol.*, 17, 2006, p. 3178.

[6]. K.-K. Liu, R.-G. Wu, Y.-J. Chuang, H. S. Khoo, S.-H. Huang, and F.-G. Tseng, Microfluidic systems for biosensing, *Sensors*, 10, 2010, p. 6623.

[7]. K. S. Kim and J.-K. Park, Magnetic force-based multiplexed immunoassay using superparamagnetic nanoparticles in microfluidic channel, *Lab Chip*, 5, 2005, p. 657.

[8]. L. Y. Yeo, H.-C. Chang, P. P. Y. Chan, and J. R. Friend, Microfluidic devices for bioapplications, *Small*, 7, 2011, p. 12.

[9]. M. Mujika, S. Arana, E. Castaño, M. Tijero, R. Vilares, J. M. Ruano-López, a Cruz, L. Sainz, and J. Berganza, Magnetoresistive immunosensor for the detection of *Escherichia coli* O157:H7 including a microfluidic network, *Biosens. Bioelectron.*, 24, 2009, p. 1253.

[10]. J.-I. Hahm, Functional polymers in protein detection platforms: Optical, electrochemical, electrical, masssensitive, and magnetic biosensors, *Sensors*, 11, 3, 2011, pp. 3327–3355.

[11]. J. B. Haun, T. J. Yoon, H. Lee and R. Weissleder, Magnetic nanoparticle biosensors, *Wiley Interdiscip. Rev.: Nanomed. Nanobiotechnol*, 2, 3, 2010, pp. 291–304.

[12]. J. Llandro, J. J. Palfreyman, A. Ionescu and C. H. W. Barnes, Magnetic biosensor technologies for medical applications: a review, *Med. Biol. Eng. Comput.*, 48, 10, 2010, pp. 977–998.

[13]. S. X. Wang and G. Li, Advances in giant magnetoresistance biosensors with magnetic nanoparticle tags: Review and outlook, *IEEE Trans. Magn.*, 44, 7, 2008, pp. 1687–1702.

[14]. D. R. Baselt, G. U. Lee, M. Natesan, S. W. Metzger, P. E. Sheehan and R. J. Colton, A biosensor based on magnetoresistance technology, *Biosens. Bioelectron.*, 13, 7–8, 1998, pp. 731–739.

[15]. I. Koh and L. Josephson, Magnetic Nanoparticle Sensors, *Sensors-Basel*, 9, 10, 2009, pp. 8130–8145.

[16]. J. Devkota, G. Kokkinis, T. Berris, M. Jamalieh, S. Cardoso, F. A. Cardoso, H. Srikanth, M. Phan and I. Giouroudi, A novel approach for detection and quantification of magnetic nanomarkers using a spin valve GMR-integrated microfluidic sensor, *RSC Advances*, 5, 2015, pp. 51169 – 51175.

[17]. G. Kokkinis, M. Jamalieh, S. Cardoso, F. A. Cardoso, F. Keplinger and I. Giouroudi, Magnetic based Biomolecule Detection using GMR Sensors, *Journal of Applied Physics*, 117, 2015, 17B731.

[18]. G. Kokkinis, S. Cardoso, F. A. Cardoso, I. Giouroudi, Microfluidics for the Rapid Detection of Pathogens using Giant Magnetoresistance Sensors, *IEEE Transactions on Magnetics*, Vol. 50, No. 11, November 2014, 4401304.

[19]. G. Kokkinis, F. Keplinger, I. Giouroudi, On-Chip Microfluidic Biosensor using Superparamagnetic Microparticles, *Biomicrofluidics*, 7, 5, 2013, S. 054117-1 - 054117-14.

[20]. I. Giouroudi, F. Keplinger., Microfluidic Biosensing Systems using Magnetic Nanoparticles, *International Journal of Molecular Sciences*, 14, 9, 2013, pp. 18535-18556.

[21]. J. Devkota, G. Kokkinis, M. Jamalieh, M. H. Phan, H. Srikanth, S. Cardoso, F. A. Cardoso, I. Giouroudi, GMR microfluidic biosensor for low concentration detection of Nanomag-D beads, *Proc. SPIE 9518, Bio-MEMS and Medical Microdevices II*, 95180W, June 1, 2015.

[22]. J. Loureiro, P. Z. Andrade, S. Cardoso, C. L. da Silva, J. M. Cabral, P. P. Freitas, Magnetoresistive chip cytometer, *Lab-on-Chip*, 11, 13, 2011, pp. 2255 – 2261.

[23]. G. Kokkinis, B. Plochberger, S. Cardoso, F. Keplinger, I. Giouroudi, Microfluidic, dual-purpose sensor for in-vitro detection of Enterobacteriaceae and biotinylated antibodies, *Lab on a Chip*, 16, 7, 2016, pp. 1261-1271.

[24]. L. Lagae, R. Wirix-Speetjens, J. Das, D. Graham, H. Ferreira, P. P. F. Freitas, G. Borghs and J. De Boeck, *Journal of Applied Physics,* 91, 2002, pp. 7445-7447. N. Sanvicens, C. Pastells, N. Pascual and M. P. Marco, Nanoparticle-based biosensors for detection of pathogenic bacteria, TrAC, *Trends Anal. Chem.,* 28, 11, 2009, pp. 1243–1252.

[25]. Diamond, E., Lee, G. Y., Akhtar, N. H., Kirby, B. J., Giannakakou, P., Tagawa, S. T. and Nanus, D. M., Isolation and characterization of circulating tumor cells in prostate cancer, *Frontiers in Oncology*, 2, 2012, pp. 1-11.

[26]. Santana, S. M., Liu, H., Bander, N. H., Gleghorn, J. P., and Kirby, B. J., Immunocapture of Prostate Cancer Cells with Anti-PSMA Antibodies in Microdevices, *Biomed Microdevices*, 14, 2, 2012, pp. 401-407.

[27]. Neuberger T., Schöpf B., Hofmann H., Hofmann M., von Rechenberg B. Superparamagnetic nanoparticles for biomedical applications: Possibilities and limitations of a new drug delivery system, *J Magnet Magnetic Mater,* 293, 2005, pp.483-96.

[28]. Jurgons R., Seliger C., Hilpert A., Trahms L., Odenbach S., Alexiou C. Drug loaded magnetic nanoparticles for cancer therapy, *J Phys: Condens Matter*, 18, 2006, pp. 2893-902.

[29]. Salgueiriño-Maceira V., Correa-Duarte M.A. Increasing the complexity of magnetic core/shell structured nanocomposites for biological applications, *Adv Mater,* 19, 2007, pp. 4131-44.

[30]. Latham A. H., Williams M. E., Controlling transport and chemical functionality of magnetic nanoparticles, *Accounts Chem.,* 41, 3, 2008, pp. 411-420.

[31]. Laurent S., Forge D., Port M., Roch A., Robic C., Elst L. V., et al., Magnetic iron oxide nanoparticles: synthesis, stabilization, vectorization, physicochemical characterizations, and biological applications, *Chem Rev.,* 108, 2008, pp. 2064-110.

[32]. Berry C. C., Progress in functionalization of magnetic nanoparticles for applications in biomedicine, *J Phys D: Appl Phys.,* 2009, 42, p. 224003.

[33]. Pankhurst Q. A., Thanh N. K. T., Jones S. K., Dobson J. Progress in applications of magnetic nanoparticles in biomedicine, *J Phys D: Appl Phys.*, 2009, 42, p. 224001.

[34]. Roca A. G., Costo R., Rebolledo A. F., Veintemillas-Verdaguer S., Tartaj P., González-Carreño T., et al., Progress in the preparation of magnetic nanoparticles for applications in biomedicine, *J Phys D: Appl Phys.*, 42, 2009, p. 224002.

[35]. Varadan, V. K., Chen, L., Xie, J., Nanomedicine: Design and Applications of Magnetic Nanomaterials, Nanosensors and Nanosystems, *John Wiley & Sons, Ltd.,* West Sussex, UK, 2008.

[36]. Guimaraes, A. P. Principles of Nanomagnetism, *Springer*, Berlin, Heidelberg, Germany, 2009.

[37]. Thorek, D., Chen, A., Czupryna, J., Tsourkas, A., Superparamagnetic iron oxide nanoparticle probes for molecular imaging, *Ann. Biomed. Eng.*, 34, 2006, pp. 23–38.

[38]. Gupta, A., Curtis, A., Surface modified superparamagnetic nanoparticles for drug delivery: Interaction studies with human fibroblasts in culture, *J. Mater. Sci. Mater. Med.*, 2004, 15, pp. 493–496.

[39]. Melancon, M., Lu, W., Li, C., Gold-based magneto/optical nanostructures: Challenges for in vivo applications in cancer diagnostics and therapy, *Mater. Res. Bull.*, 34, 2009, pp. 415–421.

[40]. Bumb, A., Brechbiel, M. W., Choyke, P. L., Fugger, L., Eggeman, A., Prabhakaran, D., Hutchinson, J., Dobson, P. J., Synthesis and characterization of ultra-small superparamagnetic iron oxide nanoparticles thinly coated with silica, *Nanotechnology,* 19, 2008, p. 335601.

[41]. Tartaj, P., Morales, M. P., Gonzalez-Carreno, T., Veintemillas-Verdaguer, S., Serna, C. J. Advances in magnetic nanoparticles for biotechnology applications, *J. Magn. Magn. Mater.*, 290–291, 2005, pp. 28 34.

[42]. Indira, T. K., Lakshmi, P. K. Magnetic nanoparticles—A review, *Int. J. Pharm. Sci. Nanotechnol.*, 3, 2010, pp. 1035–1042.

[43]. Iida, H., Takayanagi, K., Nakanishi, T., Kume, A., Muramatsu, K., Kiyohara, Y., Akiyama, Y., Osaka, T. Preparation of human immune effector T cells containing iron-oxide nanoparticles, *Biotechnol. Bioeng.*, 101, 2008, pp. 1123–1128.

[44]. Karami, H. Synthesis and characterization of iron oxide nanoparticles by solid state chemical reaction method, *J. Clust. Sci.*, 21, 2009, pp. 11–20.

[45]. Koutzarova, T., Kolev, S., Ghelev, C. H., Paneva, D., Nedkov, I., Microstructural study and size control of iron oxide nanoparticles produced by microemulsion technique, *Physica Status Solidi*, C, 3, 2006, pp. 1302–1307.

[46]. Loo, A. L., Pineda, M. G., Saade, H., Trevino, M. E., Lopez, R. G., Synthesis of magnetic nanoparticles in bicontinuous microemulsions. Effect of surfactant concentration, *J. Mater. Sci.*, 43, 2008, pp. 3649–3654.

[47]. Lu, A. H., Salabas, E. L., Schueth, F., Magnetic nanoparticles: Synthesis, protection, functionalization, and application, *Angew. Chem. Int. Ed.*, 46, 2007, pp. 1222–1244.

[48]. Li, G. X., Sun, S. H., Wilson, R. J., White, R. L., Pourmand, N., Wang, S., Spin valve sensors for ultrasensitive detection of superparamagnetic nanoparticles for biological applications. *Sens. Actuators A,* 126, 2006, pp. 98–106.

[49]. Samal, D., Kumar, P. S. A., Giant magnetoresistance, *Resonance*, 13, 2008, pp. 343–354.

[50]. Nickel, J., Magnetoresitance overview. *HP Labs,* 1995, Available online: http://www.hpl.hp.com/techreports/95/HPL-95-60.pdf (accessed on 26 July 2013).

[51]. C. D. Duarte, A. C. Fernandes, F. Arroyo Cardoso, R. Bexiga, S. Freitas Cardoso and P. J. P. Freitas, Magnetic Counter for Group B Streptococci Detection in Milk, *IEEE Transactions on Magnetics,* Vol. 51, No. 1, 2015, 5100304.

[52]. A. C. Fernandes, C. M. Duarte, F. A. Cardoso, R. Bexiga, S. Cardoso, and P. P. Freitas, Lab-on-chip cytometry based on magnetoresistive sensors for bacteria detection in milk, *Sensors*, Vol. 14, No. 8, August 2014, pp. 15496–15524.

[53]. R. Mitterboeck, G. Kokkinis, T. Berris, F. Keplinger, I. Giouroudi, Magnetic microfluidic system for isolation of single cells, *Proc. SPIE 9518, Bio-MEMS and Medical Microdevices II,* 951809, June 1, 2015, p. 2181194.

Chapter 8
On-Chip Blood Coagulation Sensor

Surya Venkatasekhar Cheemalapati and Anna Pyayt

8.1. Introduction

Measurement of the speed of blood coagulation is required for a large portion of the population. Nearly 80 % of patients with diabetes die due to excessive blood clotting [1, 2]. Malfunctioning of the coagulation system results in multiple disorders, ranging from ones with increased bleeding, such as hemophilia, to unwanted clot formation, and as a result - heart attacks and strokes [3-7]. Anticoagulant medications are taken regularly by more than 2 million patients in the U.S. [8, 9], and it currently requires periodic dose adjustments and visits to the hospitals. There is a need in a device that can be used to monitor speed of blood coagulation at home, frequently, using small volume of blood obtained using fingerprick, at low cost and easily operated by non-professional.

There are several traditional approaches to the testing of speed of blood coagulation: prothrombin time (PT) that can be done in hospital [10, 11] or at home using portable device called Coaguchek [12], partial thromboplastin time (PTT) [13] and thromboelastography (TEG) [14]. Prothrombin time test is a relatively low cost and widely used, but it provides a single data point for the whole coagulation process without any information about the dynamics. On the other hand, thromboelastography is a "gold standard" method that monitors the dynamics of coagulation by continuously measuring the strength of the clot, but it is much more expensive and slow.

Recent advances have led to new devices for coagulation analysis. These devices are based on optical [15-18], electrical [19, 20] and mechanical properties of blood [21]. Lim et al. [15] used transmission of light through blood plasma to detect the rate of coagulation but their device required pre-processing of blood before testing. Other optical devices based on surface plasmon resonance [16, 17] were used with the whole blood, but their continuous operation required sophisticated sensor pre-processing before each testing. Light scattering technique was also used for coagulation detection [18], but different effects related to presence of the cells caused decrease in sensitivity and reproducibility [22]. Magnetic detection of blood coagulation requires use of external receiver coil that

Surya Venkatasekhar Cheemalapati
Department of Chemical and Biomedical Engineering, University of South Florida, Tampa, FL, USA

cannot be integrated on-chip [21]. Change in electrical properties of blood has also been demonstrated as a mechanism for monitoring blood coagulation [19, 20]. Lei et al [20] fabricated a microfluidic sensor that required such external components as pumps and electrodes. Usually one of those limitations prevented integration of blood coagulation device on-chip. More recently, Tripathi et al. [23] utilized the change in light speckle intensity caused by cells in the whole blood. Since movement of cells becomes slower as coagulation increases this changes the rate of change of speckle intensity. However they require the use of high speed cameras making the process expensive.

The device proposed in this chapter is based on continuous monitoring of the refractive index change of whole blood during coagulation [35]. Several refractive index sensors have been developed till date. These include Brag grating sensors [24], photonic crystals [25], ring resonators [26], long period fiber grating [27] and others [28-34]. These sensors are extremely sensitive to their surrounding medium and their performance will be influenced by the presence of red blood cells. There is a need to create a new low-cost, on chip, easy to operate device that would be able to monitor coagulation of whole blood continuously near the patient.

The main sensing component for the proposed device is silicon waveguides with a silicon dioxide cladding that can be easily integrated on-chip. The device can measure the amount of light reflected back from the waveguide-whole blood interface using Fresnel's equation for reflectance. The coagulation can be monitored by observing the change in reflected power during increase of refractive index of coagulating blood. This task is usually very challenging since blood is very nonhomogeneous [36] and red blood cells absorb and scatter light [22, 37].

In this work, first the amount of noise created by the presence of red blood cells was characterized and then propose a technique to eliminate the noise generated by the presence of red blood cells by isolating them from the sensor surface using a 3D cladding filtering structure. The proposed device can be easily fabricated with the current MEMS technologies. First, layers of silicon dioxide cladding and silicon core can be deposited using one of several deposition techniques available [38]. Following this, well characterized fabrication techniques such as lithography and dry etching [39] can be utilized for fabricating low loss devices within simulated design parameters. The new proposed structure would enable the device to work with the whole blood and at the same time be miniature, low cost, and provide continuous information about the dynamics of coagulation.

8.2. Design

The blood coagulation device based on optical waveguides can be made really miniature (Fig. 8.1). During coagulation blood plasma refractive index increases while fibrin mesh is formed around blood cells. This change in refractive index can be detected by measuring amount of light reflected from the interface between the waveguide and the plasma. When refractive index of the waveguide material is known, Fresnel's equation can be used to calculate the refractive index of plasma. The Y-splitter is used to connect the light source and the detector to the sensing waveguide. Light is coupled into the waveguide and is reflected back from the waveguide-blood interface to a light detector/ power meter as shown in the Fig. 8.1. The refractive index of blood determines how much light is reflected back into the detector. By monitoring the rate of change of power reflected back to the detector speed of coagulation can be determined. The cells present in the system can also reflect some light back to the waveguide and thus introduce additional noise interfering with the measurements.

Fig. 8.1. The proposed design for the blood coagulation sensor. Light is coupled into the waveguide and is reflected back from the waveguide-blood interface into the power detector. The change in refractive is monitored by measuring the change in reflected power.

Initially different materials were considered for the waveguide. The first step for design optimization was to determine material that would provide highest signal. Fig. 8.2 (a) demonstrates reflectivity from the waveguide- uncoagulated blood interface calculated using Fresnel's equation, equation (1), where n1 is the refractive index of the waveguide material and n2 is the refractive index of the blood. The uncoagulated and coagulated blood refractive indices used in the simulations are 1.345 and 1.351 respectively [40]. Fig. 8.2 (b) shows the difference in signals for uncoagulated vs. fully-coagulated plasma.

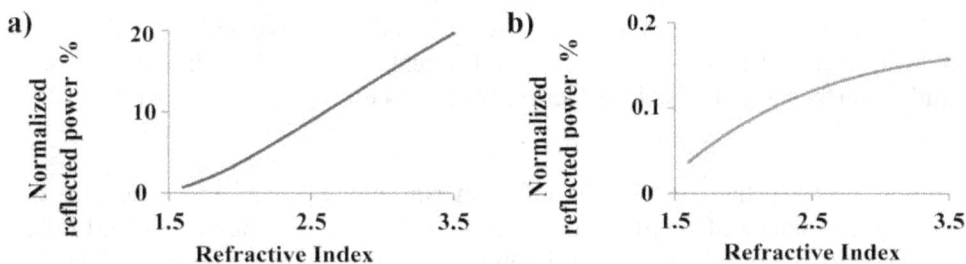

Fig. 8.2. Waveguide material selection. (a) Reflectivity from waveguide- uncoagulated blood interface depending on refractive index of the waveguide core material; (b) Difference in signals for uncoagulated and fully-coagulated plasma vs. the refractive index of the waveguide.

Both plots are normalized relative to the input intensity equal to 1 a.u. It can be noticed that a higher refractive index corresponds to a higher portion of light reflected from the interface and also to a higher contrast between coagulated and uncoagulated plasma. Therefore, for the maximum signal level, silicon with a refractive index of 3.5 is chosen for the core of the waveguide. Infrared light with the wavelength 1.5 μm was used in simulations since it can propagate in silicon waveguide with very low attenuation.

$$R = \left(\frac{n_1 - n_2}{n_1 + n_2}\right)^2 \qquad (1)$$

8.3. Results and Discussion

Three dimensional (3D) simulations were performed using commercial FDTD software. The waveguide core had square cross section with 1μm x 1μm and refractive index 3.5 corresponding to silicon. The substrate and cladding were both silicon dioxide with refractive index of 1.46. The red blood cells were modeled with refractive index 1.394 [41] and diameter 6 μm [42]. IR light was coupled into the waveguide and is reflected back from the waveguide blood interface. Initially reflected power was measured for uncoagulated and coagulated blood plasma and normalized with respect to the value for uncoagulated plasma. These two values are used as boundary conditions for the rest of the simulations in this study. All the normalized powers presented in this study are normalized with respect to the reflected power from the uncoagulated blood plasma.

The simulations were conducted in several steps. They started with characterization of the noise caused by the presence of red blood cells in the whole blood. First, orientation of the cell reflecting most of the light back into waveguide was determined. Fig. 8.3 (a) shows normalized reflected power vs angle between the cell and the waveguide interface. Fig. 8.3 (b) explains the way how angle was determined. The noise was the highest when the cell was oriented directly facing the waveguide at zero degrees. Considering this orientation of the cell the "worst case scenario" generating maximum noise, the next set simulations was conducted for this orientation.

After that the noise from the presence of the cell in front of the waveguide interface was analyzed for different distances between the cell and the waveguide. As the cell was moved away from the waveguide the total intensity, including intensity from the waveguide interface and the back-reflection from the cell, was measured at the other end of the waveguide.

Fig. 8.4 demonstrates the data for the cell that was moved away from the face of the waveguide to a distance of 10 μm with 0.5 μm per increment. The normalized reflected power is the ratio of the reflected power in presence of the cell to the power reflected from the pure uncoagulated blood plasma. It can be noticed that when the cell is close to the waveguide there is a high level of noise, and a small movement causes large signal fluctuation. The amplitude of the fluctuation is higher than the difference in signals between uncoagulated and fully coagulated plasma, what makes signal to noise ratio less than one and any detection impossible. Furthermore, the level of noise significantly decreases when the cell moves away from the interface of the waveguide. Since red blood cells can sediment at speed of several micrometers per second [43], this means that for right orientation it would take only a few seconds before the cells move out of the high noise range and not influence the detection of coagulation at all.

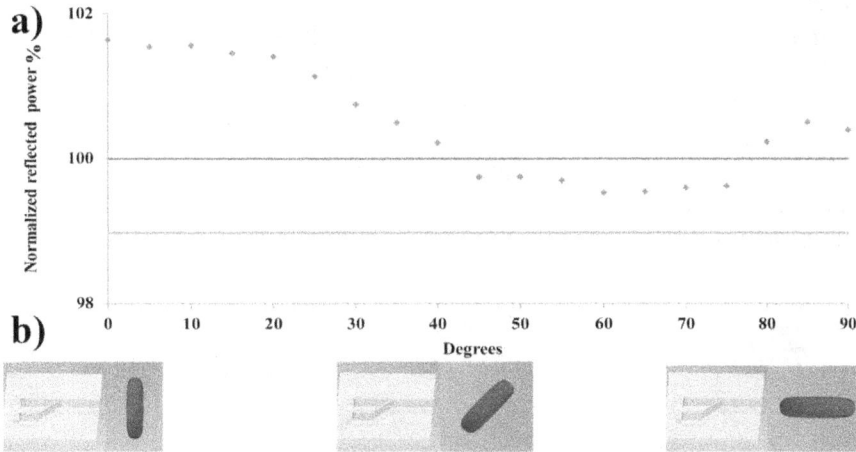

Fig. 8.3. Noise due to presence of cell at different orientations. (a) The normalized reflected power (in %) vs. angle between the cell and the interface of the waveguide. Red line is the reflected power from the uncoagulated blood. Green line is the reflected power from the coagulated blood. Blue is the reflected power in presence of the red blood cell; (b) Explanation of the orientation of the cell relative to the waveguide (not to scale).

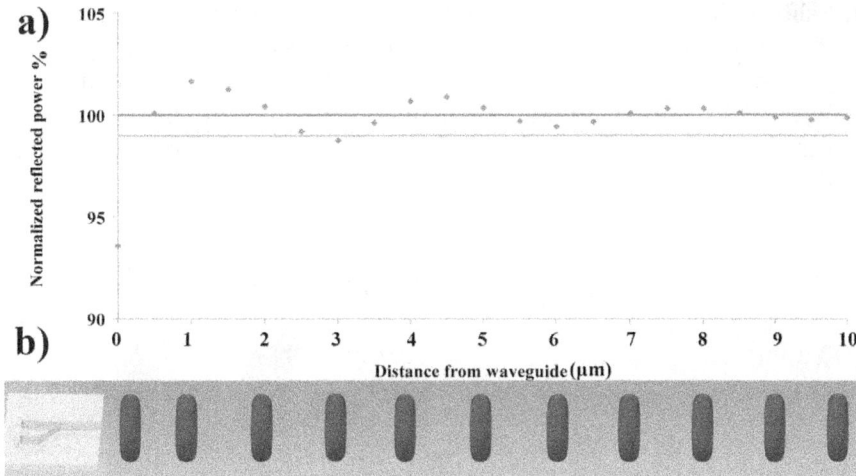

Fig. 8.4. Noise due to presence of cell at different distances from waveguide. (a) The normalized reflected power (in %) vs. distance of the cell from the interface of the waveguide. Red line is the reflected power from the uncoagulated blood, green line is the reflected power from the coagulated blood, blue dots is the reflected power in presence of the red blood cell; (b) Visualization of the cell movement from the interface of the waveguide (picture not to scale).

However this design would not be immune to noise caused by movement of the sensor as the whole what can produce random movement of cells near waveguide interface. Therefore, there is a need to completely eliminate the presence of cells from the interface of the waveguide.

In order to completely eliminate all the noise caused by the presence of the cells a new 3D cladding is proposed for the waveguide that will be filtering all the blood cells (Fig. 8.5) The cladding is 10 μm thick and contains 1μm slit above the core of the waveguide. Red and white blood cells will be filtered out and stay on the surface of the cladding, since they are larger than the size of the opening, and only plasma will be able to reach the reflective waveguide interface. Simulations were conducted for same conditions as earlier simulations. The cell is moved from the face of the waveguide to observe the influence of new design on the noise by cells.

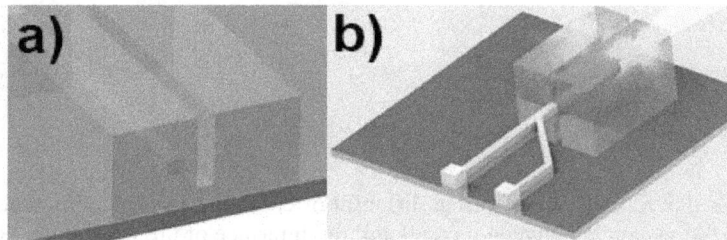

Fig. 8.5. Proposed 3D cladding design. The next iteration of the design with 3D tall cladding for filtering cells from the sensing interface of the waveguide.

Fig. 8.6 (a) shows the refractive index profile in cross section of the proposed 3D cell filtering cladding with 1 μm wide slit giving access to plasma and filtering out cells. Fig. 8.6 (b) demonstrates the mode inside the waveguide. The tail of the mode does not propagate far from the core, therefore cells on the surface of the cladding will be virtually invisible to the sensor. In order to confirm this statement additional simulations were conducted for this device configuration.

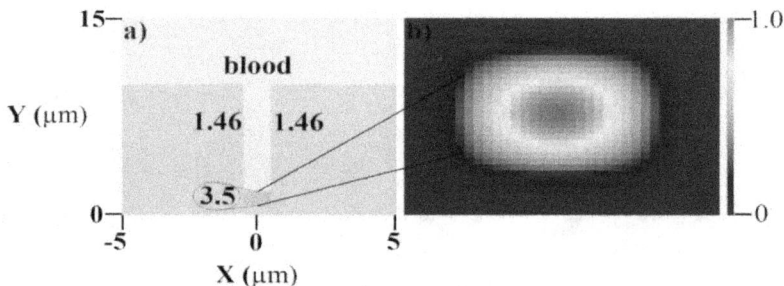

Fig. 8.6. Refractive index and mode of proposed design. (a) The refractive index plot of the blood coagulation sensor. Si, n =3.5, represented by cyan color, cladding SiO_2, n=1.46, represented by orange color and surrounded by blood, n = 1.345, represented by yellow color; (b) The mode supported by the Si waveguide.

Fig. 8.7 shows the normalized power vs the distance between the cell and the waveguide for the configurations without and with the cladding. It can be noticed that with the cladding all the noise due to the presence of the cell is completely eliminated. Since the cells are completely removed from the interface of the waveguide, the reflected power is coming only from the blood plasma completely ignoring cells.

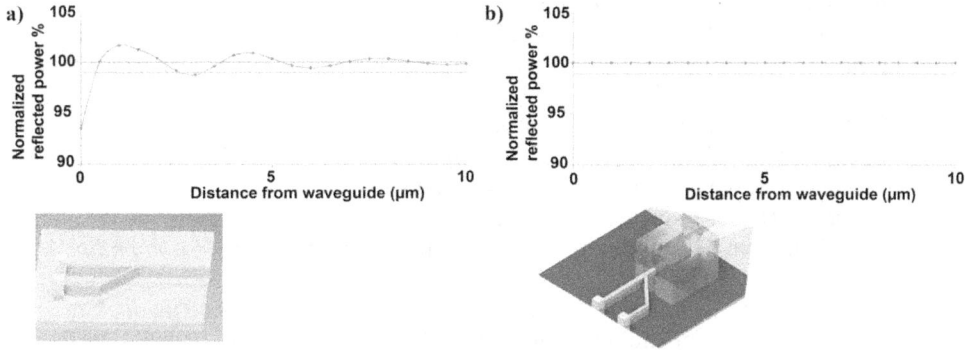

Fig. 8.7. Reduction of noise due to 3D cladding design. The normalized reflected power (in %) vs. distance of the cell from the interface of the waveguide. Red is the reflected power from the uncoagulated blood. Green is the reflected power from the coagulated blood. Blue is the reflected power in presence of the red blood cell for designs without (a), and with the 3D cladding (b).

8.4. Conclusions

Design and optimization of a new on-chip photonic sensor for continuous coagulation monitoring in whole blood was presented. Three dimensional FDTD simulations were used for the design optimization. It was observed that reflection from a waveguide interface can be used to monitor increasing reflective index of coagulating blood, but that cells would generate noise with the amplitude larger than the signal to measure. Consequently, new design has been proposed. It included incorporation of tall waveguide claddings containing small slits that can filter out the cells, but give access to the plasma. It was then demonstrated that the structure can be used to completely eliminate noise caused by the cells while accurately measuring the increase of refractive index in blood plasma.

Acknowledgements

The authors thank Harry Tuazon for the Solidworks modelliing. Surya Cheemalapati was supported by the USF Signature Research Doctoral Fellowship.

References

[1]. R. Madan, B. Gupt, S. Saluja, U. Kansra, B. Tripathi, and B. Guliani, Coagulation profile in diabetes and its association with diabetic microvascular complications, *J, Assoc. Physicians India,* Vol. 58, 2010, pp. 481-484.

[2]. O. Badulescu, C. Badescu, I. Alexa, C. Manuela, and M. Badescu, Coagulation Profile in Patients with Type 2 Diabetes Mellitus and Cardiovascular Disease, *Haematologica-the Hematology Journal,* Vol. 95, June 2010, pp. 766-766.

[3]. N. Key, M. Makris, D. O'Shaughnessy, and D. Lillicrap, Practical hemostasis and thrombosis: *Wiley Online Library,* 2009.

[4]. R. Rodvien and C. H. Mielke, Role of Platelets in Hemostasis and Thrombosis, *Western Journal of Medicine,* Vol. 125, 1976, pp. 181-186.

[5]. N. Mackman, R. E. Tilley, and N. S. Key, Role of the extrinsic pathway of blood coagulation in hemostasis and thrombosis, *Arteriosclerosis Thrombosis and Vascular Biology,* Vol. 27, August 2007, pp. 1687-1693.

[6]. J. P. Riddel, B. E. Aouizerat, C. Miaskowski, and D. P. Lillicrap, Theories of blood coagulation, *Journal of Pediatric Oncology Nursing,* Vol. 24, May-June 2007, pp. 123-131.

[7]. P. H. B. Bolton-Maggs, D. J. Perry, E. A. Chalmers, L. A. Parapia, J. T. Wilde, M. D. Williams, *et al.,* The rare coagulation disorders - review with guidelines for management from the United Kingdom Haemophilia Centre Doctors' Organisation, *Haemophilia,* Vol. 10, September 2004, pp. 593-628.

[8]. K. Nelson, I. Thethi, J. Cunanan, D. Hoppensteadt, R. Bajwa, J. Fareed, *et al.,* Upregulation of Surrogate Markers of Inflammation and Thrombogenesis in Patients With ESRD: Pathophysiologic and Therapeutic Implications, *Clinical and Applied Thrombosis-Hemostasis,* Vol. 17, June 2011, pp. 302-304.

[9]. B. Menendez-Jandula, J. C. Souto, A. Oliver, I. Montserrat, M. Quintana, I. Glch, *et al.,* Comparing self-management of oral anticoagulant therapy with clinic management - A randomized trial, *Annals of Internal Medicine,* Vol. 142, January 4, 2005, pp. 1-10.

[10]. A. J. Quick, M. Stanley-Browne, and F. W. Bancroft, A study of the coagulation defect in hemophilia and in jaundice, *American Journal of the Medical Sciences,* Vol. 190, 1935, pp. 501-511.

[11]. P. A. Owren and K. Aas, The Control of Dicumarol Therapy and the Quantitative Determination of Prothrombin and Proconvertin, *Scandinavian Journal of Clinical & Laboratory Investigation,* Vol. 3, 1951, pp. 201-208.

[12]. L. B. Lowe, S. H. Brewer, S. Kramer, R. R. Fuierer, G. G. Qian, C. O. Agbasi-Porter, *et al.,* Laser-induced temperature jump electrochemistry on gold nanoparticle-coated electrodes, *Journal of the American Chemical Society,* Vol. 125, November 26, 2003, pp. 14258-14259.

[13]. W. Korte, S. Clarke, and J. B. Lefkowitz, Short activated partial thromboplastin times are related to increased thrombin generation and an increased risk for thromboembolism, *American Journal of Clinical Pathology,* Vol. 113, January 2000, pp. 123-127.

[14]. S. M. Donahue and C. M. Otto, Thromboelastography: a tool for measuring hypercoagulability, hypocoagulability, and fibrinolysis, *Journal of Veterinary Emergency and Critical Care,* Vol. 15, March 2005, pp. 9-16.

[15]. H. Lim, J. Nam, Y. Lee, S. Xue, S. Chung, and S. Shin, Blood Coagulation study using light-transmission method, *Korea University,* 2010.

[16]. T. P. Vikinge, K. M. Hansson, P. Sandstrom, B. Liedberg, T. L. Lindahl, I. Lundstrom, *et al.,* Comparison of surface plasmon resonance and quartz crystal microbalance in the study of whole blood and plasma coagulation, *Biosensors & Bioelectronics,* Vol. 15, December 2000, pp. 605-613.

[17]. K. Hansson, T. P. Vikinge, M. Rånby, P. Tengvall, I. Lundström, K. Johansen, *et al.,* Surface plasmon resonance (SPR) analysis of coagulation in whole blood with application in prothrombin time assay, *Biosensors and Bioelectronics,* Vol. 14, 1999, pp. 671-682.

[18]. V. Kalchenko, A. Brill, M. Bayewitch, I. Fine, V. Zharov, E. Galanzha, *et al.,* In vivo dynamic light scattering imaging of blood coagulation, *Journal of biomedical optics,* Vol. 12, 2007, pp. 052002-052002-4.

[19]. A. Ur, Changes in the electrical impedance of blood during coagulation, *Nature,* 226, 1970, pp. 269 - 270.

[20]. K. F. Lei, K. H. Chen, P. H. Tsui, and N. M. Tsang, Real-Time Electrical Impedimetric Monitoring of Blood Coagulation Process under Temperature and Hematocrit Variations Conducted in a Microfluidic Chip, *Plos One,* Vol. 8, October 7, 2013, p. e76243.

[21]. L. G. Puckett, G. Barrett, D. Kouzoudis, C. Grimes, and L. G. Bachas, Monitoring blood coagulation with magnetoelastic sensors, *Biosensors and Bioelectronics,* Vol. 18, 2003, pp. 675-681.

[22]. B. Morales Cruzado, S. Vázquez y Montiel, and J. A. Delgado Atencio, Behavior of optical properties of coagulated blood sample at 633 nm wavelength, 2011, pp. 78970S-78970S-11.

[23]. M. M. Tripathi, Z. Hajjarian, E. M. Van Cott, and S. K. Nadkarni, Assessing blood coagulation status with laser speckle rheology, *Biomedical Optics Express,* Vol. 5, 2014, pp. 817-831.

[24]. W. Liang, Y. Y. Huang, Y. Xu, R. K. Lee, and A. Yariv, Highly sensitive fiber Bragg grating refractive index sensors, *Applied Physics Letters,* Vol. 86, April, 11 2005.

[25]. J. Stewart, A. Pyayt, Statistical analysis of photonic crystal spectra for the independent determination of the size and refractive index of cells, *Langmuir,* Vol. 31, June 15, 2015, pp. 7173-7177.

[26]. A. Pyayt, J. Zhou, A. Chen, J. Luo, S. Hau, A. Jen, L. Dalton, Electro-optic polymer microring resonators made by photobleaching, *Integrated Optoelectronic Devices,* February 8, 2007, pp. 64700Y-64700Y.

[27]. J.-F. Ding, L.-Y. Shao, J.-H. Yan, and S. He, Fiber-taper seeded long-period grating pair as a highly sensitive refractive-index sensor, *Photonics Technology Letters, IEEE,* Vol. 17, 2005, pp. 1247-1249.

[28]. R. Jorgenson and S. Yee, A fiber-optic chemical sensor based on surface plasmon resonance, *Sensors and Actuators B: Chemical,* Vol. 12, 1993, pp. 213-220.

[29]. J. Stewart, A. Pyayt, Photonic crystal based microscale flow cytometry, *Optics Express,* Vol. 22, 11, June 2, 2014, pp. 12853-60.

[30]. A. Pyayt, X. Zhang, J. Luo, A. Jen, L. Dalton, A. Chen, Optical micro-resonator chemical sensor, *Defense and Security Symposium,* April 27, 2007, pp. 65561D-65561D.

[31]. Y. Wang, D. Wang, M. Yang, W. Hong, and P. Lu, Refractive index sensor based on a microhole in single-mode fiber created by the use of femtosecond laser micromachining, *Optics Letters,* Vol. 34, pp. 3328-3330, 2009.

[32]. A. P. Zhang, G. F. Yan, S. R. Gao, S. L. He, B. Kim, J. Im, *et al.,* Microfluidic refractive-index sensors based on small-hole microstructured optical fiber Bragg gratings, *Applied Physics Letters,* Vol. 98, 2011.

[33]. E. Archibong, J. Stewart, A. Pyayt, Optofluidic spectroscopy integrated on optical fiber platform, *Sensing and Bio-Sensing Research,* Vol. 3, 2015, pp. 1-6.

[34]. Z. Tian, S. S. Yam, and H.-P. Loock, Refractive index sensor based on an abrupt taper Michelson interferometer in a single-mode fiber, *Optics Letters,* Vol. 33, 2008, pp. 1105-1107.

[35]. J. K. Barton, D. P. Popok, and J. F. Black, Thermal analysis of blood undergoing laser photocoagulation, *IEEE Journal of Selected Topics in Quantum Electronics,* Vol. 7, 2001, pp. 936-943.

[36]. M. Jedrzejewska-Szczerska, Measurement of complex refractive index of human blood by low-coherence interferometry, *European Physical Journal-Special Topics,* Vol. 222, October 2013, pp. 2367-2372.

[37]. R. Drezek, A. Dunn, and R. Richards-Kortum, A pulsed finite-difference time-domain (FDTD) method for calculating light scattering from biological cells over broad wavelength ranges, *Optics Express,* Vol. 6, 2000, pp. 147-157.

[38]. S. A. Campbell, The science and engineering of microelectronic fabrication, *Oxford University Press,* USA, 1996.

[39]. S. Cheemalapati, M. Ladanov, J. Winskas, and A. Pyayt, Optimization of dry etching parameters for fabrication of polysilicon waveguides with smooth sidewall using a capacitively coupled plasma reactor, *Applied Optics,* Vol. 53, 2014, pp. 5745-5749.

[40]. F. J. Barrera, B. Yust, L. C. Mimun, K. L. Nash, A. T. Tsin, and D. K. Sardar, Optical and spectroscopic properties of human whole blood and plasma with and without Y_2O_3 and $Nd_3+:Y_2O_3$ nanoparticles, *Lasers in Medical Science,* Vol. 28, November 2013, pp. 1559-1566.

[41]. B. Rappaz, A. Barbul, F. Charriere, J. Kühn, P. Marquet, R. Korenstein*, et al.,* Erythrocytes volume and refractive index measurement with a digital holographic microscope, in *Proceedings of the Biomedical Optics conference (BiOS' 07)*, 2007, pp. 644509-644509-5.

[42]. M. L. Turgeon, Clinical Hematology: Theory and Procedures, *Lippincott Williams & Wilkins,* 2005.

[43]. G. J. Tangelder, D. W. Slaaf, A. M. M. Muijtjens, T. Arts, M. G. A. O. Egbrink, and R. S. Reneman, Velocity Profiles of Blood-Platelets and Red-Blood-Cells Flowing in Arterioles of the Rabbit Mesentery, *Circulation Research,* Vol. 59, November 1986, pp. 505-514.

Chapter 9
Surface-Enhanced Raman Scattering: A Novel Tool for Biomedical Applications

Sudhir Cherukulappurath

9.1. Introduction

Surface-enhanced Raman scattering (SERS) is a powerful non-invasive spectroscopic technique that has recently gained lot of interest in molecular detection and identification owing to its high sensitivity and portability. Several applications that require biomolecular detection such as in medical biotechnology and pharmaceutical studies are now turning towards SERS as a reliable means to identify molecules. In Raman scattering, incident lights interact is made to interact with the analyte molecules resulting in a radiative scattering of the photons with not only the incident frequency (called elastic Rayleigh scattering) but also slightly shifted frequencies (termed as inelastic Raman scattering). The frequencies of Raman scattered photons can be smaller (Stokes shift) or greater (anti-Stokes shift) than that of the incident excitation light. As the probability of Raman scattering is very low (of the order of $10^{-6)}$, it is essential to have either a large concentration of molecules or a highly intense laser light for excitation in order to detect the Raman signal and proper chemical identification. However, using SERS, there is a huge enhancement of the scattered light that can be easily detected.

The first observation of enhanced Raman signals was reported in 1974 by Fleischmann et al. while studying the pyridine molecules adsorbed onto silver electrodes [1]. An explanation to the phenomenon was later given by Jeanmaire and Van Duyne in 1977 [2]. This phenomenon, later termed as SERS, is now a widely-used method in applications involving biomolecular detection of low concentration. SERS has now developed into a mature field and with the advent of state-of-the-art nanofabrication techniques, single-molecule Raman spectroscopy has been possible [3-4]. The application of SERS has moved from physics to material science, chemistry, environmental studies and more recently to biomedical applications. There are several reviews on SERS in general which form useful guide to the advancements in the field [5-10]. This chapter, an extension of

Sudhir Cherukulappurath
Physics Department, Goa University, Goa, India

the review paper [11] attempts to highlight the developments in SERS research with special reference to biomedical applications.

The first part of the chapter deals with the theory of SERS and discusses the possible mechanisms for the observed enhancement of Raman signals in SERS. Next section deals with a brief review of different substrates reported for SERS applications. SERS based biomolecular detection is currently a hot topic owing to its potential applications and will be discussed in the successive section. The sections after will deal with biomedical applications of SERS, glucose sensors based on SERS including DNA/RNA detection, immunoglobin protein detection and other relevant topics. This chapter will conclude with the basic challenges and future prospects.

9.2. Mechanism of SERS

After the first observation of enhancements in Raman scattered light by Fleischmann et al., experiments by Van Duyne group revealed that surface enhancements were due to the chemical adsorption of the molecules onto rough silver surfaces. The drastic increase in the Raman intensity of adsorbed molecules was attributed to increased surface electromagnetic fields on the metal surface. At the same time, Albrecht and Creighton had already published a report on the observation of intense Raman signals from pyridine on silver electrodes [12]. It was argued that a broadening mechanism of the excited states of the adsorbed molecules due to the presence of metal surface was responsible for the enhancements and the role of surface plasmons was speculated. Since then there has been several explanations of which some of them are based on experimental observations.

Large enhancement of Raman signals from molecules in the vicinity of metal surfaces forms the basis of SERS. While the explanation given to the first observation of Raman enhancement was based on electrochemical changes to the molecule on adhesion to the metal surface [1], now it is largely believed that surface plasmons on the metal surface has a big role [2, 13-15]. The two primary mechanisms responsible for observing SERS are (1) enhancement of local electromagnetic field due to surface plasmons and (2) chemical enhancement attributed to charge transfer mechanism.

Surface plasmons are periodic electromagnetic oscillations of the conduction electrons on a metal surface [16]. Photons can interact with surface plasmons leading to interesting effects. The field of study of surface plasmons, called plasmonics, is now very established branch of nanophotonics [17]. Noble metals such as Au, Ag are some of the most popular plasmonic metals that are used in plasmonic applications. The conduction electrons in the surface of plasmonic metals can oscillate with the same frequency as the incoming photons for a certain band of frequencies. When the frequency of the incident light is beyond a threshold frequency, these conduction electrons can no longer match the drive frequency and will tend to slow down. This usually happens for high frequency ultraviolet light and for the same reason, most plasmonic effects are pronounced in the visible part of the electromagnetic spectrum. Metal nanoparticles (usually with sizes lower than the incident wavelength) such as nanospheres have a definite geometry and thus confines the conduction electrons inside this boundary. The resonance frequency of these 'plasma' electrons largely depend on the material properties such as dielectric function of the metal as well its surrounding region as well as its geometry. This resonance is often called localized surface plasmon resonance (LSPR) and plays a crucial role in surface-enhanced processes such as SERS. The oscillating dipole nature of the plasmons create a secondary field around the metal particle often termed as 'local field' (Fig. 9.1). This in turn leads to enhanced scattering, absorption and extinction of the incident light by the plasmonic metal particle.

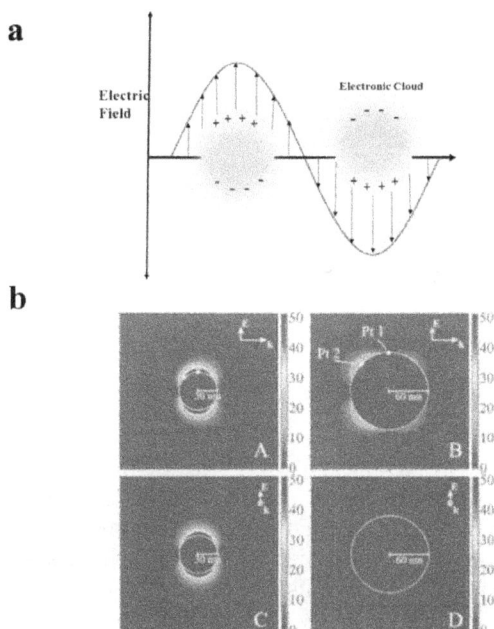

Fig. 9.1. Localized surface plasmons. (a) Schematic of localized surface plasmon oscillations on excitation with light. The electron cloud oscillates with the external electric field thereby creating a polarization of charges; (b) Simulated image of the electric field around nanoparticles of different sizes (30 nm and 60 nm radius). Reprinted with permission from [23]. Copyright (2016) American Chemical Society.

As the momentum of surface plasmons is higher than that of free photons, in order to excite and couple surface plasmons on a thin metal film with light, special schemes need to be adopted. However, this condition is not required for LSPR excitation and hence is widely used for SERS. Since most SERS platforms are based on metal nanoparticles rather than thin films, only LSPR will be discussed here.

It is now widely accepted that apart from the physical electromagnetic enhancement of the local field, a chemical enhancement process also contributes to SERS [18-20]. Several reports on experimental observation of resonant Raman scattering on molecules adsorbed onto a metal surface suggest that there is indeed a broadening of the electronic states of the adsorbate and new intermediate levels are formed due to this interaction of the analyte with the metal surface [21-22]. This 'charge transfer mechanism' is responsible to an enhancement in the scattering. Although small (of the order of 10^3), chemical enhancement do contribute to the total enhancement in SERS. In the chemical enhancement theory, it is hypothesized that the highest occupied molecular orbital (HOMO) and the lowest unoccupied molecular orbital (LUMO) tends to broaden out thereby bringing them closer to the Fermi energy level (Fig. 9.2). This facilitates charge transfer from either the molecule to the metal or vice versa when excited with light. As a result, the polarizability of the molecule is enhanced up to 1000-fold when the excitation photons are in resonance with the charge transfer energy bands.

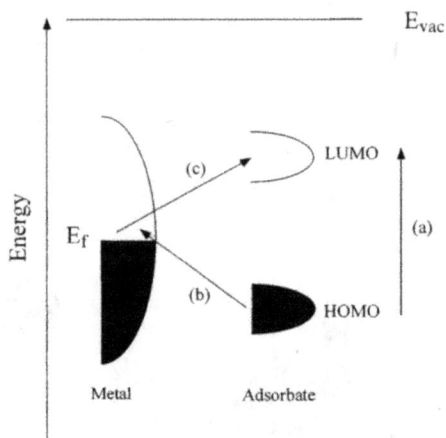

Fig. 9.2. Orbital energy diagram of a molecule adsorbed to metal surface. HOMO and LUMO levels are broadened due to the interaction thereby allowing charge transfer excitations. Reproduced from [19] with permission of The Royal Society of Chemistry.

On the other hand, electromagnetic enhancements due to surface plasmon excitations can be of several orders of magnitude larger than the chemical enhancement mechanism discussed above. At plasmonic resonance, the local electromagnetic field around the metal surface is locally enhanced thereby increasing the Raman signals [23-24]. This enhancement mechanism can be easily visualized in a classical physics point of view. The free-electron theory of metals is a good approximation in plasmonic models and can be used to explain the theory of surface plasmons without resorting to quantum mechanical treatments. The free-conduction electrons in the metal surface gets perturbed by the external electric field (usually light). This brings changes in the probability densities of the electronic wavefunction resulting in a change in the dipole moment, **P**. This induced dipole moment is directly proportional to the external electric field **E** through the constant called polarizibility α:

$$P = \alpha E \qquad (9.1)$$

It should be noted these quantities are tensors in 3D space. The local electric field interacts with the polarizability of the molecule in the vicinity of the field. This interaction leads to the inelastic scattering of incident photons collected as Raman spectrum.

Since the electric field intensity around a plasmonic nanoparticle is enhanced and the intensity of the scattered photons is related to the square of the incident intensity the overall SERS intensity is related to the induced field through [10, 25]:

$$I_{SERS} \approx |E(\omega_{inc})|^4, \qquad (9.2)$$

where I_{SERS} is the SERS intensity and ω_{inc} represents the frequency of incident excitation.

Hence if there is a 100-fold increase in the local electric field the Raman intensity will be increased by a factor of 10^8. This factor is termed 'enhancement factor' (EF) and is often quoted in SERS experiments. EF is usually determined by taking the ratio of Raman intensities with and without the plasmonic field normalized to the number of molecules on the surface.

In experiments this can obtained by comparing the SERS intensity with the Raman intensity of bulk molecules after normalizing for the number of molecules.

$$EF = \frac{I_{SERS}/N_{surf}}{I_{bulk}/N_{vol}}, \qquad (9.3)$$

where I_{bulk} represents the intensity of Raman spectrum in bulk sample while N_{surf} and N_{vol} are the respective number density of molecules.

The intensity of SERS signals decay with distance from the metallic surface. This is expected as surface plasmon fields responsible for the enhancement of Raman signals are evanescent. The molecule does not have to be in direct contact with the metal surface but need to be in the vicinity of the plasmonic field which is usually few nm from the metal surface. As the plasmonic field decays exponentially from the surface, it is necessary that the analyte molecules are close enough, usually few nanometers, to the metal surface in order to obtain SERS signal. Experimental evidences for this statement has been reported in which spacer layers of varying thickness was used between the molecules and the metal surface [26].

The experimental set up for SERS measurements can be very similar to conventional Raman systems. The excitation light, usually a low power laser, is focused onto the nanoparticle substrate which also contains the analyte molecules. The analyte molecules are either attached to the metal surface via chemical adsorption, covalent bonds or through a partition layer system. For resonant excitation, the wavelength of laser excitation should overlap with the plasmonic resonance of the nanoparticles. This optimizes the scattering process and increases the signal-to-noise ratio. The scattered signal is detected using a spectrometer system as in regular Raman spectroscopy.

It is interesting to note that SERS spectrum of a molecule can be different from the Raman signal. For example, in SERS, the intensity of higher frequency vibrations tends to lower. Overtones and overlapped bands are not observed in SERS. The resonances can be broadened and even slightly shifted when the molecule is adhered to the metal surface. Certain selection rules are either relaxed or altered and there seems to be a depolarization of the scattered light unlike bulk Raman spectra. However, the prominent peaks of Raman scattering are obtained as in conventional Raman spectra.

161

9.3. Plasmonic Nanostructures for SERS

The substrates used for SERS plays a vital role in determining the quality and quantity of Raman signals measured from the analyte molecules. The first reported observation of intense Raman scattering was on pyridine molecules adsorbed on to a simple silver electrode prepared by electrodeposition process. The inherent random roughness on the silver surface contributed to the enhanced Raman signals. Since then several different geometries of metallic nanoparticles have been studied for SERS, Ag and Au being the most popular metals used in these substrates. Single metal nanoparticles, by themselves can be used to as Raman enhancers but the most effective way is to couple nanoparticles to create stronger and larger number of electromagnetic 'hotspots'. The strong coupling of local electromagnetic field of the metal nanoparticles gives rise to intense enhancements of the local fields forming the 'hotspots'. Molecules that are in the vicinity of such coupled fields tend to show large enhancements of Raman signals.

Recent advances in nanofabrication technology has contributed to the ability of creating special geometries that can give rise to large EFs. For example, the popular metal film-over-nanospheres (FONs) that are fabricated through nanosphere lithography (NSL) process have been reported to present strong hot spots capable of observing SERS from single molecules [27-30. Van Duyne and coworkers have reported several measurements based on Ag-FONs and Au-FONs SERS substrates [31-32]. Although easy to fabricate, FON substrates presents spatial inhomogenieties, which puts other structures that show similar trend includes silica coated metallic star nanoparticles [33-34], dimers [35], nanoparticle clusters [36], shell-isolated nanoparticles (SHINERS) [37-38], Ag cage structures [39], mushrooms [40], ALD coated nanoparticles [41] and nanowire structures [42]. Some of the most popular SERS nanostructures are shown in Fig. 9.3.

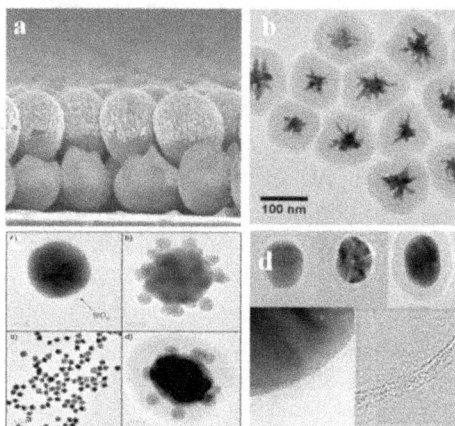

Fig. 9.3. Different nanostructures used for SERS (a) Metal film over nanostructures (FON). SEM image of AgFONs. Reprinted with permission from [29]. Copyright (2016). American Chemical Society (b) Silica coated nanostars Reprinted with permission from ([33]). Copyright (2016). American Chemical Society (c) 3D self-assembled plasmonic superstructures. From [36] with permission from John Wiley and Sons. (d) Silica shell isolated nanoparticles (SHINERS). Reprinted by permission from Macmillan Publishers Ltd: Nature [38], Copyright (2016).

It has been known that rough surfaces and random cluster of nanoparticles can give SERS signals, but a more efficient way to yield high EFs will be to properly engineer the size and distribution of the nanoparticles such that their plasmonic resonances match with the excitation light wavelength. In this regard, dimer plasmonic nanostructures are of particular interest as SERS substrates owing to the fact that a strong electromagnetic hotspot is formed at the nanogap between the two nanostructures. For example, the metal over FON structure mentioned earlier presents such highly localized hot spots thus becoming a popular substrate for SERS applications.

Tip-enhanced Raman scattering (TERS) is now becoming a hot research branch of SERS owing to its high-resolution capability and several papers on TERS for biomedical application are available [43-46]. Here, a metal coated tip is used as probe and it is illuminated with laser light to induce surface plasmon fields at the tip. The tip can then be raster scanned over the molecules under study and the scattered signals after interaction with the molecules are collected for detection. Raman-images of single molecules with super-resolution can be achieved using TERS microscopy. However, there are certain drawbacks of this techniques which includes cumbersome measurement set-up, poor reproducibility and low signal-to-noise ratio that need to be addressed.

Although Ag and Au are the most widely used for SERS, other metals such as Al, Cu, alkali metals (Li, Na, K, and Cs), Pt, Ga, In and some alloys have been tried and tested. Owing to its plasmonic resonances in the UV, Aluminum is found to be effective for UV-based SERS measurements. However, high reactivity (including large susceptibility to oxidation) and cost has restricted their use as SERS substrates.

9.4. SERS Based Biomolecular Detection

SERS has now become a well-developed and mature technique for the detection of biomolecules offering good sensitivity as well as selectivity. SERS based sensors have been reported for diagnosis and treatment of cancer, Alzheimer's and Parkinson's diseases. Large molecule sensing by SERS has shown reasonable interest owing to greater sensitivity and cost-effectiveness. In the following section, a review of some popular SERS based detection schemes are presented.

9.4.1. SERS Based Biosensing

Detection of biomolecules and bio organisms have been of great interest owing to their importance in biomedical field. In particular label-free detection of microorganisms and pathogens using SERS provide significant advantages over methods such as fluorescent marking. For example, by proper detection and study of certain pathogens, it is possible to develop pharmaceutical drugs that can be effectively used to eliminate them. In this regard, SERS has been developed as a tool for effective detection even for small concentrations of the molecules. Van Duyne el at. reported a rapid detection technique for anthrax biomarkers using SERS [47]. Calcium dipicolinate (CaDPA), considered to be a biomarker for bacillus spores, was detected using SERS based on Ag-FON samples. Spore concentrations of the order of 10^{-12} M was reported to be detected. Following this there has been an exponential growth in the number of publications on bio-detection based on SERS. Recently, several reports on detection of whole microorganisms based on SERS substrates have been published. For example, Boardman et al. have demonstrated the rapid detection of bacteria directly from blood using a novel technique based on SERS [48]. They used Au nanoparticles embedded in SiO_2 as substrate and were able to detect 17 different bacterial species separated from blood samples. In another recent report, E-coli detection was achieved by in-situ coating of the bacterial cell wall with Ag nanoparticles in water followed by SERS measurements [49]. This method offers several advantages such as high sensitivity, reduced measurement times, high portability and lower reaction volumes over conventional label-free bio-detection methods. An interesting work includes the development of lab-on-a-chip device based SERS data base for the differentiation of six mycobacteria including both tuberculosis and non-tuberculosis strains [50]. Such developments clearly aim at utilizing the potential of SERS in medical applications in the future.

9.4.2. Glucose Sensing

In recent years, there has been a concerning level of increase in the number of diabetic patients globally. The failure of insulin response mechanism in such patients cause fluctuations in glucose levels leading to further health related issues. Blood glucose levels are often monitored in diabetic patients usually by finger pricking method. Researchers have been working towards developing non-invasive glucose detection methods aiming at minimizing trauma to the patients. Raman spectroscopic method can detect glucose levels in vitro but requires strong and long laser exposures. This process can be made more efficient using SERS based detection. Ag FON (Silver film-over- nanostructures) have been successfully employed as SERS substrates for the purpose by Van Duyne, et. al [29-30]. Since glucose is not readily adsorbed on to silver film, a special partition layer is created in order to bring the glucose molecules in close vicinity of the metal surface as represented in Fig. 9.4. To achieve this, a self-assembled monolayer (SAM) of 1-decanethiol and mercapto-hexanol (DT/MH) is formed over the metal surface. This assisted in creating a glucose concentration gradient which was then detected using SERS. Several different partition layers were studied by the group but only these straight alkanethiols were found to be effective. Using this technique physiologically relevant glucose levels were detected, thus proving the utility in medical applications.

A further improvement was achieved using spatially offset Raman spectroscopy (SORS) was obtained by the same group. In this method, the scattered light is collected from different regions that are offset from the laser excitation point thereby providing an improved depth in resolution. Combining this with SERS bring a powerful tool called surface enhanced SORS (SESORS) for bio-detection. Van Duyne group successfully used this technique by injecting sensors and monitoring glucose levels through living rat´s skin [51-52]. It has been reported that this method will be a significant tool in biomedical applications in the future.

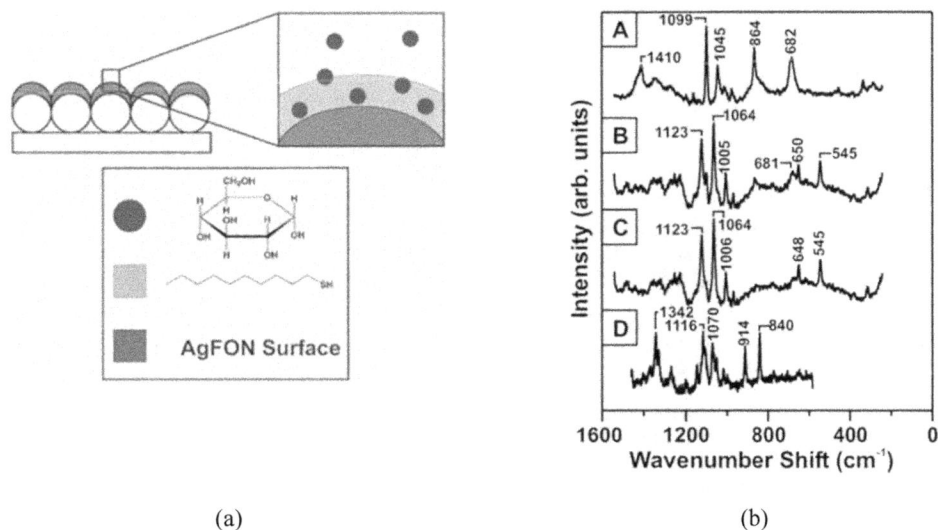

(a)   (b)

Fig. 9.4. Glucose sensing. (a) Scheme for glucose sensing using AgFON substrates. Glucose is partitioned into an alkanethiol monolayer adsorbed on the silver film substrate. (b) SERS spectra from glucose molecules. A and B represents spectra obtained without and with glucose in the partition layer respectively. C is the residual glucose spectrum obtained by subtracting A from B. D represents pure Raman spectrum of glucose for comparison. Reprinted with permission from [31]. Copyright (2016) American Chemical Society.

9.4.3. SERS Markers for DNA/RNA Detection

Recent developments in nanotechnology have facilitated sensitive and selective detection of nucleic acids. This has in turn revolutionized modern biomedical analysis as well as diagnostic tools. SERS based detection of nucleic acids offers several advantages over conventional methods such as fluorescent spectroscopy. It also forms a complementary analysis tool to other sophisticated techniques like NMR and mass spectroscopy. SERS presents better sensitivity with lower limits of detection and greater spatial resolution. Moreover, undesirable effects such as photobleaching and quenching can be reduced to a great extent. In particular, by using excitation wavelengths that spectrally match the electronic absorption bands of the biomolecules, it is possible to improve the efficiency of scattering process. This technique, often termed surface enhanced resonance Raman scattering (SERRS), has now become a potential tool of DNA detection [14].

Earlier methods of DNA detection involved immobilizing the molecules to silver or gold nanoparticles along with a Raman reporter (Fig. 9.5). In order to achieve this, surface functionalization methods were developed in order to attach the DNA strands onto the metal surface, followed by the assembly of the nanoparticles [53-54]. 13 nm Au nanoparticles were linked to oligonucleotides that were functionalized with a thiol group at their tail end. Two non-complimentary oligonucleotide solutions so prepared are then mixed together. Due to their non-complementary nature, there is no reaction. Additional linking of DNA duplex to this components and oligomerization results in the assembly of DNA-linked nanoparticles.

165

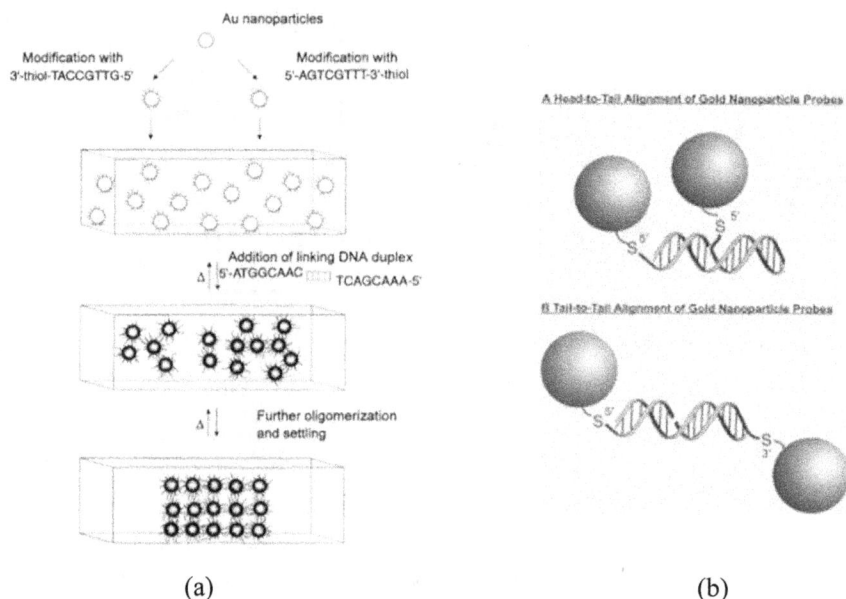

Fig. 9.5. (a) Schematic of the DNA-based nanoparticle assembly [53]. Different schemes for attaching DNA to gold nanoparticles. Reprinted with permission from [54]. Copyright (2016) American Chemical Society.

A multiplexed detection of DNA and RNA using gold nanoparticle probes that were labeled with oligonucleotides and Raman markers was reported [55]. Three component sandwich assay system was utilized in their method (Fig. 9.6). The nanoparticles were attached with cyanine3 (Cy3) thiol-capped oligonucleotides for monitoring different DNA strands. Here, Cy3 was chosen as the Raman tag as it easily hybridizes with the specific DNA strand under investigation. To further enhance the Raman signals silver hydroquinine was passed through forming silver nanoparticles along the Cy3 strands. Several different strands of DNA and RNA were detected using this method with a detection limit of 20M.

To further improve the detection mechanism, some groups have developed techniques where in a fluorescent marker molecule was also attached apart from Raman active reporter. This combined assay system provided better information regarding the DNA. For example, Fang et al. used Rhodamine-B as both Raman tag as well as fluorescent marker [56]. Single strands of DNA were detected by Fabris et al. by first hybridizing DNA with peptide nucleic acid (PNA) which was then immobilized on Ag nanoparticles that were attached with Rhodamine-6G [57]. The detection limit of this method was reported to be of the order of pM concentration (Fig. 9.7).

Multiplexed DNA detection could be achieved by using several different dyes that were excited with a single wavelength [58]. A different approach of using two different wavelengths that matched the electronic absorption of a particular oligonucleotide was also reported by the same group [59].

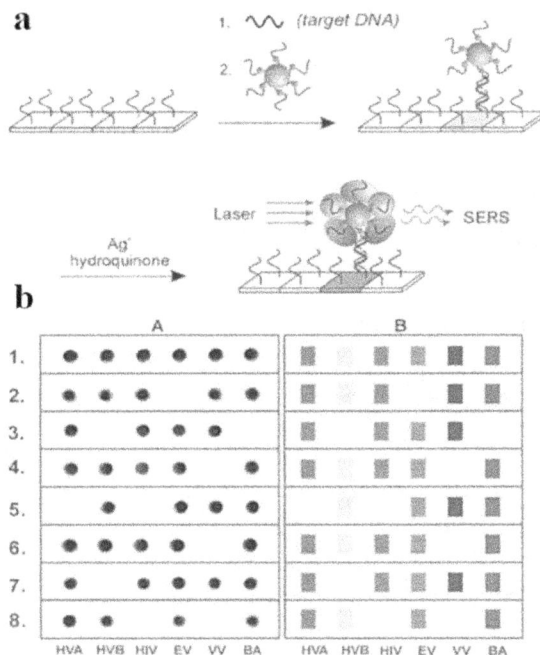

Fig. 9.6. SERRS based DNA detection using gold nanoparticles that are functionalized with dye-labelled oligonucleotide followed by a silver staining. From [55]. Reprinted with permission from AAAS.

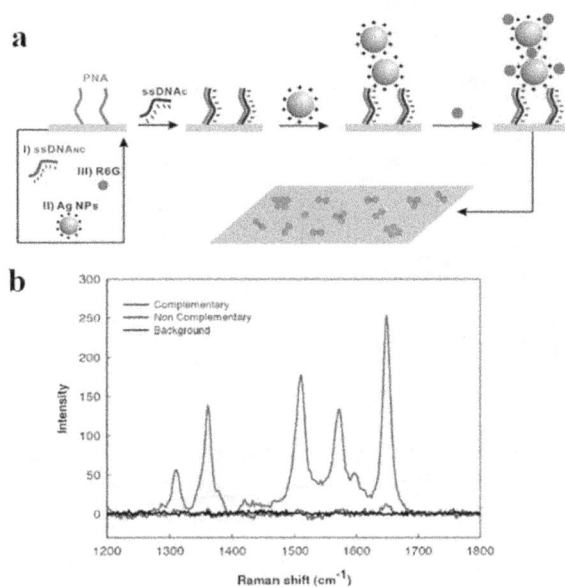

Fig. 9.7. (a) Scheme adopted by Fabris *et al.* for the detection of hybridized DNA. Peptide nucleic acid was used for hybridization and immobilized on AG colloids with Rhodamine-6G; (b) Averaged SERS signals from PNA hybridized DNA. Reprinted (adapted) with permission from [57]. Copyright (2016) American Chemical Society.

167

Another powerful and unique approach is DNA- based self-assembly of plasmonic nanoparticles to enhance Raman scattering. Originally developed by Mirkin, et al. [53-54], this method has been reported for sandwich assay with silver nanoparticles that were coated with oligonucleotides and Raman marker molecules [60]. Graham, et al. presented a controlled aggregation of DNA coated silver nanoparticles through a target-dependent sequence specific DNA hybridization assay. Maximum enhancement of Raman signals was obtained by cleverly placing the Raman scattering molecules in the interstices of the assembled metal nanoparticles.

Label-free approaches for SERS based DNA detection has gained considerable interest and simple mononucleotide detection have been reported. Bell, et al. demonstrated SERS detection of adenine, guanine, thymine, cytosine, and uracil using citrate-reduced silver colloids that were aggregated with $MgSO_4$ [61]. As in the previous method, the analyte mononucleotide can get in the hotspots of the aggregated nanoparticles thereby achieving maximum enhancement of Raman scattered light (Fig. 9.8). SERS from 2'-deoxyadenosine 5'-monophosphate (dAMP) attached to Ag colloids were obtained using this technique.

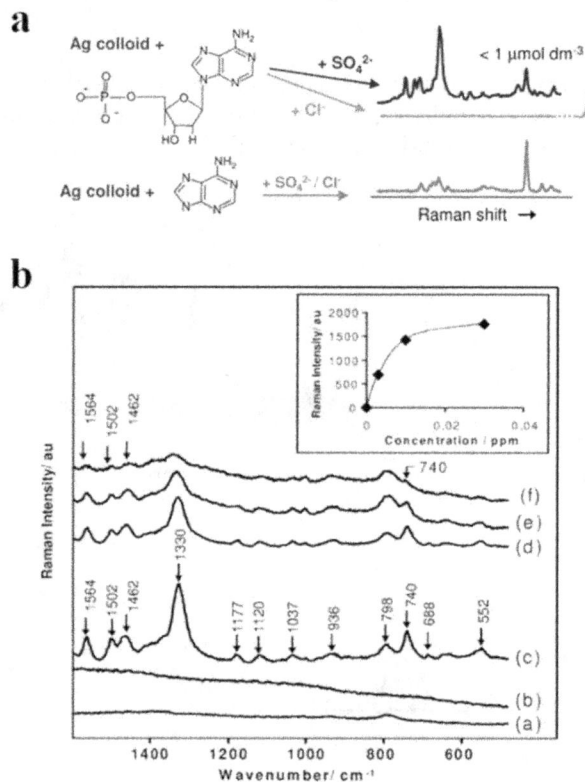

Fig. 9.8. Method for obtaining SERS spectra from DNA/RNA mononucleotide by aggregating citrate reduced Mg colloids with $MgSO_4$. SERS signal of 2'-deoxyadenosine 5'-monophosphate (dAMP) for different concentrations obtained using this method. Reprinted (adapted) with permission from [61]. Copyright (2016) American Chemical Society.

This method was also extended to single base nucleotide mismatch detections in short DNA strands [62]. Single base sensitivity of DNA bases using similar methods have been reported by several groups. Detection of DNA hybridization using label-free methods is useful in forensics and genetic studies. Barhoumi and Halas have demonstrated label-free detection of DNA in hybridized state using the plasmonic properties of Au nanoshells [63]. These Au nanoshells comprise of silica core with a thin film of Au. The dominant adenine peak at 736 cm^{-1} is removed and replaced with its isomer 2-aminopurine (Fig. 9.9). This aminopurine substituted DNA is then adsorbed onto the Au nanoshells using a thiol moiety on its ends. The ratio of intensity of peaks of adenine (at 736 cm^{-1}) and 2-aminopurine (at 807 cm^{-1}) gives a quantitative degree of hybridization.

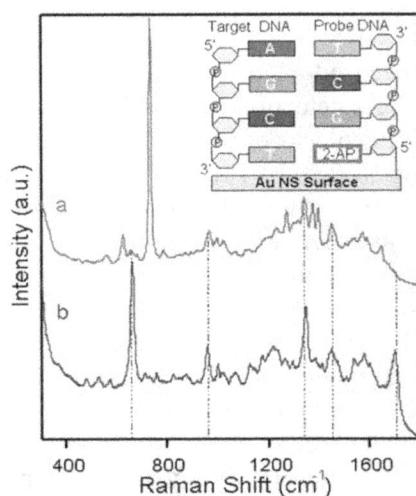

Fig. 9.9. (a) Au nanoshell based SERS spectra for a DNA sequence (a) ST20N1, containing adenine bases, and (b) ST20N2, without the adenine bases. Inset shows the schematic of DNA hybridization. Reprinted (adapted) with permission from [62]. Copyright (2016) American Chemical Society.

Improvements in label-free detection can be achieved by proper control of the plasmonic nanoparticle assembly. Dielectrophoresis has been used for assembly of nanoparticles that have DNA bases attached to them. Adenine molecules adsorbed onto Au nanoparticles were detected using a dynamic dielectrophoresis-enabled assembly of metal nanoparticles in the form of pearl chains with nanometer-sized gaps. As electrophoretic forces overcome diffusion this approach provides a rapid detection scheme with good sensitivity. Low molecular concentrations in the pM range was detected [64-65]. Magnetophoresis of magnetically active SERS nano-reporters (or plasmonically active magnetic nanoparticles) also provides a way to overcome the diffusion-limited assembly on substrates.

The number of reports on SERS based DNA/RNA detection has increased exponentially in the last few years clearly points towards the tremendous potential of the technique and the promises it holds in biomolecular detection.

9.4.4. Immunoglobin Protein Detection Based on SERS

Understanding of biomolecular process in living organisms is crucial not only in modern biology but also in medical science. In particular detection of protein plays an important role in disease diagnosis and cure. Future drug discovery is dependent on protein sensing and analysis. State-of-the-art protein detection includes immunoassay tests, fluorescence readout and microscopic methods. Raman microscopy has been used for study of proteins and their interactions [66-69]. However, conventional Raman spectroscopy suffers from low scattering cross sections, high fluorescent background and the necessity to have larger quantities of samples. Recently surface plasmon-based biosensing has gained reasonable interest owing to its improved sensitivity and portability. Indeed, owing to its advantages, SERS provide an interesting alternative to the above techniques.

Different approaches have been adopted for protein detection using SERS. The most straightforward way is the direct detection of protein molecules by collecting the SERS signal. Amino acids, the building blocks of proteins as well as smaller peptide groups have been well characterized using SERS [70-73]. In these studies, Ag colloids were used as SERS substrates and several homodipeptides that were adsorbed onto the Ag colloids were analyzed. This also gave better insights towards the orientation of adsorbed aminoacids. Stewart, et al. studied peptides and aminoacids adsorbed onto electrochemically prepared silver surface [74] while Hu, et al. used silver colloid to obtain SERS from lysosomes [75]. Water soluble proteins and dipeptides were studied by Chumanev, et al. [76]. Several other studies on small protein SERS were also reported [77-79]. Ozaki's group have studied enzymes such as lysozyme, ribonuclease B, avidin, catalase, hemoglobin, and cytochrome using SERS (Fig. 9.10). The enzymes were adsorbed onto colloidal silver after mixing acidifed sulphate, which enhanced the detection limits [80].

Large protein molecules can often show complicated SERs signal that makes identification difficult. In such cases, the complete pattern of SERS peaks is taken and analyzed instead of looking for single vibrational signatures. In addition to the direct (intrinsic) SERS measurements of proteins, it is possible to add reporter molecules to the proteins and then measure SERS (extrinsic). Some of the most common SERS reporter molecules include 5,50-dithiobis(succinimidyl-2-nitrobenzoate) (DSNB) with a peak at 1336 cm-1 shift [81], 5,5'-dithiobis(2-nitrobenzoic acid) (DTNB) with a peak at 1342 cm-1 shift [82], 4-mercaptobenzoic acid (MBA) with 1585 cm-1 peak [83], 4-nitrobenzenethiol (4-NBT) at 1336 cm-1 shift , 2-methoxybenzenethiol (2-MeOBT) with intensity monitored at 1037 cm-1 shift, 3-methoxybenzenethiol (3-MeOBT) with intensity monitored at 992 cm-1 shift, and 2-napthalenethiol (NT) with intensity monitored at 1384 cm-1 shift [84].

Sandwich immunoassay is a very common way for protein detection. An immunoassay based SERS study was first reported by Tarcha et al. where they measured SERS spectra from immunoassay of thyroid stimulating hormone (TSH) [85]. Grubisha et al. used a novel reagent consisting of reagent consists of gold nanoparticles that were modified to integrate bioselective species (e.g., antibodies) with molecular labels for the generation of strong, biolyte-selective SERS signals [86]. Gold-coated glass substrates are

170

functionalized with the target antibody and it is then exposed to the solution containing the corresponding antigens. A sandwich complex assay is formed when Raman-labelled metal colloidal solution is added. Detection of femtomolar concentration of prostate-specific antigen (PSA) using SERS was reported by authors (Fig. 9.11). This method allows in vitro early diagnosis for certain cancers in a very short time interval.

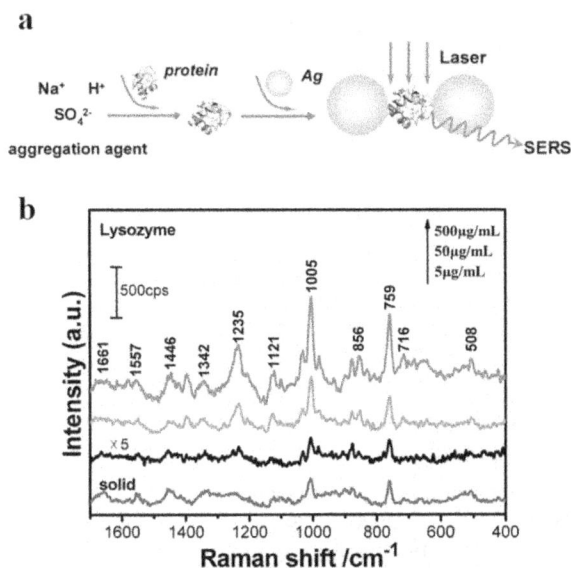

Fig. 9.10. (a) Schematic of the protocol used in the aggregation of Ag colloids for label-free protein detection. (b) SERS spectra from catalyze and control experiment spectra. Reprinted (adapted) with permission from [80]. Copyright (2016) American Chemical Society.

Raman markers were used for the detection of thrombin at subpicomolar concentrations using a protein-protein recognition system containing gold nanoparticles that were capped with a bifunctional molecule [87]. This molecule is capable of forming a covalent link with the aromatic residues of the protein moiety. Certain vibration bands of this link could be enhanced by the gold nanoparticle thereby detecting thrombin. A detection limit of 10-13 M was reported by the authors using this method. In fact, gold nanoparticles play a vital role in SERS detection systems and an extensive review of SERS nanoparticles for medical applications can be seen in reference [88]. In another interesting work, SERS based microscopy was used to image the selective localization of PSA in a prostate tissue. Gold nanostar particles were conjugated to an antibody against the tumor suppressor and white light immunization and scanning gave the Raman image of the PSA localization [89-90]. Histopathological analysis requires the localization of certain tissues using immunohistochemistry. In this work, gold nanostars that were fabricated using colloidal chemistry methods were conjugated with tumor suppressor p63, a p53 homologue. A white light source was illuminated onto the tissue for imaging. The image obtained when overlapped with the false color SERS image shows the presence of basal cells of the benign prostate. This demonstration of protein detection using SERS imaging shows the potential of this method to become a medical tool for early diagnosis of several diseases including cancer.

171

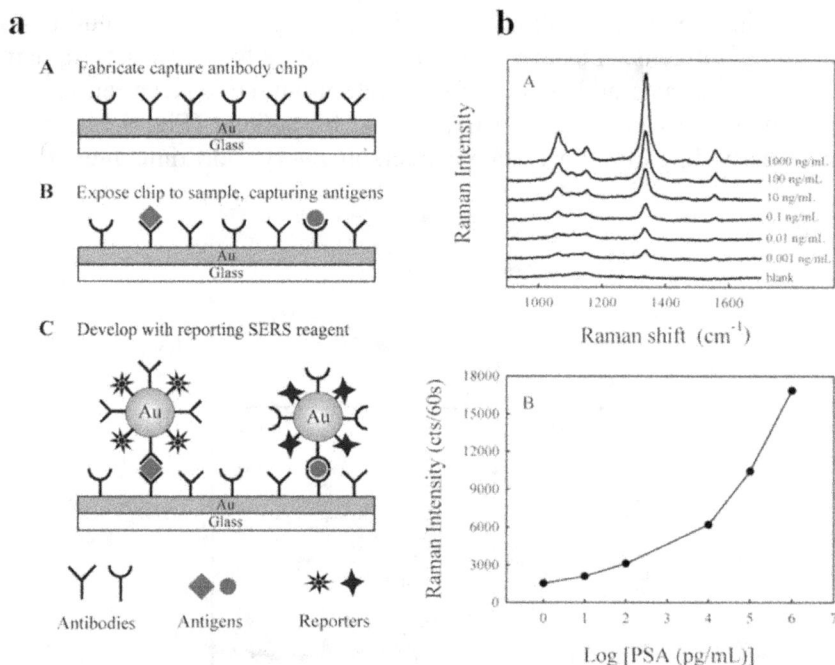

Fig. 9.11. Femtomolar detection of PSA. (a) Schematic of the steps involved in the method. (b) Evolution of SERS signal from PSA immunoassay for different concentrations and the dose-response curve for free PSA in human Serum. Reprinted with permission from [86]. Copyright (2016) American Chemical Society.

9.5. Conclusion and Future Prospects

Since the discovery of SERS, there has been tremendous increase in the number of publications in the field. Several different applications based on SERS have been demonstrated thereby bringing it to the scientific limelight in recent years. In the last decade or so, SERS has slowly developed into a useful spectroscopic tool that has potential applications in physical, chemical material as well as life sciences. In particular, the biomedical applications of SERS have motivated several researchers to develop very efficient sensing platforms based on SERS. The enhancement of Raman scattering by plasmonic fields is now well understood and researchers are now moving forward with engineering more efficient nanostructures to improve the sensitivity along with reduced fabrication costs. It is now widely believed that SERS mechanism has two contributions: the major one being electromagnetic and a minor chemical enhancement. Plasmonic field contributions plays a major role in the electromagnetic enhancement of Raman scattering process. Thus, it becomes essential to wisely engineer the plasmonic nanostructures to obtain optimum enhancements. As plasmonic fields are confined to the surface, the signal enhancement decays exponentially with the distance from the surface and hence termed evanescent. It is then imperative to have the analyte molecules very close to the nanoparticle surface to be in the vicinity of the electromagnetic field. Chemical enhancements, on the other hand, are not well understood due to the difficulty in theoretical as well as experimental observations. A theoretical description of chemical enhancement will require accurate knowledge of the vibrational and electronic states of the analyte molecule that is adsorbed onto the metal surface. A substantial amount of work, both theoretical as well as experimental, need to be carried out in order to fully understand the mechanism.

It should be mentioned that the recent developments in nanofabrication methods have contributed largely for the advancement of SERS based applications. One of the major routes of fabrication of SERS substrates is the bottom up colloidal synthesis of nanoparticles. It is now possible to obtain shape as well as size sensitive structures that can be tuned for the experimental requirements such as excitation wavelength. However, there are some drawbacks of this method based on colloidal chemistry. Firstly, inhomogeneity of the nanoparticles obtained can be an issue in quantitative studies. Different regions of the sample can present different enhancements due to structural inhomogeneity. Moreover, colloidal purity can sometimes be questionable and may require further chemical analysis. Unknown compositions in the structures can lead to spurious Raman signals that interfere with those of the studied molecules.

On the other hand, top-to-bottom approaches have some advantageous over colloidal fabrication methods. There is better control over size, shape and distribution of the nanoparticles. Good reproducibility of the shape and size of the nanostructures is possible using the top-to-bottom approach. State-of-the-art nanofabrication techniques such as electron beam lithography, nanosphere lithography, focused ion beam milling and optical lithography have been made use of for fabricating interesting nanoparticle geometries that is unachievable through colloidal chemistry synthesis. However higher cost of fabrication and longer preparation times are disadvantageous for several real-life applications.

It is clear that conventional Raman spectroscopy is now slowly being replaced by SERS in most applications. This is particularly true for biomedical applications such as disease diagnosis and bio-detection. SERS spectroscopy provides valuable information about biomolecules in living organisms and their interactions with greater sensitivity. The number of publications based on biomedical application of SERS has seen a tremendous increase in the recent years. It is now possible to use SERS for the detection of DNA/RNA and proteins. Label-free SERS in vivo as well as in vitro provide vital information on the chemical composition of these biomolecules without additional markers that can impede certain natural processes. Qualitative as well as quantitative data can be elucidated for SERS signals which can help in characterizing the molecules and the system in a complete fashion. Raman signals from single DNA as well as hybridized strands is now achievable using label-free SERRS techniques [63, 91-92]. Early stage detection of certain cancers such as prostate cancer is now possible with the help of SERS studies. Prostate cancer marker PSA can be detected in the human serum. Reports on the detection of breast cancer cell biomarkers based on SERS have been published. SERS has also been applied in other biomedical test such as the detection of calcium ions. SERS imaging forms a powerful tool for visualizing bio medically relevant processes. For example, SERS microscopy has helped in understanding protein localization in tissues that are cancerous. In a recent report, an intraoperative tumor resection based on SERS imaging in live rats was reported. These works provide promise for advanced and accurate tumor imaging and resection.

Nevertheless, there are few challenges that need to be addressed for direct application of SERS in practical biomedical applications. Fabrication of reliable and cost-effective substrates is still in the optimization stage. Although current developments in nanofabrication technology has contributed to the advancement of novel SERS substrates, it is yet not clear if ideal SERS platforms have been developed.

On the other hand, researchers are exploring the possibility of using materials beyond Ag and Au. One novel material that is now becoming a strong candidate for new generation SERS is graphene. But large-scale fabrication of graphene substrates is still a challenge. Other factors such as environmental issues of nanotechnology, cellular contamination, ethical issues are being discussed and the outcomes are not known for now. These issues need to be addressed in the near future so as to plan a viable and sustainable technology.

In conclusion, SERS is has developed into a modern technique that holds tremendous potential for biomedical applications. In the coming years, it is believed that SERS will be used in real-world applications such as disease diagnostics and biosensing and will provide valuable information of biochemical processes. SERS can provide vital information that are cannot be obtained from conventional analytical methods such as fluorescence. SERs will become portable and cost-effective tool for pharmacological applications such as drug discovery and disease diagnostics.

References

[1]. M. Fleischmann, P. J. Hendra, A. J. McQuillan, Raman spectra of pyridine adsorbed at a silver electrode, *Chemical Physics Letters,* Vol. 26, Issue 2, 1974, pp. 163-166.
[2]. David L. Jeanmaire, Richard P. Van Duyne, Surface Raman Spectroelectrochemistry: Part I. Heterocyclic, aromatic, and aliphatic amines adsorbed on the anodized silver electrode, *Journal of Electroanalytical Chemistry and Interfacial Electrochemistry,* Vol. 84, No. 1, 1977, pp. 1-20.
[3]. Shuming Nie, Steven R. Emory, Probing Single Molecules and Single Nanoparticles by Surface-Enhanced Raman Scattering, *Science,* Vol. 275, No. 5303, 1997, pp. 1102-1106.
[4]. Katrin Kneipp, Yang Wang, Harald Kneipp, Lev T. Perelman, Irving Itzkan, Ramachandra R. Dasari, Michael S. Feld, *Physical Review Letters,* Vol. 78, Issue 9, 1997, pp. 1667-1670.
[5]. Martin Moskovits, Surface-enhanced spectroscopy, *Review of Modern Physics,* Vol. 57, Issue 3, 1985, pp. 783-828.
[6]. A. Otto, Surface-enhanced Raman scattering of adsorbates, *Journal of Raman Spectroscopy,* Vol. 22, Issue 12, 1991, pp. 743-752.
[7]. Alan Campion, Patanjali Kambhampati, Surface-enhanced Raman scattering, *Chemical Society Reviews,* Vol. 27, Issue 4, 1998, pp. 241-250.
[8]. Surface-Enhanced Raman Spectroscopy: Analytical, Biophysical and Life Science Applications, S. Schlücker (Ed.), *Wiley-VCH,* Weinheim, Germany, 2011.
[9]. Kyle C. Bantz, Audrey F. Meyer, Nathan J. Wittenberg, Hyungsoon Im, Özge Kurtuluş, Si Hoon Lee, Nathan C. Lindquist, Sang-Hyun Oh and Christy L. Haynes, Recent progress in SERS biosensing, *Physical Chemistry Chemical Physics,* Vol. 13, 2011, pp. 11551-11567.

[10]. Sebastiain Schlücker, Surface-Enhanced Raman Spectroscopy: Concepts and Chemical Applications, *Angewandte Chemie, Int. Ed.,* Vol. 53, 2014, pp. 4756-4795.

[11]. Sudhir Cherukulappurath, Surface-Enhanced Raman Spectroscopy for Biomedical Applications: A Review, *Sensors & Transducers,* Vol. 197, Issue 2, February 2016, pp. 1-13.

[12]. Albrecht and Creighton, Anomalously intense Raman spectra of pyridine at a silver electrode *Journal of American Chemical Society*, 99, 1977, pp. 5215– 5217.

[13]. E. Le Ru, P. Etchegoin, Principles of Surface-Enhanced Raman Spectroscopy and Related Plasmonic Effects, *Elsevier*, Amsterdam, 2009.

[14]. Surface-Enhanced Raman Scattering: Physics and Applications, Topics in Applied Physics (Eds.: K. Kneipp, M. Moskovits, H. Kneipp), *Springer*, Berlin, Vol. 103, 2006.

[15]. R. Aroca, Surface-Enhanced Vibrational Spectroscopy, *John Wiley & Sons,* New York, 2006.

[16]. William L. Barnes, Alain Dereux, Thomas W. Ebbesen, Surface plasmon subwavelength optics, *Nature*, Vol. 424, No. 6950, 2003, pp. 824-830.

[17]. Maier Stefan Alexander, Plasmonics: Fundamentals and Applications, *Springer*, 2007.

[18]. A. Otto, J. Timper, J. Billmann, G. Kovacs, I. Pockrand, Surface roughness induced electronic Raman scattering, *Surface Science,* Vol. 92, Issue 1, 1980, pp. L55-L57.

[19]. Kambhampati Patanjali, Foster Michelle C., Campion Alan, Two-dimensional localization of adsorbate/substrate charge-transfer excited states of molecules adsorbed on metal surfaces, *The Journal of Chemical Physics,* Vol. 110, No. 1, 1999, pp. 551-558.

[20]. C. Schatz, M. A. Young, R. P. Van Duyne, Electromagnetic Mechanism of SERS, *Topics in Applied Physics,* Vol. 103, 2006, pp 19-46.

[21]. Campion A., Kambhampati P., Surface-enhanced Raman scattering, *Chemical Society Reviews,* Vol. 27, No. 4, 1998, pp. 241-250.

[22]. Lin Zhao, Lasse Jensen, George C. Schatz, Surface-Enhanced Raman Scattering of Pyrazine at the Junction between Two Ag20 Nanoclusters, *Nano Letters,* Vol. 6, No. 6, 2006, pp. 1229-1234.

[23]. K. Lance Kelly, Eduardo Coronado , Lin Zhao , and George C. Schatz, The Optical Properties of Metal Nanoparticles: The Influence of Size, Shape, and Dielectric Environment, *Journal of Physical Chemistry B,* Vol 103, Issue 3, 2003, pp. 668–677.

[24]. Bhavya Sharma, Renee R. Frontiera, Anne-Isabelle Henry, Emilie Ringe, Richard P. Van Duyne, SERS: materials, applications, and the future, *Materials Today,* Vol. 15, No. 1-2, 2012, pp. 16-25.

[25]. Paul L. Stiles, Jon A. Dieringer, Nilam C. Shah and Richard P. Van Duyne, Surface-Enhanced Raman Spectroscopy, *Annual Reviews of Analytical Chemistry,* Vol. 1, 2008, pp. 601-626.

[26]. Kennedy B. J, Spaeth S., Dickey M., Carron K. T., Determination of the distance dependence and experimental effects for modified SERS substrates based on self-assembled monolayers formed using alkanethiols, *Journal of Physical Chemistry: B,* Vol. 103, 1999, pp. 3640-3646.

[27]. Haynes C. L, Van Duyne R. P., Nanosphere lithography: a versatile nanofabrication tool for studies of size-dependent nanoparticle optics, *Journal of Physical Chemistry: B,* Vol. 105, No. 24, 2001, pp. 5599-5611.

[28]. Hicks E. M., Zhang X. Y., Zou S. L., Lyandres O., Spears K. G., Plasmonic properties of film over nanowell surfaces fabricated by nanosphere lithography, *Journal of Physical Chemistry: B,* Vol. 109, 2005, pp. 22351-22358.

[29]. X. Y. Zhang, M. A. Young, O. Lyandres, R. P. Van Duyne, Rapid Detection of an Anthrax Biomarker by Surface-Enhanced Raman Spectroscopy, *Journal of the American Chemical Society,* Vol. 127, 2005, pp. 4484-4489.

[30]. L. A. Dick, A. D. McFarland, C. L. Haynes, R. P. Van Duyne, Metal Film Over Nanosphere (MFON) Electrodes for Surface-Enhanced Raman Spectroscopy (SERS): Improvements in Surface Nanostructure Stability and Suppression of Irreversible Loss, *Journal of Physical Chemistry: B,* Vol. 106, No. 4, 2002, pp. 853-860.

[31]. Karen E. Shafer-Peltier, Christy L. Haynes, Matthew R. Glucksberg, Richard P. Van Duyne, Toward a Glucose Biosensor Based on Surface-Enhanced Raman Scattering, *Journal of the American Chemical Society,* Vol. 125, No. 2, 2003, pp. 588-593.

[32]. Chanda Ranjit Yonzon, Christy L. Haynes, Xiaoyu Zhang, Joseph T. Walsh, Jr., Richard P. Van Duyne, A Glucose Biosensor Based on Surface-Enhanced Raman Scattering: Improved Partition Layer, Temporal Stability, Reversibility, and Resistance to Serum Protein Interference, *Analytical Chemistry,* Vol. 76, 2004, pp. 78-84.

[33]. Fales A. M., Yuan H., Vo-Dinh T., Silica-coated gold nanostars for combined surface-enhanced Raman scattering (SERS) detection and singlet-oxygen generation: a potential nanoplatform for theranostics, *Langmuir,* Vol. 27, Issue 19, 2011, pp. 12186-12190.

[34]. Liu Y., Yuan H., Fales A. M., Vo-Dinh T., pH-sensing nanostar probe using surface-enhanced Raman scattering (SERS): theoretical and experimental studies, *Journal of Raman Spectroscopy,* Vol. 44, Issue 7, 2013, pp. 980-986.

[35]. Vivek V. Thacker, Lars O. Herrmann, Daniel O. Sigle, Tao Zhang, Tim Liedl, Jeremy J. Baumberg, Ulrich F. Keyser, DNA origami based assembly of gold nanoparticle dimers for surface-enhanced Raman scattering, *Nature Communications,* Vol. 5, 2014, Article 3448, pp. 1-7.

[36]. M. Gellner, D. Steinigeweg, S. Ichilmann, M. Salehi, M. Schütz, K. Kçmpe, M. Haase, S. Schlücker, 3D Self-Assembled Plasmonic Superstructures of Gold Nanospheres: Synthesis and Characterization at the Single-Particle Level, *Small,* Vol. 7, No. 24, 2011, pp. 3445-3451.

[37]. Jackson J. B., Halas N. J., Surface-enhanced Raman scattering on tunable plasmonic nanoparticle substrates, *Proceedings of the National Academy of Sciences of the USA,* Vol. 101, No. 52, 2004, pp. 17930-17935.

[38]. Jian Feng Li, Yi Fan Huang, Yong Ding, Zhi Lin Yang, Song Bo Li, Xiao Shun Zhou, Feng Ru Fan, Wei Zhang, Zhi You Zhou, De Yin Wu, Bin Ren, Zhong Lin Wang, Zhong Qun Tian, Shell-isolated nanoparticle-enhanced Raman spectroscopy, *Nature,* Vol 464, No. 7287, 2010, pp. 392-395.

[39]. Fang J., Liu S., Li Z., Polyhedral silver mesocages for single particle surface-enhanced Raman scattering-based biosensor, *Biomaterials,* Vol. 32, No. 3221, 2011, pp. 4877-4884.

[40]. Masayuki Naya, Takeharu Tani, Yuichi Tomaru, Jingbo Li, Naoki Murakami, Nanophotonics bio-sensor using gold nanostructure, *Proc. SPIE 7032, Plasmonics: Metallic Nanostructures and Their Optical Properties VI,* 70321Q, 2008.

[41]. Jon A. Dieringer, Adam D. McFarland, Nilam C. Shah, Douglas A. Stuart, Alyson V. Whitney, Chanda R. Yonzon, Matthew A. Young, Xiaoyu Zhang, Richard P. Van Duyne, Surface enhanced Raman spectroscopy: new materials, concepts, characterization tools, and applications, *Faraday Discussions,* Vol. 132, 2006, pp. 9-26.

[42]. F. De Angelis, F. Gentile, F. Mecarini, G. Das, M. Moretti, P. Candeloro, M. L. Coluccio, G. Cojoc, A. Accardo, C. Aiberale, R. P. Zaccaria, G. Perozziello, L. Tirinato, A. Toma, G. Cuda, R. Cingolani, E. Di Fabrizio, Breaking the diffusion limit with super-hydrophobic delivery of molecules to plasmonic nanofocusing SERS structures, *Nature Photonics,* Vol. 5, 2011, pp. 682-687.

[43]. R. M. Roth, N. C. Panoiu, M. M. Adams, R. M. Osgood, C. C. Neacsu, M. B. Raschke, Resonant-plasmon field enhancement from asymmetrically illuminated conical metallic-probe tips, *Optics Express,* Vol. 14, Issue 7, 2006, pp. 2921-2931.

[44]. W. H. Zhang, X. D. Cui, B. S. Yeo, T. Schmid, C. Hafner, R. Zenobi, Nanoscale Roughness on Metal Surfaces Can Increase Tip-Enhanced Raman Scattering by an Order of Magnitude, *Nano Letters,* Vol. 7, Issue 5, 2007, pp. 1401-1405.

[45]. J. Steidtner, B. Pettinger, Tip-Enhanced Raman Spectroscopy and Microscopy on Single Dye Molecules with 15 nm Resolution, *Physical Review Letters,* Vol. 100, No. 23, 2008, pp. 236101-236104.

[46]. R. Zhang, Y. Zhang, Z. C. Dong, S. Jiang, C. Zhang, L. G. Chen, L. Zhang, Y. Liao, J. Aizpurua, Y. Luo, J. L. Yang, J. G. Hou, Chemical mapping of a single molecule by plasmon-enhanced Raman scattering, *Nature*, Vol. 498, No. 7452, 2013, pp. 82-86.

[47]. Xiaoyu Zhang, Matthew A. Young, Olga Lyandres, and Richard P. Van Duyne, Rapid Detection of an Anthrax Biomarker by Surface-Enhanced Raman Spectroscopy, *Journal of American Chemical Society,* Vol. 127, 2005, pp. 4484-4489.

[48]. Anna K. Boardman, Winnie S. Wong, W. Ranjith Premasiri, Lawrence D. Ziegler, Jean C. Lee, Milos Miljkovic, Catherine M. Klapperich, Andre Sharon, and Alexis F. Sauer-Budge, Rapid Detection of Bacteria from Blood with Surface-Enhanced Raman Spectroscopy, *Analytical Chemistry*, Vol. 88, 16, 2016, pp. 8026-8035.

[49]. Haibo Zhou, Danting Yang, Natalia P. Ivleva, Nicoleta E. Mircescu, Reinhard Niessner, Christoph Haisch SERS Detection of Bacteria in Water by in Situ Coating with Ag Nanoparticles, *Analytical Chemistry,* Vol. 86, 3, 2014, pp. 1525-1533.

[50]. Anna Mühlig, Thomas Bocklitz, Ines Labugger, Stefan Dees, Sandra Henk, Elvira Richter, Sönke Andres, Matthias Merker, Stephan Stöckel, Karina Weber, Dana Cialla-May, Jürgen Popp, LOC-SERS: A Promising Closed System for the Identification of Mycobacteria, *Analytical Chemistry,* Vol. 88, 16, 2016, pp. 7998-8004.

[51]. Ma K., Yuen J. M., Shah N. C., Walsh J. T., Glucksberg M. R., Van Duyne R. P., In Vivo, Transcutaneous Glucose Sensing Using Surface-Enhanced Spatially Offset Raman Spectroscopy: Multiple Rats, Improved Hypoglycemic Accuracy, Low Incident Power, and Continuous Monitoring for Greater Than 17 Days, *Analytical Chemistry,* Vol. 83, No. 23, 2011, pp. 9146-9152.

[52]. Yuen J. M., Shah N. C., Walsh J. T., Glucksberg M. R., Van Duyne R. P., Transcutaneous Glucose Sensing by Surface-Enhanced Spatially Offset Raman Spectroscopy in a Rat Model, *Analytical Chemistry,* Vol. 82, No. 20, 2010, pp. 8382-8385.

[53]. C. A. Mirkin, R. L. Letsinger, R. C. Mucic, J. J. Storhoff, A DNA-based method for rationally assembling nanoparticles into macroscopic materials, *Nature*, Vol. 382, No. 6592, 1996, pp. 607-609.

[54]. James J. Storhoff, Robert Elghanian, Robert C. Mucic, Chad A. Mirkin, Robert L. Letsinger, One-Pot Colorimetric Differentiation of Polynucleotides with Single Base Imperfections Using Gold Nanoparticle Probes, *Journal of the American Chemical Society,* Vol. 120, No. 9, 1998, pp. 1959-1964.

[55]. Yunwei Charles Cao, Rongchao Jin, Chad A. Mirkin, Nanoparticles with Raman Spectroscopic Fingerprints for DNA and RNA Detection, *Science*, Vol. 297, No. 5586, 2002, pp. 1536-1540.

[56]. Cheng Fang, Ajay Agarwala, Kavitha Devi Buddharaju, Nizamudin Mohamed Khalid, Shaik Mohamed Salim, Effendi Widjaja, Marc V. Garland, Narayanan Balasubramanian, Dim-Lee Kwong, DNA detection using nanostructured SERS substrates with Rhodamine B as Raman label, *Biosensors and Bioelectronics,* Vol. 24, No. 2, 2008, pp. 216-221.

[57]. Laura Fabris, Mark Dante, Gary Braun, Seung Joon Lee, Norbert O. Reich, Martin Moskovits, Thuc-Quyen Nguyen, Guillermo C. Bazan, A heterogeneous PNA-based SERS method for DNA detection, *Journal of the American Chemical Society,* Vol. 129, No. 19, 2007, pp. 6086-6087.

[58]. Karen Faulds, W. Ewen Smith, Duncan Graham, Evaluation of Surface-Enhanced Resonance Raman Scattering for Quantitative DNA Analysis, *Analytical Chemistry,* Vol. 76, No. 2, 2004, pp. 412-417.

[59]. K. Faulds, F. McKenzie, W. E. Smith, D. Graham, Quantitative Simultaneous Multianalyte Detection of DNA by Dual-Wavelength Surface-Enhanced Resonance Raman Scattering, *Angewandte Chemie,* Vol. 119, Issue 11, 2007, pp. 1861-1863.

[60]. D. Graham, D. G. Thompson, W. E. Smith, K. Faulds, Control of enhanced Raman scattering using a DNA-based assembly process of dye-coded nanoparticles, *Nature Nanotechnology*, Vol. 3, No. 9, 2008, pp. 548-551.

[61]. Steven E. J. Bell, Narayana M. S. Sirimuthu, Surface-Enhanced Raman Spectroscopy (SERS) for Sub-Micromolar Detection of DNA/RNA Mononucleotides, *Journal of the American Chemical Society*, Vol. 128, No. 49, 2006, pp. 15580-15581.

[62]. Evanthia Papadopoulou, Steven E. J. Bell, Label-Free Detection of Single-Base Mismatches in DNA by Surface-Enhanced Raman Spectroscopy, *Angewandte Chemie*, Vol. 123, Issue 39, 2011, pp. 9224-9227.

[63]. Aoune Barhoumi, Naomi J. Halas, Label-Free Detection of DNA Hybridization Using Surface Enhanced Raman Spectroscopy, *Journal of the American Chemical Society*, Vol. 132, No. 37, 2010, pp. 12792-12793.

[64]. Hansang Cho, Brian Lee, Gang L. Liu, Ajay Agarwal, Luke P. Lee, Label-free and highly sensitive biomolecular detection using SERS and electrokinetic preconcentration, *Lab on a Chip*, Vol. 9, Issue 23, 2009, pp. 3360-3363.

[65]. Sudhir Cherukulappurath, Si Hoon Lee, Antonio Campos, Christy L. Haynes, Sang-Hyun Oh, Rapid and Sensitive in Situ SERS Detection Using Dielectrophoresis, *Chemistry of Materials*, Vol. 26, No. 7, 2014, pp. 2445-2452.

[66]. J. De Gelder, K. De Gussem, P. Vandenabeele, L. Moens, Reference database of Raman spectra of biological molecules, *Journal of Raman Spectroscopy*, Vol. 38, Issue 9, 2007, pp. 1133-1147.

[67]. R. Schweitzer-Stenner, Structure and dynamics of biomolecules probed by Raman spectroscopy, *Journal of Raman Spectroscopy*, Vol. 36, Issue 4, 2005, pp. 276-278.

[68]. R. Tuma, Raman spectroscopy of proteins: from peptides to large assemblies, *Journal of Raman Spectroscopy*, Vol. 36, No. 4, 2005, pp. 307-319.

[69]. Z. Wen, Raman Spectroscopy of Protein Pharmaceuticals, *Journal of Pharmaceutical Sciences*, Vol. 96, Issue 11, 2007, pp. 2861-2878.

[70]. E. Podstawka, Y. Ozaki, L. M. Proniewicz, Part I: Surface Enhanced Raman Spectroscopy of Amino Acids and Their Homodipeptides Adsorbed on Colloidal Silver, *Applied Spectroscopy*, Vol. 58, No. 5, 2004, pp. 570-580.

[71]. E. Podstawka, Y. Ozaki, L. M. Proniewicz, Part II: Surface Enhanced Raman Spectroscopy Investigation of Methionine Containing Heteropeptides Adsorbed on Colloidal Silver, *Applied Spectroscopy*, Vol. 58, No. 5, 2004, pp. 581-590.

[72]. E. Podstawka, Y. Ozaki, L. M. Proniewicz, Part III: Surface-Enhanced Raman Scattering of Amino Acids and Their homopeptide Monolayers Deposited onto Colloidal Gold Surface, *Applied Spectroscopy*, Vol. 59, 2005, pp. 1516-1526.

[73]. E. Podstawka, E. Sikorska, L. M. Proniewixz, B. Lammek, Raman and surface-enhanced Raman spectroscopy investigation of vasopressin analogues containing 1-aminocyclohexane-1-carboxylic acid residue, *Biopolymers*, Vol. 83, 2006, pp. 193-203.

[74]. S. Stewart, P. M. Fredericks, Surface-enhanced Raman spectroscopy of peptides and proteins adsorbed on an electrochemically prepared silver surface, Spectrochimica Acta. Part A: *Molecular and Biomolecular Spectroscopy*, Vol. 55, Issue 7, 1999, pp. 1615-1640.

[75]. H. Li, J. Sun, B. M. Cullum, Label-Free Detection of Proteins Using SERS- Based Immuno-Nanosensors, *NanoBiotechnology*, Vol. 2, Issue 1, 2006, pp. 17-28.

[76]. G. D. Chumanov, R. G. Efremov, I. R. Nabiev, Surface-enhanced Raman spectroscopy of biomolecules. Part I. Water-soluble proteins, dipeptides and amino acids, *Journal of Raman Spectroscopy*, Vol. 21, Issue 1, 1990, pp. 43-48.

[77]. J. D. Driskell, J. M. Uhlenkamp, R. J. Lipert, M. D. Porter, Surface-Enhanced Raman Scattering Immunoassays Using a Rotated Capture Substrate, *Analytical Chemistry*, Vol. 79, No. 11, 2007, pp. 4141-4148.

[78]. B. C. Galarreta, P. R. Norton, F. L. Labarthet, SERS detection of Streptavidin/Biotin Monolayer Assemblies, *Langmuir*, Vol. 27, Issue 4, 2011, pp. 1494-1498.

[79]. F. Domenici, A. R. Bizzarri, S. Cannistraro, Surface-enhanced Raman scattering detection of wild- type and mutant p53 proteins at very low concentration in human serum, *Analytical Biochemistry*, 2012, Vol. 421, pp. 9-15.

[80]. X. Han, G. Huang, B. Zhao, Y. Ozaki, Label-Free Highly Sensitive Detection of Proteins in Aqueous Solutions Using Surface-Enhanced Raman Scattering, *Analytical Chemistry*, Vol. 81, No. 9, 2009, pp. 3329-3333.

[81]. J. Driskell, K. Kwarta, R. Lipert, M. Porter, J. Neill, J. Ridpath, Low-Level Detection of Viral Pathogens by a Surface-Enhanced Raman Scattering Based Immunoassay, *Analytical Chemistry*, Vol. 77, No. 19, 2005, pp. 6147-6154.

[82]. Chi-Chang Lin, Ying-Mei Yang, Yan-Fu Chen, Tzyy-Schiuan Yang, Hsien-Chang Chang, A new protein A assay based on Raman reporter labeled immunogold nanoparticles, *Biosensors and Bioelectronics*, Vol. 24, Issue 2, 2008, pp. 178-183.

[83]. S. Xu, X. Ji, W. Xu, X. Li, L. Wang, Y. Bai, B. Zhao, Y. Ozaki, Immunoassay using probe-labelling immunogold nanoparticles with silver staining enhancement via surface-enhanced Raman scattering, *Analyst*, Vol. 129, 2004, pp. 63-68.

[84]. Gufeng Wang, Jeremy D. Driskell, Marc D. Porter, Robert J. Lipert, Control of Antigen Mass Transport via Capture Substrate Rotation: Binding Kinetics and Implications on Immunoassay Speed and Detection Limits, *Analytical Chemistry*, Vol. 81, No. 15, 2009, pp. 6175-6185.

[85]. T. E. Rohr, T. Cotton, N. Fan, P. J. Tarcha, Immunoassay employing surface-enhanced Raman spectroscopy, *Analytical Biochemistry*, Vol. 182, No. 2, 1989, pp. 388-398.

[86]. D. S. Grubisha, R. J. Lipert, H. Y. Park, J. Driskell, M. D. Porter, Femtomolar Detection of Prostate-Specific Antigen: An Immunoassay Based on Surface-Enhanced Raman Scattering and Immunogold Labels, *Analytical Chemistry*, Vol. 75, No. 21, 2003, pp. 5936-5943.

[87]. Anna Rita Bizzarri, Salvatore Cannistraro, SERS detection of thrombin by protein recognition using functionalized gold nanoparticles, *Nanomedicine: Nanotechnology, Biology and Medicine*, Vol. 3, Issue 4, 2007, pp. 306-310.

[88]. Lucas A. Lane, Ximei Qian, Shuming Nie, SERS Nanoparticles in Medicine: From Label-Free Detection to Spectroscopic Tagging, *Chemical Reviews*, Vol. 115, No. 19, 2015, pp. 10489-10529.

[89]. C. Jehn, B. Küstner, P. Adam, A. Marx, P. Ströbel, C. Schmuck, S. Schlücker, Water soluble SERS labels comprising a SAM with dual spacers for controlled bioconjugation, *Physical Chemistry Chemical Physics*, Vol. 11, No. 34, 2009, pp. 7499-7504

[90]. M. Schütz, D. Steinigeweg, M. Salehi, K. Kömpe, S. Schlücker, Hydrophilically stabilized gold nanostars as SERS labels for tissue imaging of the tumor suppressor p63 by immuno-SERS microscopy, *Chemical Communications*, Vol. 47, 2011, pp. 4216-4218.

[91]. Aoune Barhoumi, Dongmao Zhang, Felicia Tam, Naomi J. Halas, Surface-Enhanced Raman Spectroscopy of DNA, *Journal of the American Chemical Society*, Vol. 130, No. 16, 2008, pp. 5523-5529.

[92]. Li-Jia Xu, Zhi-Chao Lei, Jiuxing Li, Cheng Zong, Chaoyong James Yang, Bin Ren, Label-Free Surface-Enhanced Raman Spectroscopy Detection of DNA with Single-Base Sensitivity *Journal of American Chemical Society*, Vol. 137, No. 15, 2015, pp. 5149-5154.

Chapter 10
An Advanced Sensors-based Platform for the Development of Agricultural Sprayers

Paulo E. Cruvinel, Vilma A. Oliveira, Heitor V. Mercaldi, Elmer A. G. Peñaloza, Kleber R. Felizardo

10.1. Concepts of Pesticide Application Technology in Agriculture

The definition of the parameters such as the size of the drops and application volume depend directly on the ratio (target/pesticide) relationship. The liquid agricultural inputs can be applied directly to the soil or even to the plants or their leaves with lower density of drops, i.e., allowing the use of larger droplets. Therefore, this facilitates the adoption of techniques for reduction of the drift, and allows improving of security in relation to the application, as well as the use of machinery capability in relation to the use of the sprayers and its efficiency. If used properly the large droplets, they can provide adequate deposit levels. Such deposits are related with the amount of pesticide, or its volume, deposited in the target [1].

The agricultural pest management also requires controllers and adequate climatically coverage conditions. Although the use of larger droplets could be apparently indicated, in first approximation, as ideal for the operation, the truly pest management requires a better contact of the applied pesticides with the problem to be treated. Thus, by using smaller droplets become possible to reach improvements in the efficiency, i.e., since such processes are not linear, such initiatives should be associated with knowledge, intelligence and adequate application technology.

Besides, since there are either a large diversity of pests and changes in their dynamic's behavior in environment, for each different agricultural crop, and as a function of its geographical locations and climate, it is required one specific solution. In other words, in terms of pesticide application for pest control there are not only one solution for everything [2]. That's why the solution in relation to pest management in agriculture is still an open field for research, development, and innovation, mainly for the area relating

Paulo E. Cruvinel
Embrapa Instrumentation (CNPDIA), Rua XV de Novembro 1452, 13560-970 São Carlos, SP, Brazil

to pesticide application technology dedicated to food security and based on the sustainability of the natural resources.

An important feature for the definition of strategies for control of pests in relation to the application technology is the way in which the pesticides are classified in relation to plants, i.e., as having systemic movement or not after the application and its absorption, even being aerial or terrestrial application using airplanes or tractors respectively with spraying systems, This means that these products can be classified as systemic or not, i.e., if the application technology needs to provide good coverage and penetration ability of the droplets in the leaves in order to take into account all parts of the plant.

In most cases, to obtain good control there are needs to allow the coverage of the leaves and the distribution of the pesticide throughout the plants, with emphasis on the lower parts, medium parts or even in the top parts, i.e., as a function of the pest which is being treated. In the case of application directed to plants, the study of the characteristics of the targets should include the analysis of the movement of the leaves, stage of development, hairiness, roughness, and the face of the leaves. Today there are groups looking for the plant phenotyping as an indicator to define procedure related to pest control, once they are related to the plant architecture [3].

In spraying systems, nozzles break the mixture prepared mainly with pesticides plus water into droplets and form the spray pattern. In Fig. 10.1 is shown examples of nozzles used for agricultural sprayers.

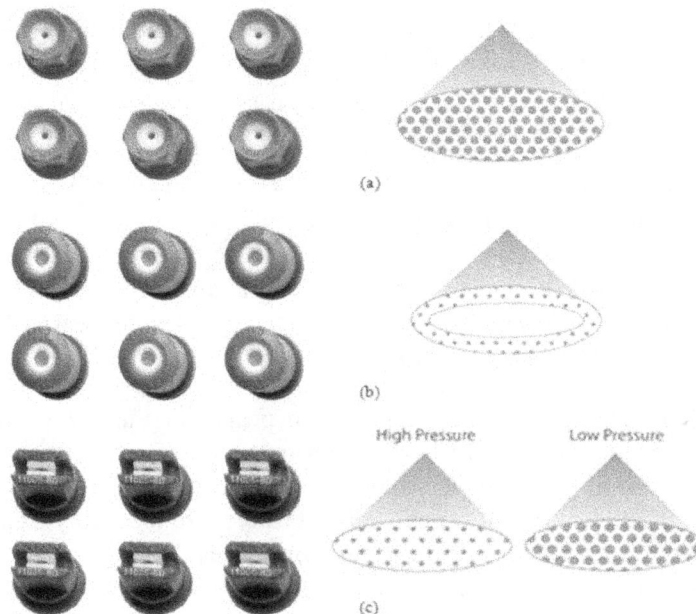

Fig. 10.1. Examples of nozzles used for agricultural sprayers: (a) Full cone spray nozzles (BD model Magnojet®); (b) Hollow cone spray nozzles (CV-IA model Magnojet®); c) Flat fan spray nozzles (CH model Magnojet®).

The determination of the correct application volume at a given operating pressure, travel speed, and spacing, as well as the size of the spray particle, are all important because they can help the determination of both efficacy and spray drift of the application of an herbicide, insecticide, or even a fungicide. The proper selection of the sensors, the embedded electronics, hydraulics components, tubes, control modality, intelligence for decision making, as well as the correct nozzles type and size are essential for proper pesticide application [4-7].

The droplet size (see Table 10.1) influences the ability class spraying to cover the target and penetrate the leaves into a plant. Smaller droplets have better coverage capacity, i.e., offering more drops/cm^2). Also, the smaller droplets provide greater penetration capability and are recommended when it is necessary good coverage and good penetration [8]. However, smaller droplets can be more sensitive to evaporation and drift processes. In productive agricultural systems in general, the large drops are preferred for application of herbicides, such as glyphosate, for example, while fine droplets are more used for insecticides, fungicides and other products of less systemicity.

Table 10.1. Based on the standard ASAE S-572 (Spray Tip Classification by Droplet Size) the droplet size classes are organized in according to the color code, which are used to identify and to select the tips. Droplet sizes are usually expressed in microns (μm). For reference tips the flow is given in liters/minute and the pressure is given in bar.

Classification category	Symbol	Color Code	Approximate Volume Median Diameter (VMD)
Very Fine	VF	Red	<100
Fine	F	Orange	100-175
Medium	M	Yellow	175-250
Coarse	C	Blue	250-375
Very Coarse	VC	Green	375-450
Extremely Coarse	XC	White	> 450

There are in the literature broad nozzles descriptions, their recommendation of use, selection of the proper nozzle type, and calibration method, also several manufactures in different counties. However, one may find little information, so far, regarding to how a sprayer for dedicated application can be developed, tested and validated for agricultural applications.

Despite the proper selection of a nozzle type and size is relevant, is also very important to take into consideration the technology to allow the occurrence of the process for application. Therefore, the whole sprayer system is involved in the application process not only in determining the amount of spray applied to an area, uniformity, and coverage but also with the target and the amount of the potential drift. Besides, during operation their nozzles allow breaking of mixture into droplets, and also propel the droplets in a proper direction. Drift can be minimized by selecting response time of the sprayers, the choice of best moment for the applications in function of the climate conditions, selection of the controllers, which are used to select the best pressure or even the best volume, as well as the nozzle that produce the required droplet size while providing adequate coverage at the intended application rate.

It is important to take into account that even when a tip predominantly produces thick drops, there is a small part of fine droplets in the applied volume. This part is sensitive and related to the drift processes. This means that a certain cutting edge does not produce any drops of same size, but a range of droplet sizes, which is called spray spectrum. Despite de climatical conditions, the risk of drift is also a function of the quality of the technology of application and the instrumentation used in the sprayer's architecture and their design.

10.2. An Advanced Sensors-Based Platform and Its Conception

Situated in the Embrapa Instrumentation (CNPDIA), and organized in partnership with the Control Laboratory of the Department of Engineering and Computer Science located at the University of São Paulo (EESC/USP), is the laboratory infrastructure for the development of research and innovation that contribute to improve agricultural sprays [9], i.e., those based on liquid agricultural inputs. Such infrastructure was designed and developed taking into account the concept of an advanced platform based on the use of sensors and actuators into a network coupled, controllers circuits, and intelligent electronics to enable project and development of sprayer systems.

Therefore, the projects of such systems can be carried out by means of monitoring and control in real-time of the main parameters and variables involved in agricultural sprayers. Further, based on digital image analysis and pattern recognition evaluation of the sprays quality, as well as the drift, can be analyzed [10].

In this context, this laboratory infrastructure has an advanced development system that enables the design of architectures involving the connections of hydraulic components and devices, mechanical pumps, electronic and computer algorithms, as illustrated by Figs. 10.2-10.4. Besides, this infrastructure allows the fusion and integration of smart sensors and advanced instrumentation with control, agricultural machinery, agronomy and precision agriculture. The term smart sensor refers to those elements containing sensing and signal processing capabilities and understanding, with objectives ranging from simple viewing to sophisticated remote sensing, surveillance, search or even track, robotics, and intelligence applications [11]. The smart sensor is expected to have the capability that functionality and architecture, as well as raw data acquisition are based the existence of a processing unit [12-14]. The advanced sensor-based platform has the potential to assist in dealing with a large amount of data that is generated by a monitoring system. On board processing at the advanced platform for agricultural sprayers development, the sensors allow a portion of the computation to be done locally on the sensors' embedded processors, with self-diagnosis and self-calibration capabilities, thus reducing that data amount of information that needs to be transmitted over a network.

Fig. 10.2. Front view of the development system for projects dedicated to the application of liquid agricultural inputs based on the advanced sensors-based platform. In the background is the panel with electrical, electronic and hydraulic components and the front boom sections with the nozzles and spray tips.

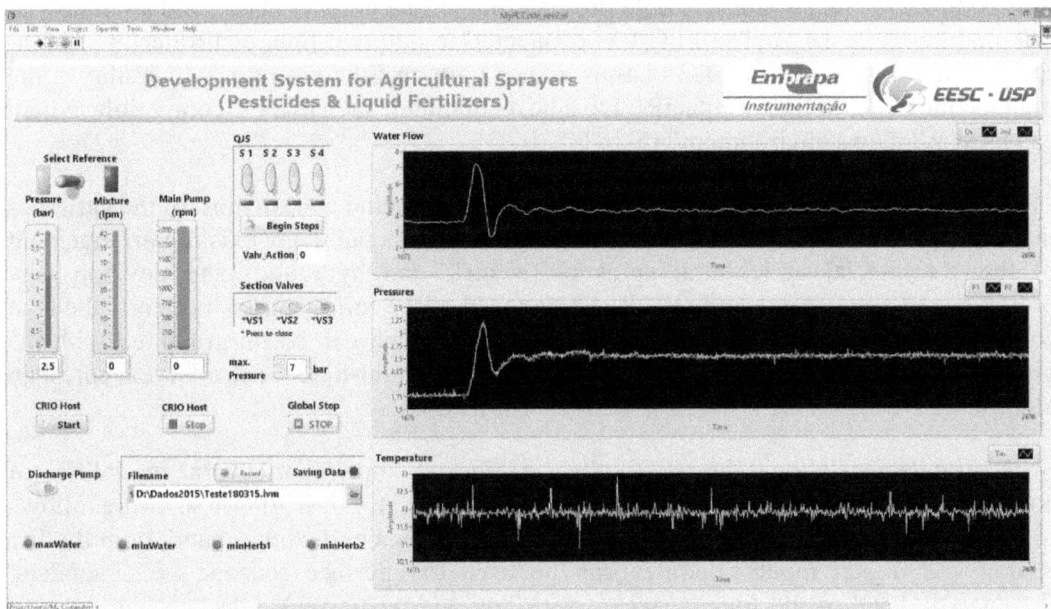

Fig. 10.3. Front view of the user interface of the development system for projects dedicated to the application of liquid agricultural inputs based on the advanced sensors-based platform.

Fig. 10.4. Front view of the stand-alone devices used on the advanced sensor-based platform, which allow interconnections in the front grille, and in the digital network: (1) frequency inverter, (2) power supplies (4) modules for automation and control of the inputs and outputs variables, (5) box with electronic circuits for signal conditioning (7) hydraulic valves, boom section and flow meter and the pressure sensor. In the rear of the grid are assembled: (3) water tank level sensor, (6) pump for mixing coupled to a three-phase motor; (8) two tanks for pesticides and level sensors (9) two piston pumps.

Fig. 10.5 shows the structure used for the automation of the advanced sensor-based platform, which took into account the use of a CompactRIO embedded controller, model cRIO-9073 manufactured by National Instruments. This controller is a rugged, reconfigurable embedded system containing three main components, which are a real-time controller, a reconfigurable field programmable gate array (FPGA), and an industrial I/O modules (Fig. 10.6). The FPGA is connected to one real-time controller by a high-speed PCI Bus. The cRIO-9073 chassis comes with an Ethernet port, which allows the digital connection of the CompactRIO to a host computer, such as a PC compatible, which can used Windows environment [15-17].

This system has the necessary instrumentation to monitor and control most of the variables that can be involved with agricultural spraying, as well as build any existing arrangement of the commercial agricultural sprayers (electric and hydraulic) and develop new prototypes of agricultural sprayers. It is composed of the following instruments: the host computer, the real-time controller, and the FPGA, which can be programmed with the graphical programming language - LabVIEW. The real-time controller executes an algorithms created in the LabVIEW real-time software.

Such arrangement allows the use of the real-time to implement several modalities of controls. In such architecture the FPGA created in the LabVIEW FPGA software, allows to read and write data from the cRIO-9073 I/O modules. On the other hand, both the I/O analog and digital inputs modules can be used to interface sensors, i.e., encoders, flowmeters, level of the liquids (water, fertilizer and pesticides), pressure, current and voltage, conductivity, response time, and others) and actuators (pumps, valves and power inverters). A host program, created in the LabVIEW PC software [18], is used to communicate with the real-time controller and also to manage the user interface for the operation and project setting on the advanced sensor based platform, including visualization, data logging and also the simulation of a broad spectra of models.

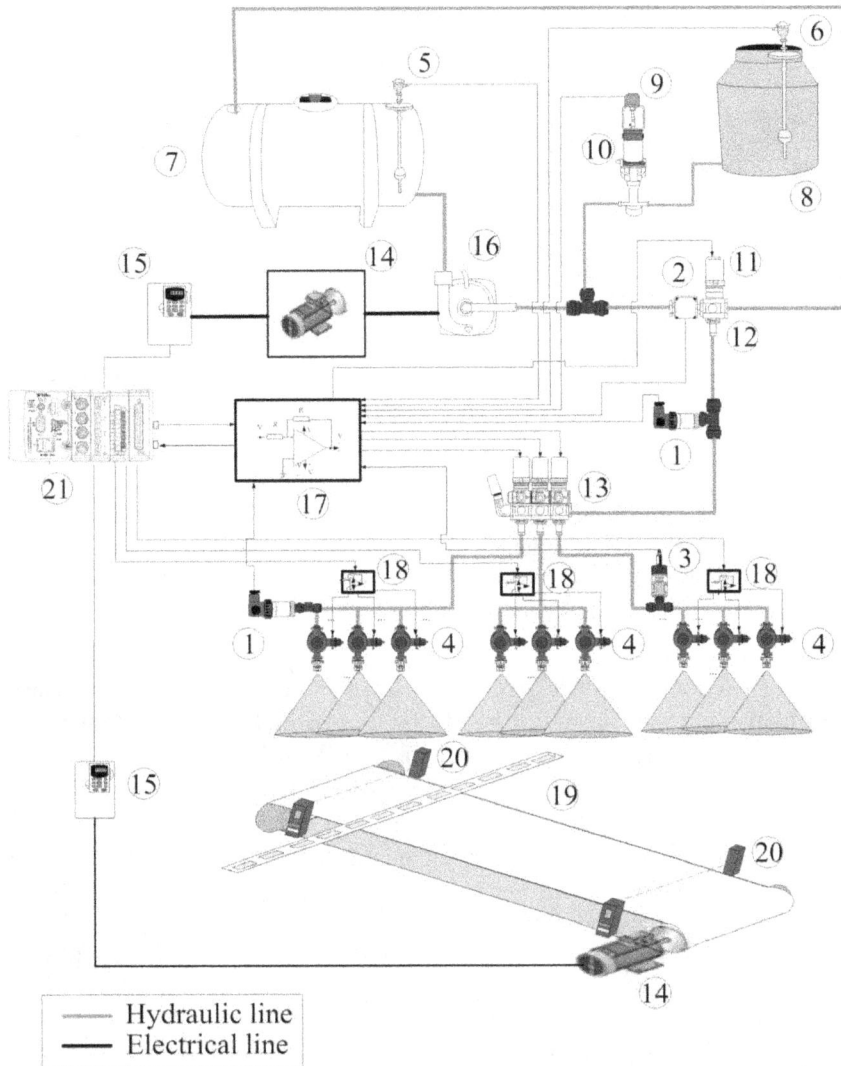

Fig. 10.5. The structure used for the automation of the advanced sensor-based platform, in which is possible to observe the following sensors and devices: (1) Pressure sensors; (2) Flow sensor; (3) Electrical conductivity sensor; (4) Smart delay time sensors; (5) Level sensor for the mixing tank; (6) Level sensor for the product tank; (7) Tank dedicated for mixing; (8) Tank used for pesticides; (9) Encoder`sensor used for the injection pump; (10) Injection pump; (11) Proportional valve encoder`sensor; (12) Proportional valve; (13) Section valves; (14) Three-phase motor; (15) Frequency inverter; (16) Spray pump; (17) Signal conditioning module; (18) Signal conditioning of the smart delay time sensors; (19) Industrial belt that simulates the tractor movement in relation to the sprayers; (20) Infrared optical sensors for position detection; (21) Programmable Automation Controller (PAC-CRIO).

187

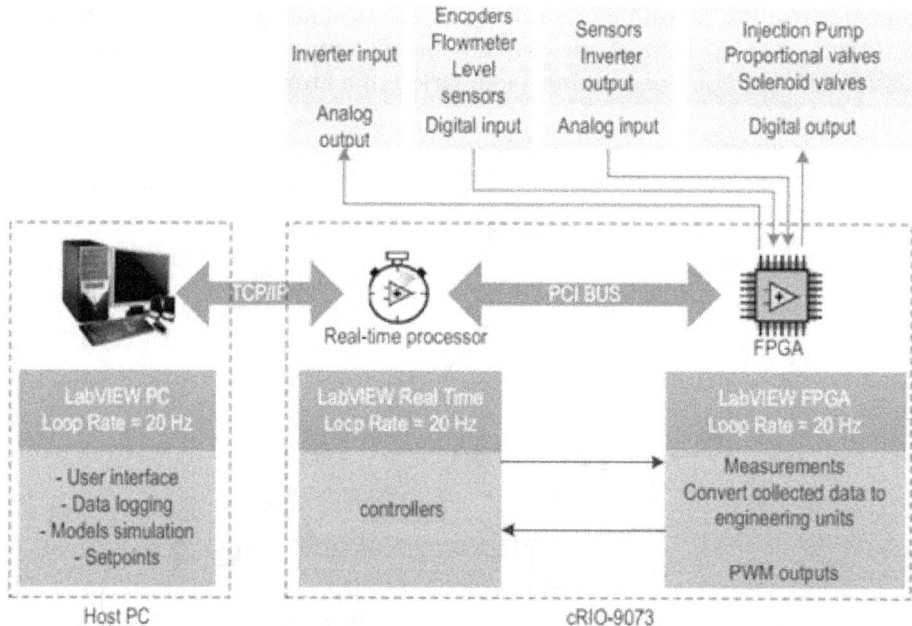

Fig. 10.6. Structure used for I/O signals in the advanced sensor-based platform, which took into account the use of a CompactRIO embedded controller manufactured by the National Instruments.

To select the pressure sensor for a specific application besides the pressure range first of all the type of pressure measurement has to be considered. Pressure sensors measure a certain pressure in comparison to a reference pressure and can be divided into absolute, gage and differential devices [19-21]. In the advanced platform for the development of sprayers are used two piezoelectric pressure sensors. One of them monitors the hydraulic system pressure in the distribution point of the mixture (water + pesticide) to the spray bars. The other monitors the pressure at the end of the spray bar. These sensors have non-linearity $< + 0.5$ % considering the best fit straight line (BFSL), i.e., along a useful pressure range, e.g., from 0 to 103.42 kPa up to 689.48 MPa; signal output in either 4-20 mA or 0-10 V; measuring deviation of the zero signal typical $\leq \pm 0.5$ % of the span; and accuracy at room temperature $\leq \pm 1$ % of span.

Flowmeters are used in fluid systems (liquid and gas) to indicate the rate of flow of the fluid. They can also control the rate of flow if they are equipped with a flow control valve. When a fluid is introduced into the tube, the float is lifted from its initial position at the inlet, allowing the fluid to pass between it and the tube wall. As the float rises, more and more fluid flows by the float because the tapered tube's diameter is increasing. Ultimately, a point is reached where the flow area is large enough to allow the entire volume of the fluid to flow past the float. This flow area is called the annular passage. The float is now stationary at that level within the tube, as its weight is being supported by the fluid forces which caused it to rise. This position corresponds to a point on the tube's measurement scale and provides an indication of the fluid's flow rate [22-24]. In the advanced platform

188

for the development of sprayers is used an electromagnetic flowmeter having a useful measuring range from 5 to 100 liters /min for pressures up to 4000 kPa (40 bar), and can operate at a maximum temperature of 50° Celsius. The flowmeter sensor is used to measure the flow rate of the mixture (water + pesticide). The flowmeter has an open collector output (0 to 12V), and presents a calibration of 600 pulses per liter (ppl), as well as accuracy in the class of 0.5 % to 1 % (full flow). The used type has no mechanical moving parts inside the pipe and performances independent from fluid density and viscosity. Besides, the part in contact with the mixture is made of polypropylene and stainless steel.

The conductivity sensor is responsible for measuring the concentration of the mixture (water + pesticide) and uses instrumentation for signal conditioning, which involves a power converter circuit to voltage (I/V). In general, the conductivity of a solution varies significantly depending on the temperature of the liquid [25-28]. When carrying out conductivity measurement, liquid temperature should be controlled as accurately as possible. The conductivity sensor used in the advanced platform for sprayer's development has four electrodes for medium and high conductivity. Also, uses a PT100 sensor built-in for automatic temperature compensation. The electrical conductivity sensor works with pressures up to 700 kPa, as well as withstands temperatures from -10 °C to 80 °C. The output of the transmitter produces current outputs in the range of 0 to 20 mA, i.e., proportional to the conductivity values (0-650 mS/cm) and the temperature in the range from 0 to 100° C. Other characteristics include a cell constant accuracy equal to ± 5 %, and a cell constant repeatability of equal to ± 2 %.

The response time for a direct injection sprayer system was defined in the literature by Peck and Roth [29] as the period from the instant the injection begins until the chemical concentration rate reaches 95 % of the equilibrium rate [30]. The rise time t_r and transport delay t_d characteristics of a sprayer proposed by these authors are shown in Fig. 10.7. A 95 % concentration rate corresponds to the chemical concentration of the spraying, which is necessary for satisfactory weed control [31].

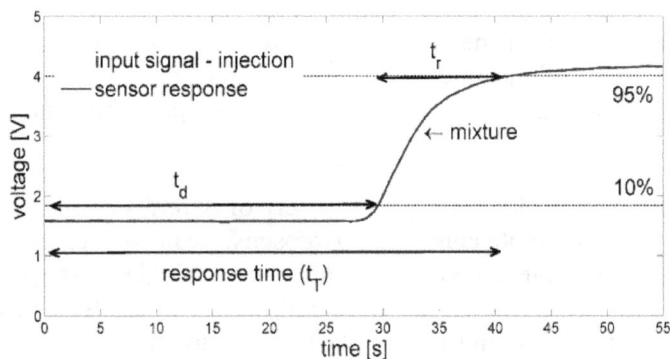

Fig. 10.7. Delay time t_d, rise time t_r and response time t_T of a typical injection system as described in [29]. The dotted line indicates the time behaviour of the concentrated mixture as a response to an injection input.

In the advanced platform for sprayers' development is used a set of intelligent sensors to measure the response time of the sprayers, which have usability in real time applications [32, 33, 34]. The decision to embed the smart sensors directly in the sprayer nozzles provide a scenario where the input data from the physical sensor could be analyzed by various knowledge-based routines. The sensor output could be raw data or preprocessed information. This information could be in the form of a flag, which shows a confidence level of the response time for pesticide applications. The use of these intelligent sensors provides more additional information than that of traditional ones, i.e., the information provided by intelligent sensors can include actual data, corrected data, validity of the data, and reliability of the sensor. The smart sensors operate in a dynamic range defined by 0.50 V ± 2 LSB < v_{output} ≤ 4.90 V ± 2 LSB which is related with the accuracy allowed by its internal ADC [35].

Liquid level sensors and switches can provide high-reliability for monitoring and detection the level of fluid media. To effectively address such a wide variety of measurement challenges, one can find a broad range of contact, non-contact, and non-intrusive liquid level sensors and switches [36-38]. These are available in multiple technology types, including magnetic reed switch-based floats, solid-state electro-optical, conductivity, capacitive, ultrasonic, and piezo-resonant. For the advanced sensor-based platform single point level switches are used, i.e., one for the liquid level of the mixing tank, and another to detect the pesticide level, i.e., installed into the pesticide tank.

Incremental encoder provides a specified amount of pulses in one rotation of a motor. The output can be a single line of pulses or two lines of pulses that are offset in order to determine rotation. The phasing between the two signals is called quadrature [39]. The typical assembly of an incremental encoder consists of a spindle assembly, a sensor array that creates just two primary signals for the purpose of position and speed, and cover.

In the advanced development platform for spraying are used the same types of encoders [40, 41] for the proportional valve and the injection pump. The encoders are connected to the DC motor spindle proportional valve, which enables incrementally 1024 pulses per revolution (ppr). This type of sensor is used in the modeling step proportional valve to obtain the angular position of the DC motor shaft, which drives plunger. Moreover, the motor angular velocity of the injection pump is also achieved through the use of an incremental encoder having 1024 pulses per revolution (ppr) either, and in the same sense, coupled to the motor spindle.

Infrared (IR) technology addresses a broad variety of wireless applications, especially in the areas of sensing and remote control. A photosensor can be used as a photointerrupter that integrates an optical receiver and emitter in a single U-shaped package. In a transmission type photointerrupter, the light emitting and detecting elements are placed facing each other. The shape and size are two of the main differentiating features of a photointerrupter [42-45]. For the advanced platform for sprayer's development are used two photosensors together the industrial belt that simulates the tractor movement in relation to nozzles sprayers. They are used to adjust by means of the frequency inverter the time line for its adequate operation. In other words, it is possible to use a constant speed of the belt or even accelerate it in order to reach different simulation conditions.

Over the belt are used a shaft that allow the use of a set of water-sensitive papers to get the spray droplet impressions or the droplets pattern for analysis. Besides, is used the e-splinke® software [46] program that allow accurate and rapid measure of spray droplet impressions from such water-sensitive paper. The process can be used to determine several useful spray drop statistics.

For instance, the percent coverage, the spray deposition rate, drift profile, single swath pattern width, and multiple pass uniformity are all determined. Droplet statistics such as VMD ($V_{(0.5)}$, Volume Median Diameter), $V_{(0.1)}$, and $V_{(0.9)}$ are automatically calculated for each drop card scanned [47]. A printout with a histogram of the drop sizes along with a graphic record is also provided by this software.

10.3. Sensors-Basis Arrangement and Examples of Operation

The spraying system with direct injection, from the viewpoint of control systems, can be represented by the block diagram of Fig. 10.8. From the diagram of Fig. 10.8, is possible to observe that two controllers are to be implemented. It is possible to consider classic controllers such as Proportional, Integral and Derivative (PID) or even advanced one's controllers, as the predictive model-based controller (MPC).

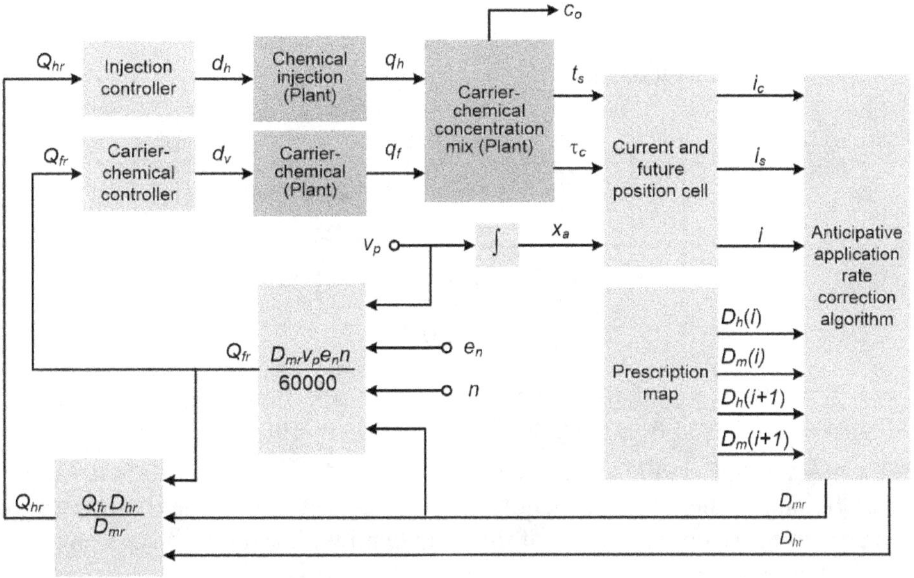

Fig. 10.8. Control loops for the variable-rate spraying system with direct injection.

10.3.1. Mathematical Models for the Injection, Spraying and Concentration Systems

The injection system consists of a piston pump and a DC motor driven by a transistorized power amplifier. This system is responsible for injecting the pesticide in the suction line

191

of the spray pump and its mathematical model describes the dynamic behavior of the flow of pesticide q_h. This state space model, described in detail in [48], is given by:

$$\dot{x} = \begin{bmatrix} -\dfrac{R_h}{L_h} & -\dfrac{K_e}{L_h} \\ \dfrac{K_t}{J} & -\dfrac{b}{J} \end{bmatrix} x + \begin{bmatrix} \dfrac{K_{pt}}{L_h} & 0 \\ 0 & -\dfrac{K_b}{J} \end{bmatrix} u,$$

$$y = [0 \quad 60000K_b]x$$

(10.1)

where $u = [d_h \quad p_h]^T$, with d_h the control signal of PWM type producing a duty cycle in the interval between 0 a 100, p_h is the pump output pressure, $y = q_h$ the output pump flow and $x = [i_h \quad \omega_h]^T$ the state vector with i_h the DC motor current and ω_h the DC motor speed. The variable ranges and parameter values in (10.1) are given in Table 10.2.

The hydraulic model of the spray system describes the dynamic behavior of the mixture flow denoted q_f and the pressure p_s of the system. This system has hydraulic and electrical components such as electric valves of the proportional type and on/off valves controlled by DC motors driven by H-bridge and PWM control, flow meter, pressure sensor, spray nozzles and hose pipes. The nonlinear state space model of the spraying system, which is described in detail in [49, 50], is given by:

$$\dot{x} = f(x,u) = \begin{bmatrix} x_2 \\ -\dfrac{1}{T_m}x_2 + \dfrac{K_m K_{ph}}{T_m}DZ(d_v) \end{bmatrix}$$

$$y = h(x,u) = \begin{bmatrix} \dfrac{Q_p\left[\sqrt{K_t\left(\alpha_0 + \alpha_1 e^{\frac{x_1}{\beta}}\right)} - \left(\alpha_0 + \alpha_1 e^{\frac{x_1}{\beta}}\right)\right]}{K_t - \left(\alpha_0 + \alpha_1 e^{\frac{x_1}{\beta}}\right)} \\ 0 \end{bmatrix},$$

(10.2)

where $u = d_v$ is the input of the H-bridge of the proportional type electrical valve, a control signal of the PWM type with duty cycle between -100 a 100, $y = q_f$ is the output flow and $x = [\theta_v \quad \dot{\theta}_v]^T$ is the state vector with θ_v and $\dot{\theta}_v$ the proportional valve needle rotation angle and the angular needle speed, respectively. The DC motor of the proportional valve has a nonlinearity of the dead band with saturation type represented in (10.2) by the variable $DZ(d_v)$:

$$DZ(d_v) = \begin{cases} D_{max}, & se\ d_v > D_{max} \\ k_+(d_v - d_+), & se\ d_+ < d_v \le D_{max} \\ 0, & se\ d_- < d_v \le d_+ \\ -k_-(d_v - d_-), & se\ d_- < d_v \le D_{min} \\ D_{min}, & se\ d_v < D_{min} \end{cases}.$$

(10.3)

The model of the dynamics of c_o for the mix concentration (water + pesticide), described in detail in [48, 49], can be approximated by a first order system with transport delay:

$$T_c \dot{c}_o + c_o = K_o c_i(t - \tau_c), \tag{10.4}$$

where $c_i = q_h/q_f$ is the input and c_o is the output of the mix concentration model, τ_c and T_c are the transport delay and time constant, respectively, which are dependent on the flow q_f as follows:

$$\tau_c = K_1 q_f^{-\varphi_1}. \tag{10.5}$$

$$T_c = K_2 q_f^{-\varphi_2} \tag{10.6}$$

where the parameter τ_c corresponds to the time the concentration output c_o remains unchanged after a change of the input c_i. From (10.5) and (10.6), the time response which is defined as the elapsed time from the time of injection until the concentration of the mixture reaches 98.2 % of its regime value at the most distant sprayer nozzle of the injection sprayer system can be estimated as:

$$t_s = 4T_c + \tau_c. \tag{10.7}$$

The values of the parameters as well as the definitions of the variables are also given in Table 10.2.

10.3.2. Proportional, Integral and Derivative Controller

The PID controllers appeared first in the 1930s, still today is considered the simplest solution for the control problems in industry [48, 52, 53]. In a spraying system with direct injection of chemicals it is necessary the design and implementation of two classic PID controllers, one for the sub-system injection and the other for the carrier-chemical sub-system. The PID controllers as illustrated in Fig. 10.9 were implemented in a platform CRIO, NI [54]. The sum of the proportional action by trapezoidal integration approximation and partial derivative is given by:

$$u(k) = u_p(k) + u_i(k) + u_d(k), \tag{10.8}$$

where

$$u_p(k) = K_c e(k)$$

$$u_i(k) = \begin{cases} u_i(k-1) + \dfrac{K_c T_{pid}}{2T_i}[e(k) + e(k-1)], & T_i > 0 \\ \\ 0, & T_i \le 0 \end{cases} \tag{10.9}$$

$$u_d(k) = -\dfrac{K_c T_d}{T_{pid}}[y(k) - y(k-1)]$$

Table 10.2. Spraying system variables and parameters.

Parameter/variable name	Value
Carrier-chemical sub-system	
Saturations limits [D_+, D_-]	[100,-100]
Pipe length L	1.35 m
Pipe internal diameter d_i	1.27×10^{-2} m
Friction factor f	8.2×10^{-3}
Mass density ρ	$1.0 \times 10^{+3}$ kg m^{-3}
Number of booms m	2
Number of nozzles per boom n	7
Sprayer pump volumetric flow rate Q_p	42.00 L min^{-1}
Proportional gain controller K_c	20
Nozzle 11003 fluid resistance K_n	2.07×10^2 kPa (L min^{-1})$^{-2}$
Nozzle 11005 fluid resistance K_n	0.75×10^2 kPa (L min^{-1})$^{-2}$
Solenoid valve fluid resistance K_{vs}	4.08×10^{-2} kPa (L min^{-1})$^{-2}$
Plumbing boom 1 until nozzle 1 fluid resistance K_{pb}	4.53×10^{-2} kPa (L min^{-1})$^{-2}$
Equivalent plumbing between nozzles fluid resistance K_{pn}	1.35×10^{-2} kPa (L min^{-1})$^{-2}$
Carrier-chemical subsystem equivalent fluid resistance K$_{eq11003}$	4.32 kPa (L min^{-1})$^{-2}$
Carrier-chemical subsystem equivalent fluid resistance K$_{eq11005}$	1.63 kPa (L min^{-1})$^{-2}$
H-bridge gain K_{ph}	1.20×10^{-1}
Proportional valve motor torque constant K_m	1.10 rad V^{-1}
Proportional valve motor time constant T_m	5.00×10^{-2} s
Proportional valve fluid resistance K_v	2.99×10^{-2} kPa (L min^{-1})$^{-2}$
Dead band limits [d_+, d_-]	[20,-20]
Gains [k_+,k_-]	[1.25,-1.25]
Coefficient α_0	2.99×10^{-2} kPa (L min^{-1})$^{-2}$
Coefficient α_1	2.81×10^{-6} kPa (L min^{-1})$^{-2}$
Coefficient β	6.53 rad
Chemical sub-system	
Armature resistance R_c	6.86×10^{-1} Ω
Armature inductance L_c	1.00×10^{-3} H
Valve motor torque constant K_t	3.75×10^{-2} N m A^{-1}
Valve motor back voltage constant K_e	3.75×10^{-2} V s rad^{-1}
Valve motor moment of inertia J	4.74×10^{-4} Kg m^2
Valve motor damping constant b	4.59×10^{-4} Kg m^2 s^{-1}
Volumetric displacement of the pump K_p	2.07×10^{-7} m^3 rad^{-1}
Power amplifier gain K_{pt}	1.20×10^{-1}
Coefficient K	5120
Carrier-chemical mix sub-system	
Coefficient K_1	158
Coefficient K_2	197
Coefficient ϕ_1	0.85
Coefficient ϕ_2	0.97
Minimum value provided by the flow meter q_-	5.00 L min^{-1}
Maximum value at the sprayer pump flow q_+	min^{-1}

where K_c is the proportional gain, T_i is the integration constant, T_d is the derivative constant, $T_{pid} = {}^{T_s}/_{60}$ is the time base of the PID controller and T_s is the sampling time. For both injection and spraying systems, the sampling period for the implementation of the digital controllers was set as $t_s = 50\ ms$ [55]. The initial setting of the controllers was performed used an autotuning relay feedback approach [56]. For the chemical carrier controller the PID parameters were adjusted to $K_c = 14$, $T_i = 3.5 \times 10^{-2}$, and $T_d = 8.0 \times 10^{-3}$. For the injection controller, the values were $K_c = 16.0$, $T_i = 6.0 \times 10^{-3}$. The implemented digital PID controller algorithm is presented in Pseudocode 10.1.

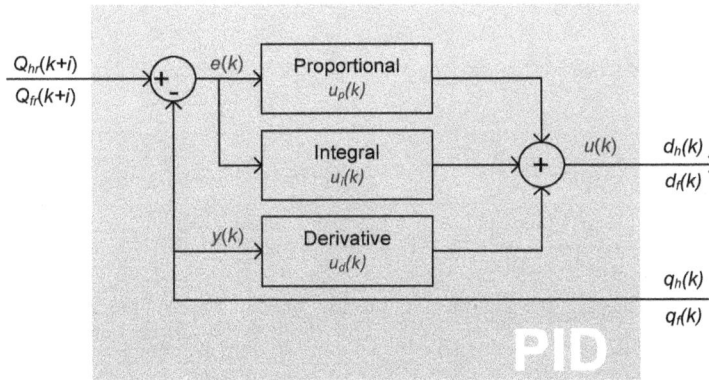

Fig. 10.9. PID based control loop for the carrier-chemical and PI for the injection controller.

Pseudocode 10.1. PID control loop.

Inputs: Q_{fr}, q_f
Outputs: d_f
BEGIN
initialization:
        $u_p \leftarrow 0; u_i \leftarrow 0;\ u_d \leftarrow 0; d_h \leftarrow 0; d_f \leftarrow 0; k \leftarrow 0;$
loop:
        READ Q_{fr}, q_f
        $y(k) \leftarrow q_f$
        $e(k) \leftarrow Q_{fr} - q_f$
        EVALUATE u_p, u_i, u_d
        $u(k) \leftarrow u_p(k) + u_i(k) +\ u_d(k)$
        $d_f \leftarrow u(k)$
        $k \leftarrow k + 1$
        WAIT T_{pid}
GOTO *loop*
END

The implemented PI and PID controllers presented satisfactory performances. The output flow reached the reference flow rates (set points) in a short period of time and with little overshoots. In the spraying system, the time taken to achieve the steady state is an important issue because it directly involves spraying efficiency. Thus, this accelerates the control action and lowers the application error. In Fig. 10.10, it is also possible to verify that the pressure values of the spraying system and spray tips were maintained in the normal operating levels, which in the case of used tips is at most 400 kPa.

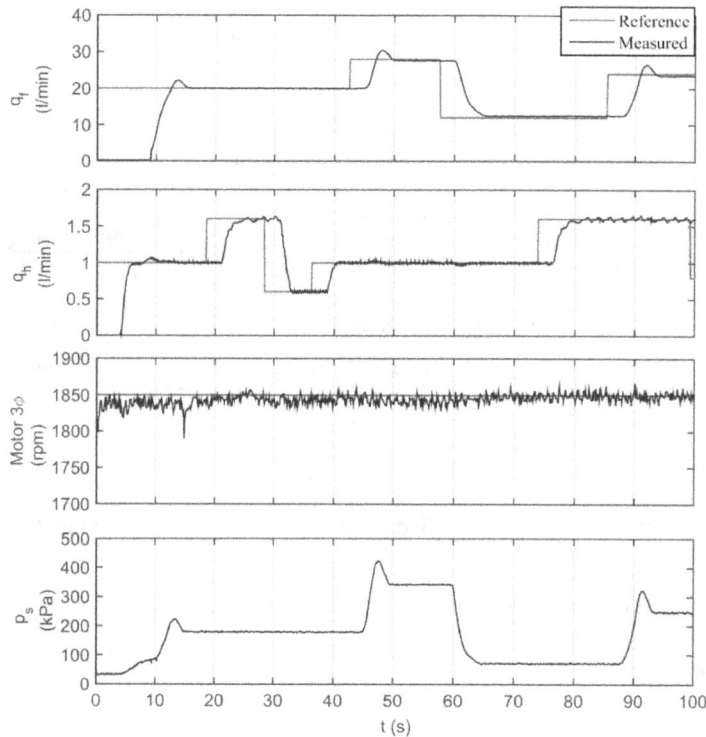

Fig. 10.10. Experimental results for the variable application rate using PI and PID controllers.

10.3.3. Model Based Predictive Controller

In general, a predictive control algorithm solves a problem of optimal control of a linear dynamic system subject to constraints. The controller predicts the future behavior of the actual system through the model of the plant over a horizon of upper and lower prediction, denoted by N_w e N_p, respectively. The optimal control input is calculated by minimizing a cost function defined over the prediction horizon. Typically, the optimization problem is formulated with a cost function specified as a sum of squared errors between the trajectory of the reference input and the predicted output and control effort as follows [51, 52]:

$$J(k) = \sum_{i=N_w}^{N_p} \|\hat{y}(k+i/k) - r(k+i/k)\|_{Q(i)}^2 + \sum_{i=0}^{N_u-1} \|\Delta u(k+i/k)\|_{R(i)}^2, \qquad (10.10)$$

subjected to:

$$\begin{aligned} u_{min} \leq u(k) \leq u_{max} \\ y_{min} \leq y(k) \leq y_{max} \\ \Delta u_{min} \leq \Delta u(k) \leq \Delta u_{max} \end{aligned} \qquad (10.11)$$

where Q(i) is the positive definite matrix weighing the error, R(i) is the positive semi-definite matrix weighing the control action, $\hat{y}(k+i/k)$ is the output prediction vector, $r(k+i/k)$ is the future reference trajectory vector and $\Delta u(k+i/k)$ is the incremental control for a horizon N_u. The command control of the current instant k is obtained by setting $u(k) = u(k-1) + \Delta\hat{u}(k|k)$. Only the actual control $\Delta\hat{u}(k|k)$ is used and the future control signal generated is discharged as at next time instant the output $y(k+1)$ will be known. The procedure is repeated and a new optimization problem is solved at the next step and the sequences actualized.

From a look up table values for the agrochemical rate of application of the mixture denoted D_{mr} and of the pesticide denoted D_{hr} were chosen and the controller references flows denoted Q_{fr} and Q_{hr} were calculated as [53, 54]:

$$Q_{fr} = \frac{D_{mr} v_p e_n n}{60000}, \qquad (10.12)$$

$$Q_{hr} = Q_{fr} \frac{D_{hr}}{D_{mr}}, \qquad (10.13)$$

where e_n is the distance between the sprayer tips, in centimeters and n is the number of active sprayer tips actives.

10.3.3.1. Implementing the MPCs

The spraying system with direct injection uses two MPCs in regulating the flows q_h and q_f. The MPC algorithm was developed in the LabVIEW RT with the Toolkit MPC Controller tool LabVIEW Control Design and was run on the embedded system cRIO-9073 at a deterministic sampling rate of 50ms [49, 50]. For regulating the flows q_h and q_f the diagrams of Fig. 10.11 and Fig. 10.12 were implemented, respectively.

For regulating q_h, the state space model described in (10.1), for $p_h = 0$, since the injection of agrochemical is made in the suction line of the sprayer pump, was discretized using the c2d Matlab function with 50 ms one sampling period. Therefore, considering the values of the parameters shown in Table 10.2, the following discrete model was obtained for the injection system:

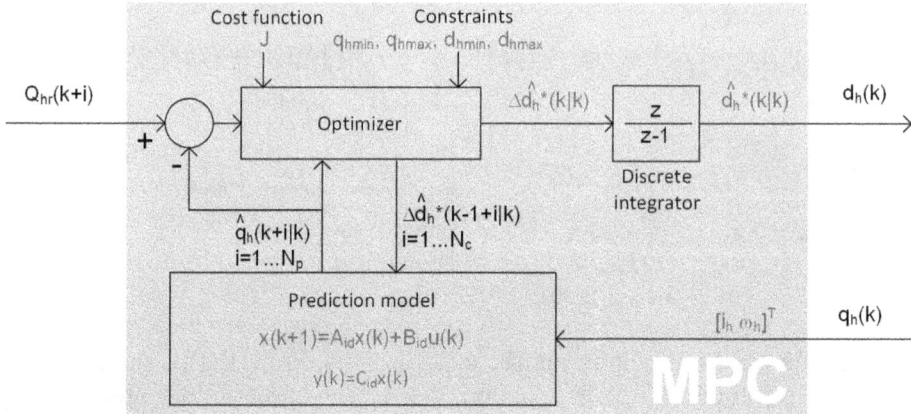

Fig. 10.11. MPC based control loop for injection controller.

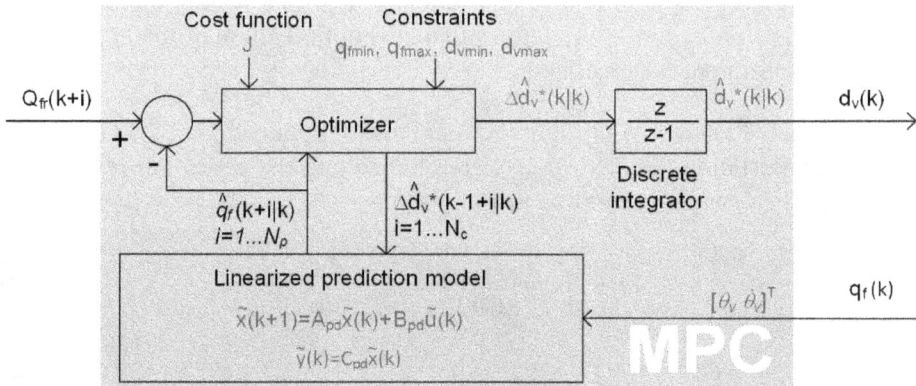

Fig. 10.12. MPC based control loop for carrier-chemical controller.

$$x(k+1) = \begin{bmatrix} -0.004938 & -0.04248 \\ 0.08963 & 0.7711 \end{bmatrix} x(k) + \begin{bmatrix} 0.1432 \\ 0.5956 \end{bmatrix} u(k)$$
$$y(k) = [0 \quad 0.01241]x(k)$$

(10.14)

where $u(k)$ is the control signal denoted d_h, and $y(k)$ is the output signal denoted q_h. The MPC tuning was initially carried out by simulation on the PC LabVIEW and the obtained parameters were used as a starting point for controller tuning directly in the sensor-based platform using the cRIO-9073. The final values of the parameters of the injection MPC were: $N_p = N_w = 2$, $N_c = 1$, $Q = 12$, $R = 0.15$, $u_{min} = 0$, $u_{max} = 65$, $y_{min} = 0$, $y_{max} = 2$, $\Delta u_{min} = -1$ e $\Delta u_{max} = 1$.

For regulating the flow q_f, the non-linear model represented by (2) was linearized using the fsolve function of Matlab around the equilibrium $\theta_v = 65$ rad and $\dot{\theta}_v = 0$ rad/s which corresponds to the midpoint of the work range of the proportional valve. Then, the linearized model is discretized using a 50 ms sampling. Therefore, considering the values of the parameters shown in Table 10.2, the following discrete model [55] was obtained for the spraying system:

$$x(k + 1) = \begin{bmatrix} 1 & 0.03161 \\ 0 & 0.3679 \end{bmatrix} x(k) + \begin{bmatrix} 0.002152 \\ 0.07396 \end{bmatrix} u(k)$$
$$y(k) = [0.55 \quad 0]x(k)$$

(10.15)

where $u(k)$ is the control signal denoted d_v and $y(k)$ is the output signal denoted q_f. The tuning of the MPC was done in the same way as for regulating the flow q_h. The final values of the parameters of the carrier-chemical controller MPC were: $N_p = N_w = 50$, $N_c = 1$, $Q = 15$, $R = 0.015$, $u_{min} = -90$, $u_{max} = 90$, $y_{min} = 0$, $y_{max} = 34$, $\Delta u_{min} = -1$ e $\Delta u_{max} = 1$. The implemented digital MPC controller algorithm is presented in Pseudocode 10.2.

Pseudocode 10.2. MPC control loop for the carrier-chemical.

Parameters: $A_{pd}, B_{pd}, C_{pd}, N_p, N_w, N_c, Q, R, u_{min}, u_{max}, y_{min}, y_{max}, \Delta u_{min}, \Delta u_{max}$
Inputs: Q_{fr}, q_f
Outputs: d_v
BEGIN
initialization:
      $\Delta \hat{d}_v^* \leftarrow 0; \hat{d}_v^* \leftarrow 0; d_v(\text{k}) \leftarrow 0; \hat{q}_f \leftarrow 0;$
loop:
      READ Q_{fr}, q_f
      EVALUATE $\hat{q}_f$ (prediction model)
      EVALUATE $\Delta \hat{d}_v^*$ (optimal control sequence)
      $d_v(\text{k}) \leftarrow d_v(\text{k} - 1) + \Delta \hat{d}_v^*(\text{k}|\text{k})$
      $k \leftarrow k + 1$
      WAIT mpc sample rate period
GOTO *loop*
END

10.3.4. Experimental Results for Variable Rate Application Based on Infestation Maps

An experiment was conducted in the spraying sensor-based platform using tap water in the water tank. The agrochemical tank was filled with 16 liters of tap water containing 50 g of sodium chloride (NaCl). The application map considered featured a hypothetical example of a soybean infested by weeds. The culture contained $i = 7$ cells of length of $L_c = 300$ m. The agrochemical rate of application D_{hr} between the cells ranged from 2 to 12 L/ha and the application rate of the mix denoted D_{mr} was 230 L/ha. The recommendations of the application rates were based on the label of the herbicide glyphosate athanor 48. The chosen v_p application speed was 10 km/h with random variations of ± 10 % in some places [49]. Fig. 10.13 shows the responses of the instantaneous flow rates and q_f in function of speed variations.

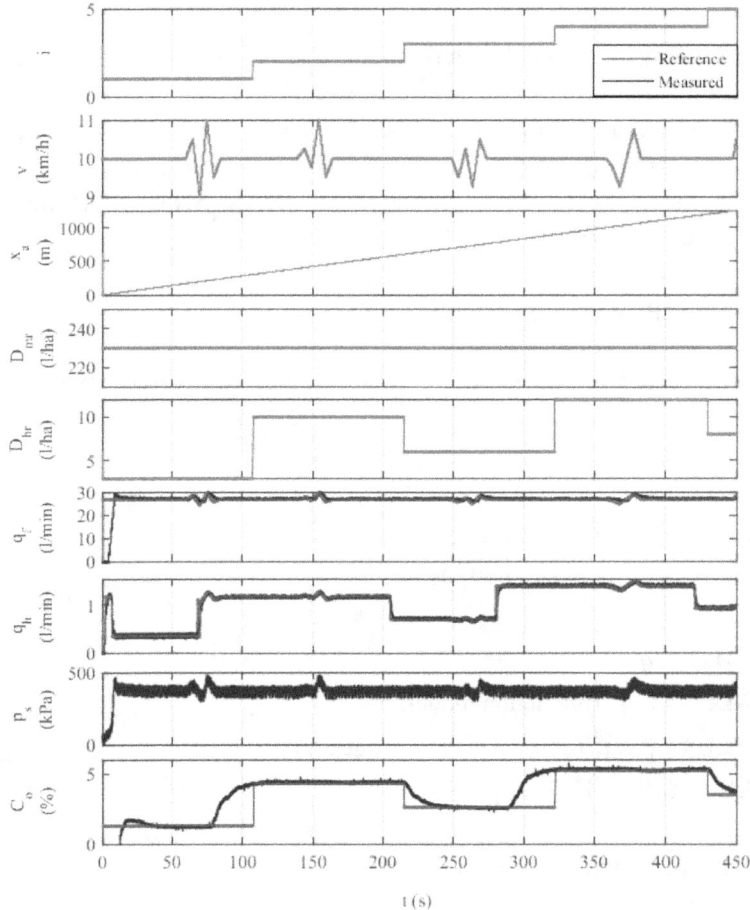

Fig. 10.13. Experimental results of the application rate based in maps.

Notice that the system behavior showed no peaks or excessive variations in the values of instant flow rates over time, due to the speed variation. The changes in flow rates do not exceed 5 % relative to the average, indicating an acceptable behavior of the MPCs [56]. It is also noticeable that the flows reached the reference values in an acceptable time and with little overshoot.

The change in the mixture concentration due to application rate changes and/or speed variations occurred during a short time and the peaks were below 5 % due to simultaneous adjustment of q_h and q_f flows performed by the MPCs.

It was also noticed that an anticipation strategy of the application rates can be performed with success in cases where the application rates increases from one cell to another. This strategy avoids subapplication cases and reduces the error of the mixture concentration as it ensures the application of the agrochemical in the desired mixture concentration in cells where the application rate is greater than the application rate of the previous cell.

10.4. Conclusions

The models and real-time optimization methods customized for agricultural sprayers can be in time embedded into a microcomputer by means of dedicated software development using very high-level languages and validated in an environment which allows virtual instruments, development, tests, calibration, and data analysis for support on the decision making processes.

The goal of such initiative, i.e., the organization of an advanced sensor-based platform for the development of agricultural sprayers took into account programming and sensor-based instrumentation and their accessibility to engineers, programmers, technical personnel, students, researchers and other experts, which seek, for instance, the rapid prototyping and flexible programmable architectures for the development of agricultural sprayers.

Based on the above, one may conclude that this infrastructure is adequately in establish a laboratorial facility dedicated for learning, design and construction of agricultural sprayers for the liquid agricultural inputs, which uses the sensors, actuators, and control systems. Besides, the presented laboratory infrastructure can also be identified as high-level programmable functions, which included a set of I/O ports, not only analogical gates but also digital, which allow a complete design in compliance with the applicable requirements and standards, i.e., used to reach both spray quality and efficiency in the agricultural pest control.

Acknowledgements

The authors would like to thank the Brazilian Agricultural Research Corporation (Embrapa) by grants MP2: 02.07.11.025.00.00, and MP1: 01.09.01.002.00.00, as well as the National Council for Scientific and Technological Development (CNPq) by grants 143452/2008-8, 479306/2008-7, 304985/2009-0, and 306477/2013-0.

References

[1]. M. Heidary, J. P. Douzals, C. Sinfort, A. Vallet, Influence of spray characteristics on potential spray drift of field crop sprayers: A literature review, *Crop Protection*, Guildford, Vol. 63, 2014, pp. 120-130.
[2]. H. C. Simmons, Development of instrumentation for spray drop-size research, *Pesticide Formulations and Application Systems*, Vol. 15, *ASTM STP 915*, L. D. Spicer and T. M. Kaneko, Eds., American Society for Testing and Materials, *Philadelphia*, 1986, pp. 108-113.

[3]. S. Paulus, J. Behmann, A. K. Mahlein, L. Plumer, H. Kuhlmann, Low-Cost 3D Systems: Suitable Tools for Plant Phenotyping, *Sensors*, 2014, Vol. 14, 3001-3018.

[4]. Standard Practice for Determining Data Criteria and Processing for Liquid Drop Size Analysis, Annual Book of ASTM® Standards, *General Methods and Instrumentation*, Vol. 14, 02, 1996, pp. 535-539.

[5]. Standard Terminology relating to liquid particle statistics. 1996 Annual Book of ASTM Standards, *General Methods and Instrumentation*, Vol. 14, 02, 1996, pp. 810-812.

[6]. R. G. Prinn, P. G Simmonds, R. A. Rasmussen, A. J. Crawford, R. D. Rosen, F. N. Alyea, C. A. Cardelino, D. M. Cunnold, P. J. Fraser, J. E. Lovelock, The atmospheric lifetime experiment. 1: Introduction, instrumentation, and overview, *Journal of Geophysical Research*, Vol. 88, C13, 1983, pp. 8353–8367.

[7]. A. J. Prata, R. P. Cechet, I. J. Barton, D. T. Llewellyn Jones, The along-track scanning radiometer for ERS-1: Scan geometry and data simulation, *IEEE Transactions on Geoscience and Remote-sensing*, Vol. 28, 1990, pp. 3–13.

[8]. B. K. Fritz, W. C. Hoffmann, J. A. Bonds, M. Farooq, Volumetric collection efficiency and droplet sizing accuracy of rotary impactors, *Transaction of America Society of Agricultural and Biological Engineers*, 2011, pp. 57-63.

[9]. K. P. Gillis, D. K. Giles, D. C. Slaughter, D. Downey, Injection mixing system for boomless, target-activated herbicide spraying. *Transaction of ASAE*, 46, 4, 2003, pp. 997-1008.

[10]. A. Osterman, T. Godesa, M. Hocevar, B. Sirok, M. Stopar, Real-time positioning algorithm for variable-geometry air-assisted orchard sprayer, *Computers and Electronics in Agriculture*, Vol. 98, 2013, pp. 175–182.

[11]. R. Taymanov, K. Sapozhnikova, Metrological selfcheck and evolution of metrology, *Measurement*, Vol. 43, Issue 7, 2010, pp. 869-877.

[12]. K. L. Hughes, A. R. Frost, A review of agricultural spray metering, *Journal of Agricultural Engineering Research*, Vol. 32, Issue 3, 1985, pp. 197-207.

[13]. S. Y. Yurish, Sensors: Smart vs. Intelligent, *Sensor & Transducers*, Vol. 114, Issue 3, March 2010, pp. 1-6.

[14]. R. C. Michelini, R. P. Razzoli, G. M. Acaccia, R. M. Molfino, Mobile robots in greenhouse cultivation: inspection and treatment of plants, *ACTA Horticulture*, Vol. 453, 1998, pp. 52-60.

[15]. O. Mencer, ASC: A stream compiler for computing with FPGAs, *IEEE Transaction on Computer-Aided Design of Integrated Circuits and Systems*, Vol. 15, No. 9, 2006, pp. 1603–1617.

[16]. A. Krikelis, C. Weems, Associative processing and processors, *Computer*, Vol. 27, No. 11, 1994, pp. 12–17.

[17]. N. Ertugrul, New era in engineering experiments: an integrated interactive teaching/ learning approach and real time visualisations, *International Journal of Engineering Education*, 1998, Vol. 14-5, pp. 344-355.

[18]. N. Ertugrul, Towards virtual laboratories: a survey of LabVIEW-based teaching/learning tools and future trends, *International Journal of Engineering Education*, Vol. 16-3, 2000, pp. 1-10.

[19]. W. P. Eaton, J. H. Smith, Micromachined pressure sensors: review and recent developments, *Smart Materials and Structures*, Vol. 6, Number 5, 1997, pp. 530-539.

[20]. R. P. Benedict, Fundamentals of temperature, pressure and flow measurements, Third Edition, *John Wiley & Sons*, New York, 1984.

[21]. W. C. Dunn, Introduction to instrumentation, sensors, and process control, *Artec House Inc.,* London, 2006.

[22]. K. Minemura, K. Egashira, K. Ihara, H. Furuta, K. Yamamoto, Simultaneous measuring method for both volumetric flow rates of air-water mixture using a turbine flowmeter,

Transaction on ASME and Journal of Energy Resources Technology, Vol. 118, 1996, pp. 29-35.

[23]. S. Y. Yurish, Digital Sensors and Sensor Systems: Practical Design, *International Frequency Sensor Association (IFSA) Publishing*, 2011.

[24]. D. W. Spitzer, Industrial Flow Measurement, *Research Triangle Park, ISA*, 1991.

[25]. B. X. J. R. Chen, J. Yi, Working principle and improvement of testing methods of thermal conductivity sensor, *Chemical Engineering & Equipment*, Vol. 2, 2010, pp. 64-66.

[26]. F. Rastrello, P. Placidi, A. Scorzoni, Measurements, FEM simulation and spice modeling of a thermal conductivity detector, in *Proceedings of the IEEE Instrumentation and measurements Technology Conference*, Hangzhou, China, 2011, pp. 651-655.

[27]. L. Parra, S. Sendra, J. Lloret, I. Bosch, Development of a Conductivity Sensor for Monitoring Groundwater Resources to Optimize Water Management in Smart City Environments, *Sensors*, 2015, No. 15, pp. 20990-21015.

[28]. L. Parra, V. Ortuño, S. Sendra, J. Lloret, Low-Cost Conductivity Sensor Based on Two Coils, in *Proceedings of the 1st International Conference on Computational Science and Engineering (CSE'13)*, Valencia, Spain, 2013, pp. 107–112.

[29]. D. R. Peck, L. O. Roth, Field sprayer induction system development and evaluation, *Transaction of the ASAE*, 1975, Paper No. 75-1541.

[30]. K. A. Bennet, R. B. Brown, Direct nozzle injection and precise metering for variable rate herbicide application, *Transaction of the ASAE*, 1997, Paper No. 97-1046.

[31]. K. Ogata, Modern Control Engineering, 4th Edition, *Prentice Hall PTR*, 2001.

[32]. H. V. Mercaldi, C. H. Fujiwara, E. A. G. P. Penaloza, V. A. Oliveira, P. E. Cruvinel, Smart and customized electrical conductivity sensor for measurements of the response time from sprayers based on direct injection, *Sensors & Transducers*, Vol. 193, Issue 10, October 2015, p. 1-10.

[33]. E. A. Anglund and P. D. Ayers, Field evaluation of response times for a variable rate (pressured-based and injection) liquid chemical applicator, *Applied Engineering in Agriculture*, Vol. 19, Issue 3, 2003, pp. 273-282.

[34]. M. S. Miller, D. B. Smith, A review of application error for sprayers, *Transactions of the ASAE*, Vol. 35, Issue 3, 1992, pp. 787-791.

[35]. M. Villa, F. Tian, P. Cofrancesco, J. Hal·mek, M. Kasal, High resolution digital quadrature detection, *Review of Scientific Instruments*, Vol. 67, No. 6, 1996, pp. 2123-2129.

[36]. E. Virginia Ebere, O. O. Francisca, Microcontroller based automatic water level control system, *International Journal of Innovative Research in Computer and Communication Engineering*, Vol. 1, Issue 6, 2013, pp. 1390-1396.

[37]. E. J. Chern, B. B. Djordjevic, Nonintrusive ultrasonic low-liquid-level-sensor, *Materials Evaluation*, 48, 4, 1990, pp. 481–485.

[38]. N. B. Manik, S. C. Mukherjee, A. N. Basu, Studies on the propagation of light from a light-emitting diode through a glass tube and development of an optosensor for the continuous detection of liquid level, *Optical Engineering*, 40, 12, 2001, pp. 2830–2836.

[39]. T. Dogša, M. Solar, B. Jarc, Delaying analogue quadrature signals in Sin/Cos encoders, *Journal of Microelectronics, Electronic Components and Materials*, Vol. 44, No. 1, 2014, pp. 69 – 74.

[40]. Z. Jun, W. Z. Gong, H. Q. Sheng, X. Jie, Optimized design for high-speed parallel BCH encoder, in *Proceedings of the IEEE International Workshop on VLSI Design and Video Technology*, Suzhou, China, 2005, pp. 97-100.

[41]. J. Lee, S. Shakya, Implementation of Parallel BCH Encoder Employing Tree-Type Systolic Array Architecture, *International Journal of Sensor and Its Applications for Control Systems*, Vol. 1, No. 1 2013, pp. 1-12.

[42]. P. M Novotny, N. J. Ferrier, Using infrared sensor and the Phong illumination model to measure distances, in *International Conference on Robotics and Automation*, Detroit, MI, USA, Vol. 2, 1999, pp. 1644-1649.

[43]. G. Benet, F. Blanes, J. E. Simo, P. Perez, Using infrared sensors for distance measurement in mobile robots, *Journal of Robotics and Autonomous Systems*, Vol. 10, 2002, pp. 255-266.

[44]. A. M. Flynn, Combining sonar and infrared sensors for mobile robot navigation, *The International Journal of Robotics Research*, Vol. 7, 6, 1988, pp. 5-14.

[45]. P. Sharma, D. Joshi, H. Raghuvanshi, L. Yogi, Remote Operated Master Switch via Infrared Technology, *International Journal of Emerging Research in Management &Technology*, Vol. 4, Issue-5, 2015, pp. 102-106.

[46]. P. E. Cruvinel, S. R. Vieira, S. Crestana, E. R. Minatel, M. L. Mucheroni, A. T. Neto, Image processing in automated measurements of raindrop size and distribution, *Computers and Electronics in Agriculture*, 23, 3, 1999, pp. 205-217.

[47]. G. A. Matthews, Pesticide Application Methods, 3rd Edition, *CRC Press*, Ames, 2000.

[48]. C. C. Yu, Autotuning of PID Controllers: Relay Feedback Approach, *Springer-Verlag*, 1999.

[49]. K. R. Felizardo, H. V. Mercaldi, P. E. Cruvinel, V. A. Oliveira, B. L. Steward, Modeling and model validation of a chemical injection sprayer system, *Applied Engineering in Agriculture*, Vol. 32, 2016, pp. 285-297.

[50]. K. R. Felizardo, Modelagem e controle preditivo de um sistema de pulverização com injeção direta (in Portuguese), Doctoral Dissertation, *Universidade de São Paulo*, 2013.

[51]. K. R. Felizardo, H. V. Mercaldi, V. A. Oliveira, P. E. Cruvinel, Modeling and predictive control of a variable-rate spraying system, in *Proceedings of the 8th EUROSIM Congress on Modelling and Simulation*, Cardiff, United Kingdom, 202-207, 2013.

[52]. J. Maciejowski, Predictive Control with Constraints, 1st Ed., *Prentice Hall*, 2002.

[53]. E. F. Camacho, C. Bordons, Model Predictive Control, 2nd Ed., *Springer*, 2007.

[54]. K. R. Felizardo, H. V. Mercaldi, V. A. Oliveira, P. E. Cruvinel, Sistema embarcado de tempo real para automação de uma bancada de testes de pulverizadores (in Portuguese), in *Proceedings of the 10th IEE/IAS International Conference on Industry Applications*, Fortaleza, Brazil, 2012, pp. 5-7.

[55]. C. H. Houpis, G. B. Lamont, Digital Control Systems, 2nd Ed., *McGraw-Hill Higher Education*, 1992.

[56]. O. Katsuiko. Modern Control Engineering 4th Edition, *Prentice Hall*, 2003.

Chapter 11
Design, Implementation and Characterization of Time-to-Digital Converter on Low-Cost FPGA

Dadouche F., Turko T., Malass I., Skilitsi A., Léonard J. and Uhring W.

11.1. Introduction

Nowadays, numerous applications require a precise measurement of time duration separating two or several physical events. Examples of these applications include: Time of Flight sensors, Time resolved spectroscopy and fluorescence, positron emission tomography instruments as well as pulse position demodulators [1].

3D scanners or 3D console games represent typical applications using Time of Flight (TOF) measurements to reconstitute a three-dimensional scene. In such systems, the light is emitted by a laser or light-emitting diode and detected by suitable light sensors after reflection by the illuminated object. The light pulse TOF is proportional to the distance traveled by the latter. The measurement is made independently by several pixels allowing the reconstitution of the 3D scene [2, 3].

To measure this duration, one uses devices capable of converting extremely low time resolution (some tens of picoseconds) into digital values understandable for downstream processing and conditioning chain. These devices are commonly known as Time-to-Digital Converters (TDCs) [4]. They have been largely used for several years in numerous smart sensor systems, particle and high-energy physics applications as well as measurement and instrumentation applications such as digital scopes and logic analyzers [4, 5].

The several techniques of time-to-digital conversion proposed in the literature can be regrouped in two families: the first generation family based on ADC converters and the second generation family which is fully digital [4]. In order to design a TDC, a method

Wilfried Uhring
ICube, UMR 7357, Université de Strasbourg and CNRS, Strasbourg, France

based on Time-to-Amplitude Conversion (TAC) can be used. This technique is based on controlling a capacitor discharge. The time elapsed between start and stop signals is determined by the capacitor voltage before and after discharging. However, the TACs need very fast analog switches as well as high precision current sources and ADCs. So this technique is not well suited for FPGA use [6].

As an alternative fully digital techniques are proposed. The simplest way to achieve this objective is to use a digital counter starting and stopping by the measured signal. Nevertheless, the resolution of such systems is limited by the operating frequency. Indeed, to achieve a resolution of 100 ps, a frequency of 10 GHz is theoretically required. With the current clock frequency limitation, this technique is considered to be restricted for time resolution above 1 ns in most of the FPGA based applications [7].

The solution which seems to be interesting for digital FPGA devices is the direct time-to-digital conversion which allows to measure both long and short time intervals. Some of the digital techniques that can be readily identified are [4, 7, 8]: Tapped Delay Lines (TDL), Delay Locked Loop (DLL), Vernier Delay Line (VDL), Multilevel TDC, etc. All of these time-to-digital conversion techniques are usually designed as Application-Specific Integrated Circuits (ASICs) [9]. ASIC TDCs offer high performances but suffer from a high cost, slow time to market and limited reconfiguration possibilities. It is also worth noting that the ASIC solutions are not suitable for integration into reconfigurable digital designs mostly described in Hardware Description Languages (HDL). As a result, numerous solutions for implementing TDCs on FPGA circuits have emerged [10-15]. However, the most significant limitation of these architectures is the difficulty to predict the placement and routing delays as well as the time delay of the logic gates itself. The consequence of this inevitable hardware restriction is a non-stable resolution of the designed TDC [10].

In this chapter we aim to bring together and extend our previous works [16-18] in a more complete and unique document dedicated to introduce the design, the implementation as well as the characterization of a high performance TDC on a low cost FPGA. It is organized as follows: Section 11.2 overviews TDC's families and introduces the structure of the studied TDC; Section 11.3 presents the proposed design methodology; Section 11.4 gives a first case study showing the implementation of a 42 ps TDC on a cyclone IV FPGA; Section 11.5 deals with TDC characterization using Poisson Process Events as well as the presentation of a fast and efficient correction method; Section 11.6 illustrates with a second case study the use of the designed TDC for a fluorescence lifetime measurement based on TCSPC (Time-Correlated Single Photon Counting) and demonstrates the efficiency of the correction method at improving the transfer function linearity; Section 11.7 provides some final observations.

11.2. Time-to-digital Converters Families and their Functioning Principle

A TDC is an electronic system that measures the time interval between two occurring events of a given signal. Its main purpose is to convert temporal information to binary sequence understandable for a downstream processing chain. But there are different ways to perform this operation that leads to different kinds of architectures which can be divided into two families: the first generation family based on ADC converters and the second generation family which is fully digital. This section will present the general operation principles of these techniques and introduces the chosen structure to be implemented and characterized in the following sections.

11.2.1. First Generation of TDCs or Analog Time-to-Digital Converters

The first generation TDCs convert the time into digital values in two steps. The first step consists of converting the time interval into an amplitude (voltage) using the Time-to-Amplitude Converters (TAC). In the second step the obtained voltage is digitized by an Analog-to-Digital Converter (ADC). The operation principle is illustrated by Fig. 11.1 bellow. To digitize the time interval to be measured (t_m), two signal events are used to generate a pulse with the same time as t_m [4]. The generated pulse is applied to an integrator which converts it to a voltage which is then digitized by an ADC.

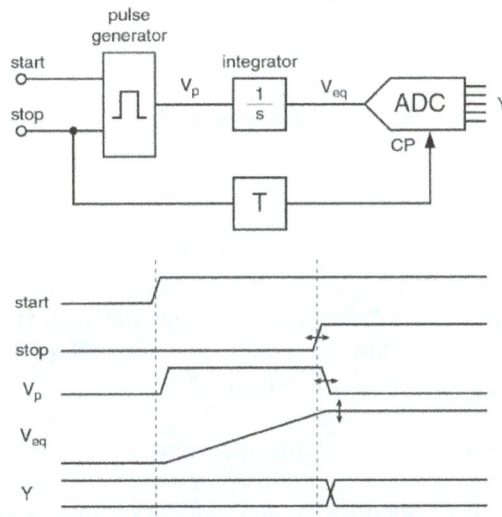

Fig. 11.1. Block and signal diagram of basic analog time-to-digital converter, e.g. [4].

The time-to-amplitude conversion is introduced in 1942 by Bruno Rossi to measure muons life time with a good precision but for military raisons it was withheld until 1946 [19] [20]. This operation is carried out firstly by using vacuum tubes before to move to ferrite cores and finally CMOS circuits.

The performances of Analog Time-to-Digital Converters strongly depends on that of the included ADC and on the integrator output noise. Indeed, to measure short time durations with a good precision it is necessary to use a high resolution ADC. However measuring long durations needs to increase further the ADC dynamic range. As a result, it is very difficult to digitize long time durations with a high precision that leads to a bad TDC resolution for high dynamic range time measurements [7]. This is the main weakness of

5

the first generation TDCs using a TAC associated with an ADC. We will see in the following subsection that fully digital TDCs don't suffer from this limitation.

11.2.2. Second Generation of TDCs or Fully Digital TDCs

The need for increasingly high time resolution resulted in shifting to fully digital TDC architectures. The first idea is to use a counter with high clock frequency. The counting process is triggered by the signal to be measured. It starts by the first raising edge of the signal and it stops with its falling edge. Consequently, the time resolution is determined by the counter clock frequency. The resolution is as high as the clock frequency. However, the higher the frequency, the more energy-intensive the circuit. Moreover, the standard CMOS technology is not suitable for very high frequencies. For example, the maximum frequency achievable in 65 nm CMOS technology is around 5 - 10 GHz, leading to a time resolution of 100 ps at best.

To overcome this limitation, one interesting and commonly used technique is the Nutt method allowing to measure, at the same time, long as well as short time intervals. Its principle is introduced in the following.

11.2.2.1. Functional Principle of Digital TDCs or Nutt Method

The technique presented by Ronald Nutt in 1968 is used to measure long interval times (up to 1 millisecond) with very high resolution (200 ps) [21]. For this purpose the Nutt method combines different time measurements with appropriate scale for each one. More precisely, it is composed of three blocks: two fine measurement blocks and a coarse one. The coarse one counts the number (N) of clock periods between start and stop signals, and the fine blocks evaluate the uncertainties, which are shorter than the clock period, on both sides of the time interval, as illustrated in Fig. 11.2.

The time interval to be measured (T_m) is thus a combination of three individual durations: (1) T_{Coarse}, which represents an integer number of clock periods, (2) T_{Fine1}, representing the time between the START signal rising edge and the first following clock rising edge, and (3) T_{Fine2} which is the time between the falling edge of the measured interval and the following clock rising edge. According to this timing diagram, the measured time will be expressed as follows:

$$T_m = T_{Fine1} + T_{Coarse} - T_{Fine2},$$ (1)

however, given the fact that:

$$T_{Coarse} = N \cdot T_{clk},$$ (2)

we obtain the following expression:

$$T_m = T_{Fine1} + N \cdot T_{clk} - T_{Fine2},$$ (3)

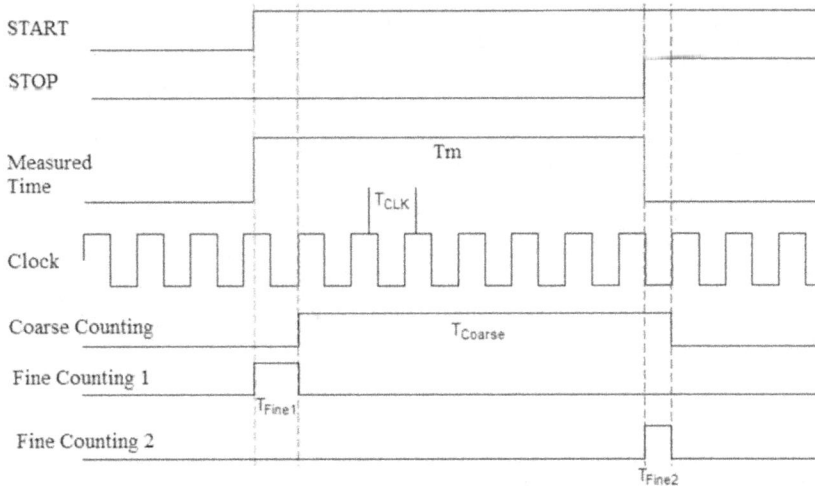

Fig. 11.2. Functioning principle of a generic TDC.

In most situations the start signal can be synchronized with the main clock. As an example, in fast imaging systems we need to know the delay separating a photon emission by a laser diode and the detection of that photon by a Single-Photon Avalanche Diode (SPAD). So the START signal triggering the laser diode can be synchronized with the coarse counter clock. The STOP signal announces the detection of the photon by the SPAD. In those cases, the whole TDC can be reduced to a coarse counter associated to a fine TDC measuring T_{Fine2}. Consequently, the measured interval time will be given by:

$$T_m = N \cdot T_{clk} - T_{Fine2}, \qquad (4)$$

Since the coarse block is a simple counter incremented by the system clock, we will focus in the following sections on the implementation of the fine TDC.

11.2.2.2. Structure of the Studied TDC

As mentioned previously, there are different techniques of designing TDCs. In this work, we focus on the commonly used Tapped Delay Lines (TDL) architecture depicted in Fig. 11.3.

A TDL TDC consists of N cascaded delay elements whose inputs are stored in D Flip Flops (DFFs). We would then have as many DFFs as there are delay elements. Therefore, each delay element associated with its DFF forms an elementary cell of the TDC.

The number (N) of these elementary cells depends on the common DFF clock frequency ($1/T_{clk}$), as well as the propagation time of the delay element (t_d). This is given by the ratio of clock period to propagation time t_d. Since the value of t_d is not provided, it is determined experimentally.

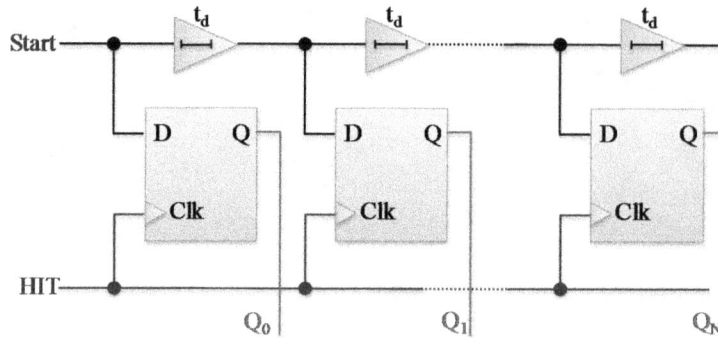

Fig. 11.3. Tapped delay line TDC.

11.3. Design of the Studied TDC

11.3.1. Design and Validation of the Elementary Cell

To implement the structure of the TDL TDC introduced above, the first idea that comes in mind is to design an elementary cell which will be replicated according to the desired line size. To do so, we can use a simple invertor as a delay element associated with a DFF as illustrated by Fig. 11.4.

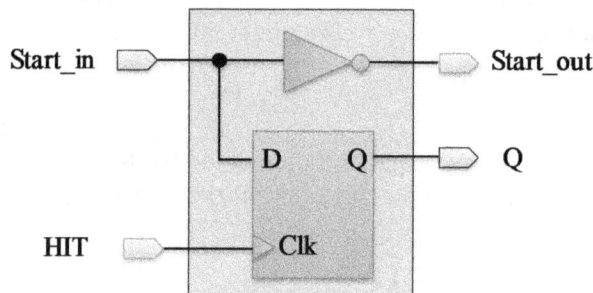

Fig. 11.4. Simple TDC elementary cell.

However, implementing a TDC chain on an FPGA by replication of this cell leads to a simplified circuit entirely different from the desired function. Indeed, if the input signal Start_in and the output of the second TDC elementary cell signal Start_out have the same logic equation, the logic synthesis tool used (the software Quartus II in our case) will simplify the logical equation giving the output versus the input so that it saves place and time. The consequence is a bypass of the delay element of the TDC cell.

To illustrate this phenomenon, we represent in Fig. 11.5 the RTL (Register Transfer Level) view resulting from the implementation of a simple TDC chain composed of four elementary cells.

It can readily be seen that, in spite of the presence of inverters, the software has simplified the logical equations. Consequently, all the inverted signals are grouped independently of the non-inverted ones. It is thus evident that this method is not suitable for designing a TDC on FPGA. Nevertheless, it is worth to notice that, if it is not possible to prevent the Quartus II software to optimize the data path, it is instead possible to create this path manually by operating directly on the logical resources of the FPGA. Indeed, the Quartus II Chip Planner tool allows physical access to logical resources available on the chip.

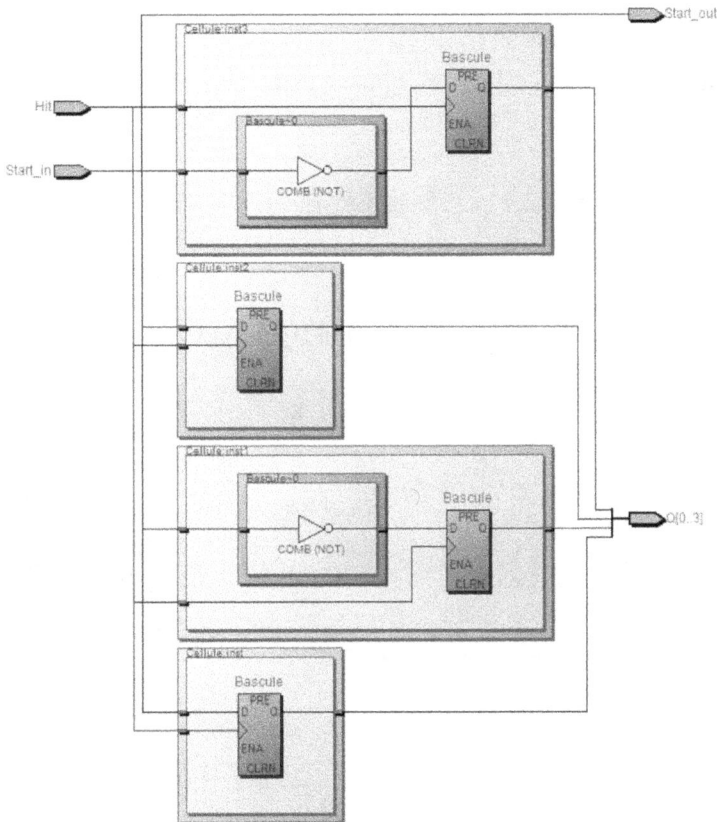

Fig. 11.5. RTL view of the implementation of a simple 4-cells TDC.

Using this tool, one can customize the configuration of the logic elements and impose the data path. However, the manual configuration of logic elements is tedious and time consuming in particular for systems with a certain complexity such as TDCs.

Even if we can use this technique to implement a TDC on an FPGA, given the large number of logic elements to be configured individually, it is still difficult to set up. Moreover, the TDC chain size can vary from an application to another; it will be therefore preferable to automate the configuration so that the solution will be generic and adaptable. Hence, we propose an appropriate design methodology in the following section.

11.3.2. Proposed Design Methodology

In order to provide solutions to the issues raised above, in this section we suggest an alternative approach to implement a TDC structure fulfilling the following needs:

- Avoid the software data path simplification;

- Increase TDC resolution by reducing the propagation time through delay elements;

- Automate the elementary cells set-up process to optimize the design time and enable the development of generic and adaptable structures;

- And use a low cost FPGA target to implement the TDC.

This method relies on i) using adders as delay elements, ii) using the Carry Chain Logic of the FPGA, and iii) using the Chip Planner tool.

11.3.2.1. Using Adders and Carry Chain Logic

The implementation of digital circuits on FPGA targets depends on the architecture of the logical resources of the target. In this work, we are aiming to use a low cost FPGA from Altera Cyclone family. The selected target is the Cyclone IV (EP4CE55F23C8) based on the logic element shown by Fig. 11.6 [22].

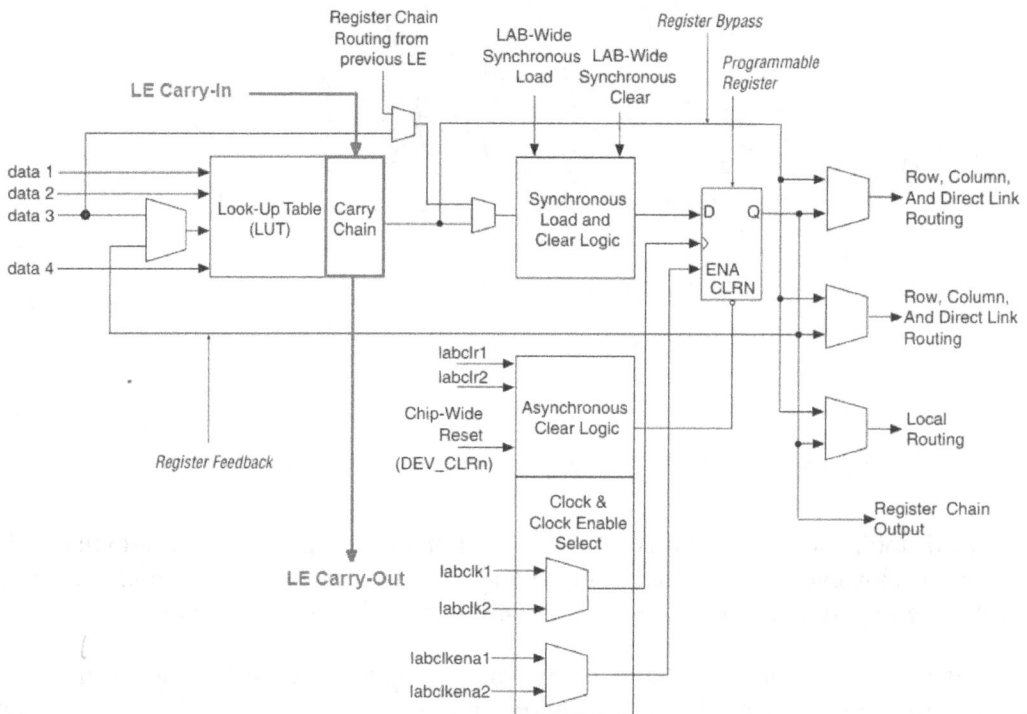

Fig. 11.6. Cyclone IV logic element structure.

The Cyclone IV logical element, provides a dedicated path for fast carry propagation. The role of this carry chain is to use specific fast paths for carry propagation instead of general-purpose routing network. Doing so, allows us to drastically optimize the propagation time. This is ideal for the enhancement of the TDC resolution. Moreover, it allows harmonizing the delays of the TDC elementary cells.

The problem is that customized handling of carry chains is reserved to high performance FPGAs such as the Stratix family from Altera whose cost is outstandingly high. However, it is possible to configure the Quartus II synthesis tool to optimize speed. In this case, the synthesis tool uses the carry chain logic automatically when synthetizing an HDL model involving adders.

It is therefore possible to use the carry chain logic to minimize and harmonize propagation delays for components involving adders. It is precisely the idea that is exploited here to design TDC elementary cells based on simple adders. This was done by developing a simple behavioral VHDL model for an adder with a customizable number of elementary cells. The number of cells depends on the data length modeled by a generic parameter called DATA_WIDTH. The whole model is given by Fig. 11.7.

```vhdl
entity signed_adder is
    generic ( DATA_WIDTH : natural := 127 );

    port (  a      : in signed   ((DATA_WIDTH-1) downto 0);
            b      : in signed   ((DATA_WIDTH-1) downto 0);
            cin    : in std_logic;
            cout   : out std_logic;
            result : out signed ((DATA_WIDTH-1) downto 0) );
end entity;

architecture rtl of signed_adder is

begin

PROCESS (cin, a, b)
    VARIABLE  s : signed ((DATA_WIDTH) downto 0);
    BEGIN
        s := ('0' & a) + ('0' & b) + ('0'&cin);
        result<=s((DATA_WIDTH-1) downto 0);
        cout <= s(DATA_WIDTH);
    END process;
end rtl;
```

Fig. 11.7. Adder VHDL model.

The fine TDC using adders can be performed by: (1) applying the TDC input signal STOP to the carry input signal (cin) of the adder, and (2) choosing values for adder operand inputs (a and b) so that an output carry is generated (cout='1') if input carry is equal to '1'. The output carry is then an exact replication of the input carry delayed by a transmission time through the cell. To do so, we simply set all the bits of the first operand to '1' and the bits of the second operand to '0'. For each bit (i) the arithmetic sum $a(i)+b(i)$

gives '1'. When the input carry is activated (cin='1') by the TDC input signal (STOP), the arithmetic sum a+b+cin gives '0' and the carry output moves to '1'.

Fig. 11.8 illustrates the implementation of one elementary cell of a TDC by a logic element of the Cyclone IV target. The adder cell is obtained by the look up table (LUT) and the DFF by the sequential configurable output register.

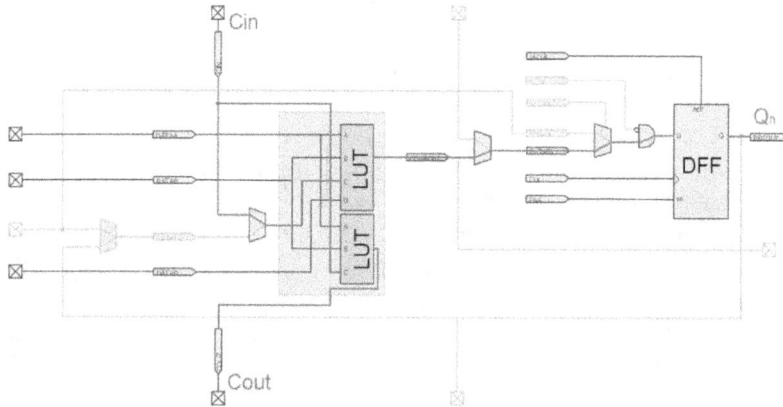

Fig. 11.8. Implementation of a TDC elementary cell by a logic element.

Theoretically, to obtain a TDC chain similar to the TDL structure shown by Fig. 11.3, it is sufficient to duplicate the structure of Fig. 11.8 as often as necessary to reach the desired number of cells. However, when implementing such a chain on the FPGA, some DFFs of the TDC elementary cells are dissociated of their corresponding 1-bit adder cells even if the data path is perfectly respected. This phenomenon occurs randomly and leads to the placing of the DFF and the delay element of the same TDC's elementary cell in different logic elements, as shown in Fig. 11.9.

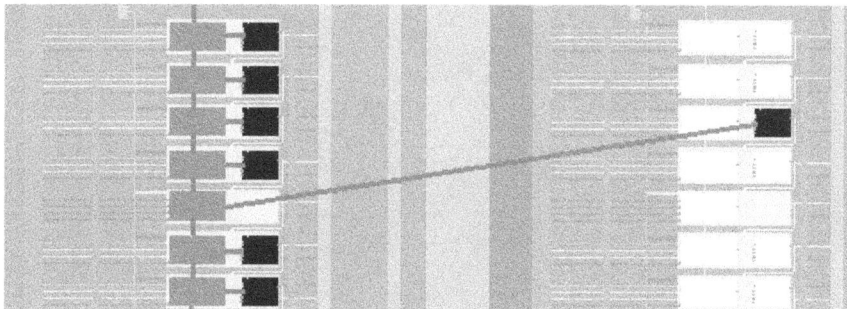

Fig. 11.9. Random placing of DFFs on the chip.

The direct consequence of component misplacing is that the time delay is no longer identical for all cells. This inevitably generates unpredictable artifacts. To ensure a reliable operation, it is necessary to overcome this problem by constraining the placement tool to

214

bring together the components of the same cell in the same logic element. This is the purpose of the next section.

11.3.2.2. Using Chip Planner

Fast imaging systems require TDC's capable of measuring very short time durations. It is therefore necessary to master all of the signal propagation delays through the cells as well as the routing network.

As we have seen in the previous section, unconstrained automatic implementation of a TDC on an FPGA usually leads to an inhomogeneous and irreproducible structure. Consequently, the measurement results are tainted by these uncertainties. Therefore, it is necessary to control the exact physical location of TDC cells on the chip.

This could be achieved by using the Chip Planner tool provided by Altera which, according to the user's needs, allows to specify given implementation regions on the chip for blocks constituting the whole system. In addition, it supports incremental compilation to preserve the well-implemented parts and reduce the compilation time. This operation takes place in three distinct steps:

- Creating Design Partitions: the first step consists of dividing the design in individual partitions according to system complexity as well as user needs.

- Defining logic regions: after partitioning the design, it is necessary to define logical zones that will be associated to the partitions. This allows individual compiling and optimizing of each region. The tool used to perform this operation is LogicLock Region (LLR) within Chip Planner.

- Physical assignment of logic regions: in order to physically preserve the logic regions defined in the previous step, by means of the LLR tool, physical regions of the chip are assigned to implemented partitions.

The physical delimitation of regions permits to constrain the placing and root tool to put partitions in their specified regions defined by the user. This allows not only avoiding the random placement of certain DFFs away from their associated delay elements, but also implementing the concerned partitions as close as possible to input signal pins (HIT and STOP). The purpose of the latter operation is to reduce the propagation delays of input signal to the intended blocks. For illustrative purposes, we represent on Fig. 11.10 the assignments of physical allocations of the partitions defined using LLR and a close-up view of the layout of a 16-cells fine TDC on Fig. 11.11.

The TDC fits perfectly within the reserved region that would be assigned to it. Consequently, the DFF and the delay element of each TDC elementary cell are now implemented by the same logic element. The transmission delays are then identical for all cells.

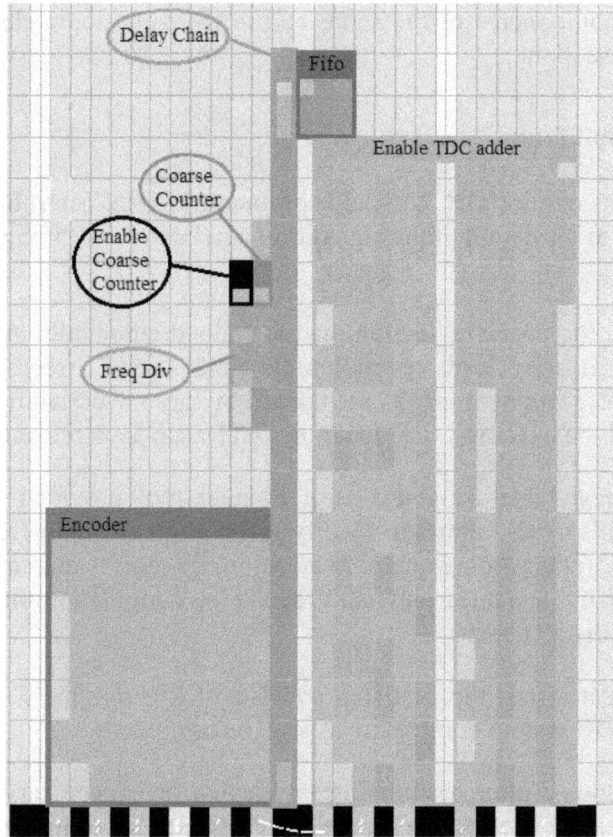

Fig. 11.10. Layout of implemented partitions of a TDC.

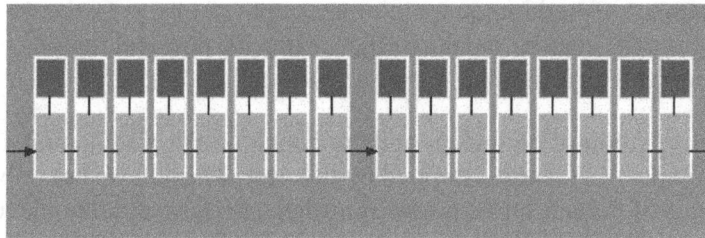

Fig. 11.11. Physical implementation of 16-cells TDC.

11.4. Implementation of a 42 ps TDC on a Cyclone IV FPGA

In this section we apply the above methodology to a realistic case. After introducing the experimental conditions, we present a robust encoder capable of detecting failure sequences, than we discuss the effect of constraining cell placement by using the Chip Planner tool, and finally we develop improved results obtained by overcoming a hardware limitation of the used FPGA design kit.

11.4.1. Experimental Measurement Conditions

The proposed TDC design has been implemented within the Cyclone IV (EP4CE55F23C8) FPGA target. The coarse counter clock is 200 MHz, i.e., the clock period is 5 ns. The delay line for the fine TDC, based on carry chain adder architecture, comprises 128 cells in order to cover a dynamic range of more than 5 ns. The signal that needs to be measured propagates through the delay chain, until the FPGA clock disables the DFFs to block their outputs and then memorizes their states. The value of these DFFs describes the time spent between the signal STOP and Clock.

The data is then transmitted to a USB port via an FTDI FT232H operating in parallel mode with transfer rates reaching up to 40 Mbyte per second. To acquire data measurements, we developed a specific application using the LabVIEW software.

In order to reduce the size of the data transmitted to the USB port, we developed a VHDL model of a specific encoder converting the 128 bits to a one byte data. Moreover it filters potential errors. The functioning of the latter is described in the Section 11.4.2.

All the blocks are summarized in Fig. 11.12 showing the synoptic view of the whole system.

The TDC has been characterized on its whole dynamic range, i.e. from 0 to 640 ns with a 5 ps step. A Stanford research DG 645 digital delay generator has been used to generate the START and STOP signals with a temporal jitter below 25 ps rms.

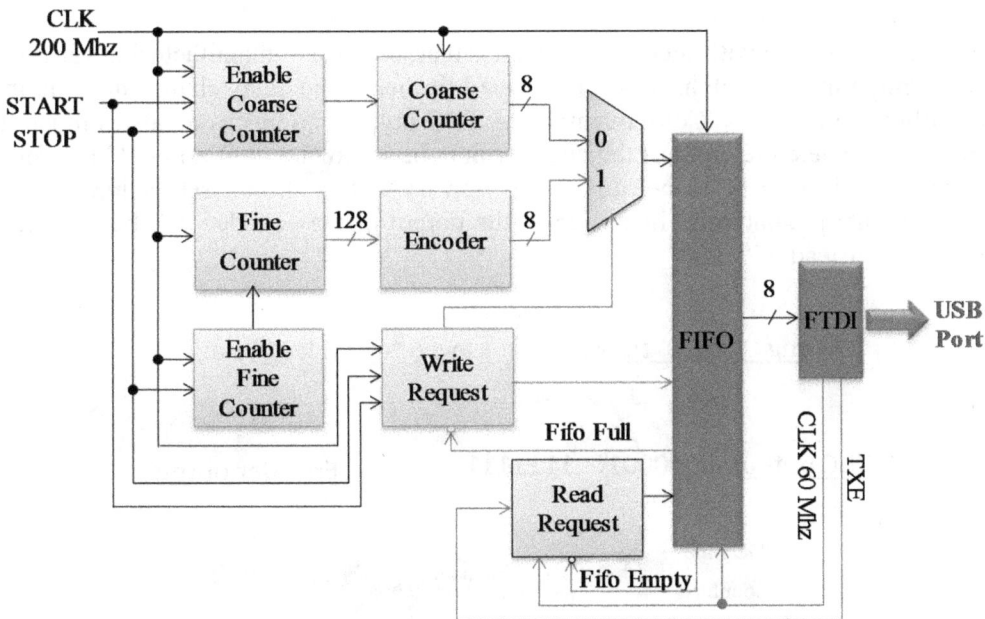

Fig. 11.12. Synoptic view of the implemented TDC system.

11.4.2. Encoding Fine Counter Output Binary Stream

As described previously the output binary data stream of the fine counter, representing the measured time, is applied to an encoder. The latter prepares the data before saving it into a FIFO (First in First Out) memory. Indeed, at the output of the delay chain, data are presented as a string made of zeros ('0') on the left and ones ('1') on the right. The encoder's role is to count the number of consecutives bits switched to '1' and generate the corresponding 8-bit binary code. So the first idea is to use a simple priority encoder to detect the position of the most significant bit moved to '1' and gives the corresponding binary code. To illustrate this we show on Fig. 11.13 the example of a 16-cell TDC output when half of the cells were crossed by the measured signal. In this case the input of the encoder is "0000000011111111", thus yielding "1000" at the output.

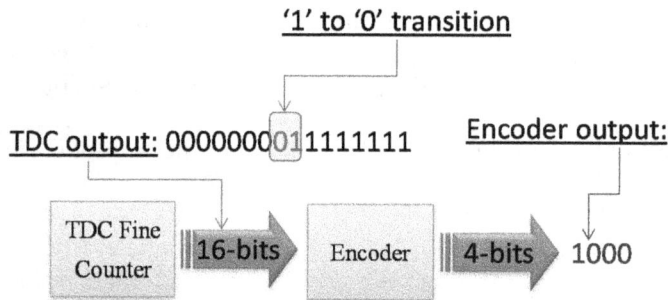

Fig. 11.13. The theoretical string FOR 16-cells TDC.

However, the problem with such an approach is that, because of manufacturing variations, such as setup time mismatch, flip-flops located further in the delay chain can sometimes react earlier or vice versa. Consequently, the TDC output can be erroneous as illustrated by Fig. 11.14 where the input of the encoder may be set to "0000001011111111" for the same delay as in Fig. 11.13 because a flip flop presents a shorter setup time and thus detects the data prematurely. In this case, the output of the encoder will be wrong, i.e. "1010" here instead of "1000".

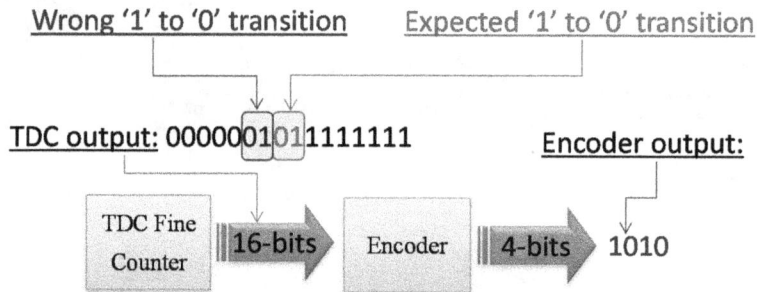

Fig. 11.14. Illustration of potential errors in the data string.

218

To overcome such errors we designed a robust encoder which, in addition to encoding the 128 bits in one byte data, it detects failed measures due to the flip-flops setup and hold times mismatches. The method adopted here is to detect '011' sequence instead of '01'. Hence all the sequences including '1' between two or more zeros ('0') are identified as wrong behavior of the corresponding DFF, and thus ignored. This method is more robust, and generates the correct output even if a wrong code appears as illustrated in Fig. 11.15.

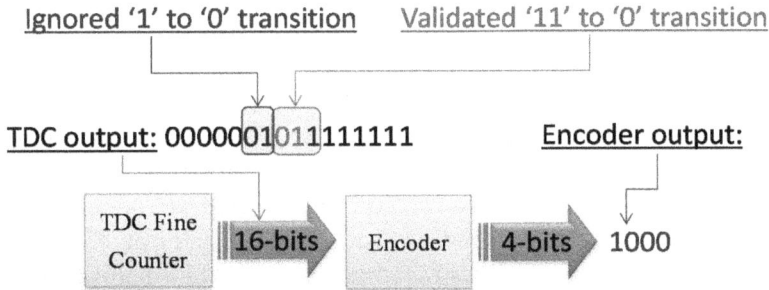

Fig. 11.15. Correction of potential errors by the encoder.

11.4.3. Effect of Constraining Cell Placement by the Chip Planer Tool

To show the effect of constraining cell placement by the Chip Planner tool we report on Fig. 11.16 the detail of the unconstrained and constrained fine TDC measurements between two reference clock edges, on a range of 5 ns.

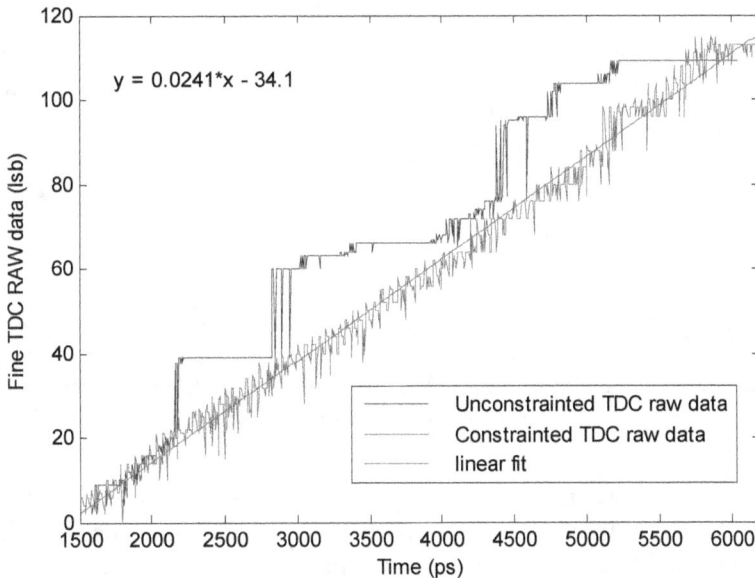

Fig. 11.16. Responses of Fine unconstrained and constrained TDC.

The unconstrained fine TDC response (blue) shows a large discrepancy of the LSB (Last Significant Bit) value indicating that some DFFs have been randomly placed. The resulting large steps make the unconstrained fine TDC unusable for sub nanosecond timing. Consequently, the use of the Chip Planner tool as described in section III is mandatory to obtain the behavior of the constrained fine TDC represented by the green curve. A linear fit is then used to assess the LSB value of the fine TDC which is given by the inverse of the linear fit slope, i.e. 41.5 ps in this study case.

11.4.4. Jitter, INL and DNL Evaluation

The noise in the fine TDC response shown in Fig. 11.16 is due to the jitter which adds uncertainty on each measurement. It can be evaluated by computing the standard deviation of a set of measurements at a given fixed delay between the START and STOP signals. The jitter depicted in Fig. 11.17 has been characterized for different delays corresponding to a given signal propagation along the fine TDC line. As each fine TDC elementary cell adds its own jitter [23], the global jitter will then increase as a square root of the number N of cells as given by the following expression:

$$\sqrt{\alpha^2 + \beta^2 . N} \, , \tag{5}$$

where α is the initial jitter present at the input of the first cell and β the single cell jitter.

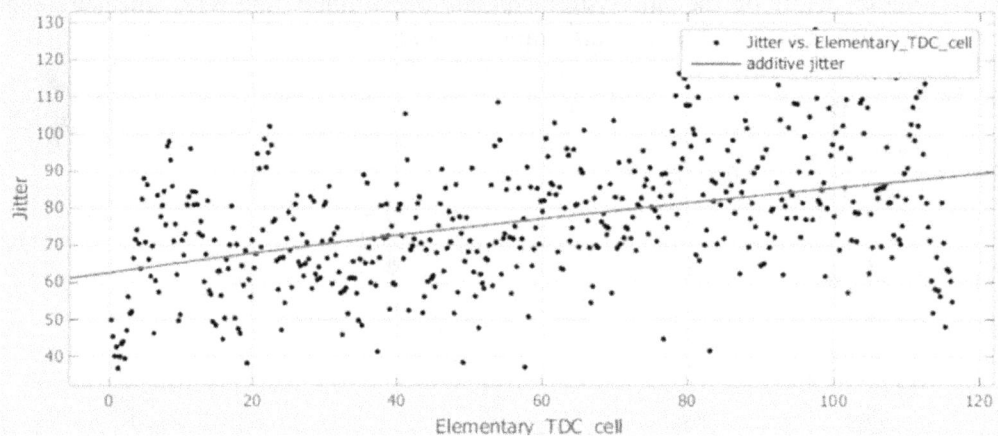

Fig. 11.17. Jitter measurement according to the elementary level, the jitter increases as the signal propagates along the fine TDC cells.

A curve, following this law is fitted on the jitter profile to underline the jitter's variation relationship in the delay line. The extraction of this parameters leads to an initial jitter α of 62 ps rms and a single cell jitter β of 5.8 ps rms. The accumulated jitter across the fine TDC delay line leads to a mean jitter of 90 ps rms. Thus, the line length (N) has to be kept as low as possible in order to obtain the best accuracy. This can be done by using the fastest achievable frequency for the coarse counter.

The Integral Non Linearity error (INL) and the Differential Non Linearity (DNL) have been measured over the entire range of the TDC. For illustrative purposes, the results from a delay of 0 to 160 ns are represented by Fig. 11.18 hereafter. It can be seen that, the implemented system shows an INL of 132 ps rms and a DNL of 50 ps rms. The static periodicity of the INL is arising from the Tapped Delay Line, but the additional jitter noise is still observable on this measurements and has a strong negative impact on the INL and DNL errors.

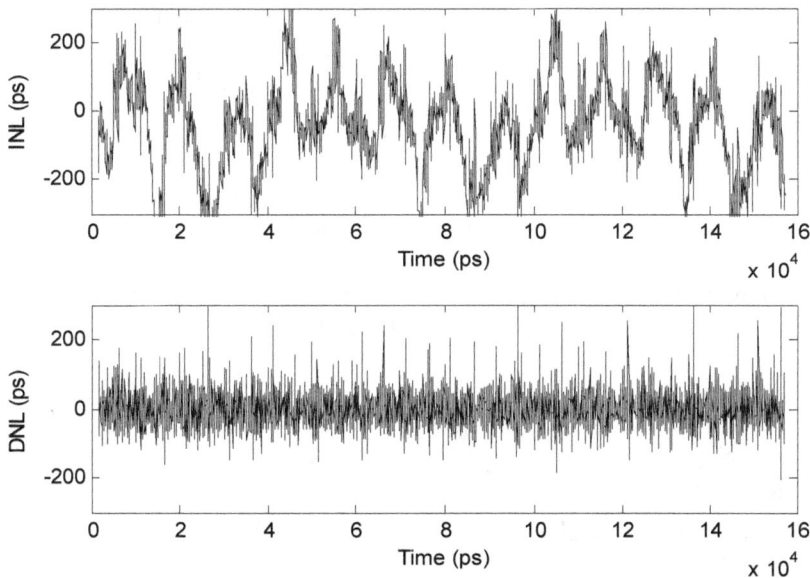

Fig. 11.18. INL and DNL errors of the implemented TDC over a range of 160 ns.

To determine the origin of the high jitter we investigated the effect of the FPGA input/output buffer delays, the FPGA oscillator frequency and the USB communication interface. Finally, we find that most of the jitter arises from an electrical noise present on the 1.2 volt core FPGA power supply. Indeed, to generate this voltage the used FPGA kit uses a switching DC/DC converter which presents a periodic noise of about 10 millivolts rms at a frequency of a tens of kHz. A modification of the power supply of the FPGA macrocell is affecting its propagation time and in this case, the propagation time modification is amplified by the number of chained elements at a given position of the tapped delay line. To overcome this limitation the provided power supply has been unsoldered and replaced by an external linear regulated power supply generator. The new measurements of jitter, INL and DNL are reported on Fig. 11.19 and Fig. 11.20.

It can readily be seen that the mean jitter is improved and reaches 26 ps rms which is mostly due to the delay generator Stanford DG 645 used to make the measurement. The periodic behavior of the jitter is due to the transition between two slices of the tapped delay line. The jitter is no longer increasing with the position of the delay cell within the tapped delay line indicating that the elementary delay cell jitter β is negligible in regards of the initial jitter α of 26 ps rms of the Stanford generator. As a result, we can conclude that thank to a clean main FPGA core supply voltage a jitter performance much better than the bin size can be obtain, insuring a very good measurement.

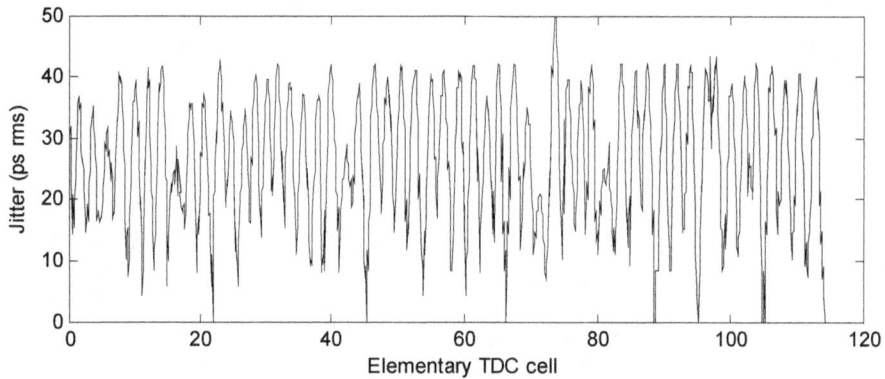

Fig. 11.19. New Jitter measurement after replacing the switching DC/DC power supply core by a DC regulated power supply.

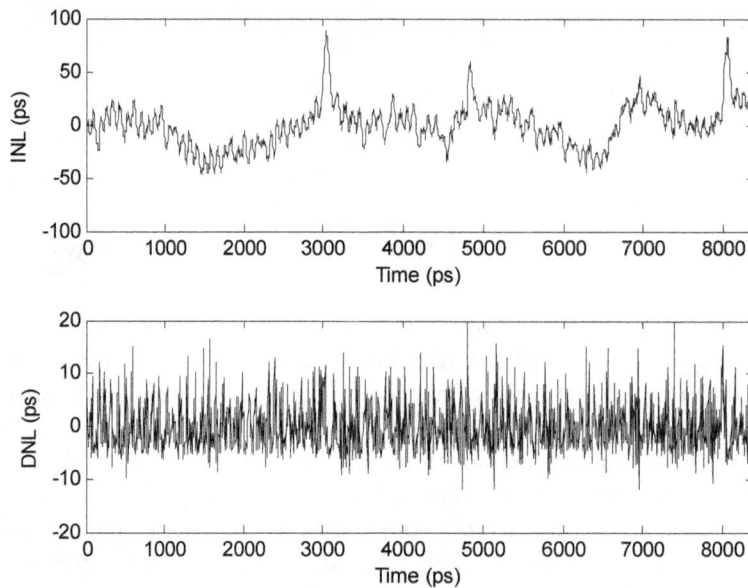

Fig. 11.20. New INL and DNL errors of the implemented TDC over a range of 8 ns after cleaning the FPGA core supply voltage.

The INL and the DNL are reduced to 22 ps rms and 13 ps rms, respectively. The spikes visible in the INL at 3100 and 8100 ns correspond to the toggle of a coarse counter bit. Once again, these measurement are showing a periodicity due to the taped delay line length of 5 ns.

11.5. TDC Characterization and Correction with Poisson Process Events

In this section we aim to present a correction method that improves the transfer function linearity of any TDC independently of its structure and its material implementation. So it is evident that this technique focuses on downstream TDC data post processing. The first step of this technique is the characterization of the TDC thus we begin by developing the characterization approach before presenting the proposed correction method.

11.5.1. TDC Characterization

The TDC characterization with Poisson Process Events approach consists in measuring a perfectly uniform distribution of temporal events in order to evaluate the accuracy of all the TDC bins. By bin accuracy, we mean the temporal size of each bin. Indeed, if a bin of the TDC is temporally larger than it should be, it gathers more events from the uniform distribution and thus the total count of this bin is higher. Conversely, a smaller bin shows a total count proportionally lower than expected. A perfect generator of random uncorrelated events is a SPAD that detects photons generated by a "continuous wave" (CW) light source such as a battery-powered Light-Emitting Diode (LED). Indeed, the single photon detection is well known to be an ideal Poisson Process which is completely uniform and uncorrelated. As a reminder The Poisson distribution is a discrete probability distribution that expresses the probability of events occurring in a fixed interval of time if these events occur with a known average rate and independently of the time since the last event.

In an experiment where the start events are given by the single-photon detection events provided by a SPAD under weak CW illumination and the stop event are given by a regular independent clock signal (of arbitrary frequency), an ideal TDC should display the exact same photon counting probability per time bin. Hence the corresponding distribution of detection times with respect to the reference clock should yield a flat histogram. If a default appears on the TDC chain, the error is temporally static. This means, that if the measurement is done several times the same error will occur every time and a Fixed Pattern Noise (FPN) will appear on the measured histogram. More precisely, in a given experimental realization, due to the Poisson statistics, the effective number N of counts per bin is affected by the so-called photon shot noise which is a noise of amplitude sqrt(N) rms leading to a relative noise on the bin value of:

$$\frac{\sqrt{N}}{N} = \frac{1}{\sqrt{N}}, \tag{6}$$

Thus one must ensure that the detected FPN is not due to the random noise of the measurement by checking that its value is well above sqrt(N) .

For example, the histogram displayed in Fig. 11.21 results from a 10 seconds acquisition under a photon counting rate of 1800 kHz with the FPGA-based TDC described in [17] with a temporal resolution of 89 ps over a range of 40 ns. The measured FPN is obtained by the detection of a large number of non-correlated Poisson events generated by a SPAD illuminated with CW light. This signal should be flat for an ideal TDC. In the specific case of this TDC, the fine counter is a Tapped Delay Line made of 56 elementary delay cells of 89 ps for a total length of 5 ns. This fine TDC counter is thus periodically reinitialized every 5 ns when the coarse counter is incremented. Hence the overall FPN of the TDC is the periodic repetition of the FPN characterizing the 5-ns-long tapped delay line used for the fine TDC counter. Fig. 11.22 displays a zoom in this 5-ns long FPN motif.

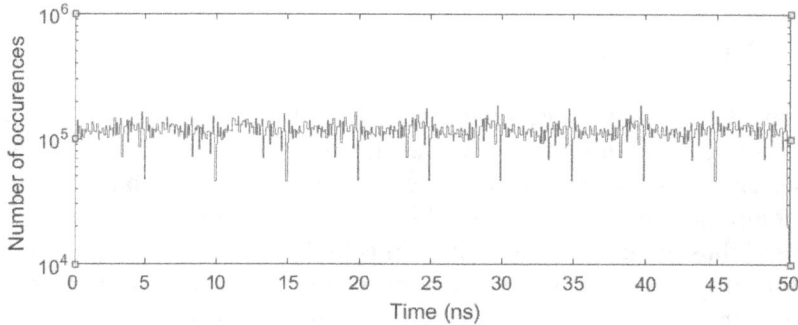

Fig. 11.21. TDC FPN measured over a range of 50 ns.

The overall photon accumulation time is large enough that the uncertainty sqrt(N) due to Poisson statistic on the average number N of photons per bin becomes much less than the detected variation of N from bin to bin, i.e. the FPN amplitude (see Fig. 11.22). Hence, the observed FPN comes from the delay mismatch of the tapped delay line used for the fine TDC counter.

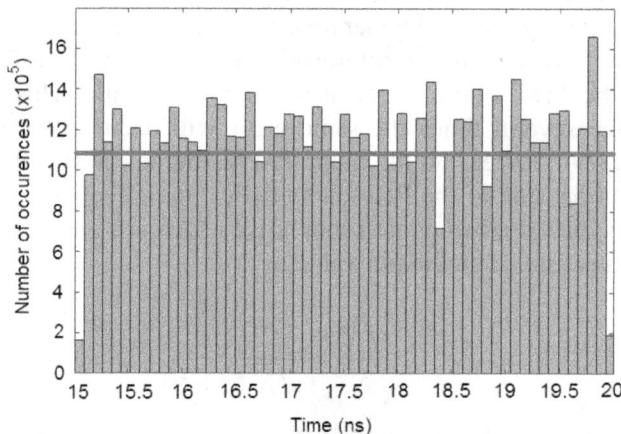

Fig. 11.22. Zoom on the 5 ns time laps corresponding to the 15-20ns time window of Fig. 11.21.

Another method for Poisson process events generation using a SPAD is to exploit the photodetector dark count rate (DCR) originating from thermal activation at ambient temperature. Fig. 11.23 displays the FPN histogram characterizing the fine counter, upon accumulating dark counts over 10 minutes when the SPAD is maintained in complete dark. The dark count rate is 250 Hz in this case. We note that the FPN is very similar to that of Fig. 11.22, as expected, since it is a property of the TDC, independent of the process (light or dark counts) generating the Poisson distributed events. The slight differences are consistent with the physical shot noise of about 100 count rms in the second measurement. Consequently, the FPN characterization with a CW light source is much faster (higher counting rate) and therefore preferred. Other light sources like day light and neon light (100 Hz frequency) have been tested yielding the same FPN histogram.

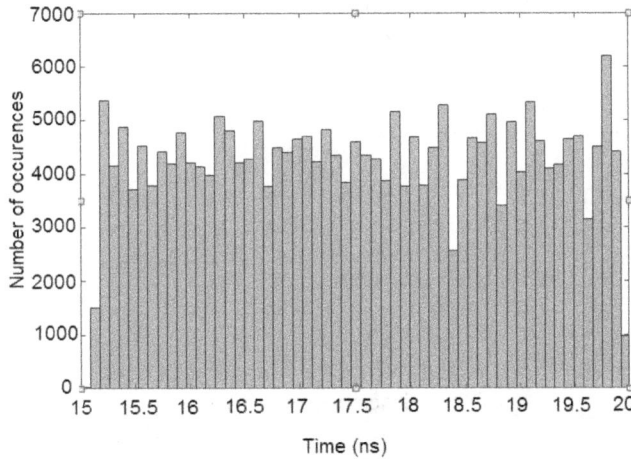

Fig. 11.23. TDC FPN histogram measured from the dark counts generated by the SPAD.

The above FPN measurement can be used to retrieve the TDC transfer function without the need of a complex delay generator. As mentioned before, because we detect Poisson process events, if a bin of the TDC is temporally larger than the mean bin width, its number of counts is increased in the same proportion. Consequently, the absolute time TDC transfer function can be extracted from the number of counts in the TDC bins with:

$$TDC(n) = \sum_{i=1}^{n} T_i = \frac{T}{\sum_{i=1}^{M} bin_i} \sum_{i=1}^{n} bin_i , \qquad (7)$$

where T_i is the actual duration of each individual TDC bin, bin_i is the number of events detected in bin number i, T is the total time range and M is the total number of TDC bins in the time range T. The extracted transfer function of the TDC over the 5-ns time range of the fine counter is presented in Fig. 11.24. The linear fit allows characterizing the Integral Non Linearity of the TDC, which is the residual of the fit, shown in Fig. 11.24. The results obtained with this method are consistent with the measurements reported in [17].

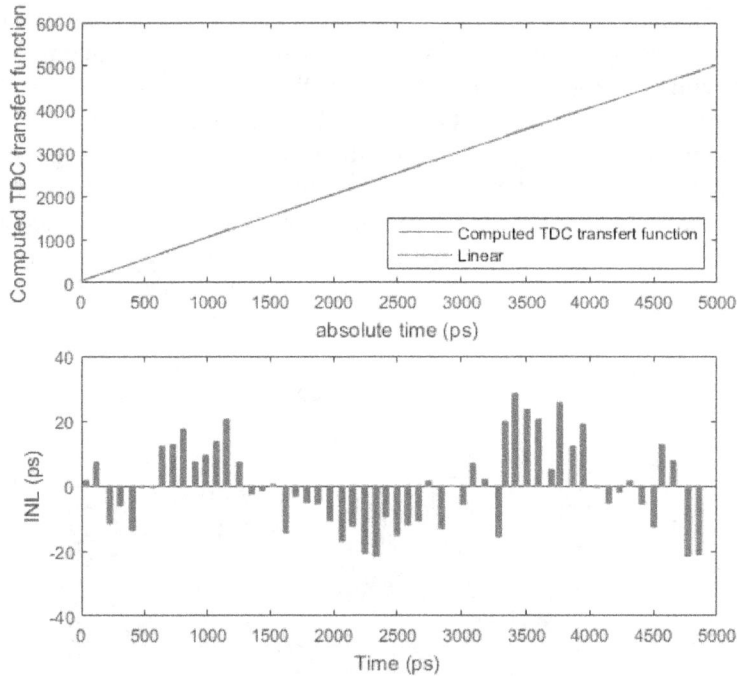

Fig. 11.24. Transfer function and INL of the 5-ns-long TDL of the TDC fine counter described in [17].

11.5.2. Correction Methodology

The FPN evidenced in Fig. 11.21 being a systematic bias, it will also affect histograms recorded in TCSPC experiments as well as in any other application of the TDC. However, it can be used to define a scaling factor for each bin so as to correct, by post-processing, the TDC transfer function. This factor is simply the deviation of the count number of each bin_i during the calibration process described previously of the reference FPN histogram relative to the average value (red line in Fig. 11.22), such that the corrected bin count $binc_i$ is given by:

$$binc_i = \frac{\frac{1}{M}\sum_{i=1}^{M}bin_i}{bin_i}.binm_i,$$ (8)

where M is the total number of TDC bins and $binm_i$ is the raw data without correction. The correction applied to the FPN histogram leads to a completely flat histogram, by construction. Note that this method is not applicable with the missing code of the TDC that are reporting a null or even very close to zero density as the division will lead to infinity or to an amplification of the noise of the close to zero bin.

11.6. TCSPC Measurement of Fluorescence Lifetime

In the following, the characterized TDC is used in a TCSPC experiment to measure the fluorescence lifetime of fluorescein dissolved in water buffered at pH = 7.4. A picosecond, 405-nm laser pulse [24] excites the fluorescence. The fluorescence signal is detected by a SPAD and the arrival time of individual photons relative to the previous laser pulse is measured and stored by the TDC on the FPGA board. Fig. 11.25 shows the raw data, therefore displaying the same FPN evidenced above: the imperfect transfer function of the TDC creates a static pattern (FPN) at the origin of the periodic glitches in the raw histogram. The 5-ns periodicity of the FPN is clearly seen also in this data.

The same data are post-processed by applying to each bin count the correction factor introduced above, and the corrected data are plotted in Fig. 11.26. The FPN is suppressed, and the extracted fluorescence lifetime is 4.19±0.02 ns, in very good agreement with the expected value for fluorescein at pH=7.4 [25].

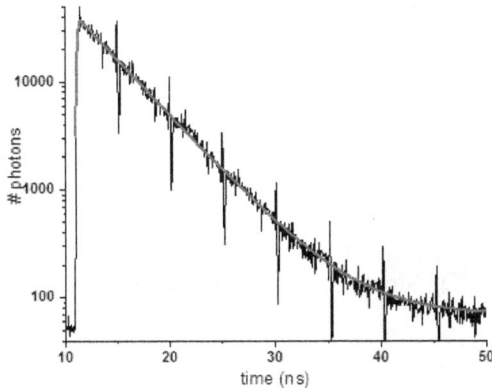

Fig. 11.25. Fluorescence decay without correction.

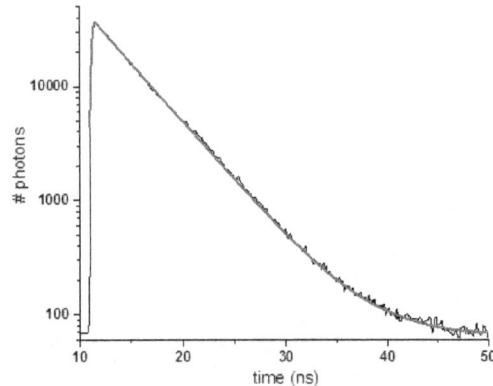

Fig. 11.26. Fluorescence decay with correction.

11.7. Conclusion

This chapter proposes a global methodology to design and implement Time-to-Digital Converters on low-cost FPGA targets. It presents how to use different tools to enhance the TDC resolution by reducing propagation delays through the connection network as well as the logic gates themselves. First, the use of adders as delay elements, to benefit from a dedicated carry chain logic path, is presented. Then the chip planner is taken advantage of, to constrain the placing and root tool to put the partitions of the system in user specified physical regions. This is central in mastering of propagation delays and consequently improving the resolution and the stability of the TDC.

The proposed methodology is applied to design and implement a TDC with a resolution of about 42 ps on a Cyclone IV FPGA. The implemented TDC presents a jitter of only 26 ps rms, and the DNL and INL has been measured to be 22 and 13 ps rms, respectively.

In addition, we introduces a fast and efficient correction method to improve the transfer function linearity of either ASIC or FPGA Time-to-Digital Converters. The approach is based on the measurement of a large number of Poisson process events generated by a simple SPAD lightened by a continuous wave light source. Importantly, it allows measuring the TDC transfer function without the need of any expensive delay generator to calibrate the device. The proposed correction is a post process operation, and thus can be implemented for any type of TDC. Its efficiency is demonstrated in a real TCSPC experiment. The function transfer correction method is simple, fast, efficient, and does not require hardware modification of the TDC.

The highlighted results in this chapter are very promising, not only because they are suitable for domains requiring high performances, but also because they are achieved by using a low-cost FPGA family which opens the door to a broader use in a great amount of fast application fields.

References

[1]. M. Fishburn, L. H. Menninga, C. Favi and E. Charbon, A 19.6 ps, FPGA-Based TDC With Multiple Channels for Open Source Applications, *IEEE Transactions on Nuclear Science*, Vol. 60, Issue 3, 2013, pp. 2203-2208.

[2]. E. Charbon, M. Fishburn, R. Walker, R. Henderson and C. Niclass, SPAD-based sensors TOF Range-Imaging Cameras, F. Remondino and D. Stoppa (Eds.), *Springer-Verlag*, Berlin Heidelberg, 2013, pp. 11–38.

[3]. L. Li, Time-of-flight camera – an introduction, SLOA190B – Techniqueal White Paper, *Texas Instruments*, January 2014, revised May 2014.

[4]. S. Henzler, Time-to-Digital Converters, *Springer Science+Business Media B. V.*, 2010.

[5]. S. Y. Yurish, Smart Optoelectronic Sensors and Intelligent Sensor Systems, *Sensors & Transducers*, Vol. 14-1, Special Issue, 2012, pp. 18-31.

[6]. W. M. Henebry and A. Rasiel, Design Features of a Start-Stop Time-to-Amplitude Converter, *IEEE Transactions on Nuclear Science*, Vol. 13, Issue 2, 1966, pp. 64-68.

[7]. J. Kalisz, Review of methods for time interval measurements with picoseconds resolution, *Metrologia*, Vol. 41, Issue 1, 2004, pp. 17–32.

[8]. C. S. Hwang, P. Chen and H. W. Tsao, A high-precision Time-to-Digital converter using a two-level conversion scheme, *IEEE Transactions on Nuclear Science*, Vol. 51, Issue 4, 2004, pp. 1349–1352.

[9]. I. Malasse, W. Uhring, J. Le Normand, N. Dumas and F. Dadouche, 10-ps Resolution Hybrid Time to Digital Converter in a 0. 18 μm CMOS Technology, in *Proceedings of the 12th IEEE International New Circuits and Systems Conference (NEWCAS)*, Trois-Rivières, Canada, 22-25 June, 2014, pp. 105-108.

[10]. J. Kalisz, R. Szplet, J. Pasierbinski and A. Poniecki, Field-programmable-gate-array-based time-to-digital converter with 200-ps resolution, *IEEE Trans. Instrum. Meas.*, Vol. 46, Issue 1, 1997, pp. 51–55.

[11]. I. Vornicu, R. Carmona-Galán and Á. Rodríguez-Vázquez, Wide range 8ps incremental resolution time interval generator based on FPGA technology, in *Proceedings of the IEEE*

International Conference on 21*st* Electronics, Circuits and Systems (ICECS), Marseille, France, 7-10 December 2014, pp. 395-398.

[12]. M. Lin, G. Tsai, C. Liu and S. Chu, FPGA-Based high area efficient Time-to-Digital IP design, in *Proceedings of the 10th Region Conference TENCON IEEE 10*, Hong Kong, China, 14-17 November 2006, pp. l-4.

[13]. R. Narasimman, A. Prabhakar and N. Chandrachoodan, Implementation of a 30 ps resolution Time-to-Digital Converter in FPGA, in *Proceedings of the International Conference on Electronic Design, Computer Networks & Automated Verification (EDCAV)*, Shillong, India, 29-30 January 2015, pp. 12-17.

[14]. A. Aloisio, P. Branchini, R. Giordano, V. Izzo and S. Loffredo, High-precision Time-to-Digital converter in a FPGA device, in *IEEE Nuclear Science Symposium Conference Record*, Orlando Florida, USA, 25-31 October 2009, pp. 290-294.

[15]. S. S. Junnarkar, P. O'Connor, and R. Fontaine, FPGA based self calibrating 40 picosecond resolution, wide range Time-to-Digital converter, in *IEEE Nuclear Science Symposium Conference Record (NSS '08)*, Dresden, Germany, 19-25 October 2008, pp. 3434–3439.

[16]. F. Dadouche, T. Turko, W. Uhring, I. Malass, J. Bartringer, and J-P. Le Normand, Design Methodology of TDC on Low Cost FPGA Targets (Case Study: Implementation of a 42 ps Resolution TDC on a Cyclone IV FPGA Target), in *Proceedings of the 9th International Conference on Sensor Technologies and Applications (SENSORCOMM'15)*, Venice, Italy, 23-28 August 2015, pp. 29-34.

[17]. F. Dadouche, T. Turko, W. Uhring, I. Malass, N. Dumas and J-P. Le Normand, New Design-methodology of High-performance TDC on a Low Cost FPGA Targets, *Sensors & Transducers*, Vol. 193, Issue 10, October 2015, pp 123-134.

[18]. T. Turko, A. Skilitsi, W. Uhring, J-P. Le Normand, N. Dumas, F. Dadouche, J. Léonard, Time to Digital Converter Transfert Function Improvement using Poisson process event, in *Proceedings of the 1st International Conference on Advances in Signal, Image and Video (SIGNAL'16)*, Lisbon, Portugal, June 26-30 June 2016, pp. 50-53.

[19]. B. Rossi, N. Nereson, Experimental Arrangement for the Measurement of Small Time Intervals between the Discharges of Geiger-Müller Counters, *Review of Scientific Instruments*, 1946, pp. 17-65.

[20]. G. W. Clark, The Contributions of Bruno B. Rossi to Particle Physics and Astrophysics, in *Attidel XXV Congresso Nazionale di Storia della Fisica e dell'Astronomia*, Milano, Italy, 10-12 November 2005 (Milano: SISFA, 2008), pp. R1.1-R1.16.

[21]. R. Nutt, Digital Time Intervalometer, *Review of Scientific Instruments*, Vol. 39, 1968, pp. 1342-1345.

[22]. Cyclone IV Device Handbook, *Altera Corporation*, April 2014, Vol. 1, Chapter 2.

[23]. M. Zlatanski, W. Uhring, J.-P. Le Normand, and D. Mathiot, A Fully characterizable asynchronous multiphase delay generator, *IEEE Transactions on Nuclear Science*, Vol. 58, Issue 2, 2011, pp. 418-425.

[24]. W. Uhring, V. Zint, J. Bartringer, A low-cost high-repetition-rate picosecond laser diode pulse generator, in *Proceedings of the SPIE - The International Society for Optical Engineering*, 2004, Vol. pp. 583-590.

[25]. J. Léonard, N. Dumas, J-P Caussé, S. Maillot, N. Giannakopoulou, S. Barrea and W. Uhring, High-throughput time-correlated single photon counting, *Lab on a Chip*, 2014, Vol. 14, Issue 22, 2014, pp. 4338-4343.

Chapter 12
New Approaches to Extend Lifetime in Wireless Sensor Network Based on Optimal Placement of Sensor Nodes and Using Duty Cycle Technique

Diery Ngom, Pascal Lorenz, Bamba Gueye

12.1. Introduction

Wireless Sensor Networks (WSN) are kind of wireless networks including many sensors node which can be deployed rapidly and cheaply over a geographical region of interest, and thereby they can be used for different purposes such as environment monitoring, wildlife habitat monitoring, security surveillance, industrial diagnostic, agricultural of precision, improve health care, etc.

Optimizing the network lifetime, minimizing the number of active nodes, maintaining full coverage of the monitored region, and providing optimal network connectivity are critical issues in WSN. These issues are usually conflicting and complementary in many WSN applications such as monitoring critical region, wildlife habitat monitoring, agricultural application. For these applications, a full coverage of the monitoring region and good network connectivity are mandatory as well an energy-awareness network lifetime.

We present in this chapter a new Medium Access Control Scheduling Algorithm (MAC-SA) to optimize these four issues simultaneously. Therefore, the geographic distribution of sensor nodes takes into account coverage and network connectivity constraints. The optimal placement of sensors based on square grids, and the ON/OFF scheduling approaches based on duty cycle techniques enable to reduce the energy consumed by sensors nodes. Furthermore, MAC-SA algorithm allows a full coverage of the monitored region and ensures optimal network connectivity. Firstly, we design and validate MAC-SA analytically. Secondly, by extensive simulations we show that MAC-SA significantly

Diery Ngom
University Alioune Diop of Bambey, Senegal

reduces the number of powered ON sensors, and thus the energy consumed during data transport by up to 30 %.

12.2. Related Works

Network lifetime, placement methods, coverage, and network connectivity problem are important issues in WSN. A lot of works have been done in recent years by the researchers for addressing these issues. In this Section, we present and discuss some of the most important works proposed.

Akewar, *et al.* [1] discuss the different deployment strategies such as forces, computational geometry and pattern based deployment. These surveys are good references to have an overall view of coverage and connectivity issues in WSN. However, they don't address the lifetime issues in their study.

With the same goal, Ankur, *et al.* [2] presents different placement strategies of sensors nodes in WSN taking into account the lifetime issues. They note that the most objective of placement techniques have focused on increasing the area coverage, obtaining strong network connectivity and extending the network lifetime.

A more study of coverage and connectivity issues in WSN are presented in a survey by Khou, *et al.* [3]. In this survey, the authors motivate their study by giving different use cases corresponding to different coverage, connectivity, latency and robustness requirements of the applications considered. They present also a general and detailed analysis of deployment problems in WSN. In their analysis, different deployment algorithms for area coverage, barrier coverage, and coverage of points are studied and classified according to their characteristics and properties. Note that, this survey is good references to have an overall view of coverage and connectivity issues in WSN. However, note that in their survey the network lifetime problem are not addressed while this problem is often in conflict with the coverage and connectivity problems.

Zhu, *et al.* [4] address the issues of coverage, connectivity, and lifetime in WSN; and they distinguish two coverage problems: static coverage and dynamic coverage. After the study of coverage problem, they propose a scheduling mechanism for sensors activities in order to reduce the energy consumption in the network and they analyze at the same time the relationship between coverage and network connectivity. Nevertheless, note that placement problem is not study and take account in their proposal. With the same goal, another approach which take account the sensors placement method based on territorial predator scent marking behavior is proposed by Abidin, *et al.* [5]. The main goals of their proposal are: to achieve maximum coverage, to reduce the energy consumed and to guaranty network connectivity. However, note that in their approach the full coverage of the monitored region is not guaranteed. Also in this context,

Mulligan, *et al.* [6] present different coverage protocols that try to maximise the number of sensor which put into sleep mode while guaranteeing k-coverage and network connectivity.

Singaram, *et al.* [7] present also a recent study in which they propose a self-scheduling algorithm that extends the network lifetime while minimizing the number of active sensors. Note that in these two studies, connectivity issues are also not addressed by these authors. A recent survey for sensors lifetime enhancement techniques in WSN is presented by Ambekar, *et al.* [8]. Nevertheless, as some previous authors in the related works, coverage and connectivity issues are not addressed by these authors. In the same purpose, existing surveys introduce basic concepts related to coverage and connectivity.

Ghosh, *et al.* [9] classify coverage problems as coverage based on exposure and coverage exploiting mobility. Area coverage, point coverage and barrier coverage is another classification proposed in detailed by respectively Fan, *et al.* [10] and Wang, *et al.* [11]. With the same goal, Zhu, *et al.* [12] distinguish two coverage problems: static coverage and dynamic coverage. They also propose a study of sleep scheduling mechanisms to reduce energy consumption and analyze the relationship between coverage connectivity. However, placement strategies of SN and lifetime problems are often missed in their surveys.

With the same goal for optimizing the network lifetime in WSN by scheduling the sensors activities, more energy efficient MAC protocols based on duty cycle are developed. In fact, the duty cycle approach is the main feature of synchronous and asynchronous MAC protocols where any node can alternate between active and sleep states in order to save its energy. In this approach, nodes can only communicate when they are in active state. In so doing, several MAC protocols such as "S-MAC", "T-MAC", "B-MAC", "X-MAC" and "RI-MAC" based on duty cycle approach were proposed in [13-15] by respectively Kaur, *et al.*, Kakria, *et al.* and Ullah, *et al.*

In S-MAC (Sensor MAC) [13], nodes alternates between active and sleep periods. During active periods, the node radios are turned ON to communicate and during sleep periods the node radios are turned OFF to save energy. Nodes establish and maintain synchronization in order to choose common fixed active periods. The active period is divided into two sub-periods for exchanging synchronization packets (SYNC packets) and DATA packets. Each node is assigned a radio ON/OFF schedule. A node, after deploying, waits for one cycle of active and sleep period to receive existing network schedule. If a SYNC packet is found then it accepts the schedule carried by the SYNC packet otherwise it uses its own schedule. S-MAC saves energy by reducing idle listening with sleep schedules. However this protocol has some limitations. Firstly, nodes broadcast their schedule to all neighbor nodes using the SYNC packet; so that this mechanism is not efficient in energy consumption. Secondly, all the border nodes incorporate the schedules and keep their radios ON during all of the active periods. Thirdly, predefined and constant sleep and listen periods is a reason for reduced efficiency of S-MAC under variable traffic.

T-MAC (Time-out MAC) [13-14] extends S-MAC and provides several improvements. In T-MAC, the S-MAC limitations were overcome by including an adaptive duty cycle when the length of the active period is varied according to traffic. Each node predicts channel activity during an active period so that it can adjust the length of its current active period. Another improvement consists to maintain node in active state during a time-out in order the node can continue to transmit packets in a burst. T-MAC significantly increases the network lifetime by downsizing the length of the active periods and by using traffic indicators at the beginning of the active periods, nodes determine when to remain active or to switch in sleep period. However, such as S-MAC in this protocol nodes broadcast their schedule to all communication neighbors using the SYNC packet. Thus this mechanism is not efficient in energy consumption and is not suitable in a network with redundancy coverage. Another default of this protocol is the over-listening problem as a node, even if he is not involved in the communication must remains active for a period of time-out.

B-MAC (Berkeley MAC) [14] adopts the famous technical LPL (Low Power Listening). In this technical the nodes periodically switches between active and sleep state. The active state is usually very short, just allows the node to sampling the channel. When a node wakes up, he lights his radio and checks the state of the channel (CCA: Clear Channel Assessment). If there is no activity, then it goes back to sleep state. Otherwise, it remains active to receive packets. After the reception, the node returns to the sleep mode. For the transmitter, each transmission of a packet is preceded by the transmission of a long preamble. The size of the preamble should be longer than the wake up interval to make sure it can be detected by a receiver (next hop). In this way, the receiver is notified to receive the data packet. B-MAC provides good energy efficiency and the active period of each receiver may be adjusted depending on the load of the transmitter. It is therefore with dynamic duty-cycle and self-adapting to the change of the traffic. B-MAC also provides a high level interface for reconfiguring the sleep interval to find a good compromise between power and network throughput. Since B-MAC uses CSMA/CA for the medium access, it suffers flow problem at the high load due to the collisions and the random backoff periods necessary to avoid these collisions. Such as S-MAC, another problem of B-MAC is the over-listening of the preamble by all neighbor nodes because even if the packet is intended only for a particular node (next hop), all other neighbor nodes must still active to listen preamble; so that, a lack of efficiency is noted in term of energy consumed.

X-MAC [15] is an improvement of B-MAC to solve the over-listening problem. Instead of transmitting a long preamble, X-MAC divides it into a series of small packets preamble, each of them containing the receiver's address packet to be transmitted. Time intervals are inserting between these packets preamble and thus allow the destination node to send an acknowledgment (ACK) when it receives one of these preambles packets. Once the transmitter receives the ACK, it knows that the next hop node is awakened and stops sending suites preamble packets and immediately sends the packet to the receiver. As B-MAC, X-MAC also provides self-adjustment of the sleep interval according to variation of the traffic. Compared to B-MAC, X-MAC improves energy efficiency and reduces the time using the shortcut preamble. However as explained above, X-MAC may choose only one next hop (router) to move the packet to its destination, even if there are multiple paths in the network whose exploitation could make robustness in the transmission. Another limitation of X-MAC is the low flow problem. Indeed when the load is high this remains no resolved due to the use of CSMA/CA mechanism for the medium access.

In RI-MAC (Receiver-Initiated MAC) [14], it's the receivers which initiate data transmission technique. In this transmission technique, the sender remains active and waits silently until the receiver explicitly signifies when to start data transmission by sending a short beacon frame. As only beacon frame and data transmissions occupy the medium in RI-MAC, with no preamble transmissions as in LPL technical used in B-MAC protocol; occupancy of the medium is significantly decreased, so that other nodes can exchange data. The receiver-initiated design in the RI-MAC not only substantially reduces overhearing, but also achieves lower collision probability and recovery cost. Therefore, RI-MAC significantly improves throughput and packet delivery ratio, especially when there are contending flows such as bursty traffic or transmissions from hidden nodes. In this protocol, the nodes are scheduled to wake up periodically to verify if any data packets are intended for them. They send out a beacon frame, which is picked up by an awakened sensor node that has pending data packets to send. After receiving the beacon the sender node starts transmitting the data packets. On the reception of these, the receiver node sends an ACK beacon. The ACK beacon plays a dual role; first to acknowledge the reception of the data packet and second to ask for more data packets if any from the same node. In RI-MAC, medium access control among senders that want to transmit data frames to the same receiver is mainly controlled by the receiver. This design of RI-MAC makes it more efficient in detecting collisions and recovering data frames that are lost than B-MAC and X-MAC where the senders are hidden to each other. As a receiver expects incoming data only RI-MAC reduces overhearing within a small window after beacon transmission. With the lower cost for recovering lost data frame and detecting collisions, RI-MAC has higher power efficiency even when the load of network increases. However, as the previous MAC duty cycle protocols presented, RI-MAC suffers some default. Indeed, where there are several transmitters, the collision can occur in this protocol.

Even if these MAC protocols are efficient in term of energy consumption, they suffer some common limitations. Indeed, in almost of these protocols, the nodes broadcast their schedule to all neighbor nodes using the synchronization packet; so that a lack of efficiency is noted in term of energy consumed. Note also that the scheduling approaches used in almost the MAC duty cycle protocols described above are not suitable in a network with redundancy coverage; due mainly to the broadcast of synchronization and data packets by the senders to all communication neighbors' nodes and the retransmission packets.

Furthermore, Boulis [16] proposes the "TunableMAC" protocol based also on the duty cycle approach. As in other MAC protocols note that, in TunableMAC the CSMA/CA mechanism is used for the medium access. It is worth noticing that with this protocol all the nodes are not aligned in their active period, so that each sender transmit an appropriate train of beacon frames to wake up potential receivers before transmitting each data packet. Thus with respect to this mechanism, all neighbour that act as potential receivers of a given sender will be awakened when they received the beacon frame from the sender. Therefore, a lack of efficiency is noted in term of energy consumed. However, TunableMAC is very flexible and can be used to make comparisons with new MAC algorithms developed for WSN.

12.3. Proposals for New Strategies to Optimize Lifetime in WSN with Respect to Coverage and Network Connectivity Constraints

In this Section, we present in the first time different analytical models such as coverage, communication channel, sensing, connectivity and energy consumption in WSN based on the related work. Note that, these models are adapted according to our study. Secondly, we have made proposals that are articulated in two levels:

1. The placement of sensor nodes, we have set up a new deterministic placement strategy for sensor nodes that reduces the number of sensors necessary sensors to ensure the full coverage of a given region, and provides optimal network connectivity. This placement model is presented in Section 12.3.2.

2. The MAC layer, we have implemented a new distributed scheduling algorithm called MAC-SA (Medium Access Control Scheduling Algorithm), based on our placement model and using duty cycle technique. MAC-SA optimizes the energy consumption of sensor nodes in the whole network during the communications phase (TX and RX), thereby optimizes the network lifetime. In addition, the full coverage of the monitored region and good network connectivity obtained through our placement model is saved at any time of the network lifetime during the execution of MAC-SA algorithm. This algorithm is presented in Section 12.3.3.

12.3.1. Modeling Components and Parameters of the WSN

12.3.1.1. Network Model

We represent the WSN by a graph:

$$G = (V, E),$$ (12.1)

where V represents all vertices (nodes of the network) and $E \subseteq V^2$ represents the set of edges giving all possible communications. There is an ordered pair $(u, v) \in E^2$ if the sensor node u is physically capable to transmit messages to the sensor node v. In this case, sensor node v is located in the communication range of sensor node u. Thus, each node u has its key communication range noted $R_C(u)$ that allows it to communicate with others sensor nodes. We assume that all sensor nodes have equal communication ranges noted R_C. Thus, for two given sensor nodes u and v such that $u \neq v$ which their communication ranges are respectively $R_C(u)$ and $R_C(v)$ we have:

$$R_C(u) = R_C(v) = R_C$$ (12.2)

Each sensor node u also has a sensing range noted $R_S(u)$ that allows it to sense and capture data from the environment. We also assume that all sensor nodes have the same sensing ranges noted R_S. Therefore, for two given sensor nodes u and v such that $u \neq v$ which their sensing ranges are respectively $R_S(u)$ and $R_S(v)$ we have:

$$R_S(u) = R_S(v) = R_S \qquad (12.3)$$

The entire sensor node v located inside the communication range of a given sensor node u are called neighbour nodes of sensor node u and are noted $N(u)$. A bidirectional wireless link exists between a sensor node u and every neighbour node $v \in N(u)$ and is represented by the directed edges (u,v) and $(v,u) \in E$. Note that all the neighbour nodes can communicate directly each other.

In the following we note respectively A and $M = \{S_1, S_2, ..., S_M\}$ the surface of the monitored region where the SN are deployed and the set of SN in the WSN. We note also $N = |M|$ the cardinality of the set M that also represents the number of sensor nodes in the WSN. On the other hand, we assume in our study that all the sensor nodes transmit their captured data to a Sink node which is the only receiver of the application packets. Fig. 12.1 illustrates an example of this network model.

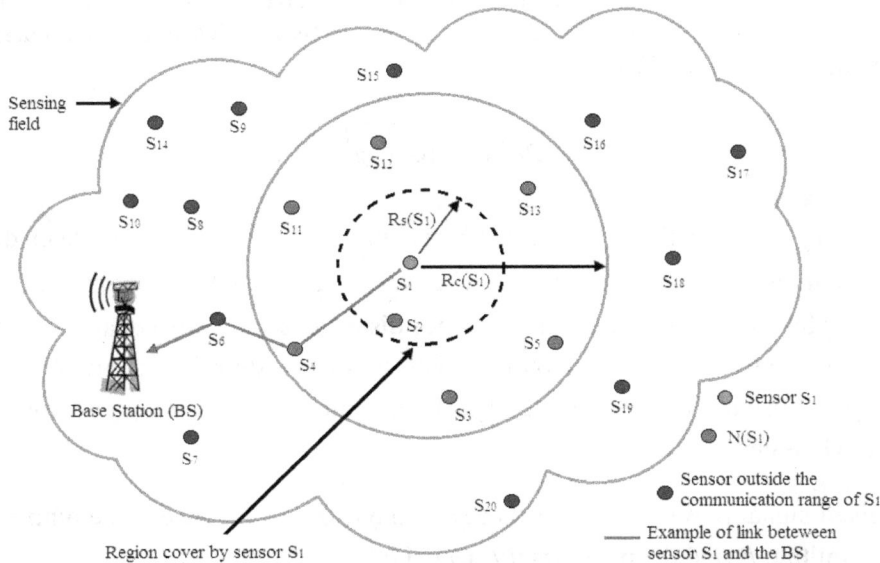

Fig. 12.1. Illustration of the network model.

12.3.1.2. Modeling the Wireless Communication

The performances of a wireless communication system are determined based on the communication channel in which it operates [23]. In WSN, modelling communication is very difficult because the nodes communicate in low power, and therefore radio links nodes are very unreliable. The unit disk model is the simplest deterministic models of communication that illustrates a unidirectional link between two SN. This model assumes that each node is able to transmit its data to any node being in its communication range. The communication range of each node is in correlation with its power transmission. Therefore, we can say that two sensor nodes u and v can communicate each other if and only if the Euclidean distance noted $d(u,v)$ between the two sensor nodes is less than their communication range R_C. Thus, two nodes $u,v \in M$ can communicate if:

$$d(u,v) \leq R_C \qquad (12.4)$$

Therefore, the communication between SN is based on geometrical considerations. Note that even if the unit disk model is widely used for analytical models, it suffers some limitations. One of these limitations is that, this model is considered to be ideal as it assumes that the messages are still received with no mistake, i.e. it suppose the conditions of the MAC layer as ideal.

Another model which takes important aspect for the wireless channel is the log-normal shadowing model. This model enables to estimate the average path loss between two sensors nodes, or in general, two points in space. For WSN, where the separation of nodes is a few meters to a few hundred meters, this model is the most used to provide accurate estimates for the average path loss. The formula below enables to estimate the path loss in decibel (dB) depending on the distance between two nodes and other parameters described in the following [23].

$$PL_0(d) = PL(d_0) + 10\eta Log\left(\frac{d}{d_0}\right) + X_\sigma, \qquad (12.5)$$

where $PL_0(d)$ is the path loss at a distance d that represents the Euclidean distance between the transmitter and the receiver. The parameter $PL(d_0)$ represents the path loss known at a reference distance d_0. This reference distance is generally equal to 1 meter (m) or 1 km. The parameter η is the exponent path loss depending in the environment and whose value is usually in the range [2~4]. The parameter X_σ is a random variable with mean zero Gaussian standard deviation σ.

The received signal power P_r at a distance d is the difference between the output power of the transmitter P_t and the path loss $PL_0(d)$, i.e.

$$P_r = P_t - PL_0(d) = P_t - \left(PL(d_0) + 10\eta Log\left(\frac{d}{d_0}\right) + X_\sigma\right) \qquad (12.6)$$

In this formula all the powers are expressed in dB. With this formula, we can control and estimate the communication range of the SN.

We consider in this chapter these two models.

Given the graph $G = (V, E)$ defined in the Section 12.3.1.1 and the communication range R_C of the SN, the unit disk model defines the set $E \subseteq V^2$ of edges which represent also the communication link between the SN by:

$$\mathrm{E} = \left\{ (u, v) \in V^2 \middle| u \neq v \wedge d(u, v) \leq R_C \right\} \tag{12.7}$$

Thus, based on Equation (12.6) and (12.7), we can determine all SN which are in the transmission range of another given SN by computing their communication ranges. In so doing, we use Equations (12.5), (12.6) and others radio parameter defined in [16] to compute the radio range (communication range) of SN. Afterwards, based on our assumptions, we can compute also the sensing range of SN and the grid length of our placement model. We detailed all these calculations in Section 12.4.1.

On the other hand, we use the well-known IEEE 802.11 as MAC layer and CSMA/CA (Carrier-Sense Multiple Access/with Collisions Avoidance) as medium access protocol.

12.3.1.3. Modelling the Coverage

Coverage is an important performance metric in WSN, which reflects how well a sensing field is monitored [11]. We may interpret the coverage concept as a nonnegative mapping between the space points of a sensing field and the sensors of a WSN [17]. There exist many type of coverage: area coverage (coverage of region), barrier coverage, and coverage of points [3]. We consider in our study the area coverage and coverage of points. Thus, we say that a sensor node S_i covers a point $q \in A$ if and only if:

$$d(S_i, q) = R_S \tag{12.8}$$

A coverage of surface (sensing coverage) means the total surface lying below the range of capture of data at least of a given sensor node. Let $S_i \in M$ a sensor node and note $C(S_i)$ the surface cover by the sensor node S_i, then:

$$C(S_i) = \left\{ q \in A \middle| d(S_i, q) \leq R_S \right\} \tag{12.9}$$

The surface covered by a subset of sensor nodes $S_{MC} = \{S_1, S_2, ..., S_C\} \subseteq M$ is then:

$$C(S_{MC}) = \bigcup_{i=1}^{|S_{MC}|} C(S_i) \tag{12.10}$$

An area is said to be covered if and only if each location of this area is within the sensing range of at least one active sensor node. For the coverage of area, we say that a sensor node S_i covers a region A if and only if for each point $q \in A$ then:

$$d(S_i, q) = R_S \tag{12.11}$$

Aera coverage is one of the fundamental problems in wireless sensor networks [11]. In the area coverage problem, the goal is to cover the whole area of the network. Depending on the application requirements, full or partial coverage is required. However, full coverage provides the best surveillance quality of the region [3]. There are two types of coverage: simple coverage and k-coverage defined as multiple coverage and depending on the degree of robustness required by the application [3]. Multiple coverage is defined as an extension of simple coverage. This type of coverage is suitable to applications such as security surveillance, distributed detection, mobility tracking, monitoring in high security areas, agricultural of precision, and military intelligence in a battlefield. In many kind of WSN related above, it is necessary to ensure full coverage of the monitored area, optimal network connectivity while deploying the minimum number of sensor nodes. This can be satisfied by covering every location in the field using at least one sensor node. Many studies aim to optimize the number of sensor nodes deployed while ensuring a high level of coverage and optimal network connectivity. So that, data captured in this location by the SN should be reported to the sink. The Fig. 12.2 and Fig. 12.3 below illustrate respectively the mechanisms of simple coverage and multiples coverage.

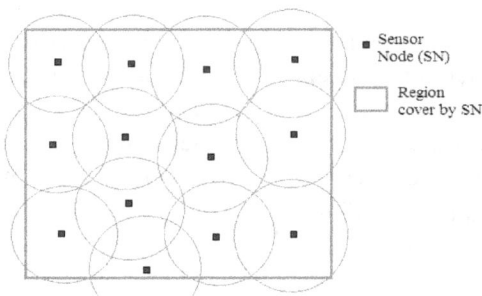

Fig. 12.2. Illustration of simple coverage.   **Fig. 12.3.** Illustration of multiple coverage.

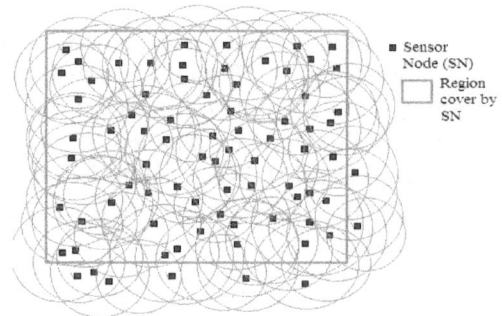

The problem of coverage area consists to apply scheduling mechanism of sensors activities to decide what sensor must be made in active mode (radio ON) or sleep mode (radio OFF) while maintaining a full coverage of the monitoring region. As we say below one degree of coverage is not sufficient for many applications of WSN related above. So that to schedule the sensors activities for these kinds of applications while ensuring full coverage of the monitored region, i.e. to ensure that if an event takes place at any geographic point of this area, it is detected by at least one sensor, it is necessary to guarantee multiple coverage when placing the SN in the interest region. In this case, an area may be covered by many sensors at the same time; this is due to overlapping coverage area of neighbour sensors. Therefore an event can be detected and reported by several sensors; this is inefficient in term of energy consumption, as some sensors will dissipate energy unnecessarily in the capture, processing and transmission. So that to reduce the energy consumption and optimize the network lifetime, it is necessary to apply scheduling strategies after planning optimal placement of SN; and while guaranteeing at the same time full coverage of the monitored region and optimal network connectivity. In this

240

chapter we will use distributed strategies based on an optimal placement of sensors to schedule the SN activities while maintaining full coverage and network connectivity. The following Fig. 12.4 illustrates a scheduling mechanism of SN activities in order to reduce the energy consumed in the WSN while ensuring the entire coverage of the monitored region and optimal network connectivity. This scheduling mechanism is based on ON/OFF scheduling approach and must allow to all SN which are in active mode (ON) to ensure the network functionally while maintaining full coverage of the monitored region and optimal network connectivity.

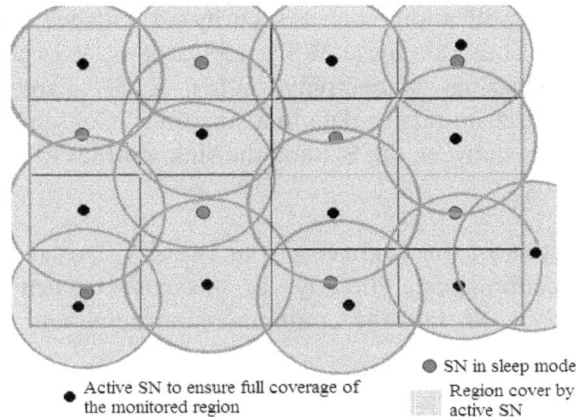

● SN in sleep mode

● Active SN to ensure full coverage of the monitored region

Region cover by active SN

Fig. 12.4. Illustration of an ON/OFF scheduling mechanism to reduce energy consumed by SN while ensuring full coverage of a monitored region.

12.3.1.4. Modelling the Network Connectivity

Two SN are said to be connected if and only if they can communicate directly (one-hop connectivity) or indirectly (multi-hop connectivity) [3]. In WSN, the network is considered to be connected if there is at least one path between the sink and each sensor node in the considered area. Connectivity is important issue in WSN. The connectivity essentially depends on the existence of routes. It is affected either by the topology changes due to mobility of SN, or the failure of sensors nodes, or malicious sensors nodes, etc. The results are the loss of communication links, the isolation of nodes, the network partitioning, thus the coverage of the monitored area can be degrade and/or the network lifetime can be decrease. Therefore, connectivity problem must be study and take into account in the design and the deployment of many WSN applications in order to guarantee coverage constraints and to ensure robustness in communication.

There are two types of network connectivity: full network connectivity and intermittent network connectivity [3]. Full network connectivity can also be either simple (1-connectivity) or multiple (k-connectivity). Full connectivity is said to be simple if there is a single path between any SN to the sink node; and it is said to be multiple if there are multiple disjoint paths between any SN and the sink node. In addition, full connectivity can be maintained during the deployment strategy of SN or it can be provided only when SN have been deployed in the monitored region. In the following, we use connectivity to represent full connectivity. Note that, in some WSN applications, it is not necessary to ensure full connectivity at any given time in the monitored region. For these WSN applications, it is sufficient to guarantee intermittent connectivity by using a mobile sink that moves and collects data from disconnected sensor nodes. There are two types of intermittent connectivity: the first one uses only one or several mobile sinks and the second uses a mobile sink and multiple throw boxes (Cluster heads).

In our study, we consider static sensors node, so that we consider only full connectivity. As we say in previous sections connectivity is often conflicting and complementary to coverage for many WSN applications such as security surveillance, agricultural precision, habitat monitoring, etc. Thus, for these WSN applications, it is not enough to ensure coverage without considering connectivity. When a SN captures data from the environment, it must be transmit these data to a sink node. Consequently, it is necessary to ensure the connectivity between the SN and the sink in order to guarantee the transfer of information to the sink.

Referring to the definition of the connectivity of two SN, two sensor nodes u and v are connect if they can communicate directly. In this case we say that these SN are communication neighbour. So that the communication neighbour of a sensor node u noted $N(u)$ is define by:

$$N(u) = \left\{ v \in V \mid v \neq u \wedge d(u,v) \leq R_C \right\} \tag{12.12}$$

A graph of a network which is connected is call a graph connected. Referring to this definition a graph is called k-connected if there is at least k disjoint path between two nodes of this graph. As we say above coverage is often related to connectivity in WSN. So that, to deal with the full coverage and the optimal network connectivity and to ensure the coverage and connectivity conditions, we consider in this chapter that the communication range R_C is twice the sensing range R_S, thus:

$$R_C = 2R_S \tag{12.13}$$

From equation (12.12) and (12.13), connectivity condition between two SN S_i and S_j of the WSN according to our model becomes:

$$d(S_i, S_j) \leq 2R_S \tag{12.14}$$

We use equation (12.14) to give mathematical proofs the optimal network connectivity. This evidence is presented in paragraph 3.3.2.2.

The graph of a connected network is a connected graph. A WSN is said k-connected if the graph associated with this network is k-connected, that is to say, there are at least k disjoint paths between two different nodes of the graph. The topology associated with such a network is thus called a k-connectivity topology. For example, Fig. 12.5 illustrates a k-connectivity topology. The dashed lines depict the links of connectivity between the sensor nodes.

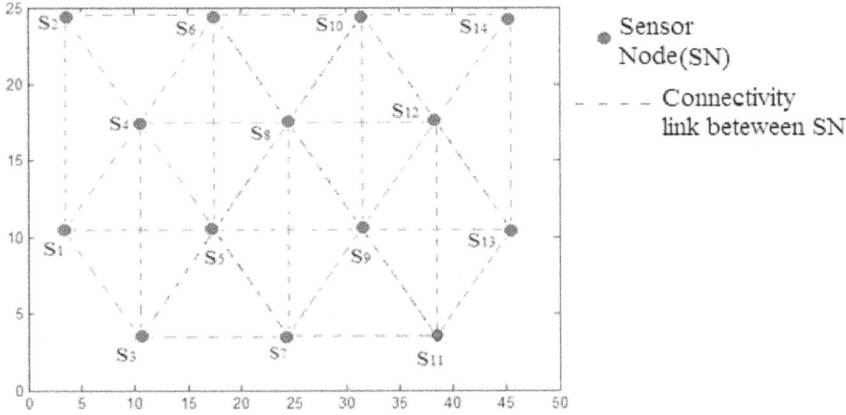

Fig. 12.5. Illustration of graph of connectivity.

12.3.1.5. Sensing Model

In our study, we consider the physical sensing model defined in [16]. However, this model is adapted according to our study. As defined in [16], in this sensing model, each SN has a physical process module which manages the detection of physical phenomena. This model is based on a random number of sources that generate physical phenomena and diffusing them in the space. A given point source may change in time and space, so that, the physical phenomena generate by this source may change in time and space. A phenomenon that occurs at any given point p of a region A covered by a set of N sensor nodes noted $M_N = \{S_1, S_2, ..., S_M\}$ is additive and its value at that point p of space and at a time t, and the value noted $V_{M_N}(p,t)$ is given by:

$$V_{M_N}(p,t) = \sum_{S_i \in M_N} \frac{V_{S_i}(t)}{\left(K * d(S_i, p)(t) + 1\right)^a} + N(0, \sigma) \qquad (12.15)$$

In Equation (12.15), $V_{S_i}(t)$ represents the value of the physical process captured by the source S_i at the time t. $d(S_i, p)(t)$ is the Euclidean distance between the source S_i and the point p at time t. The parameters K and 'a' determine how a value of a specific physical phenomena is broadcasted in the space. Finally, $N(0, \sigma)$ is a Gaussian random variable with mean zero and standard deviation σ. In [16], the values of parameters K, 'a' and σ are respectively fixed at 0.25, 1.0 and 0.2.

12.3.1.6. Models of Energy Consumption

The energy consumption is the most important issue in WSN. The energy consumed by a sensor node is mainly due to the following: capture, processing and data communication [21].

The energy of capture is dissipated by the SN to perform the following tasks: sampling, A/D conversion and activation of the capture probe. The cost of this energy depends on the specific sensor types (image, sound, temperature, etc.) and previous tasks assigned to him. In general, the energy of capture represents a small percentage of the total energy consumed by a SN.

The processing energy corresponds to the energy consumed by a sensor node during activation of its data processing unit (operations, read/write in memory). It is divided in two parts: switching energy and leakage energy. The switching energy is determined by the supply voltage and the total capacitance switched at the software level (by executing software). The leakage energy is the energy consumed when the computer unit performs no processing. In general, the processing energy is low relative to that required for communication.

The energy of communication is divided into two parts: the reception energy (energy consumed in RX mode) and the transmission energy (energy consumed in TX mode). This energy is determined by the amount of data to be communicated and the transmission distance, as well as by physical properties of the radio module. The scope of transmission of a signal depends on its transmission power (TX power). When the TX power is high, then the signal will have a large scope and the energy consumed will be higher. Note that the energy of communication represents the largest portion of the energy consumed by a SN.

The cost of the energy consumed by a sensor node must also depend on the activity status of this sensor (TX, RX, Idle and Sleep). These activities modes (or states) represent the different modes of operation of a SN [22]. Note that in the idle mode the sensor node can listen to the wireless channel without accessing to this wireless channel, while in sleep mode, the radio module is OFF and no communication is possible. It should be added that the transition between these different modes induces a cost on energy consumption even if it is small compared to other costs of energy consumption in the other modes.

For a given SN, the cost of consumed energy respectively in the TX, RX, Idle, Sleep, and Transition (Sw) states are respectively noted: $E_{Tx}(k, P_{out}), E_{Rx}(k), E_{Idle}, E_{Sw}$ where k represents the message length in bytes and P_{out} represents the TX power. If the consumed energy is expressed in Joules (J), it's regarded as the product of the voltage in Volt (V) applied to the circuit, the intensity of the current in Ampere (A) following through it, and the elapsed time in seconds (s) to perform the operation. So that, the cost of consumed energy in the different states described above can be expressed by the following equations:

$$E_{Tx}(k, P_{out}) = k.C_{Tx}(P_{out}).V_B.T_{Tx} \tag{12.16}$$

$$E_{Rx}(k) = k.C_{Rx}.V_B.T_{Rx} \tag{12.17}$$

$$E_{Idle} = C_{Idle}.V_B.T_{Idle} \tag{12.18}$$

$$E_{Sleep} = C_{Sleep}.V_B.T_{Sleep} \tag{12.19}$$

$$E_{Sw} = C_{Sw}.V_B.T_{Sw},\qquad(12.20)$$

where V_B represents the tension provided by the battery. C_{Tx}, C_{Rx}, C_{Idle}, C_{Sleep}, and C_{Sw} represent respectively the intensity of the current in the four states: TX, RX, Idle, Sleep, and transition. T_{Tx}, and T_{Rx} denoted respectively the TX and the RX time for one byte (with $T_{Tx} = T_{Rx}$). T_{Idle} is the time between the end of one communication (TX or RX) and the beginning of the next communication. T_{Sleep} is the interval time spent by a SN in the Sleep mode, and T_{Sw} is the switching time between two different modes. In this paper, we will use the radio type CC2420 [20] for the validation of our proposal by simulations and we will consider only the TX, RX, and Sleep modes. The transition time (in ms) and the transition power (in mA) between the considered modes are respectively given by the Delay transition matrix (described in Table 12.1) and the Power transition matrix (described in Table 12.2).

Table 12.1. Delay transition matrix.

	RX	TX	Sleep
RX	-	0.01	0.194
TX	0.01	-	0.194
Sleep	0.05	0.05	-

Table 12.2. Power transition matrix.

	RX	TX	Sleep
RX	-	62	62
TX	62	-	62
Sleep	1.4	1.4	-

Another model of energy consumption is presented in [24] by Heinzelman et al. For this model, to transmit a message of length k bits between a transmitter and a receiver which Euclidean distance is equal to d (in meter), the radio emitter expends the amount of energy (in Joules):

$$E_{Tx}(k,d) = E_{Tx\text{-elec}}(k) + E_{Tx\text{-amp}}(k,d)\qquad(12.21)$$

To receive this message, the radio expends the amount of energy (in Joules):

$$E_{Rx}(k) = E_{Rx\text{-elec}}(k),\qquad(12.22)$$

where $E_{Tx\text{-elec}}$ and $E_{Rx\text{-elec}}$ represent respectively the Transmitter electronic and the Receiver electronic, and ε_{amp} represent the Transmit amplifier. In the other hands, we have:

$$E_{Tx\text{-elec}} = E_{Rx\text{-elec}} = E_{elec} = 50\text{nJ / bits}\qquad(12.23)$$

$$\varepsilon_{amp} = 100pJ / bits / m^2 \tag{12.24}$$

Based to equations (21), (22), (23) and (24), we have:

$$E_{Tx}(k,d) = E_{elec} * k + \varepsilon_{amp} * k * d^2 \tag{12.25}$$

$$E_{Rx}(k) = E_{elec} * k \tag{12.26}$$

For these parameter values, receiving a message is not a low cost operation; the protocols should thus try to minimize not only the transmit distances but also the number of transmit and receive operations for each message. We make the assumption that the radio channel is symmetric such that the energy required to transmit a message from a node S_i to a node S_j is the same as the energy required to transmit a message from node S_j to node S_i for a given SNR (Signal to Noise Ratio).

For our simulations presented in Section 12.3.4, we also assume that all SN are sensing the environment at a fixed rate and thus always have data to send to the sink node. We assume also that, all SN send their captured data (with constant packets payload equal to k bits) to a given sink node which is the only receiver of these data packets.

12.3.2. Proposal of a New Placement Method Sensor Nodes Based on Grids

We consider the sensor placement model described in the following Fig. 12.6.

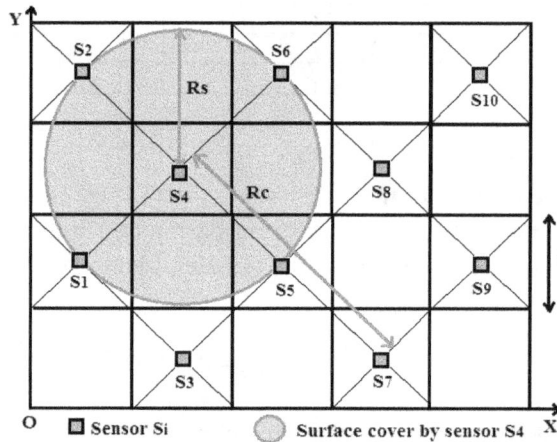

Fig. 12.6. Sensors placement model based on grids.

According to the deployment of sensors described in the Fig. 12.6 above, the geographical region of interest is partitioned into contiguous square grids having the same dimensions that equal to c. Each SN S_i is placed at a given area of a grid such that the entire area of the monitored region is covered and the number of necessary sensors is minimized. Our geographic placement model of SN presents the following advantages:

246

1. The number of sensors needed to cover the whole area is minimized.

2. The position and the surface cover by each SN are known and can be respectively determined by its coordinates (x, y) and its sensing range R_S.

3. A full coverage of the monitored region and an optimal network connectivity are ensured.

4. It exist an overlapping area with respect to the sensing coverage of SN that will be exploited by our MAC-SA algorithm for ON/OFF scheduling of SN.

Now, the optimal length c of the grid to ensure full coverage and network connectivity of our network model can be determined based on the sensing range R_S. The sensing range depends on the communication range R_C based on our assumptions $R_C = 2R_S$.

Finally, based on the geometric properties of the obtained squares and the diamonds formed by the position of SN (Fig. 12.6), the length c of the grid can be computed by using Pythagoras theorem. Thus, we can use the following equation to compute c:

$$R_S^2 = c^2 + c^2 \qquad (12.27)$$

$$\Rightarrow R_S = c\sqrt{2} \qquad (12.28)$$

$$\Rightarrow c = \frac{R_S}{\sqrt{2}} \qquad (12.29)$$

Based on our assumptions, we have:

$$R_C = 2R_S \Leftrightarrow R_S = \frac{R_C}{2} \qquad (12.30)$$

Thus, according to (12.29), and (12.30), we have:

$$c = \frac{R_C}{2\sqrt{2}} \qquad (12.31)$$

Note $M = \{S_1, S_2, ..., S_M\}$ the set of SN deployed according to our placement model described in the Fig. 12.6. Note also that each SN S_i has (x, y) in the coordinate system (O, X, Y) as shown in the Fig. 12.6 where O, (OX) and (OY) denote respectively the origin, the X axis and the Y axis of this coordinate system. Thus, using this coordinate system and according to our placement model, we can express the coordinate (x, y) of each SN function of the length c of the grids. For example as it is shown below:

$$S_1\left(\frac{c}{2},\frac{3c}{2}\right),\ S_2\left(\frac{c}{2},\frac{7c}{2}\right),\ S_3\left(\frac{3c}{2},\frac{c}{2}\right),\ S_4\left(\frac{3c}{2},\frac{5c}{2}\right),\ S_5\left(\frac{5c}{2},\frac{3c}{2}\right),\ S_6\left(\frac{5c}{2},\frac{7c}{2}\right),\ S_7\left(\frac{7c}{2},\frac{c}{2}\right),$$

$$S_8\left(\frac{7c}{2},\frac{5c}{2}\right),S_9\left(\frac{9c}{2},\frac{3c}{2}\right),\text{ and }S_{10}\left(\frac{9c}{2},\frac{7c}{2}\right).$$

Note $S_{MC}=\{S_1,S_2,...,S_C\}\subseteq M$ a subset of SN deployed in the WSN according to our placement model. Then referring to the definition of the surface covered by a subset of sensor nodes described in Equation (12.10), we show that according to our placement method, an area may be covered by many sensor nodes at the same time; this is due to overlapping coverage area of neighbours sensor nodes. Therefore, to save energy consumed in the network and to maximize network lifetime, it is necessary after the final deployment to schedule sensor nodes activities by applying Sleep/wake-up strategies (e.g. redundant nodes for full coverage, useless nodes for partial coverage) while ensuring full coverage of the monitored region and optimal network connectivity.

Note that the scheduling activity for SN differs from deployment method of SN, because existing sensor nodes are only switched ON or OFF but are not moved. In the following section we'll present the MAC-SA algorithm which is based on our geographic placement method and which enables to schedule the SN activities and optimize the network lifetime while maintaining full coverage of the monitored region and network connectivity.

12.3.3. Presentation and Analytical Evaluations of MAC-SA

12.3.3.1. Overview of MAC-SA

MAC-SA algorithm (illustrate above) considers our geometric placement model and is a distributed scheduling mechanism for SN activities.

Inputs :

1 : "c" represents the length of a given grid according to our placement model

2 : "M = $\{S_1,S_2,...,S_N\}$" represents the set of sensors deployed in the WSN according to our placement model

3 : "d (S_i,S_j)" represents the euclidean distance between two given sensors $S_i,S_j \in M$

4 : "Neighbor_Table$[S_i]$" represents the neighbor table of a given sensor $S_i \in M$

5 : "ID (S_i)" represents the ID (identifiant) of a given sensor $S_i \in M$

6 : "B_{ik}" represents a beacon frame sent by a given source $S_i \in M$

Outputs :

- A set of active sensors to relay packets send by a given source S_i and which ensure a full coverage of the monitored region and optimal network connectivity

- A set of sensors which are put in sleep mode to save their energy

7 : **for** each sensor $S_i \in M$ **do**

8 :   **for** each sensor $S_j \in M \wedge S_j \neq S_i$ **do**

9 :   **if** $d(S_i, S_j) \leq 2c\sqrt{2}$ **then**

10 :   Insert $\left(ID(S_j), \text{Neighbor\_Table}[S_i] \right)$

11 :   **end if**

12 :   **end for**

13 : **end for**

14 : **for** each sensor $S_i \in M$ which broadcast a train of beacon frames : $\{B_{i1}, B_{i2}, ..., B_{iN}\}$ **do**

15 :   **if** $S_j \in M$ receives $B_{ik} \in \{B_{i1}, B_{i2}, ..., B_{iN}\} \wedge ID(S_j) \in \text{Neighbor\_Table}[S_i]$ **then**

16 :   **if** $d(S_i, S_j) \leq c\sqrt{2}$ **then**

17 :   Make S_j in sleep state until it receives a next beacon B_{ik}

18 :   **else**

19 :   Make S_j active mode to relay packets transmit by the sender S_i

20 :   **end if**

21 :   **end if**

22 : **end for**

It enables to minimize the energy consumed by the overall network while maintaining a full coverage and network connectivity with respect to all SN. The MAC-SA algorithm exploits the redundancy of sensing coverage due to our geographic placement method. Indeed, according to TunableMAC protocol, each sender should transmit a train of beacons frames in order to wake up its entire neighbourhood before sending any data. However, according to our MAC-SA deployment of SN, where we have a sensing coverage redundancy due to our placement strategy of SN, we do not need to wake up all a given SN's neighbourhood. It is worth noticing that in TunableMAC, the set of SN has equal sleep interval and equal listening interval. Put simply, MAC-SA wakes up only few nodes among a well-chosen SN's neighbourhood in order to reduce the energy consumed during transmission and reception as well as mitigates the number of collisions between SN.

According to MAC-SA, each SN uses a neighbourhood's table that contains the *ID* of neighbour's nodes which is determined by the communication range R_C. Also, the SN has different sleeping and listening intervals. MAC-SA addresses the following two issues noted in previous studies:

1. The set of sender's neighbours that should wake up according to its neighbour's table;

2. The scheduling of sleeping and listening intervals according to the parameters of the duty cycle.

In order to select the best potential neighbours that enable to minimize the energy consumption during transmission and reception modes while ensuring a full coverage and network connectivity, and to taking into account the two issues raised above, we consider two types of neighbours for each node: "*close neighbours*" located at a maximum distance of $c\sqrt{2}$ from the sender and "*remote neighbours*" located at a distance strictly greater than $c\sqrt{2}$. For a given sender, its neighbour's receivers are only its remote neighbours. Therefore, remote neighbours must be woken up and all the remaining nodes within its close neighbourhood must be set in sleeping mode (line 14 to line 22 of MAC-SA algorithm). If they receive other beacons frame, they can decide whether they should wake up again to relay packets. Our algorithm allows the following benefits:

1. Save the energy consumed in the network, so that the network lifetime will be improved;

2. Save full coverage and network connectivity at every time of the network lifetime;

3. Balance energy consumption in the network;

4. Reduce collisions that may be due to the CSMA/CA mechanism, so that the rate of received packets by the Sink will be improved.

As shown in the illustration of MAC-SA, the lines 1 to 13 enable to compute the neighbour table of each SN $S_i \in M$ by inserting the entire ID of its neighbour $S_j \in M \wedge S_j \neq S_i$. After this step, each SN $S_j \in M$ neighbour of a given sender $S_i \in M$ will decide if it will be switched in Active or Sleep mode based on the beacon frame received by this sender (which precede the data transmission of the source) and its neighbour table (lines 14 to 22). Therefore, the SN which will usually switch in Sleep mode will save more energy; so that the network lifetime will be improved. Note that the full coverage and network connectivity will be preserved during all the network lifetime. We will give in the following part the analytical proof of the full coverage and the network connectivity.

12.3.3.2. Analytical Evaluations of MAC-SA

In this section, we evaluate firstly the complexity of MAC-SA algorithm to demonstrate its effectiveness whatever the load and the network size. Secondly, the full coverage of the monitored region and the network connectivity, are analyzed.

12.3.3.2.1. Evaluation of the Complexity of MAC-SA Algorithm

This section is devoted to the study of computational complexity of DSMAC in order to prove its effectiveness in the worst case. To do this, we will first present some formal tools that allow us to calculate the complexity of algorithms.

Are respectively n and $T(n)$ the average size of data (in bits) and the cost of the time complexity of the MAC-SA algorithm. This algorithm consists of two major parts:

1. A first portion extending from line 7 on line 13, that allows each SN to build its neighbors table by inserting all the *ID* of its one hop communications neighbors nodes.

2. A second portion from line 14 on line 22, that allows each SN S_j belonging to the neighbor table of a given source S_i to decide whether to switch into active mode in order to relay packets from that source or to be put in sleep mode to save energy. As illustrated in MAC-SA algorithm, the decision to be put in sleep or active state is based on the Euclidean distance between each sender S_i and each SN of its neighbors communication, and secondly on beacon frame which are synchronization packets that is broadcasted by the source before it's transmission. Note respectively $T^A(n)$ and $T^B(n)$ the costs complexities of the first part and the second part of MAC-SA described above. The total cost of the MAC-SA complexity is:

$$T(n) = T^A(n) + T^B(n)$$

(12.32)

Now let calculate $T^A(n)$ and $T^B(n)$. In so doing, consider initially the first part of MAC-SA algorithm described above. Note $T_2^A(n)$ the cost of the complexity of the loop (line 8 to line 12). Then we have:

$$T^A(n) = \sum_{k=1}^{N} T_2^A(n)$$

(12.33)

Note $T_1^A(n) = C_1 \in IR$, the cost of the complexity of elementary instruction in line 10. Referring to calculate the cost of complexity to a sequence of instructions and taking account the fact that each sensor node $S_i \in M$ has at most 12 communication neighbors (show Fig. 12.7), we have:

$$T_2^A(n) \le \sum_{k=1}^{N} \frac{T_2^A(n)}{12} \le \frac{1}{12}\sum_{k=1}^{N} C_1 = \frac{N*C_1}{12}$$

(12.34)

Note $C_2 = \frac{N*C_1}{12}$. Since $C_1 \in IR$, then $C_2 \in IR$. Thus, according to (12.33) and (12.34), we have:

$$T^A(n) = \sum_{k=1}^{N} T_2^A(n) \le \sum_{k=1}^{N} C_2 = N*C_2$$

(12.35)

Now calculate $T^B(n)$. To do this, note respectively $T_{11}^B(n) = C_3 \in IR$ and $T_{12}^B(n) = C_4 \in IR$, the cost of the complexity of elementary instructions in line 17 and line 19. Afterward, note $T_1^B(n)$ the cost of the complexity of the instructions from line 16 on line 20. Then, we have:

$$T_1^B(n) = \max\left(T_{11}^B(n), T_{12}^B(n)\right) = \max\left(C_3, C_4\right)$$

(12.36)

According, to (12.36), we have:

$$T^B(n) = \sum_{k=1}^{N} T_1^B(n) = \sum_{k=1}^{N} \max\left(C_3, C_4\right) = N * \max\left(C_3, C_4\right) \qquad (12.37)$$

Based on (12.32), (12.35) and (12.37), we have:

$$T(n) = T^A(n) + T^B(n) = N * C_2 + N * \max\left(C_3, C_4\right) = N * \left(C_2 + \max\left(C_3, C_4\right)\right) \quad (12.38)$$

According to (12.38), $T(n) = N*(C_2 + C_3)$ if $C_3 \geq C_4$, $T(n) = N*(C_2 + C_4)$ either. Since $C_2, C_3, C_4 \in IR$ and the number N of nodes in the WSN is a integer, then according to the definitions of classes of algorithm complexity, the complexity MAC-SA is linear. On the other hand, since a linear complexity is one of the best classes of computational complexity, then we can deduce that MAC-SA is an efficient algorithm.

12.3.3.2.2. Analytical Evaluations of Coverage and Connectivity in MAC-SA Algorithm

In this section, we state two theorems which deal with the complete coverage of the surveillance zone and on the connectivity of WSN in our investment model and according DSMAC algorithm. These two theorems are proved in the sequel.

- Theorem of the full coverage according to MAC-SA algorithm

Let A the area covered by a set of N sensors node $M = \{S_1, S_2, ..., S_N\}$ deployed in a monitored region according to the placement model described in Section 12.3.2 and our MAC-SA algorithm. So that, at any time of the network lifetime, A remains full covered by all the active sensors in the WSN.

- Demonstration

Let us consider a sender $S_i(x,y) \in M$. As we said that in the description of our MAC-SA algorithm, before this SN transmits data packets, it broadcasts a train of beacons frames noted $B_{i1}, B_{i2}, ..., B_{ik}$ in order to wake up all the sensor nodes S_j belonging to its neighbour table and located at a distance strictly greater than $c\sqrt{2}$. Based on our placement model described in Fig. 12.6 and according to the Fig. 12.7 that illustrates the coordinate of remote and close neighbour for a given sender $S_i(x,y)$, the coordinate of this sender's remote neighbour are function of the grid length c, x, y, and are expressed as follows:

$$(x, y-2c), (x, y+2c), (x-2c, y-2c), (x-2c, y), (x-2c, y+2c),$$
$$(x+2c, y-2c), (x+2c, y), (x+2c, y+2c)$$

According to Fig. 12.6 and Fig. 12.7, the coordinates of other close neighbours of the sender $S_i(x, y)$ that can be put in sleep mode are expressed as follows:

$$(x-c, y-c), (x-c, y+c), (x+c, y-c), (x+c, y+c)$$

Now, let us consider the sensor node $S_7(x, y)$ shown in Fig. 12.7. Its neighbourhood's table contains the ID of the set of the following SN:

$$\{S_1, S_2, S_3, S_4, S_5, S_6, S_8, S_9, S_{10}, S_{11}, S_{12}, S_{13}\}$$

If the sensor $S_7(x, y)$ wants to transmit, then the set of sensors located to its neighbourhood table which must wake up after receiving the beacon frames sent by the SN S_7 are: $\{S_1, S_2, S_3, S_6, S_8, S_{11}, S_{12}, S_{13}\}$, and the following SN: $\{S_4, S_5, S_9, S_{10}\}$ should be put in sleeping mode. According to the Fig. 12.7, S_4, S_5, S_9 and S_{10} are in sleeping modes at the same time whereas other SN belonging to S_7's neighbour table are in active mode (powered ON) and maintain a full network coverage. We show that the areas covered by the following SN S_4, S_5, S_9, S_{10} which are in sleep mode, and the one covered by the four active SN located at the vicinity of these sleeping SN are fully covered by the active SN. Let us consider the SN S_4 which is in sleep mode (Fig. 12.7), then according to the definition of the sensing coverage of this sensor noted $C(S_4)$, we have:

$$C(S_4) = \{q \in A \,|\, d(S_4, q) \leq c\sqrt{2}\} \qquad (12.39)$$

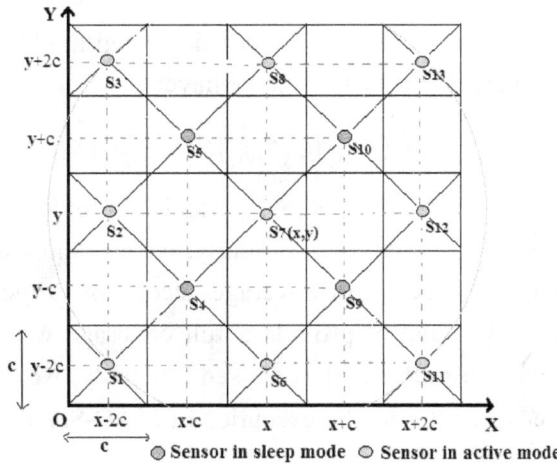

Fig. 12.7. Illustration of close and remote neighbours of $S_7(x, y)$

According to the active SN S_1, S_2, S_7 and S_6 which are around the sensor S_4, the sensing coverage of each SN is:

$$C(S_1) = \{q \in A \mid d(S_1, q) \le c\sqrt{2}\} \tag{12.40}$$

$$C(S_2) = \{q \in A \mid d(S_2, q) \le c\sqrt{2}\} \tag{12.41}$$

$$C(S_7) = \{q \in A \mid d(S_7, q) \le c\sqrt{2}\} \tag{12.42}$$

$$C(S_6) = \{q \in A \mid d(S_6, q) \le c\sqrt{2}\} \tag{12.43}$$

Note $S_C = \{S_1, S_2, S_7, S_6\}$.

Based on the coverage area of a subset of SN described in (12.10), we have:

$$C(S_C) = C(S_1) \bigcup C(S_2) \bigcup C(S_7) \bigcup C(S_6) \tag{12.44}$$

On the other hand, if we compute the Euclidean distance between the SN S_4 and each SN $S_j \in C(S_C)$, we have:

$$d^2(S_4, S_1) = \left[(x-c)-(x-2c)\right]^2 + \left[(y-c)-(y-2c)\right]^2 = 2c^2 \Rightarrow d(S_4, S_1) = c\sqrt{2} \tag{12.45}$$

$$d^2(S_4, S_2) = \left[(x-c)-(x-2c)\right]^2 + \left[(y-c)-y\right]^2 = 2c^2 \Rightarrow d(S_4, S_2) = c\sqrt{2} \tag{12.46}$$

$$d^2(S_4, S_7) = \left[(x-c)-x\right]^2 + \left[(y-c)-y\right]^2 = 2c^2 \Rightarrow d(S_4, S_7) = c\sqrt{2} \tag{12.47}$$

$$d^2(S_4, S_6) = \left[(x-c)-x\right]^2 + \left[(y-c)-(y-2c)\right]^2 = 2c^2 \Rightarrow d(S_4, S_6) = c\sqrt{2} \tag{12.48}$$

Thus according to (12.45), (12.46), (12.47) and (12.48), we have:

$$d(S_4, S_1) = d(S_4, S_2) = d(S_4, S_7) = d(S_4, S_6) = c\sqrt{2} \tag{12.49}$$

Based on the sensing coverage of SN S_1, S_2, S_6, S_7 described in (12.40), (12.41), (12.42) and (12.43); according to (12.44) and (12.49), we have:

$$C(S_4) \subseteq C(S_1) \bigcup C(S_2) \bigcup C(S_7) \bigcup C(S_6) \tag{12.50}$$

Hence according on Equation (12.50), S_1, S_2, S_6, and S_7 provide a full coverage with respect to the area covered by the SN S_4. Similarly, we can show that S_2, S_3, S_8, and S_7 (resp. S_6, S_{11}, S_{12}, and S_7) provide a full coverage according to the area covered by S_5 (resp. S_9). Finally, S_8, S_{13}, S_{12}, and S_7 provide a full coverage with respect to the area covered by S_{10}. Since the sensor $S_7(x, y)$ is chosen randomly, we can conclude that the network remains fully covered during the execution of MAC-SA algorithm.

- Theorem of the optimal network connectivity according DSMAC algorithm

Let A be the area covered by a subset of N sensors node $M = \{S_1, S_2, ..., S_N\}$ deployed in a monitored region according to the placement model described in Section 12.3.2 and MAC-SA algorithm. So that, at any time of the network lifetime, the subset of active sensors node remain k-connected (with $k \ge 4$).

- Demonstration

In fact, based on our assumptions and the modelling of the network connectivity presented in Section 12.3.1.4, two SN S_i and S_j are connected if and only if:

$$d\left(S_i, S_j\right) \leq 2c\sqrt{2} \tag{12.51}$$

In order to demonstrate the network connectivity, it is sufficient to show that all active neighbours of a given sender $S_i\left(x, y\right)$ are connected to this sender. The remote neighbours of the SN $S_i\left(x, y\right)$ noted $R\_Neighbor\_S_{x,y}$ are:

$$R\_Neighbor\_S_{x,y} = \{S_{N1}\left(x, y-2c\right), S_{N2}\left(x, y+2c\right), S_{N3}\left(x-2c, y-2c\right), S_{N4}\left(x-2c, y\right),$$
$$S_{N5}\left(x-2c, y+2c\right), S_{N6}\left(x+2c, y-2c\right), S_{N7}\left(x+2c, y\right), S_{N8}\left(x+2c, y+2c\right)\}$$

If we compute the Euclidian distance between the sensor $S_i\left(x, y\right)$ and each of the sensor nodes $S_j \in R\_Neighbor\_S_{x,y}$, we have:

$$d\left(S_i, S_j\right) \leq 2c\sqrt{2} \tag{12.52}$$

For instance:

$$d^2\left(S_i, S_{N1}\right) = \left(x-x\right)^2 + \left(y-\left(y-2c\right)\right)^2 = \left(2c\right)^2 \Rightarrow d\left(S_i, S_{N1}\right) = \sqrt{\left(2c\right)^2} = 2c \leq 2c\sqrt{2}$$

Thus, $d\left(S_i, S_{N1}\right) \leq 2c\sqrt{2}$

Therefore, according to (12.52) and based to (12.51) which illustrates the connectivity condition between two sensors, all sensors $S_j \in R\_Neighbor\_S_{x,y}$ are connected to SN $S_i\left(x, y\right)$. Since the SN $S_i\left(x, y\right)$ is chosen randomly, then all active sensors will be connected during the execution of our MAC-SA algorithm. In addition, according to the definition of a graph which is k-connected, the network is at least 4-connected; therefore, optimum routing topology exists in this network. However, we will not discuss the routing aspect in this chapter.

12.4. Evaluations of MAC-SA by Simulation

We validated our proposal by extensive simulations done with "*Castalia.3.0*" framework [16]. Castalia is a WSN simulator for Body Area Networks (BAN) and generally networks of low-power embedded devices. It is based on the OMNeT++ platform [19] and can be used by researchers and developers who want to test their distributed algorithms and/or protocols in realistic wireless channel and radio models, with a realistic node behavior especially relating to access of the radio.

12.4.1. Description of Simulation Parameters

We consider a field 2D of size equal to $200m \times 200m$. The sensor nodes are deployed monitored region according to our placement model presented in Section 12.3.2. In our simulations, we considered different network scenarios with size respectively equal to 40, 80, 120, 160, 200 nodes in addition to the base station (which is the only receiver of the application packets). We choose the type of radio module CC2420 [20].

The different simulation parameters and their values are described in Table 12.3. All these parameters are taken from [16] and adapted according to our context. However, the sensor communication range is calculated based on the modeling of the wireless channel described in paragraph 3.1.2. This calculation detailed above is done based on the basis of two parameters which are the transmission power (assumed equal for all the SN) and the average path loss $PL(d_0)$.

To compute the communication range of SN, we use the Equation (12.6) presented in paragraph 3.1.2. According to this equation:

$$P_r = P_t - PL_0(d) = P_t - \left(PL(d_0) + 10\eta Log\left(\frac{d}{d_0}\right) + X_\sigma \right) \tag{12.53}$$

In Equation (12.53), the parameter d represents the Euclidean distance between Emitter and Receiver, i.e. the communication range of SN. So that, we can calculate this communication range d as a function of other parameters define in (12.53). Thus, from equation (12.53), we have:

$$Log\left(\frac{d}{d_0}\right) = \frac{P_t - P_r - PL(d_0) - X_\sigma}{10\eta} \tag{12.54}$$

$$\Rightarrow \ln\left(\frac{d}{d_0}\right) = \left(\frac{P_t - P_r - PL(d_0) - X_\sigma}{10\eta}\right) * \ln(10) \tag{12.55}$$

$$\Rightarrow \frac{d}{d_0} = e^{\left(\frac{P_t - P_r - PL(d_0) - X_\sigma}{10\eta}\right) * \ln(10)} \tag{12.56}$$

$$\Rightarrow d = d_0 * \left(10^{\left(\frac{P_t - P_r - PL(d_0) - X_\sigma}{10\eta}\right)} \right) \tag{12.57}$$

In Equation (12.57), P_t and P_r represent respectively the transmission power and the reception power. P_r in Decibel (dB) is determined by [16] according to the receiver sensitivity (*ReceiverSensibility*) and *Noisefloor* parameter using the following equation (12.58).

$$P_r = \max\left(\text{Re}\,ceiverSensibility, NoiseFloor + 5dB\right) \tag{12.58}$$

Based to Equation (12.57) and (12.58); and according to (12.29), (12.31), we calculate respectively the value of the communication range R_C, the sensing range R_S and the grid range c, and we have:

$R_C \approx 20m, R_S \approx 10m$ and $c \approx 7m$.

The Table 12.3 shows the different values of the simulation parameters.

Table 12.3. Simulation parameters and settings.

Parameter	Value
Field sixe	$200m \times 200m$
Number of node considered during each simulation	40, 80, 120, 160, 200
Deployment type	Static
Simulation time	400 s
Communication range (R_C)	20 m
Sensing range (R_S)	10 m
Grid range(c)	~7 m
Radio type	CC2420
Transmission power	0 dB
Power consumed respectively during TX, RX and sleep states	62 mW, 62 mW, 1.4 mW
Initial energy	18720 J
Energy consumed per sensing	0.02 mJ
Data rate	250 kbps
Modulation type	PSK
Bit per symbol	4
Bandwidth	20 MHz
Noise Bandwidth	194 MHz
NoiseFloor	-100 dB
ReceiverSensibility	-95 dB
Path loss exponential (η)	2.765
Average initial path loss ($PL(d_0)$)	55
Reference distance (d_0)	1 m
Gaussian mean variable (X_σ)	4.0
Application considered	ThrouputTest [16]

The sensor nodes are placed according to their coordinates which depend on the length of the grid. For example, Fig. 12.8 illustrates the network topology for 14 sensor nodes deployed in the region of interest according to our placement model described in Section 12.3.2, and for a grid length equal to 7 m.

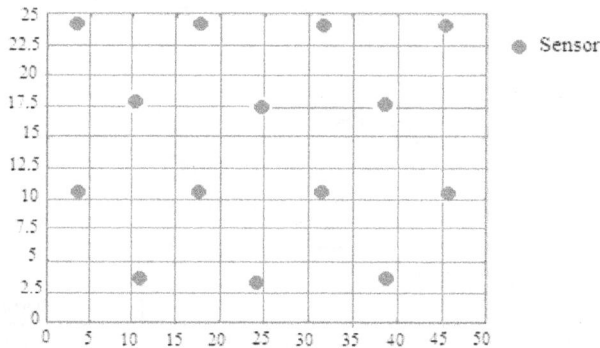

Fig. 12.8. Illustration of the topology of 14 sensors node deployed in a 2D region according to MAC-SA placement model

12.4.2. Performance Evaluations of MAC-SA

In our simulations, we used an application available in Castalia simulator called "ThroughputTest" [16]. This application allows nodes sensors to send packets of constant size, each with a load equal to 2000 bytes. These packets captured by the different network nodes of the sensors are sent to the base station with a transmission rate of 5 packets per second. The metrics measured performance is the average energy consumed, the average residual energy, the number of packets received by the base station and the level of average packet latency.

After assessments of these different metrics, comparisons between MAC-SA and TunableMAC protocol are made considering the same scenarios and assessing the same performance metrics. The different simulation results are presented in Section 12.4.2.1.

12.4.2.1. Simulation Results

12.4.2.1.1. Evaluation of Energy Consumption between MAC-SA and TunableMAC

The curves illustrated in Fig. 12.9 and Fig. 12.10 show respectively the average of energy consumed in Joules (J) and the average of remaining energy in J for the both algorithms. MAC-SA outperforms TunableMAC with respect to the energy consumed. Indeed, with MAC-SA only few senders' neighbours woke up in contrast to TunableMAC where the entire set of node's neighbours are awakened. Therefore, more actives nodes exist and thus the energy consumed is increased. The average of energy consumed in the network is roughly equal to 19.18 J (resp. 27.09 J) for MAC-SA (resp. TunableMAC). According to MAC-SA SN can save up to 30 % of their energy compared to TunableMAC. As shown that in Fig. 12.7 the average remaining energy in the network is roughly equal to 18700.803 J (resp. 18692.914 J) for MAC-SA (resp. TunableMAC).

Thus, the network lifetime time is improved in MAC-SA relative to TunableMAC.

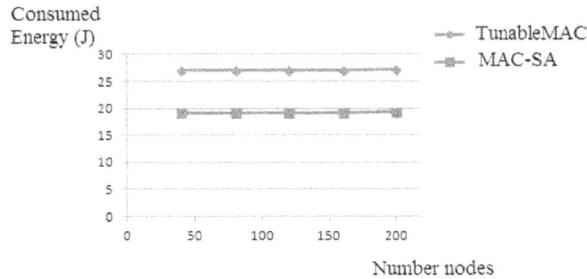

Fig. 12.9. Average consumed energy in MAC-SA and TunableMAC.

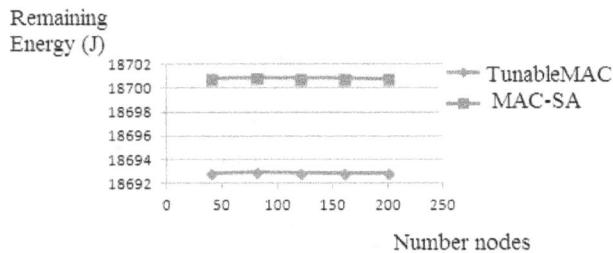

Fig. 12.10. Average remaining energy in MAC-SA and TunableMAC.

12.4.2.1.2. Evaluation of received packets between MAC-SA and TunableMAC

Fig. 12.11 (resp. Fig. 12.12) shows the average packets received by the Sink (resp. the average packets failed due to interferences). Fig. 12.11 illustrates that MAC-SA outperforms TunableMAC according to the number of packets received by the Sink. The main reason is due to the fact that MAC-SA algorithm mitigates the number of collisions.

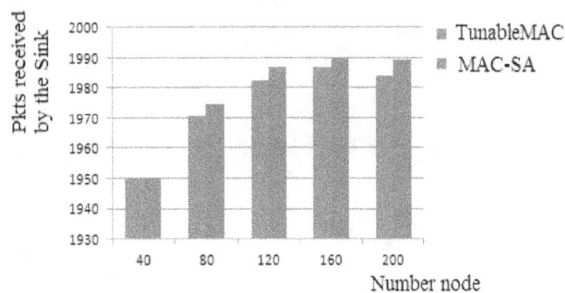

Fig. 12.11. Average packets received by the Sink in MAC-SA and TunableMAC.

Furthermore, Fig. 12.12 shows the average packets failed due to interferences. The gap between both algorithms is more important. Indeed, the average number of packets failed with interferences is roughly equals to 310.33 (resp. to 24.21) for TunableMAC (resp. MAC-SA).

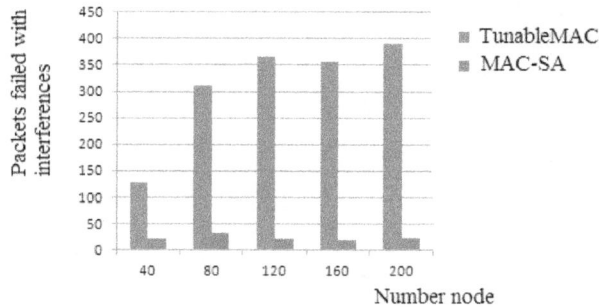

Fig. 12.12. Average packets failed with interferences in MAC-SA and TunableMAC.

12.4.2.1.3. Evaluation of application level latency between MAC-SA and TunableMAC

Fig. 12.13 shows the application level latency for both algorithms. As shown in this figure the performance of TunableMAC is lightly upper than MAC-SA but the upper level latency in these two algorithms is less than 166.9 ms, thus the level latency is reasonable in MAC-SA regarding to the most applications for WSN.

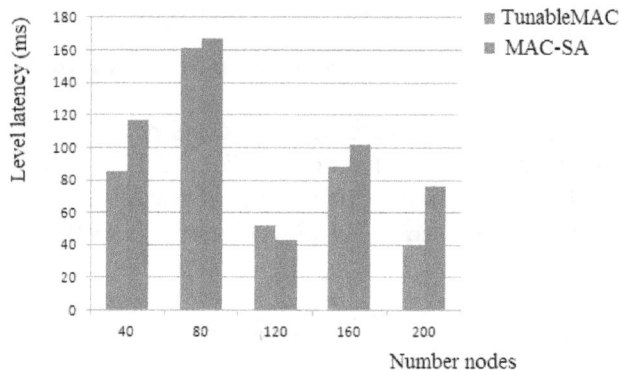

Fig. 12.13. Average application level latency in MAC-SA and TunableMAC.

12.5. Conclusion and Perspectives

In this chapter we have presented a new Medium Access Control Scheduling Algorithm (MAC-SA) based on a deterministic placement strategy for sensor nodes that reduces the number of sensors necessary sensors to ensure the full coverage of a given region, and provides optimal network connectivity.

Firstly, we study different analytical models such as coverage, communication channel, sensing, connectivity and energy consumption in WSN based on the related work. Then, we choose and adapted some of these models according to our study.

Secondly, we implement the MAC-SA algorithm and evaluate it analytically and by simulations. The simulation results show that MAC-SA optimizes the network lifetime, the average of received packets by the Sink, and minimized the number collisions in the network.

In addition, we demonstrated by the analytical evaluations that the full coverage of the monitored region and optimal network connectivity obtained through our placement model is saved at any time of the network lifetime during the execution of MAC-SA algorithm.

As future work, we plan to take into account the path loss and temporal variations of the wireless channel by proposing a more realistic modelling of the wireless communication. We also intend to show that MAC-SA enables an optimum routing based on given topology.

References

[1]. M. C. Akewar, N. V. Thkur, A Study of Wireless Mobile Sensor Network Deployment, *International Journal of Computer and Wireless Communication*, Vol. 2, No. 4, 2012, pp. 533- 541.
[2]. P. Ankur, M. A. Rizvi, Analysis of Strategies of Placing Nodes in Wireless Sensor Networks, *International Journal of Advanced Research in Computer Science and Software Engineering*, Vol. 4, Issue 2, February 2014, pp. 398-402.
[3]. I. Khou, P. Minet, A. Laouiti, S. Mahfoudh, Survey of Deployment Algorithms in Wireless Sensor Networks: Coverage and Connectivity Issues and Challenges, *International Journal of Autonomous and Adaptive Communications Systems*, 2014, pp. 1- 24.
[4]. C. Zhu, C. Zheng, L. Shu, G. Han, A survey on coverage and connectivity issues in wireless sensor networks, *Journal of Networks and Computer Applications*, Vol. 35, No. 2, 2012, pp. 619-632.
[5]. H. Z. Abidin, N. M. Din, N. A. M. Radzi, Deterministic Static Sensor Node Placement in Wireless Sensor Network based on Territorial Predator Scent Marking Behavior, *International Journal of Computer Network and Information Security*, Vol. 5, No. 3, December 2013, pp. 186-191.
[6]. R. Mulligan, Coverage in Wireless Sensor Networks: A Survey, *Networks Protocols and Algorithms*, Vol. 2, No. 2, 2010, pp. 35- 41.
[7]. M. Singaram, D. S. Finney, N. K. Sathish, V. Chandraprasad, Energy Efficient Self-Scheduling Algorithm For Wireless Sensor Networks, *International Journal of Scientifique & Technology Research*, Vol. 3, Issue 1, 2014, pp. 75-78.
[8]. D. V. Ambekar, A. D. Bhoi, R. D. Kharadkar, A Survey on Sensors Lifetime Enhancement Techniques in Wireless Sensor Networks, *International Journal of Computer Applications*, Vol. 107, No. 19, December 2014, pp. 22-25.
[9]. A. Ghosh, S. K. Das, Coverage and Connectivity Issues in Wireless Sensor Networks: A survey', *Pervasive and Mobile Computing*, Vol. 4, Issue 3, 2008, pp. 303-334.

[10]. G. Fan, S. Jin, Coverage Problem in Wireless Sensor Networks: A Survey, *Journal of Networks*, Vol. 5, No. 9, 2010.

[11]. B. Wang, Coverage Problem in Sensor Networks: A Survey, *ACM Computing Surveys*, Vol. 43 No. 4, 2011.

[12]. C. Zhu, C. Zheng, L. Shu, G. Han, A survey on coverage and connectivity issues in wireless sensor networks, *Journal of Network and Computer Applications*, Vol. 35, No. 2, 2012, pp. 619-632.

[13]. J. Kaur, T. C. Aseri, A. Kakria, A survey of locally synchronized MAC protocols in Wireless sensor network, *International Journal of Computer Applications*, Vol. 91, No. 11, April 2014, pp. 37- 41.

[14]. A. Kakria, T. C. Aseri, A Survey on Asynchronous MAC protocols in Wireless Sensor Networks, *International Journal of Computer Applications*, Vol. 108, No. 9, December 2014, pp. 19-22.

[15]. A. Ullah, G. Kim, J. S. Ahn, Performance Analysis of X-MAC protocol with Collision Avoidance Algorithm, in *Proceedings of the IEEE INFOCOM*, 2013, pp. 211-212.

[16]. A. Boulis, Castalia. A simulator for Wireless Sensor Networks and Body Area Networks, Version 3.2, *User's Manual*, March 2011.

[17]. H. Xu, J. Zhu, B. Wang, On the Deployment of a Connected Sensor Network for Confident Information Coverage, *Sensors*, Vol. 15, No. 5, 2015, pp. 11277-11294.

[18]. M. Varposhti, M. Dehghan, R. Safabakhsh, A. Distributed Homological Approach to Location-Independent Area Coverage in Wireless Sensor Networks, *Wireless Personal Communications*, Vol. 83, No. 4, 2015, pp. 3075-3089.

[19]. OPNET Technologies, Inc. Opnet modeler wireless suite - ver. 15. 0. A. PL1. http://www.opnet.com

[20]. Chipcon, CC2420 Data Sheet. http://www.chipcon.com

[21]. W. Heinzelman, A. Chandrakasan, H. Balakrishnan, Energy-Efficient Communication Protocol for Wireless Micro sensor Networks in *Proceedings of the 33rd Hawaii International Conference on System Sciences,* Vol. 8, January 2000, pp. 8020-8030.

[22]. M. Younis, T. Nadeem, Energy efficient MAC protocols for wireless sensor Networks, *Technical Report, University of Mryland baltimre County*, USA, 2004.

[23]. C. T. Kone, Design of the architecture for a large Wireless Sensor Network, Thesis Network and Telecommunications, *Henri Poincare University, Nancy I*, 2011, pp. 27-31.

[24]. W. R. Heinzelman, A. Chandrakasan, and H. Balakrishnan, Energy-Efficient Communication Protocol for Wireless Microsensor Networks, in *Proceedings of the 33rd Hawaii International Conference on System Sciences*, 2000, pp. 1-10.

[25]. D. Ngom, P. Lorenz, B. Gueye, A Distributed Scheduling Algorithm to Improve Lifetime in Wireless Sensor Network based on Geometric Placement of Sensors with Coverage and Connectivity Constraints, in *Proceedings of the 9th International Conference on Sensor Technologies and Applications (SENSORCOMM'15)*, Venice, Italy, 23-28 August 2015, pp. 57-63.

[26]. D. Ngom, P. Lorenz and B. Gueye, A MAC Scheduling Algorithm for Optimizing the Network Lifetime in Wireless Sensor Networks Based on Coverage and Connectivity Constraints, in *Sensors & Transducers,* Vol. 194, Issue 11, November 2015, pp. 1-14.

Chapter 13
Distributed Algorithm for Multiple Target Localization in Wireless Sensor Networks Using Combined Measurements

Slavisa Tomic, Marko Beko, Rui Dinis, Milan Tuba

13.1. Introduction

Accurate localization of people and objects is a very important task in countless wireless structures nowadays. Hence, recently the area of localization has attracted much attention in the research society. Typically, wireless localization systems rely on distance measurements [1], extracted from the time-of-arrival, time-difference-of-arrival, received signal strength (RSS), angle-of-arrival (AoA) information, or a combination of them, depending on the available hardware [2]. Due to limited energy resources, a real challenge is to develop localization algorithms that are fast, scalable and abstemious in their computational and communication requirements.

Several hybrid localization algorithms that combine distance and angle measurements are available in the literature today. To solve the non-cooperative target localization problem in a 3-D scenario, the authors in [3] proposed two estimators: linear least squares (LS) and optimization based. The LS estimator is a relatively simple and well known estimator, while the optimization based estimator was solved by Davidson-Fletcher-Powell algorithm [4]. In [5], the authors derived an LS and a maximum likelihood (ML) estimator for a hybrid scheme that combines RSS difference (RSSD) and AoA measurements. Non-linear constrained optimization was used to estimate the target's location from multiple RSS and AoA measurements. Both LS and ML estimators in [5] are λ-dependent, where λ is a non-negative weight assigned to regulate the contribution from RSS and AoA measurements. A selective weighted LS (WLS) estimator for RSS/AoA localization problem was proposed in [6]. The authors determined the target location by exploiting weighted ranges from two nearest anchor measurements, which were combined with the serving base station AoA measurement. A WLS estimator for a 3-D RSSD/AoA non-cooperative localization problem when the transmitted power is unknown was presented

Slavisa Tomic
ISR/IST, LARSyS, Lisbon, Portugal

in [7]. However, the authors in [7] only investigated a small-scale WSN, with extremely low noise power. An estimator based on semidefinite programming relaxation technique for cooperative target localization problem was proposed in [8]. The authors in [8] obtained angle measurements from a triplet of points. However, due to the consideration of triplets of points, the computational complexity of their approach increases rather substantially with the network size.

All of the above approaches solve a centralized hybrid range/angle localization problem. Although centralized approaches are stable, in large-scale networks, a central processor with enough computational capacity might not be available. Hence, a distributed solution is of practical interest.

In this work, we study a hybrid RSS/AoA localization problem in a large-scale cooperative wireless sensor network (WSN), where the transmit power, P_T, of the nodes is not known. Moreover, the scenario where the path loss exponent (PLE) is different for each link and not perfectly known is examined. For such a challenging localization problem, we propose a distributed solution based on second-order cone relaxation (SOCR) technique. The proposed algorithm does not involve a central processor and has a computation-free initialization. Information exchange is permitted between two incident sensors solely and data processing is performed locally by each node.

Throughout the work, upper-case bold type, lower-case bold type and regular type is used for matrices, vectors and scalars, respectively. $\mathbb{R}^n$ and $\mathbb{C}^n$ respectively denote the n dimensional real and complex Euclidean space. The operators $(\cdot)^T$ and $(\cdot)^H$ denote transpose and Hermitian, respectively. The cardinality of a set $\mathcal{V}$, *i.e.*, the number of elements in the set is denoted by $|\mathcal{V}|$. The normal (Gaussian) distribution with mean μ and variance σ^2 is denoted by $\mathcal{N}(\mu, \sigma^2)$. The N-dimensional identity matrix is denoted by I_N and the $M \times N$ matrix of all zeros by $\mathbf{0}_{M \times N}$ (if no ambiguity can occur, subscripts are omitted). $\|x\|$ denotes the vector norm defined by $\|x\| = \sqrt{x^H x}$, where $x \in \mathbb{C}^n$. For Hermitian matrices A and B, $A \succcurlyeq B$ means that $A - B$ is positive semidefinite.

The remainder of this work is organized as follows. In Section 13.2, the RSS and AoA measurement models are introduced and the target localization problem is formulated. Section 13.3 presents the development of the proposed distributed estimator. In Section 13.4 we provide an analysis about the computational complexity, while in Section 13.5 we discuss the performance of the proposed algorithm. Finally, Section 13.6 summarizes the main conclusions.

13.2. Problem Formulation

Consider a large-scale network with $|\mathcal{A}| = N$ anchors and $|\mathcal{T}| = M$ target nodes, randomly deployed over a region of interest. The considered WSN can be seen as a connected graph, $\mathcal{G}(\mathcal{V}, \mathcal{L})$, with $|\mathcal{V}| = M + N$ vertices and $|\mathcal{L}|$ links (connections). The known locations of the anchor and the unknown locations of the target nodes are denoted by $a_1, a_2, \cdots, a_N$, and $x_1, x_2, \cdots, x_M$ ($x_i, a_j \in \mathbb{R}^3, \forall i \in \mathcal{T}$ and $\forall j \in \mathcal{A}$), respectively.

Due to battery limitations, it is assumed that all nodes have limited communication range, R. Therefore, two nodes, i and j, are linked if and only if they are within the communication range of each other. The sets of all target/anchor and target/target edges are defined as $\mathcal{L}_{\mathcal{A}} = \{(i,j): \|x_i - a_j\| \leq R, \forall i \in \mathcal{T}, \forall j \in \mathcal{A}\}$ and $\mathcal{L}_{\mathcal{T}} = \{(i,j): \|x_i - x_j\| \leq R, \forall i,j \in \mathcal{T}, i \neq j\}$, respectively.

To determine the unknown locations of the target nodes, we engage a hybrid system that combines distance and angle measurements (see Fig. 13.1). In Fig. 13.1, $x_i = [x_{i1}, x_{i2}, x_{i3}]^T$ and $a_j = [a_{j1}, a_{j2}, a_{j3}]^T$ represent the unknown coordinates of the i-th target and the known coordinates of the j-th anchor, respectively, while $d_{ij}^{\mathcal{A}}$, $\phi_{ij}^{\mathcal{A}}$ and $\alpha_{ij}^{\mathcal{A}}$ respectively denote the distance, azimuth angle and elevation angle between the i-th target and the j-th anchor.

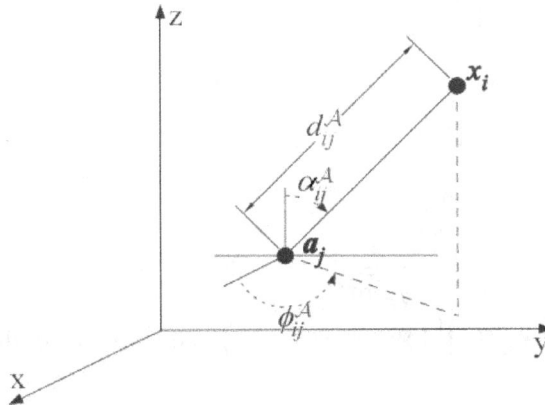

Fig. 13.1. Illustration of a target and anchor locations in a 3-D space.

It is assumed here that the distances are withdrawn from RSS measurements exclusively, because RSS-based ranging does not require any additional hardware [9]. The noise-free RSS between nodes i and j can be modeled as [10, Ch. 3]

$$P_{ij} = P_T \left(\frac{d_0}{d_{ij}}\right)^{\gamma} 10^{-\frac{L_0}{10}}, \forall (i,j) \in \mathcal{L}_{\mathcal{A}} \cup \mathcal{L}_{\mathcal{T}}, \tag{1}$$

where P_T is the transmit power of a node, L_0 is the reference path loss value measured at a short reference distance d_0 ($d_0 \leq d_{ij}$) from the transmitting antenna, γ is the path loss exponent, and d_{ij} is the distance between nodes i and j. By using the relationship L_{ij} (dB) $= 10 \log_{10} \frac{P_T}{P_{ij}}$, the RSS model in (1) can be replaced by the following path loss model

$$L_{ij} = L_0 + 10\gamma \log_{10} \frac{d_{ij}}{d_0} + n_{ij}, \forall (i,j) \in \mathcal{L}_{\mathcal{A}} \cup \mathcal{L}_{\mathcal{T}}, \tag{2}$$

where $n_{ij} \sim \mathcal{N}\left(0, \sigma_{n_{ij}}^2\right)$. Not knowing P_T corresponds to not knowing L_0.

In order to realize the AoA measurements (both azimuth and elevation angles), we assume that anchors are appropriately equipped (*e.g.* with directional antenna or antenna array [3] or with video cameras [11]). Hence, azimuth and elevation anlge measurements can be modeled respectively as [3]

$$\phi_{ij}^{\mathcal{A}} = \tan^{-1}\left(\frac{x_{i2} - a_{j2}}{x_{i1} - a_{j1}}\right) + m_{ij}, \forall(i,j) \in \mathcal{L}_{\mathcal{A}}, \tag{3}$$

and

$$\alpha_{ij}^{\mathcal{A}} = \cos^{-1}\left(\frac{x_{i3} - a_{j3}}{\|x_i - a_j\|}\right) + v_{ij}, \forall(i,j) \in \mathcal{L}_{\mathcal{A}}, \tag{4}$$

where $m_{ij} \sim \mathcal{N}\left(0, \sigma_{m_{ij}}^2\right)$ and $v_{ij} \sim \mathcal{N}\left(0, \sigma_{v_{ij}}^2\right)$ are the measurement errors of azimuth and elevation angles, respectively.

Given the observation vector $\boldsymbol{\theta} = [\boldsymbol{L}^T, \boldsymbol{\phi}^T, \boldsymbol{\alpha}^T]^T$ ($\boldsymbol{\theta} \in \mathbb{R}^{3|\mathcal{L}_{\mathcal{A}}|+|\mathcal{L}_{\mathcal{T}}|}$), where $\boldsymbol{L} = [L_{ij}]^T, \boldsymbol{\phi} = [\phi_{ij}^{\mathcal{A}}]^T, \boldsymbol{\alpha} = [\alpha_{ij}^{\mathcal{A}}]^T$, the conditional probability density function is given as

$$p(\boldsymbol{\theta}|\boldsymbol{\chi}) = \prod_{i=1}^{3|\mathcal{L}_{\mathcal{A}}|+|\mathcal{L}_{\mathcal{T}}|} \frac{1}{\sqrt{2\pi\sigma_i^2}} \exp\left(-\frac{(\theta_i - f_i(\boldsymbol{\chi}))^2}{2\sigma_i^2}\right), \tag{5}$$

where $\boldsymbol{\chi} = [x_1^T, x_2^T, \cdots, x_M^T, L_0]^T$ ($\boldsymbol{\chi} \in \mathbb{R}^{(3M+1)\times 1}$) is the vector of all unknown variables, and

$$f(\boldsymbol{\chi}) = \begin{bmatrix} \vdots \\ L_0 + 10\gamma \log_{10}\frac{d_{ij}}{d_0} \\ \vdots \\ \tan^{-1}\left(\frac{x_{i2} - a_{j2}}{x_{i1} - a_{j1}}\right) \\ \vdots \\ \cos^{-1}\left(\frac{x_{i3} - a_{j3}}{\|x_i - a_j\|}\right) \\ \vdots \end{bmatrix}, \boldsymbol{\sigma} = \begin{bmatrix} \vdots \\ \sigma_{n_{ij}} \\ \vdots \\ \sigma_{m_{ij}} \\ \vdots \\ \sigma_{v_{ij}} \\ \vdots \end{bmatrix}.$$

The ML estimate of the unknown variables, $\hat{\boldsymbol{\chi}}$, is obtained by maximizing the logarithm of the likelihood function in (5) with respect to $\boldsymbol{\chi}$ [12, Ch. 7], as

$$\hat{\boldsymbol{\chi}} = \arg\max_{\boldsymbol{\chi}} \ln\{p(\boldsymbol{\theta}|\boldsymbol{\chi})\} = \arg\max_{\boldsymbol{\chi}} \ln\left\{\prod_{i=1}^{3|\mathcal{L}_{\mathcal{A}}|+|\mathcal{L}_{\mathcal{T}}|} \exp\left\{-\frac{(\theta_i - f_i(\boldsymbol{\chi}))^2}{2\sigma_i^2}\right\}\right\} \tag{6}$$

$$= \arg\min_{\boldsymbol{\chi}} \sum_{i=1}^{3|\mathcal{L}_{\mathcal{A}}|+|\mathcal{L}_{\mathcal{T}}|} \frac{1}{\sigma_i^2}(\theta_i - f_i(\boldsymbol{\chi}))^2.$$

The ML estimator in (6) does not have a closed form solution and is not convex. Therefore, recursive methods (*e.g.* gradient descent method) can get trapped into local minimum leading to poor estimation accuracy. Hence, in the remaining of this work, we will show that the ML problem in (6) can be solved in a distributed fashion by applying certain approximations. More precisely, a convex relaxation technique leading to a distributed SOCP estimator which can be solved efficiently by interior-point algorithms [13] will be proposed.

13.3. Distributed Localization

In large-scale WSNs, it might be virtually impossible to solve (6); thus, in large-scale networks, a distributed solution of (6) such that each target node determines its own location is of interest. By assigning initial target location estimates, $\hat{x}_i^{(0)}, \forall i \in \mathcal{T}$, the problem in (6) can be divided into local sub-problems, *i.e.*, it can be solved by each target node individually. Accordingly, target node i updates its location estimate in each iteration, t, by solving the following local ML problem

$$\hat{x}_i^{(t+1)} = \arg\min_{x_i, L_0} \sum_{j=1}^{3|\mathcal{L}_{\mathcal{A}_i}| + |\mathcal{L}_{\mathcal{T}_i}|} \frac{1}{\sigma_j^2} \left(\theta_j - f_j(\chi_i) \right)^2, \forall i \in \mathcal{T}, \tag{7}$$

where $\mathcal{L}_{\mathcal{A}_i} = \{j : (i,j) \in \mathcal{L}_{\mathcal{A}}\}$ and $\mathcal{L}_{\mathcal{T}_i} = \{j : (i,j) \in \mathcal{L}_{\mathcal{T}}\}$ represent the set of all anchor and all target neighbors of the target i respectively, and the first $|\mathcal{L}_{\mathcal{A}_i}| + |\mathcal{L}_{\mathcal{T}_i}|$ elements of $f_j(\chi_i)$ are given as

$$f_j(\chi_i) = L_0 + 10\gamma \log_{10} \frac{\|x_i - \hat{a}_j\|}{d_0}, \text{for } j = 1, \dots, |\mathcal{L}_{\mathcal{A}_i}| + |\mathcal{L}_{\mathcal{T}_i}|,$$

and

$$\hat{a}_j = \begin{cases} a_j, \text{if } j \in \mathcal{A}, \\ x_j^{(t)}, \text{if } j \in \mathcal{T}. \end{cases}$$

Having $\hat{x}_i^{(0)}, \forall i \in \mathcal{T}$ at hand, and assuming that the noise is sufficiently small, from (2) we have

$$\lambda_{ij} \|x_i - \hat{a}_j\| \approx \eta d_0, \forall (i,j) \in \mathcal{L}_{\mathcal{A}} \cup \mathcal{L}_{\mathcal{T}}, \tag{8}$$

where $\lambda_{ij} = 10^{-\frac{L_{ij}}{10\gamma}}$ and $\eta = 10^{-\frac{L_0}{10\gamma}}$.

Similarly, from (3) and (4) we get

$$c_{ij}^T (x_i - a_j) \approx 0, \forall (i,j) \in \mathcal{L}_{\mathcal{A}}, \tag{9}$$

and

267

$$\mathbf{k}^T(\mathbf{x}_i - \mathbf{a}_j) \approx \|\mathbf{x}_i - \mathbf{a}_j\| \cos(\alpha_{ij}^{\mathcal{A}}), \forall (i,j) \in \mathcal{L}_{\mathcal{A}}, \tag{10}$$

where $\mathbf{c}_{ij} = [-\sin\phi_{ij}^A, \cos\phi_{ij}^A, 0]^T$ and $\mathbf{k} = [0,0,1]^T$. Following the LS criterion, from (8), (9) and (10), we obtain the following optimization problem

$$\hat{\mathbf{x}}_i^{(t+1)} = \arg\min_{\mathbf{x}_i, \eta} \sum_{j \in \mathcal{E}_{\mathcal{A}_i} \cup \mathcal{E}_{\mathcal{T}_i}} \left(\lambda_{ij}\|\mathbf{x}_i - \hat{\mathbf{a}}_j\| - \eta d_0\right)^2 + \sum_{j \in \mathcal{E}_{\mathcal{A}_i}} \left(\mathbf{c}_{ij}^T(\mathbf{x}_i - \mathbf{a}_j)\right)^2$$
$$+ \sum_{j \in \mathcal{E}_{\mathcal{A}_i}} \left(\mathbf{k}_{ij}^T(\mathbf{x}_i - \mathbf{a}_j) - \|\mathbf{x}_i - \mathbf{a}_j\| \cos(\alpha_{ij}^{\mathcal{A}})\right)^2. \tag{11}$$

Although the problem in (11) is non-convex, it can be readily transformed into a convex one, by applying appropriate relaxation. To do so, introduce auxiliary variables $r_{ij} = \|\mathbf{x}_i - \hat{\mathbf{a}}_j\|, \forall i \in \mathcal{T},$ for $j = 1, \dots, |\mathcal{L}_{\mathcal{A}_i}| + |\mathcal{L}_{\mathcal{T}_i}|$ and $\mathbf{z} = [z_{ij}]$, $\mathbf{g} = [g_{ij}]$ and $\mathbf{h} = [h_{ij}]$ where $z_{ij} = \lambda_{ij}r_{ij} - d_0, \forall i \in \mathcal{T},$ for $j = 1, \dots, |\mathcal{L}_{\mathcal{A}_i}| + |\mathcal{L}_{\mathcal{T}_i}|$, $g_{ij} = \mathbf{c}_{ij}^T(\mathbf{x}_i - \mathbf{a}_j), \forall i \in \mathcal{T},$ for $j = 1, \dots, |\mathcal{L}_{\mathcal{A}_i}|$, and $h_{ij} = \mathbf{k}_{ij}^T(\mathbf{x}_i - \mathbf{a}_j) - \|\mathbf{x}_i - \mathbf{a}_j\| \cos(\alpha_{ij}^{\mathcal{A}}), \forall i \in \mathcal{T},$ for $j = 1, \dots, |\mathcal{L}_{\mathcal{A}_i}|$. Then, problem in (11) can be rewritten as

$$\underset{\mathbf{x}_i, \eta, \mathbf{z}, \mathbf{r}, \mathbf{g}, \mathbf{h}}{\text{minimize}} \|\mathbf{z}\|^2 + \|\mathbf{g}\|^2 + \|\mathbf{h}\|^2$$

subject to

$$r_{ij} = \|\mathbf{x}_i - \hat{\mathbf{a}}_j\|, \forall i \in \mathcal{T}, \text{for } j = 1, \dots, |\mathcal{L}_{\mathcal{A}_i}| + |\mathcal{L}_{\mathcal{T}_i}|,$$
$$z_{ij} = \lambda_{ij}r_{ij} - \eta d_0, \forall i \in \mathcal{T}, \text{for } j = 1, \dots, |\mathcal{L}_{\mathcal{A}_i}| + |\mathcal{L}_{\mathcal{T}_i}|, \tag{12}$$
$$g_{ij} = \mathbf{c}_{ij}^T(\mathbf{x}_i - \mathbf{a}_j), \forall i \in \mathcal{T}, \text{for } j = 1, \dots, |\mathcal{L}_{\mathcal{A}_i}|,$$
$$h_{ij} = \mathbf{k}_{ij}^T(\mathbf{x}_i - \mathbf{a}_j) - r_{ij}\cos(\alpha_{ij}^{\mathcal{A}}), \forall i \in \mathcal{T}, \text{for } j = 1, \dots, |\mathcal{L}_{\mathcal{A}_i}|.$$

Introduce epigraph variables ξ, ρ and κ, to obtain the constraints $\xi = \|\mathbf{z}\|^2$, $\rho = \|\mathbf{g}\|^2$ and $\kappa = \|\mathbf{h}\|^2$ from the objective function. Apply SOCR technique to relax the constraints $r_{ij} = \|\mathbf{x}_i - \hat{\mathbf{a}}_j\|$ and $\xi = \|\mathbf{z}\|^2$ into conic constraints $r_{ij} \geq \|\mathbf{x}_i - \hat{\mathbf{a}}_j\|$ and $\xi \geq \|\mathbf{z}\|^2$, respectively. Then, the problem in (12) is transformed into the following second-order cone programming (SOCP) problem [13, Ch. 4]

$$\underset{\mathbf{x}_i, \eta, \mathbf{z}, \mathbf{r}, \mathbf{g}, \mathbf{h}, \xi, \rho, \kappa}{\text{minimize}} \xi + \rho + \kappa$$

subject to

$$\|\mathbf{x}_i - \hat{\mathbf{a}}_j\| \leq r_{ij}, \forall i \in \mathcal{T}, \text{for } j = 1, \dots, |\mathcal{L}_{\mathcal{A}_i}| + |\mathcal{L}_{\mathcal{T}_i}|,$$
$$z_{ij} = \lambda_{ij}r_{ij} - \eta d_0, \forall i \in \mathcal{T}, \text{for } j = 1, \dots, |\mathcal{L}_{\mathcal{A}_i}| + |\mathcal{L}_{\mathcal{T}_i}|,$$
$$g_{ij} = \mathbf{c}_{ij}^T(\mathbf{x}_i - \mathbf{a}_j), \forall i \in \mathcal{T}, \text{for } j = 1, \dots, |\mathcal{L}_{\mathcal{A}_i}|, \tag{13}$$
$$h_{ij} = \mathbf{k}_{ij}^T(\mathbf{x}_i - \mathbf{a}_j) - r_{ij}\cos(\alpha_{ij}^{\mathcal{A}}), \forall i \in \mathcal{T}, \text{for } j = 1, \dots, |\mathcal{L}_{\mathcal{A}_i}|,$$
$$\left\|\begin{bmatrix} 2\mathbf{z} \\ \xi - 1 \end{bmatrix}\right\| \leq \xi + 1, \left\|\begin{bmatrix} 2\mathbf{g} \\ \rho - 1 \end{bmatrix}\right\| \leq \rho + 1, \left\|\begin{bmatrix} 2\mathbf{h} \\ \kappa - 1 \end{bmatrix}\right\| \leq \kappa + 1.$$

Let us define $\mathcal{C}$ as the set of colors (numbers) of the nodes in order to manage the network and avoid message collision [14]. We summarize the proposed SOCP algorithm for unknown L_0 in Algorithm 13.1. The algorithm is explained in three parts. First part includes lines $1 - 12$, where we estimate the targets' locations by solving the proposed SOCP estimator in (13), S times. Lines $7 - 9$ are introduced in order to minimize the oscillations in the location estimates. Second part involves lines $14 - 18$, where in the S-th iteration we use the last location updates to find the estimate of L_0, $\hat{L}_{0i}$, at each target node. Owing to our assumption that L_0 is identical for all nodes, we execute an average consensus algorithm in order to in order to get the average estimated value of L_0, $\hat{L}_0$. To do so, the local-degree weights method [15] is exploited, since it is particularly suitable for distributed implementation and guarantees convergence (provided that the graph is not bipartite). The final part of our algorithm covers lines $1 - 12$, starting at $(S + 1)$-th iteration. In this part, we take advantage of the $\hat{L}_0$ estimate to update the targets' location estimates by solving (13) as if L_0 is known. To solve (13) for known L_0, we just have to compute $\hat{\eta} = 10^{-\frac{\hat{L}_0}{10\gamma}}$ and plug in this estimated value into (13). We refer to the proposed distributed algorithm as "SOCP" in the remaining text.

Algorithm 13.1. The proposed distributed SOCP algorithm

Require: $\hat{x}_i^{(0)}, \forall i \in \mathcal{T}, a_j, \forall j \in \mathcal{A}, \mathcal{C}, S, T_{\max}$

  1. **Initialize:** $t \leftarrow 0$
  2. **repeat**
  3.   **for** $c = 1, \dots, \mathcal{C}$ **do**
  4.   **for all** $i \in \mathcal{C}_c$ (in parallel) **do**
  5.   Collect $\hat{a}_j, \forall j \in \mathcal{E}_{\mathcal{A}_i} \cup \mathcal{E}_{\mathcal{T}_i}$
  6. $\hat{x}_i^{(t+1)} \leftarrow \begin{cases} \text{solve (13),} & \text{if } t \leq S \\ \text{solve (13) using } \hat{L}_0, & \text{if } t > S \end{cases}$
  7.   **if** $\frac{\left\| \hat{x}_i^{(t+1)} - \hat{x}_i^{(t)} \right\|}{\hat{x}_i^{(t)}} > 1$ **then**
  8. $\hat{x}_i^{(t+1)} \leftarrow \hat{x}_i^{(t)}$
  9.   **end if**
  10.   Broadcast $\hat{x}_i^{(t+1)}$ to $\hat{a}_j, \forall j \in \mathcal{L}_{\mathcal{A}_i} \cup \mathcal{L}_{\mathcal{T}_i}$
  11.   **end for**
  12.   **end for**
  13. $t \leftarrow t + 1$
  14. **if** $t = S$ **then**
  15.   **for all** $i \in \mathcal{T}$ (in parallel) **do**
  16. $\hat{L}_{0i} \leftarrow \dfrac{\sum_{j \in \mathcal{L}_{\mathcal{A}_i} \cup \mathcal{L}_{\mathcal{T}_i}} L_{ij} - 10\gamma \log_{10} \frac{\left\| \hat{x}_i^{(t)} - \hat{a}_j \right\|}{d_0}}{\left| \mathcal{L}_{\mathcal{A}_i} \right| + \left| \mathcal{L}_{\mathcal{T}_i} \right|}$
  17.   **end for**
  18. **end if**
  19. **begin consensus**
  20. $\hat{L}_0 \leftarrow \frac{1}{M} \sum_{i=1}^{M} \hat{L}_{0i}$
  21. **end consensus**
  22. **until** $t < T_{\max}$

13.4. Complexity Analysis

The worst case computational complexity of an SOCP algorithm is given as [16]

$$\mathcal{O}\left(\sqrt{U}\left(m^2\sum_{n=1}^{U}u_n + \sum_{n=1}^{U}u_n + m^3\right)\right),\tag{14}$$

where U is the number of the second-order cone constraints, m is the number of the equality constraints, and u_n is the dimension of the n-th second-order cone. In (13), $U = N_i + M_i + 3$, $m = 3N_i + M_i$, and $u_n = p + 2$ (for $n = 1, \ldots, N_i + M_i$), $u_n = N_i + M_i + 2$ (for $n = N_i + M_i + 1, \ldots, 2(N_i + M_i)$) and $u_n = N_i + 2$ (for $n = 2(N_i + M_i) + 1, \ldots, 3N_i + 2M_i$), where p represents the space dimension, $N_i = |\mathcal{L}_{\mathcal{A}_i}|$ and $M_i = |\mathcal{L}_{\mathcal{T}_i}|$. Therefore, according to (14), the worst case computational complexity of the proposed algorithm is of the order: $T_{\max} \times M \times \mathcal{O}\left(\max_i\{(N_i + M_i)^{3.5}\}\right)$. From this result, one can conclude that the computational complexity of distributed algorithms is dependent on the size of the neighborhood fragments, and not on the size of the network.

13.5. Performance Results

This section presents a set of results obtained through computer simulations to obtain the performance of the proposed approach in terms of estimation accuracy and convergence. To the best of authors' knowledge, no distributed hybrid range/angle algorithms are available in the literature, and modification of the existing centralized algorithms is not straightforward to distributed setting. To show the advantage of merging two radio measurements versus traditional localization systems, we include also the performance results of the proposed method when only RSS measurements are engaged, called here "SOCP2$_{\text{RSS}}$". All of the presented algorithms were solved by using the MATLAB package CVX [17], where the solver is SeDuMi [18].

As the main performance metric, we opted to use the normalized root mean square error (NRMSE), defined as

$$\text{NRMSE} = \sqrt{\frac{1}{MM_c}\sum_{i=1}^{M_c}\sum_{j=1}^{M}\|x_{ij} - \widehat{x}_{ij}\|^2},$$

where $\widehat{x}_{ij}$ denotes the estimate of the true location of the j-th target, x_{ij}, in the i-th Monte Carlo (M_c) run. We consider a random deployment of M targets and N anchors inside a cube region of length B in each M_c run. This setting is interesting because the algorithms are tested against different network topologies, and their robustness to different scenarios is examined. To make the comparison of the considered approaches as fair as possible, we first obtained $M_c = 500$ targets' and anchors' locations, as well as noise realizations between two sensors $\forall(i, j) \in \mathcal{L}_{\mathcal{A}} \cup \mathcal{L}_{\mathcal{T}}$ in each M_c run. Also, we made sure that the network graph is connected in each M_c run, and the localization problem was solved for

270

those settings. In all simulations presented here, the reference distance was set to $d_0 = 1$ m, the reference path loss to $L_0 = 40$ dB, the communication range of a sensor to $R = 6.5$ m, and the PLE was fixed to $\gamma = 3$. However, in practice it is almost impossible to perfectly estimate the value of the PLE. Thus, to account for a realistic measurement model mismatch and test the robustness of the considered approaches to imperfect knowledge of the PLE, the true PLE for each link was drawn from a uniform distribution on an interval $\gamma_{ij} \in [2.7, 3.3], \forall(i,j) \in \mathcal{L_A} \cup \mathcal{L_T}$. Finally, we assumed that the initial guess of the targets' locations, $\hat{\boldsymbol{x}}_i^{(0)}, \forall i \in \mathcal{T}$, is in the intersection of the big diagonals of the cube area.

Fig. 13.2 illustrates the NRMSE versus t performance comparison for different N. One can see from the figure that the performance of all methods improves as t and/or N increases. This behavior is anticipated since the algorithm is expected to progress gradually with t, and more realiable information is available in the network with the increase of N. Moreover, a saturation of the SOCP curve can be noticed at $t = 3$; hence, at this point we obtain an estimate of L_0, and we resume our algorithm as if L_0 is known. This action explains a sudden performance boost in the SOCP curve after $t = 3$. Also, for $N = 20$ at $t = 3$, we can observe that the proposed hybrid algorithm has marginally worse performance than the traditional one which uses RSS measurements only. Apart from the fact that SOCP2$_{\text{RSS}}$ uses the true value of L_0, this anomaly can be explained to some extent by the fact that the hybrid algorithm is affected by relativey high noise level for angle measurements, and that M is significantly larger than N. We can see that for $N = 30$ this peculiarity disappears. Finally, it can be argued that the new algorithm performs exceptional, achieving its lower bound, given by the results of the proposed approach when the real value of L_0 is used (labeled here as "SOCP2"), for all N.

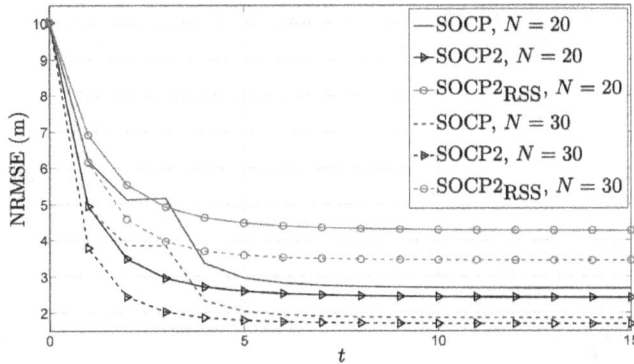

Fig. 13.2. NRMSE versus t comparison, when $M = 60, R = 6.5$ m, $\sigma_{n_{ij}} = 3$ dB, $\sigma_{m_{ij}} = 6$ deg, $\sigma_{v_{ij}} = 6$ deg, $\gamma_{ij} \in [2.7, 3.3], \gamma = 3, B = 20$ m, $L_0 = 40$ dB, $d_0 = 1$ m, $M_c = 500$.

Fig. 13.3 illustrates the NRMSE versus t performance for different M. From the figure, it can be seen that the proposed algorithm requires a slightly higher number of iterations to converge when M is increased, as predicted. Nevertheless, the performance of the new algorithm does not deteriorate as M is increased, and once again it reaches its lower bound for all settings.

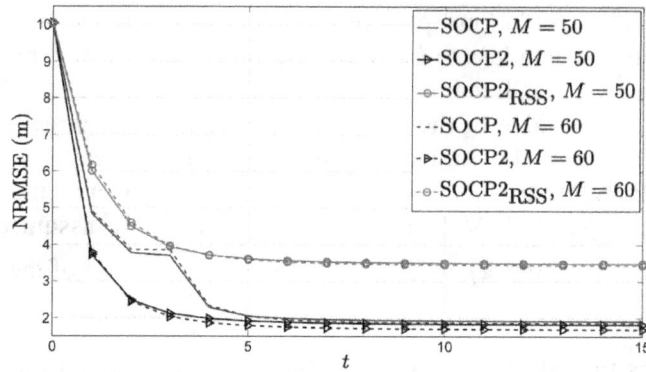

Fig. 13.3. NRMSE versus t comparison, when $N = 30, R = 6.5$ m, $\sigma_{n_{ij}} = 3$ dB, $\sigma_{m_{ij}} = 6$ deg, $\sigma_{v_{ij}} = 6$ deg, $\gamma_{ij} \in [2.7, 3.3], \gamma = 3, B = 20$ m, $L_0 = 40$ dB, $d_0 = 1$ m, $M_c = 500$.

Figs. 13.4, 13.5 and 13.6 respectively illustrate the NRMSE versus $\sigma_{n_{ij}}$ (dB), $\sigma_{m_{ij}}$ (deg) and $\sigma_{v_{ij}}$ (deg) performance for $T_{\max} = 15$. In these figures, it can be observed that the performance of the proposed algorithm exacerbates with the quality of a certain measurement, as foreseen. We can also see that the quality of the RSS measurements has the most significant impact on the performance of the proposed algorithm, while the error in the azimuth and elevation angle measurements have marginal influence on the performance. This is not surprising, since the AoA measurements are acquired at the anchors only, while RSS measurements were obtained by all nodes in the network. Even so, one can see from Fig. 13.4 that the performance loss is lower than 15 % for the proposed algorithm, which is relatively low for the considered noise range. Finally, we can see from Figs. 13.4, 13.5 and 13.6 that the performance of the proposed algorithm is excellent in all scenarios, performing very close to its lower bound.

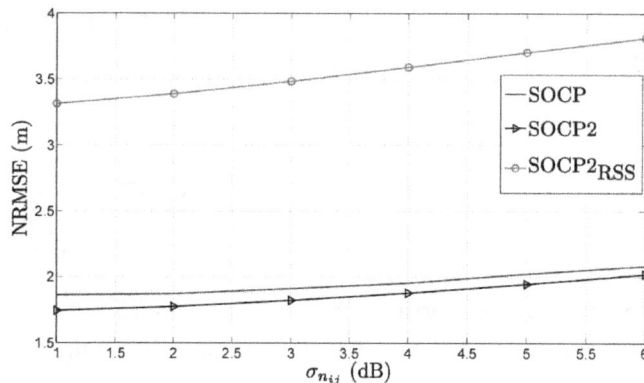

Fig. 13.4. NRMSE versus $\sigma_{n_{ij}}$ (dB) comparison, when $N = 30, M = 50, R = 6.5$ m, $\sigma_{m_{ij}} = 6$ deg, $\sigma_{v_{ij}} = 6$ deg, $\gamma_{ij} \in [2.7, 3.3], \gamma = 3, B = 20$ m, $L_0 = 40$ dB, $d_0 = 1$ m, $T_{\max} = 15, M_c = 500$.

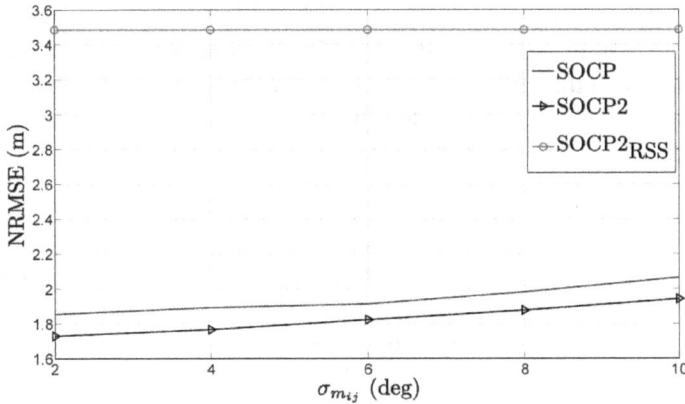

Fig. 13.5. NRMSE versus $\sigma_{m_{ij}}$ (deg) comparison, when $N = 30, M = 50, R = 6.5$ m, $\sigma_{n_{ij}} = 3$ dB, $\sigma_{v_{ij}} = 6$ deg, $\gamma_{ij} \in [2.7, 3.3], \gamma = 3, B = 20$ m, $L_0 = 40$ dB, $d_0 = 1$ m, $T_{\max} = 15, M_c = 500$.

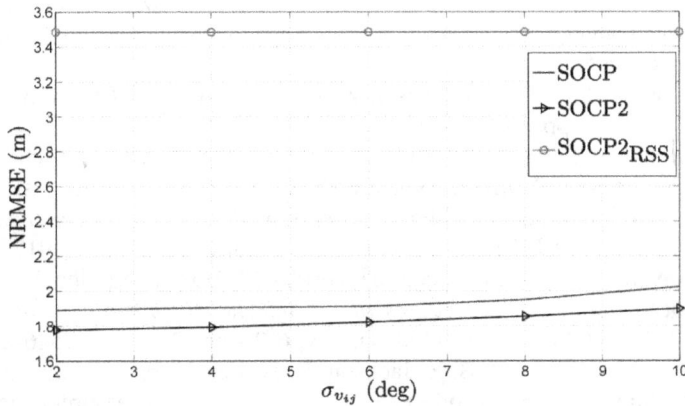

Fig. 13.6. NRMSE versus $\sigma_{v_{ij}}$ (deg) comparison, when $N = 30, M = 50, R = 6.5$ m, $\sigma_{n_{ij}} = 3$ dB, $\sigma_{m_{ij}} = 6$ deg, $\gamma_{ij} \in [2.7, 3.3], \gamma = 3, B = 20$ m, $L_0 = 40$ dB, $d_0 = 1$ m, $T_{\max} = 15, M_c = 500$.

13.6. Conclusions

In this work, we addressed a distributed algorithm for hybrid RSS/AoA localization problem for the case where the transmit powers are identical for all nodes and not known. To solve this very challenging problem, we have derived first a novel non-convex LS estimator, which tightly approximates the ML one for small noise levels and is suitable for distributed implementation. We have showed that, by applying appropriate SOCR technique, the derived estimator can be transformed into a convex one. After a certain number of iterations, we took advantage of the obtained location estimates to calculate the ML estimate of the transmit powers at each node. The local degree weights method was

273

then applied in order reach a consensus about the transmit power estimates between the nodes, and our algorithm was continued as if the transmit powers were known. The obtained results confirm that the new approach efficiently solves the localization problem, offering accurate localization in just a few iterations.

Acknowledgements

This work was partially supported by Fundação para a Ciência e a Tecnologia under Projects UID/EEA/00066/2013, PEst-OE/EEI/LA0008/2013 (IT pluriannual founding and HETNET), and PTDC/EEITEL/6308/2014-HAMLeT, as well as the grants SFRH/BPD/108232/2015 and SFRH/BD/91126/ 2012.

References

[1]. D. C. Popescu, M. Hedley, Range Data Correction for Improved Localization, *IEEE Wirel. Commun. Letters*, Vol. 4, Issue 3, 2015, pp. 297-300.

[2]. N. Patwari, J. N. Ash, S. Kyperountas, A. O. Hero III, R. L. Moses, N. S. Correal, Locating the Nodes: Cooperative Localization in Wireless Sensor Networks, *IEEE Sig. Process. Mag.*, Vol. 22, Issue 4, 2005, pp. 54-69.

[3]. K. Yu, 3-D Localization Error Analysis in Wireless Networks, *IEEE Trans. Wirel. Commun.*, Vol. 6, Issue 10, 2007, pp. 3473-3481.

[4]. R. Fletcher, Practical Methods of Optimization, *John Wiley & Sons*, Chichester, UK, 1987.

[5]. S. Wang, B. R. Jackson, R. Inkol, Hybrid RSS/AOA Emitter Location Estimation Based on Least Squares and Maximum Likelihood Criteria, in *Proceedings of the Biennial Symposium on Communications (IEEE QBSC'12)*, Kingston, Ontario, 28-29 May 2012, pp. 24-29.

[6]. L. Gazzah, L. Najjar, H. Besbes, Selective Hybrid RSS/AOA Weighting Algorithm for NLOS Intra Cell Localization, in *Proceedings of the Wireless Communications and Networking Conference (IEEE WCNC'14)*, Istanbul, Turkey, 6-9 April 2014, pp. 2546-2551.

[7]. Y. T. Chan, F. Chan, W. Read, B. R. Jackson, B. H. Lee, Hybrid Localization of an Emitter by Combining Angle-of-Arrival and Received Signal Strength Measurements, in *Proceedings of 27th Canadian Conference on Electrical and Computer Engineering (IEEE CCECE'14)*, Toronto, Ontario, 4-7 May 2014, pp. 1-5.

[8]. P. Biswas, H. Aghajan, Y. Ye, Semidefinite Programming Algorithms for Sensor Network Localization Using Angle of Arrival Information, in *Proceedings of the Asilomar Conference on Signals, Systems, and Computers*, Pacific Grove, California, 30 Oct - 02 Nov 2005, pp. 220-224.

[9]. S. Tomic, M. Beko, R. Dinis, RSS-based Localization in Wireless Sensor Networks Using Convex Relaxation: Noncooperative and Cooperative Schemes, *IEEE Trans. Vehic. Technol.*, Vol. 64, Issue 5, 2015, pp. 2037-2050.

[10]. T. S. Rappaport, Wireless Communications: Principles and Practice, *Prentice-Hall*, Upper Saddle River, New Jersey, 1996.

[11]. M. B. Ferreira, J. Gomes, J. P. Costeira, A Unified Approach for Hybrid Source Localization Based on Ranges and Video, in *Proceedings of the IEEE International Conference on Acoustics, Speech and Signal Processing (IEEE ICASSP'15)*, South Brisbane, Queensland, 19-24 April 2015, pp. 2879-2883.

[12]. S. M. Kay, Fundamentals of Statistical Signal Processing: Estimation Theory, *Prentice-Hall*, Upper Saddle River, New Jersey, 1993.

[13]. S. Boyd, L. Vanderberghe, Convex Optimization, *Cambridge University Press*, Cambridge, United Kingdom, 2004.

[14]. S. Tomic, M. Beko, R. Dinis, Distributed RSS-Based Localization in Wireless Sensor Networks Based on Second-Order Cone Programming, *Sensors*, Vol. 14, Issue 10, 2014, pp. 18410-18432.

[15]. L. Xiao, S. Boyd, Fast Linear Iterations for Distributed Averaging, *System & Control Letters*, Vol. 53, Issue 2004, 2004; pp. 65-78.

[16]. I. Pólik, T. Terlaky, Interior Point Methods for Nonlinear Optimization, *Nonlin. Optim.*, G. Di Pillo, F. Schoen, (Eds.), *Springer*, 1st Edition, 2010.

[17]. M. Grant, S. Boyd, CVX: Matlab Software for Disciplined Convex Programming, Version 1.21. (http://cvxr.com/cvx).

[18]. J. F. Sturm, Using SeDuMi 1.02, a MATLAB Toolbox for Optimization Over Symmetric Cones, *Optim. Meth. Softw.*, 1998.

Chapter 14
Environmental Data Recovery Techniques and its Applications using Polynomial Regression in the Sensor Network Systems

Noboru Ishihara, Yoshihiro Yoneda, Koji Kurihara, Takashi Suganuma, Hiroyuki Ito, Kunihiko Gotoh, Koichiro Yamashita and Kazuya Masu

14.1. Introduction

In the Internet of Things (IoT) era [1], the wireless sensor networks (WSNs) play an important role in aggregating the data. The WSNs use a plenty of wireless sensor nodes to monitor environmental parameters, such as temperature, humidity, pH, light, air pressure and so on. WSNs have many possible applications, ranging from structural health monitoring to field monitoring. Thanks to the progress in microelectronics based on the integrated circuit technology, small wireless sensor nodes with low power consumption have been achieved and suitable for the IoT systems. Thus, several companies and universities around the world are increasingly focused on the development of WSNs toward IoT era [2-7]. However, problems exist with data loss owing to data collision between the sensor nodes and electromagnetic noise. As the interval of aggregate data is not fixed in the time and space domains, digital signal processing using Fourier or wavelet transforms cannot be applied directly to the aggregated data. Moreover, noise degrades the data accuracy. Because the environmental characteristics have various waveforms, data reliability cannot evaluated by signal analysis. To overcome these problems, various techniques, such as data collection timing [9], redundant system [10], data recovery [11], have been used to increase data reliability.

In this chapter, polynomial regression for environmental data recovery based on the correlations among the environmental data is applied [12, 13]. Environmental characteristics are recovered from aggregated data of the sensor nodes using polynomial regression. Thus, data loss is tolerated, and the data can be analyzed easily. Basic sinusoidal environmental variations are assumed to evaluate the data recovery function

Noboru Ishihara
Tokyo Institute of Technology, Nagatsutacho 4259, Midori-ku, Kanagawa, 226–8503, Japan

277

with polynomial regression. If the sinusoidal characteristics can be modeled appropriately, arbitrary waveform characteristics, such as single-shot, periodic and non-periodic waveforms, can also be modelled theoretically. The recovered data accuracy is evaluated by comparing the recovered and source characteristics.

It is also proposed a data reliability evaluation flow that does not rely on signal analysis [14]. It is clarified the relation between the accuracy of the recovered characteristics and the polynomial regression order, and the effects of data loss and number of sensor nodes is analyzed. Furthermore, it is shown that the use of polynomial regression has the advantage of low-pass filtering that enhances the signal-to-noise ratio (SNR) of the environmental characteristics. In addition, it is shown that polynomial regression can recover arbitrary environmental characteristics.

In Section 14.2, the environmental data recovery technique based on polynomial regression is introduced. In Section 14.3, the reliability of the recovered data is discussed. The frequency domain characteristics are evaluated in Section 14.4. In Section 14.5, we confirm the ability of polynomial regression to recover arbitrary environmental characteristics. Application results of data recovery for measured environmental characteristics are described in Section 14.6. Finally, the conclusions are drawn in Section 14.7.

14.2. Environmental Data Recovery Using Polynomial Regression

The aggregated data analysis is shown in Fig. 14.1. If the interval of the aggregated sampled data is fixed, the environmental data characteristics can be analyzed directly using Fourier or wavelet transforms. However, when the interval of the data is not fixed, the data cannot be directly analyzed. Therefore, continuous environmental characteristics are recovered from the aggregated data, and then, the fixed interval data are resampled from the recovered characteristics.

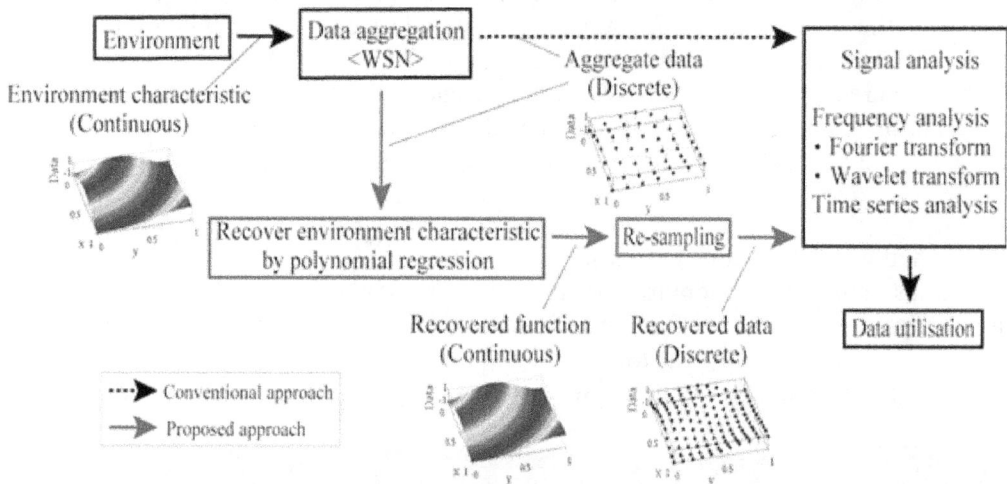

Fig. 14.1. WSNs system using polynomial regression.

14.2.1. Environmental Data Recoveries Using Polynomial Regression

There are several ways of expressing the recovered characteristics, e.g., Fourier series expansion, polynomial regression, interpolation and so on. Polynomial regression is simple and suitable for expressing continuous characteristics as it tolerates data loss [15-17]. However, polynomial regression is not good at expressing characteristics with many inflection points. If the frequency band is limited, the environmental characteristics at the limited bandwidth can be expressed using polynomial expressions. Therefore, polynomial regression is used in environmental data gathering and recovery. When one-dimensional data are $t=[t_1,\cdots,t_N]^T$ and the environmental data are $d=[d_1,\cdots;d_N]^T$, the environmental source characteristics function $f_W(t)$ can be recovered and recovered function $f_R(t)$ is

$$f_R(t)=\sum_{i=0}^{m}a_i t^i \,,\tag{14.1}$$

where $a=[a_0,\cdots,a_m]^T$ is the coefficient vector. The value of a is obtained using at least-squares methods. m is the order of polynomial equation. For two-dimensional data obtained by sensor nodes arranged in coordinates $(x_1,y_1),\ldots,(x_N,y_N)$ and coordinates $x=[x_1,\cdots,x_N]^T$ and $y=[y_1,\cdots,y_N]^T$, the recovered function is

$$f_R(x,y)=\sum_{\substack{j,k\geq 0 \\ j+k\leq m}}a_{jk}x^j y^k \,,\tag{14.2}$$

where the coefficient vector a is the column vector. It's size is $\dfrac{(m+1)(m+2)}{2}\times 1$.

14.2.2. Data Reliability Evaluation Flow

14.2.2.1. Evaluation Flow

The reliability evaluation flow is shown in Fig. 14.2. Two-dimensional data are assumed in the evaluation. The environmental source characteristic function is $f_W(x_i,y_i)$. To consider the effect of noise and data loss, the sensor node model is defined. When the noise is expressed as $f_N(x_i,y_i)$, the sampled data with noise $f_S(x_i,y_i)$ can be expressed as

$$f_S(x_i,y_i)=f_W(x_i,y_i)+f_N(x_i,y_i)\,.\tag{14.3}$$

To evaluate the effect of data loss, the following function is added.

$$f_O(x_i,y_i)=\begin{cases}f_S(x_i,y_i)\ (\textit{without data loss})\\ 0\ (\textit{with data loss})\end{cases},\tag{14.4}$$

where $f_O(x_i, y_i)$ represents the sampled data considering the effect of noise and data loss. The amount of data in $f_O(x_i, y_i)$ decreases compared with the number of $f_S(x_i, y_i)$. The error of the recovered data is defined as

$$f_E(x, y) = f_R(x, y) + f_W(x, y). \tag{5}$$

The data accuracy that is sampled at fixed intervals using the above continuous functions is compared with the accuracy of the evaluated data. The root mean-square error (RMSE) at each comparison point is defined as data reliability. RMSE is given by

$$RMSE = \frac{1}{\sigma_W} \sqrt{mean(f_E(x, y)^2)} \times 100(\%). \tag{14.6}$$

In the evaluation, we carry out 1000 iterations to minimize the effect of noise variation and data loss.

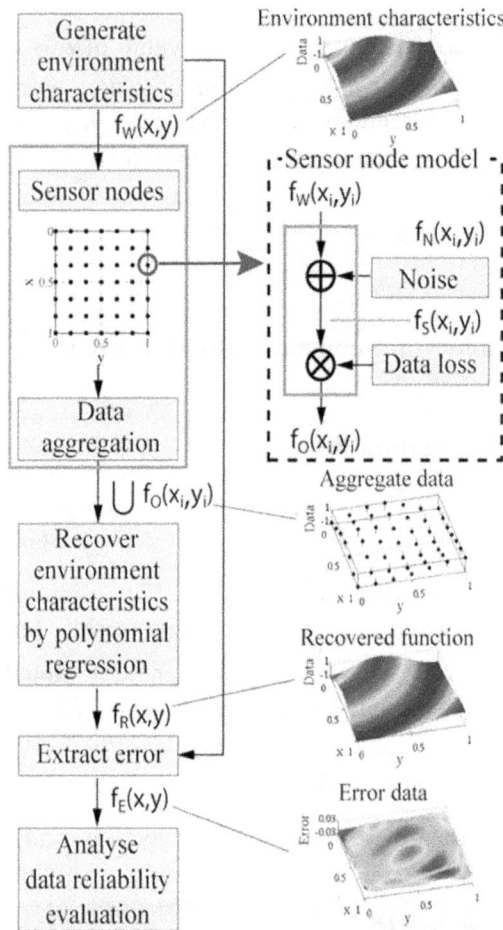

Fig. 14.2. Data reliability evaluation flow.

14.2.2.2. Parameter Setting

To evaluate the data recovery reliability, the following conditions are considered.

14.2.2.2.1 Sinusoidal Environmental Characteristics

The correlations among the actual environmental data are complex. However, to determine a generalized index, it is preferable to use simple data characteristics. In this study, a sinusoidal wave is assumed as the basic environmental characteristic because any arbitrary characteristic can be expressed as a linear combination of a sinusoidal wave. The following equations are the sinusoidal functions used for one- and two-dimensional data.

$$f_W(t) = \frac{A_{PP}}{2} \sin(2\pi \frac{t}{L} + \theta) \ . \tag{14.7}$$

$$f_W(x,y) = \frac{A_{PP}}{2} \sin(2\pi \frac{\sqrt{(x-x_0)^2 + (y-y_0)^2}}{L} + \theta) . \tag{14.8}$$

where (x_0, y_0) is showing the position of the wave generation source, A_{pp} is the peak-to-peak amplitude, L is the wavelength and θ is the phase.

14.2.2.2.2 Observation Region

The observation region is the region where the polynomial regression is applied. The observation region is partitioned and then polynomial regression is applied to each partition. Each data recovery function is joined to express the characteristics of the observation region. The partitions of the region are determined by the cycle (wavelength) of the highest frequency of the environmental characteristics.

14.2.2.2.3 Number of Sensor Nodes

The sensor nodes are arranged at equal intervals in the observation region, including the upper boundary. In the case of two-dimensional structures, the sensor nodes are set on a grid. The density of the sensor nodes is represented by the number of sensor nodes N in the observation region. When the analysis is in the time domain, the one-dimensional coordinate axis is evaluated with respect to the time axis. In this case, the number of sensor nodes in the observation region represents the number of sampled data.

14.2.2.2.4 Noise

Electromagnetic noise generated at the sensor interface consisting of an amplifier and an analogue-to-digital converter and electromagnetic noise in the environment mainly contribute to data noise. The SNR is defined by following equation.

$$SNR = 10\log_{10}\frac{\text{var}(f_S)}{\text{var}(f_N)} \quad . \tag{14.9}$$

14.2.2.2.5. Data Loss

Data loss occurs because of data collisions or intermittent failures in the wireless communication. To simulate the effect of data loss, we use a pseudorandom data generation technique in the evaluation.

14.3. Data Reliability of Periodic Characteristics

The reliability of recovered data that are resampled from the recovered function is evaluated by comparing with the environmental source characteristics function. The data reliability is evaluated using the conditions described in the previous section. Fig. 14.3 shows the sinusoidal signal that is assumed as the environmental characteristics of one- and two-dimensional conditions.

The SNR at each sensor node is set to be 40 dB. Therefore, the reference position of the RMSE is determined as 1.0 %. When the number of sensor nodes is increased, the reference position of the RMSE is 0.5 %. Without sensor node noise, the reference position of the RMSE is 0.25 %.

14.3.1. Application Range of the Polynomial Regression

Firstly, the relation between the partition region cycle and RMSE was analyzed by changing the order of the polynomial without the sensor node noise. The number of sensor nodes is ten for the one-cycle partition region in the one-dimensional case and 10 x 10 for the one-cycle partition region in the two-dimensional case. We also examined the five-cycle partition region and monitored the maximum error. Results for the one- and two-dimensional sinusoidal signals (Fig. 14.3) are shown in Fig. 14.4. For 0.25 % error and one-cycle partition region, the order of the polynomial should be higher than seven for one- and two-dimensional signals.

14.3.2. Effect of Sensor Node Number

The number of sensor nodes is thought to strongly affect the data reliability. The relation between the number of sensor nodes and RMSE was analyzed when the case of SNR is 50, 40 and 30 dB at each sensor node. And ninth-order polynomial was used in the analysis. The results for the one- and two-dimensional cases are shown in Fig. 14.5. Obviously, the errors are reduced with the number of sensor nodes.

The increasing number of sensor nodes reduced the RMSE owing to noise. Fig. 14.6 shows the results for the required SNR at each sensor node when the RMSE is 0.25 %, 0.5 % and 1 %. The precision of each sensor node is improved by increasing the number of sensor nodes.

282

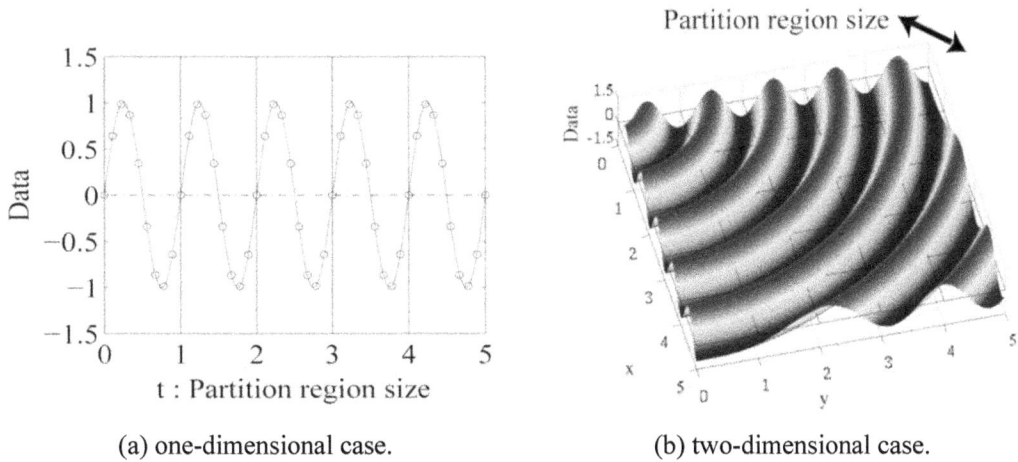

(a) one-dimensional case.

(b) two-dimensional case.

Fig. 14.3. Examples of the recovery function.

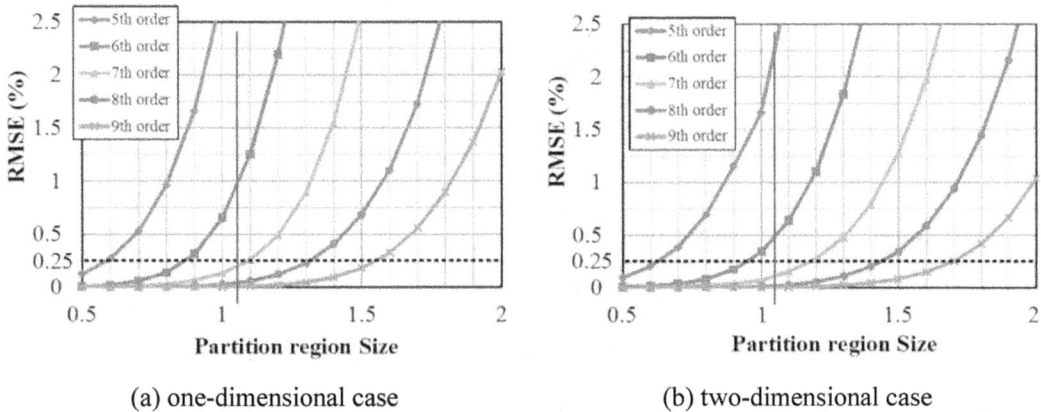

(a) one-dimensional case

(b) two-dimensional case

Fig. 14.4. Precision of polynomial regression.

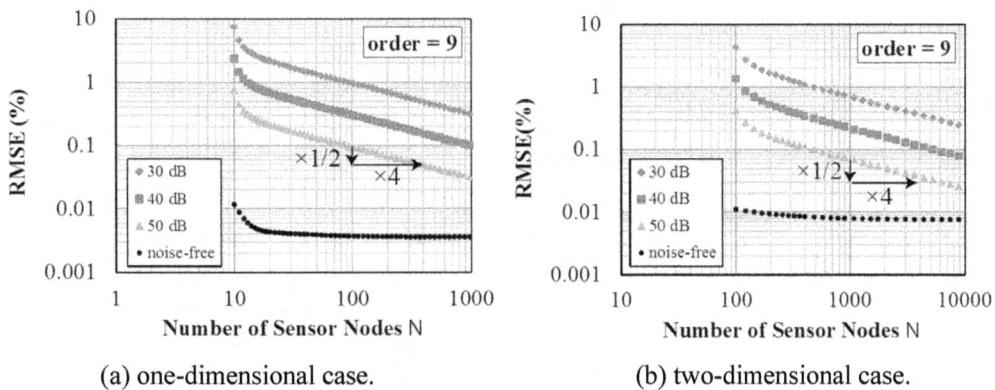

(a) one-dimensional case.

(b) two-dimensional case.

Fig. 14.5. Required number of sensor nodes.

283

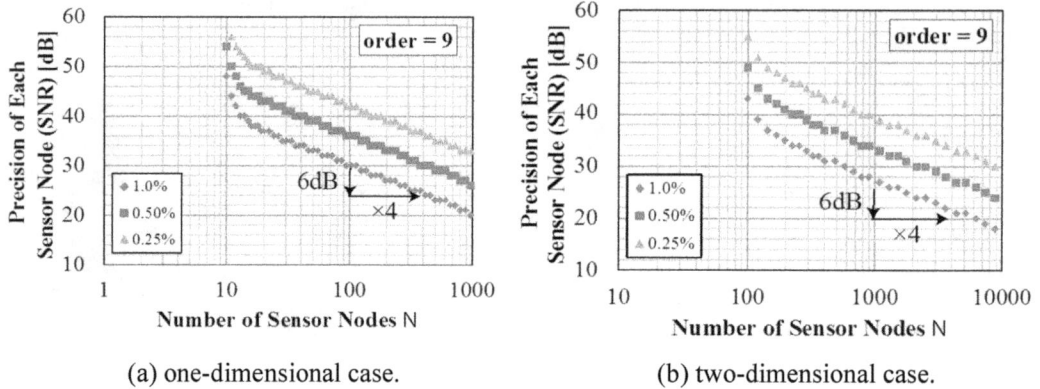

(a) one-dimensional case.   (b) two-dimensional case.

Fig. 14.6. Sensor accuracy.

There are two ways to improve the data reliability. The first is to increase the number of sensor nodes and the second is that sensor nodes should be high SNR. If the number of sensor nodes is increased four times, the RMSE decreases by 50 %. If the SNR of each sensor node is improved by 6 dB, the RMSE decreases by 50 %.

14.3.3. Effect of Data Loss

The relation between data loss rate and the RMSE was analyzed. A ninth-order polynomial and 40-dB SNR at each sensor node was assumed. The number of sensor nodes was selected to satisfy the RMSE of 0.5 % and 0.25 %. In the one-dimensional case, 36 and 149 nodes were selected for the analysis. In the two-dimensional cases, 225 and 841 nodes were selected in the evaluation.

The results for the one- and two-dimensional cases are shown in Fig. 14.7.

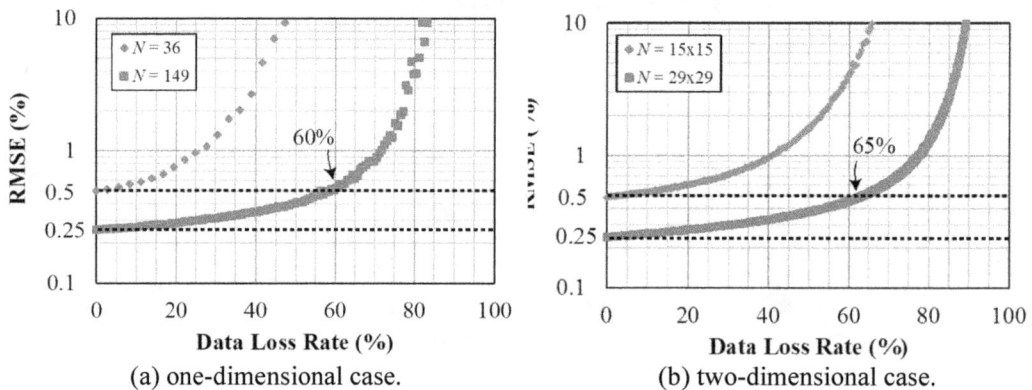

(a) one-dimensional case.   (b) two-dimensional case.

Fig. 14.7. Data loss robustness.

The RMSE increases with data loss rate, of course. However, by increasing the number of sensor nodes, the RMSE decreases. The number of sensor nodes satisfies the RMSE of 0.25 % adequately, whereas the data loss rate is 60 % for RMSE of 0.5 % in the one-dimensional case and 65 % in the two-dimensional case. These results suggest that a redundant system can enhance the data reliability by increasing the number of sensor nodes.

14.4. Reliability in the Frequency Domain

The reliability of the recovered data using polynomial regression was also evaluated in the frequency domain [18]. The fast Fourier transform (FFT) was applied to the recovered data.

14.4.1. FFT Analysis

The one-dimensional sinusoidal environmental characteristics are assumed to be the same as in the previous sections. Recovered data at fixed intervals are obtained by sampling the data recovered by polynomial regression. FFT is applied to the recovered data. The signal-to-noise and distortion ratio (SNDR) and spurious-free dynamic range (SFDR) were evaluated. The SFDR is used to discuss the effect of harmonic distortion.

When the fundamental frequency is f_0 , the number of FFT points is FFT_{POINT} and the sampling frequency is F_s, f_0 is given by

$$f_0 = \frac{F_S}{FFT_{POINT}} .$$

(14.10)

Therefore, the input signal frequency f_{in} and wavelength of the input signal per division λ_{in} is

$$f_{in} = k \cdot f_0 .$$

(14.11)

$$\lambda_{in} = \frac{k}{D_n} ,$$

(14.12)

where D_n is the number of divisions and k is an integer number.

14.4.2. Reliability in the FFT Analysis

14.4.2.1. Effect of Input Signal Cycle (wavelength)

The relation between signal cycle (wavelength) in the polynomial regression and the evaluation indices of gain, SNDR and SFDR was analyzed using FFT. In the analysis, a ninth-order polynomial, 40-dB SNR at each sensor node, 32 divisions dividing FFT points into partition region and 1024 of FFT points are assumed. Evaluation conditions are summarized in Table 14.1.

The results are shown in Fig. 14.8. This is showing the relation between input frequency cycle (wavelength) for polynomial regression and the evaluation indexes. Decreases of 1 dB are tolerated by the SNDR and SFDR and for wavelength with the maximum partition of 1.6 cycles. Above 1.6 cycles, the partition region signals are filtered out. The gain is flat up to the three-cycle partition region. Thus, the polynomial regression acts as a low-pass filter. This means that the SNDR and SFDR are improved because the polynomial regression limits the bandwidth of the environmental signals.

Table 14.1. Conditions of FFT analysis.

Parameters	Condition
Noise	40 [dB]
Polynominal regression order	9
FFT point	1024
Divisions	32

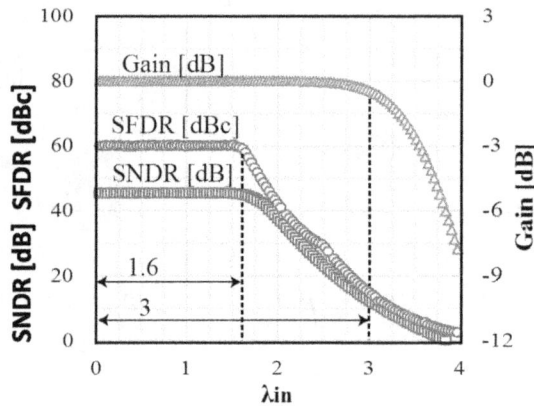

Fig. 14.8. SNDR, SFDR and gain vs. input wavelength.

14.4.2.2. Effect of the Number of Sensor Node

The number of sensor nodes per partition region is evaluated. The results are shown in Fig. 14.9. The FFT results for the source environmental signals were 40-dB SNDR and 59-dB SFDR. For 13 sensor nodes, the SNDR is the same as the result of the source environmental signals. For 25 sensor nodes, the SNDR is the same as the result of source environmental signals. Higher SNDR and SFDR are possible by increasing the number of sensor nodes. By increasing the number of sensor nodes four times, both SNDR and SFDR improved by 6 dB.

14.4.2.3. Frequency Spectrum

The frequency spectrum is evaluated by FFT. The results are shown in Fig. 14.10. Based on the results of Figs. 14.8 and 14.9, the 1.6-cycle (wavelength) input signal region and 25 sensor nodes per partition region were assumed. Compared with the spectrum of the source environmental signal, the noise level of the high-frequency region is filtered out. Table 14.2 summarizes the result of FFT analysis. By limiting the observation region in the polynomial regression, both SNDR and SFDR are improved.

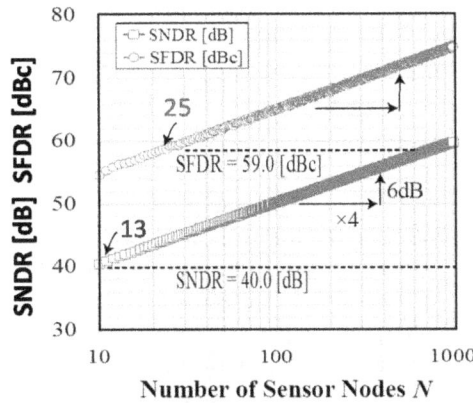

Fig. 14.9. SNDR and SFDR vs. number of sensor nodes.

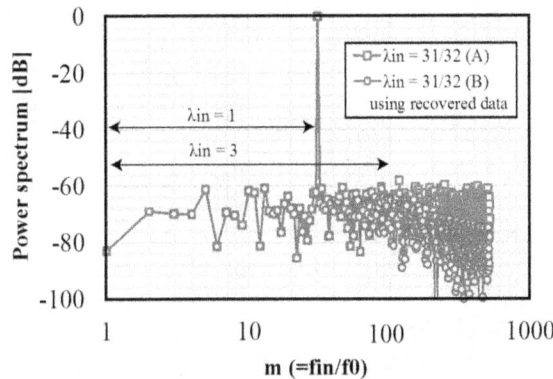

Fig. 14.10. Frequency spectrum with and without recovered data.

Table 14.2. Result of FFT analysis.

	SNDR [dB]	SFDR [dBc]
(a) FFT	40.0	59.0
(b) FFT (using recovered data)	45.2	59.9
Differences ((b)-(a))	+5.2	+0.9

14.5. Recovery for Environmental Arbitrary Characteristics

In Sections 14.3 and 14.4, it was clarified that polynomial regression can recover the sinusoidal environmental characteristics. Polynomial regression for arbitrary characteristics is also validated by selecting the order of the polynomial equation for each partition region. Scale-space filtering (SSF) [19] and Akaike's information criterion (AIC) [20-22] were used to select the partition region and the order of the polynomial regression based on aggregate data. Recovery of arbitrary data is done by following steps.

[Step 1]: Select the partition region for polinoninal regression by using SSF.

The SSF detects extreme values by the convolution of the Gaussian function. The partition region is obtained as the region between the extreme points.

[Step 2]: Determine the polinoninal order for the partition region data.

The AIC is a statistical measure that estimates the quality of the environmental source characteristics from aggregated data, including noise and data loss. The order of the polynomial regression for the partition region is obtained. In the calculation, practical c-AIC [22] is used. Fig. 14.11 is showing an example of Polynomial order extraction using c-AIC. c-AIC is applied for the data of $f_W(t) = \sin(t)$ $(0 \leq t \leq 1)$ with 40 dB SNR. When the order is 7th, c-AIC value is minimum as shown in Fig. 14.11(a). This is the appropriate order for polynomial regression. Recovered characteristic using 7th order polynominal regression is shown in Fig. 14.11 (b).

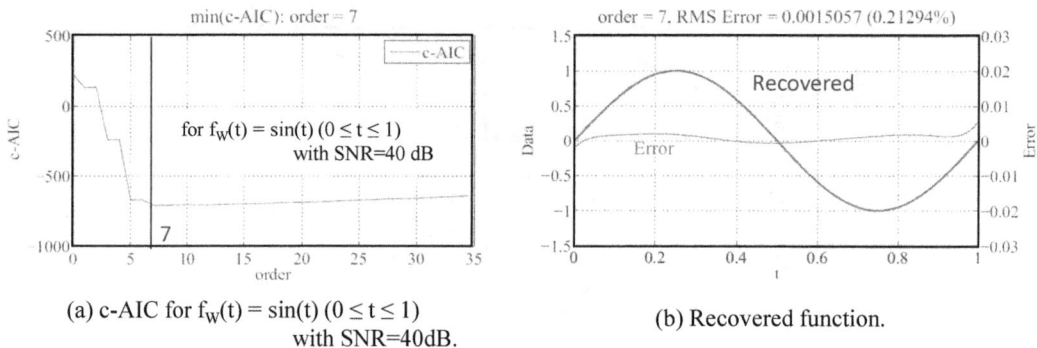

(a) c-AIC for $f_W(t) = \sin(t)$ $(0 \leq t \leq 1)$ with SNR=40dB.

(b) Recovered function.

Fig. 14.11. Polynomial order extraction by using c-AIC.

By detecting the extreme values by SSF and estimating the quality of source characteristics between the appropriately selected extreme points by c-AIC, arbitrary characteristics can be recovered using polynomial regression [23].

Fig. 14.12 shows extreme values of arbitrary environmental characteristics with and without noise of 40-dB SNR detected by SSF. Time domain transient characteristic that have two peaks was supposed as the arbitrary environmental characteristics. Partition regions can be obtained by using the extreme values as shown in Fig. 14.12. Without noise, partition regions can be divided clearly. When there are noises, observation region is divided into smaller regions.

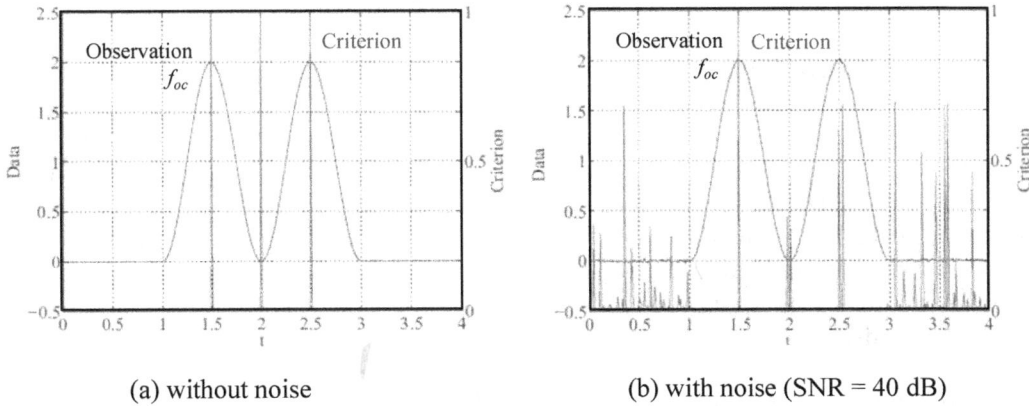

(a) without noise   (b) with noise (SNR = 40 dB)

Fig. 14.12. Extreme values extraction using SSF.

For those partition regions, the order of polynomial regression is thus optimized. The regions divided by the criterion of extreme values are the partition regions, and the order of the polynomial regression is set at each partition region.

Fig. 14.13 shows the recovered data from arbitrary characteristics with and without noise of 40-dB SNR using the SSF, c-AIC and polynomial regression. The RMSE is under 0.1 % even when SNR is 40 dB; thus, the polynomial regression can obviously recover the arbitrary environmental characteristics.

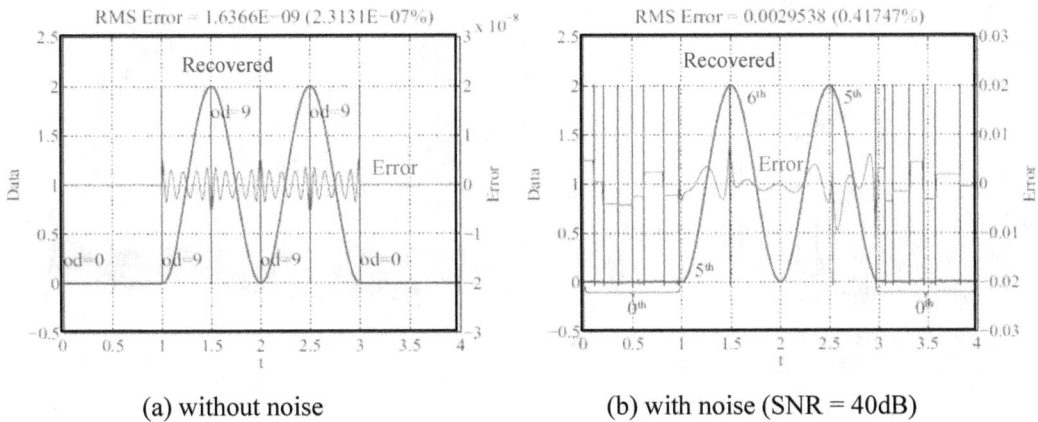

(a) without noise   (b) with noise (SNR = 40dB)

Fig. 14.13. Recovered characteristics using c-AIC.

14.6. Application Results for Actual Measured Data

In Section 14.5, it was clarified that polynomial regression can recover the arbitrary environmental characteristics. Therefore, polynomial regression was applied to the actual measured data. Two examples are shown in the section.

14.6.1. Temperature Data in Time Domain

Example of measured temperature data is shown in Fig. 14.14. The data was sampled at every 10 minutes from Jul. 17th, 2015 to Aug. 16th, 2015 (1 month) using thermos recorder placed in a farm. The temperature change which is day by day is observed. Polynomial regression techniques previously described were applied to the measured data of every 12 hours. The polynomial order is the same for the each partition regions, and the polynomial order was optimized to be the minimum value of the RMSE in each the polynomial order.

The recovered results are shown in Fig. 14.15. Table 14.3 summarizes the conditions for the temperature data recovery. Measured characteristics in Fig. 14.14 were recovered with the RMSE less than about ± 1 %. This result shows the data recovery with the polynomial regression is applicable for the environmental data that changes successively at random.

The effect of sampling period in the measurement was also evaluated. RMSE values are calculated, when sampling periods were 10, 20, 30, 40 and 60 minutes by thinning out the data. Table 14.4 is showing the results. As increase of sampling period, RMSE is increased and required polynominal order is decreased. If RMSE of 3 % is tolerable, 60-minites sampling period is enough for measurement. This measuring time period can be feedback to measurement system conditions.

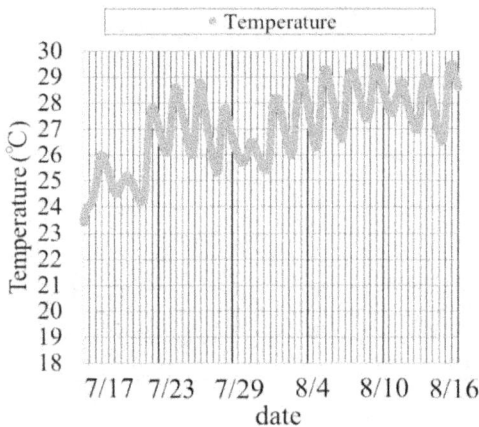

Fig. 14.14. Measured temperature characteristics.

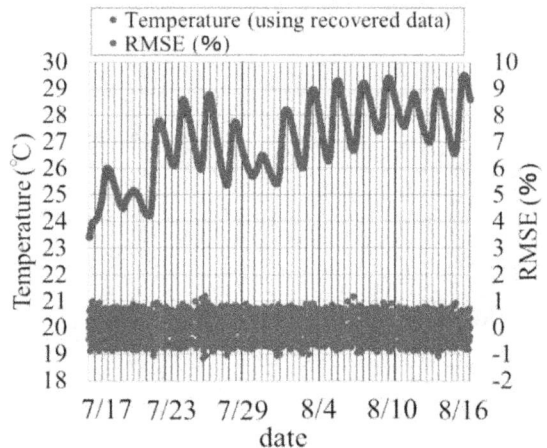

Fig. 14.15. Recovered temperature characteristics.

Table 14.3. Data recovery conditions for measured temperature.

Parameters	Conditions
Term	from Jul. 17th, 2015 to Aug. 16th, 2015
Interval of Sampling	10 minutes
Number of the partition region	60
Number of samples (1/one partition region)	36
Polynomial order	8

Table 14.4. Measurement time period dependency on and RMSE and polynominal order.

Time period	10 min.	20 min.	30 min.	40 min.	60 min.
RMSE(%)	1.97	2.07	2.32	2.95	2.95
Polynomial order	8	10	5	5	4

14.6.2. Acceleration Data in Frequency Domain

Recovery of measured acceleration sensor data was also examined with the polynomial regression techniques. Recovered frequency spectrum was compared with measured characteristics. Data recovery conditions for measured acceleration data is shown in the Table 14.5. Acceleration sensor data was measured with 10-Hz signal vibrator supposing person's health monitoring. The sampling frequency of acceleration sensor was 50 Hz. Polynomial regression was applied to every 32 data of the acceleration sensor data measured.

Table 14.5. Data recovery conditions for measured acceleration.

Parameters	Conditions
Input frequency	10 [Hz]
Sampling frequency	50 [Hz]
Number of the partition region	32
Number of samples (1/one partition region)	32
Polynomial order	13

The results are shown in Fig. 14.16. The LPF effect is observed in the data recovered. And in the low frequency region including the 10-Hz signal, it can be seen that there is no difference between the power spectrums measured and power spectrums recovered.

Table 14.6 is showing the calculation results of SNDR and SFDR, SNDR is improved by 3.09 dB by the LPF effect.

The effect of the polynomial regression in the frequency domain was confirmed with the acceleration sensor data measured.

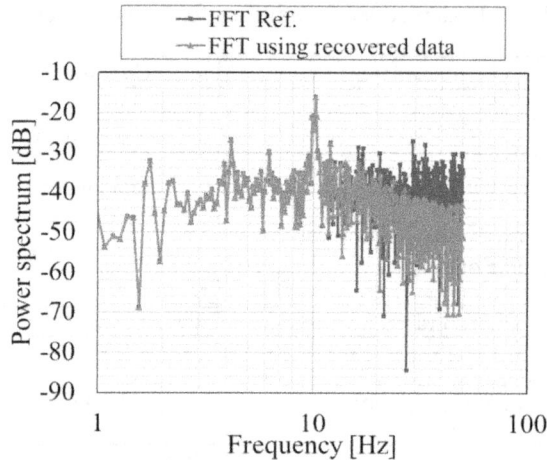

Fig. 14.16. Data recovery for measured acceleration.

Table 14.6. SNDR and SFDR for the acceleration data.

FFT	SNDR(dB)	SFDR(dBc)
Ref.	-5.69	10.66
Using polynominal regression	-2.6	10.66

14.7. Conclusions

In this chapter, environment data recovery techniques using polynomial regression for sensor networks have been examined and discussed.

(1) A data reliability evaluation procedure for WSNs was proposed with polynomial regression.

(2) The recovered data reliability depends on the order of the polynomial regression; the number of sensor nodes, the effect of noise and data loss were quantified.

(3) From FFT analysis, it is seen that polynomial regression act as a low-pass filter. Data recovery using polynomial regression enhances the SNDR or SFDR in the WSNs system.

(4) Polynomial regression can recover arbitrary environmental characteristics and can be used with SSF and c-AIC.

(5) Proposed polynomial regression techniques were successfully applied to two kinds of measured data; temperature and acceleration.

In conclusion, environmental data recovery using polynomial regression can be useful for practical sensor networks.

Acknowledgements

Part of this work was a joint research project of Tokyo Institute of Technology and Fujitsu Laboratories Limited. We would like to express our thanks to all who supported this project.

References

[1]. Embracing the Internet of Everything To Capture Your Share of $14.4 Trillion, White Paper, *CISCO,* 2013.

[2]. J. Hill, R. Szewczyk, A. Woo, S. Hollar, D. Culler, and K. Pister, System architecture directions for networked sensors, in *Proceedings of the ACM International Conference on Architectural Support for Programming Languages and Operating Systems*, November 2000, pp. 93–104.

[3]. Wireless sensors from UC Berkeley and Intel researchers help conservation biologists monitor elusive seabird in Maine, *UC Berkeley,* http://berkeley.edu/news/media/releases/2002/08/05 snsor.html

[4]. J. L. Hill and D. E. Culler, MICA: A Wireless Platform for Deeply Embedded Networks, *IEEE Micro,* Vol. 22, No. 6, 2002, pp. 12–24.

[5]. I. F. Akyildiz, W. Su, Y. Sankarasubramaniam, and E. Cayirci, A Survey on Sensor Networks, *IEEE Communications Magazine,* Vol. 40, No. 8, Aug. 2002, pp. 102–114.

[6]. A. Ogawa, T. Yamazato, and T. Ohtsuki, Information and Signal Processing for Sensor Networks, *IEICE Transactions on Fundamentals of Electronics, Communications and Computer Sciences,* Vol. E87-A, No. 10, Oct. 2004, pp. 2599–2606.

[7]. A. Boukerche and S. Samarah, In-Network Data Reduction and Coverage-Based Mechanisms for Generating Association Rules in Wireless Sensor Networks, *IEEE Transactions on Vehicular Technology,* Vol. 58, No. 8, Oct. 2009, pp. 4426–4438.

[8]. L. Doherty and D. A. Teasdale, Towards 100 % reliability in wireless monitoring networks, ACM international Workshop on Performance Evaluation of Wireless Ad Hoc, *Sensor and Ubiquitous Networks*, Oct. 2006, pp. 132–135.

[9]. Sivrikaya, F., et al., Time synchronization in sensor networks: a survey, *IEEE Network,* 18, 4, 2004, pp. 45-50.

[10]. K. Yamashita, et al., Implementation and Evaluation of Architecture Search Simulator Including Disturbance for Wide-range Grid Wireless Sensor Network., in *Proceedings of the Multimedia, Distributed, Cooperative, and Mobile Symposium,* 2014, pp. 1368-1377.

[11]. Doherty, L., et al., Algorithms for position and data recovery in wireless sensor networks., Diss. Department of Electrical Engineering and Computer Sciences, *University of California at Berkeley,* 2000.

[12]. Kohei Ohba, Yoshihiro Yoneda, Koji Kurihara, Takashi Suganuma, Hiroyuki Ito, Noboru Ishihara, Kunihiko Gotoh, Koichiro Yamashita and Kazuya Masu, Environmental Data Recovery Using Polynomial Regression for Large-scale Wireless Sensor Networks., in *Proceedings of the SENSORNETS'16*, No. 11, 2016, pp. 161-168.

[13]. Kohei Ohba, Yoshihiro Yoneda, Koji Kurihara, Takashi Suganuma, Hiroyuki Ito, Noboru Ishihara, Kunihiko Gotoh, Koichiro Yamashita and Kazuya Masu, Polynomial Regression Techniques for Environmental Data recovery in Wireless Sensor Networks., *Sensors & Transducers,* Vol. 199, Issue 4, April 2016, pp. 1-9.

[14]. Y. Yoneda, R. Hasama, K. Kurihara, T. Suganuma, H. Ito, K. Gotoh and K. Masu, A study on Data Reliability Evaluation Index of Wireless Sensor Network for Environmental Monitoring., in *Proceedings of the 41st SICE Symposium on Intelligent Systems,* 2014.

[15]. R. L. Eubank and P. Speckman, Curve Fitting by Polynomial-Trigonometric Regression, *Biometrika*, Vol. 77, No. 1, Mar. 1990, pp. 1–9.

[16]. Wang, G., Cao, J., Wang, H. and Guo, M., Polynomial regression for data gathering in environmental monitoring applications, in *Proceedings of the Global Telecommunications Conference (GLOBECOM'07)*, 2007, pp. 1307-1311.

[17]. T. Sato, H. Ueyama, N. Nakayama, and K. Masu, Determination of Optimal Polynomial Regression Function to Decompose On-die Systematic and Random Variations, in *Proceedings of the IEEE Asia and South Pacific Design Automation Conference*, March 2008, pp. 518–523.

[18]. Kohei Ohba, Yoshihiro Yoneda, Koji Kurihara, Takashi Suganuma, Hiroyuki Ito, Noboru Ishihara, Kunihiko Gotoh, Koichiro Yamashita and Kazuya Masu, A method of recovering environment data using polynomial regression for large-scale wireless sensor networks., *IEICE Technical Report*, ASN2015-72, 2015.

[19]. A. P. Witkin, Scale-Space Filtering: A New Approach to Multi-Scale Description, in *Proceedings of the IEEE International Conference on Acoustics, Speech and Signal Processing*, Vol. 9, 1984, pp. 150–153.

[20]. H. Akaike, A New Look at the Statistical Model Identification, *IEEE Transactions on Automatic Control*, Vol. AC-19, No. 6, 1974.

[21]. C. M. Hurvich and C.-L. Tsai, Bias of the Corrected AIC Criterion for Underfitted Regression and Time Series Models, *Biometrika*, Vol. 78, No. 3, Sep. 1991, pp. 499–509.

[22]. N. Sugiura, Further Analysis of the Data by Akaike's Information Criterion and the Finite Corrections, *Communications in Statistics - Theory and Methods*, Vol. 7, No. 1, 1978, pp. 13–26.

[23]. K. Haze, et al., Modeling home appliance power consumption by interval-based switching Kalman filters, *Technical Report of IEICE*, 112.31, 2012, pp. 39-44.

Chapter 15
Robust Header Compression for the Internet of Things

Pekka Koskela, Mikko Majanen and Mikko Valta

15.1. Introduction

Due to the extensive growth of Internet of Things (IoT), the number of wireless devices connected to the Internet is increasing and will continue to increase remarkably in the near future. For example, Gartner estimates that the IoT, which excludes PCs, tablets and smartphones, will grow to 26 billion units installed in 2020, representing an almost 30-fold increase from 0.9 billion in 2009 [1].

In wireless IoT networks, the available bandwidth and energy is often restricted. Therefore, all means for saving those resources are welcome. The biggest resource consumption source is radio communication, including both transmission and reception [2]. One efficient way to control transmission and reception is the utilization of the duty-sleep cycle control of the radio with a MAC protocol [2]. Another effective way to reduce radio transmission, which supplements the previous approach, is the utilization of packet compression. The packet compression can be targeted to cover only header or payload part, or both parts of the packet.

IoT devices, e.g., sensors, periodically report their current data values to the cloud services in the Internet. Thus, the transmitted data may be only couple of bytes, whereas the protocol headers of the packet (MAC, IP, TCP/UDP, etc.), are many tens of bytes. This big header overhead is the motivation behind the compression of packet headers and the design of lightweight protocols with small headers. For instance, the header of the traditional application layer protocol, Hypertext Transfer Protocol (HTTP), can take easily over 40 bytes, whereas replacing HTTP with Constrained Application Protocol (CoAP) [3] can drop the header size to less than 10 bytes.

In this chapter, we summarize our previous work related on header compression [4-6] and discuss further on the topic. We first give a brief description of packet compression

Pekka Koskela
VTT Technical Research Centre of Finland Ltd., Espoo, Finland

techniques in Section 15.2, which is followed by a more careful presentation of the evolution of header compression and a comparison between different header compression methods in Section 15.3. In Section 15.4, we present the CoAP profile for Robust Header Compression (ROHC) and study its performance in fast (LTE, WLAN) and low speed (XBEE) radio networks. Finally, Section 15.5 concludes this work.

15.2. Packet Compression Techniques

There are several compression techniques, see for example recent surveys in [6-9]. In this section, we shortly summarize the main compression techniques that are depicted in Fig. 15.1.

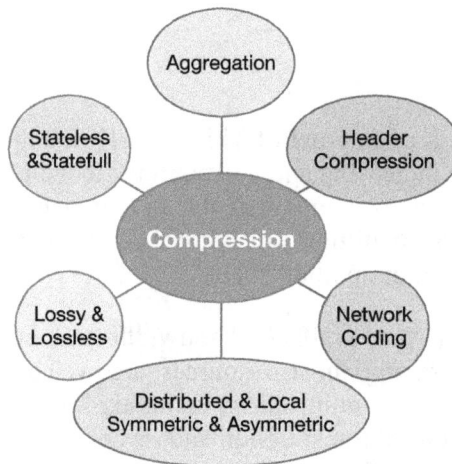

Fig. 15.1. Compression techniques.

Aggregation is an efficient method to compress data whenever it is suitable for the function in question. The problem of using aggregation is that it will reduce information like by averaging, and the network structure and routing has to support aggregation. The other shortage is that if packet loss exists, then losing one aggregated packet means losing several original packets, especially when no packet caching is used. During decades, numerous aggregation schemes have been proposed, in which the most recent ones, e.g. [10] and [11], consider also security issues.

Network coding: Depending on the information, there may be a possibility to define sequences, which are repeating inside the message(s). By utilizing the repeated information, it is possible to describe the same information in a shorter way and to achieve compression in that way.

In traditional routing networks, packets are cached and forwarded separately downstream, even if they have the same destination. In network coding, the messages are merged and the code and the accumulated result are forwarded to the destination. After receiving the accumulated message, it is decoded at the destination. Network coding techniques for wireless sensor networks (WSN) are discussed more carefully in [12].

There are several methods with varied complexity for describing the repeated information. Depending on the method, the decompression will return exactly the original data or it may have some losses or mistakes. In general, the method with more complexity will provide more lossless compression and better compression rate.

Distributed vs. local and symmetric vs. asymmetric: Compression operation can be carried out locally at the node or it can be distributed to several nodes in order to share the load between the nodes. Load sharing may help especially nodes that have scarce of resources. Usually, capabilities of network devices vary. Terminal nodes, like sensor nodes, have least resources whereas core network devices, like servers, can share their resources. Often a distribution is an asymmetric system, where the most capable devices take care of the highest load and thus save the resources of the terminal nodes.

Lossy & lossless: Depending on the data reconstruction after the decompression, compression methods can be divided into lossy and lossless techniques. Lossless techniques aim to return exactly the original data, whereas lossy techniques give only an approximation of the original data. Generally, lossy algorithms provide higher compression, but also higher data loss. Which approach will fit best depends on the requirements of the application. For instance, in video and voice applications, lossy compression may be accepted, but data loss or a wrong value may cause serious problems in the case of control measurements.

Stateless or stateful: Stateless compression does not require any per-flow state, which could be corrupted during the change in wireless connection. The idea of the stateless compression is based on the assumption that some content between the sender and the receiver is well-known and thus can be assumed or extrapolated from the received data. That kind of data is for instance the static parts of the protocols' header data, like the network prefix, which can be compressed into a single bit. If similar data is already available in several protocol headers, like checksum, then that data can be removed.

In the case of stateful compression (or shared-context), there will always be negotiation between the sender and the receiver. During the negotiation, the sender and the receiver agree on the semantics how the compression will be performed. When the compression is used, the sender and the receiver have to agree from time to time that the compression state is still valid.

The advantage of the stateful approach compared to the stateless is that it will allow much higher compression rate than the stateless. For instance, in a data flow, almost all header information can be compressed under one ID flag. Another advantage of the stateful approach is that it is more dynamic because there is no need to assume anything before the negotiation and so the packet formats can change freely.

Because of the nature of wireless communication, there will always be some packet loss and bit errors during messaging. If the losses and errors are significant, then stateless approach will outperform stateful, because the stateless approach does not need any pre-configuration before compression. In the case of stateful compression, if an error or a loss happens, then pre-configuration must be redone, which causes extra control traffic. So, maintaining the flow states will be difficult [13]. This makes pure stateful solution, despite it has a better compression rate, applicable only in good link conditions.

Resource-constrained devices must also consider memory usage and computational complexity. From this point of view, the stateless approach is more efficient, because it does not need any state establishment or management, and it has a simpler source code.

Non-adaptive and adaptive: In general, an adaptive compression can adjust system to environmental changes, which can be, for example, changes in the data type or connection performance. The connection can be improved, for instance, by chancing the communication interface or operation of the link layer (L2), or utilizing multipath routing to get better connection. Adaptivity makes the system more complex, but at the same time, it provides better performance in changing circumstances.

Header Compression: The network packet can be divided into two parts: the header and the payload part. From the compression point of view, the header part is interesting because there is redundant data among different protocol headers and, especially, between consecutive packets that belong to the same flow. This kind of redundant data can be elided [8, 14]. For example, many header fields remain constant or change according to a known pattern between packets. Over a single link, not all that data is needed and part of it can be temporarily removed, i.e., the full packet will be re-created on the receiving side of the link. In the next section, we will take a closer look to different header compression solutions.

15.3. Development of Header Compression

Header compression is not a new idea, but compression standards are still evolving. The first header compression scheme, CTCP (i.e., VJ compression) [15], compresses TCP/IPv4 headers and it was presented in 1990. The evolution continued when IPHC [16] and CRTP [17] were presented in 1999 with wider protocol support (UDP, RTP and IPv6) and improvements in packet loss handling. The next evolution step was presented in ROHC [18] in 2001 and 6LoWPAN [19] in 2007. ROHC presents a robust compression scheme with modular protocol support, i.e., protocol profiles. 6LoWPAN presents a compact solution aimed for IoT environment. In the next subsections, we will discuss more carefully the above mentioned header compression solutions and compare them to each other.

15.3.1. Van Jacobson Header Compression (CTCP)

CTCP header compression defines compression for TCP/IP(v4) datagrams, which was specifically designed to improve TCP/IP performance over slow serial links, with bit rate around 300 bps. The primary idea behind the compression was to define per-packet information that likely to stay constant over the connection. In the compression, the constant bits can be omitted and the changed bits can be indicated with a bitmask. The bitmask tells the difference between the previous and the current packet and that way only the differences in the changing fields are sent rather than the whole fields themselves. In practice, this is done by saving the states of the TCP connections at the both ends of the link, and sending only the differences in the header fields that changed. Van Jacobson compression reduces the normal 40 byte TCP/IPv4 packet headers down to 3-4 bytes in average, see Fig. 15.2. Because the scheme is designed only for low bandwidth connections, where bit errors and packet loss are not a big issue, it works well there. When the bit error rate (BER) increases over 10^{-4}, the scheme does not perform well [20]. This is mainly due to that the scheme does not have its own feedback and recovery mechanism concerning packet loss, instead it relies on TCP's own recovery mechanisms. Nowadays, it is well known that TCP's recovery mechanisms for packet loss do not perform well in wireless connections [21].

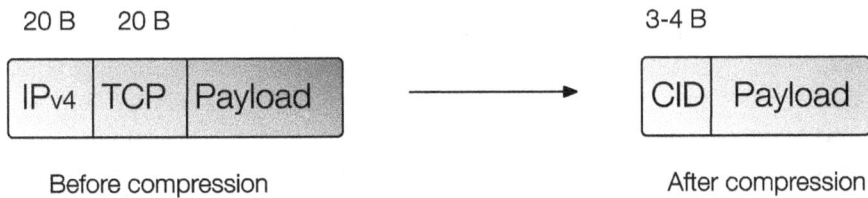

Fig. 15.2. CTCP header compression.

15.3.2. IP Header Compression (IPHC)

In general, IPHC is based on a similar compression and decompression idea as CTCP, where known data is compressed and decompression is done based on the saved context information of the compression. The main development step that IPHC brought in was that it supports UDP, IPv6 and extension headers in addition to TCP. Moreover, IPHC improved error recovery mechanisms (important especially in lossy links) and supported multiple IP header compression (in case of tunneling of IP packets) and allowed extensions for multi-access links and multicast. There also described two additional mechanisms that increase the efficiency of TCP header compression over lossy links. When many packet streams (several hundreds) traverse the link, a phenomenon known as context ID (CID) thrashing can occur. In CID thrashing, headers cannot be matched with an existing context and have to be sent uncompressed or as full headers.

15.3.3. Compressing RTP Headers (CRTP)

CRTP uses similar compression approach like CTCP and IPHC. As a new thing, it provides support for RTP protocol. As additional features, CRTP brings in a method for reporting packet loss information. The information contains a health report of packets and compression level that sources are capable to adapt. CRTP replaces the IPv4, UDP, and RTP headers (40 bytes) with a 2-4-byte context ID (CID), as depicted in Fig. 15.3.

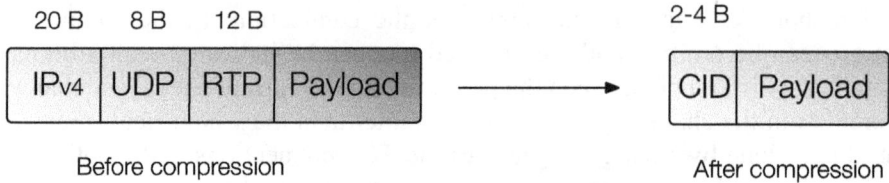

Fig. 15.3. CRTP header compression.

15.3.4. Robust Header Compression (ROHC)

ROHC has similar idea like above, where compressed packets are decompressed based on the saved context data in the decompression side. One main difference for previous compression schemes is that the whole compression process is divided into different states depending on the link performance. The states are maintained within two finite state machines: one at a compressor and the other at a decompressor. The compressor states are Initialization & Refresh (IR), First Order (FO) (i.e., partial compression), and Second Order (SO) (i.e., full compression). The state machine aided compression makes compression process more robust and takes advantage of the link quality, but on the other hand, it increases complexity. The other remarkable difference is that new protocol headers can be presented as profiles, which makes a modular base for protocol implementation and development. Currently, ROHC supports, e.g., RTP, UDP, UDP-Lite, TCP, ESP, and IP protocols [22-24]. CoAP compression profile for ROHC was introduced in our earlier work [5]. Fig. 15.4 presents an example of CoAP/UDP/IPv4 packet header compression, where the headers are compressed from 37 bytes to 5 bytes.

15.3.5. IPv6 Over Low Power Wireless Area Networks (6LoWPAN)

6LoWPAN compression approach differs from previous ones as being a stateless compression. It uses only known or assumed headers in the compression process and thus there is no need to create decompression tables in the receiver side. This simplifies the compression process and enables stateless compression, which has a clear advantage in lossy communication like wireless communication with resource scare devices. 6LoWPAN is not only about header compression but also involves fragmentation, neighbor discovery, under mesh addressing, stateless address configuration, and management information base (MIB). 6LoWPAN is defined over IEEE 802.15.4 devices. Based on the protocol stack, 6LoWPAN header compression involves the following header compression:
• Application layer header compression (MIP);
• Transport layer header compression (TCP, UDP);
• Network layer header compression (IPv6 unicast and multicast, routing and other extension headers, ICMP);
• Data link layer (fragmentation).

OSI - layers

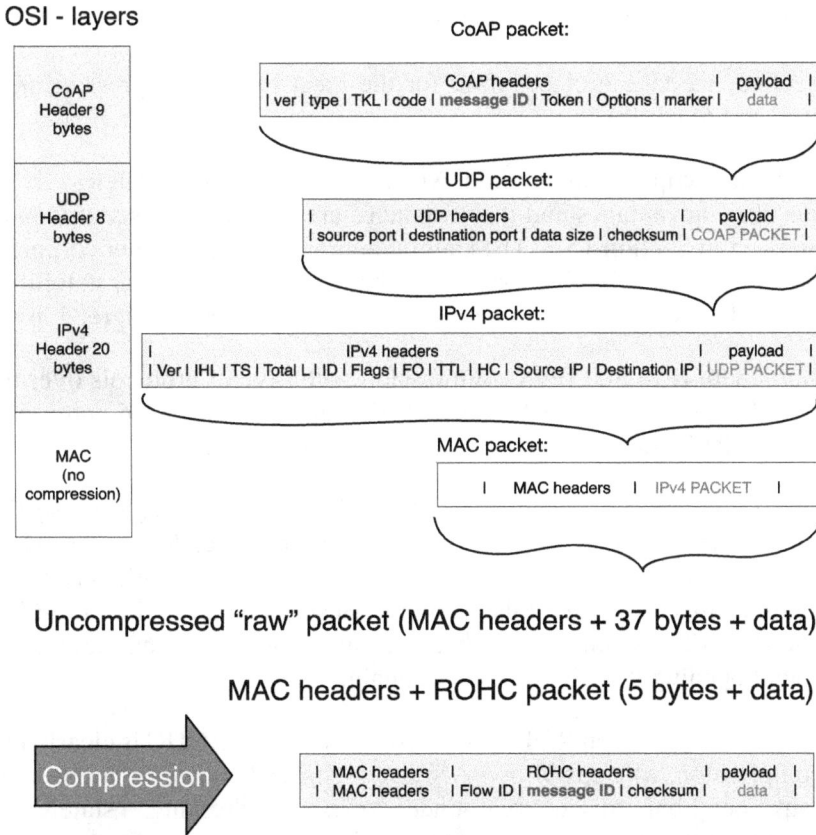

Fig. 15.4. ROHC header compression with CoAP/UDP/IPv4 profile.

6LoWPAN compresses IPv6/UDP headers (48 bytes) to 7 bytes compressed header (CH) and, respectively, IPv6/TCP headers (60 bytes) to 7-31 bytes compressed header, as depicted in Fig. 15.5.

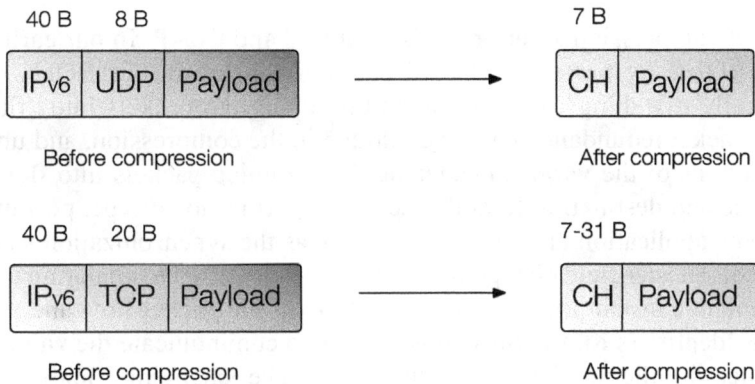

Fig. 15.5. 6LoWPAN header compression.

15.3.6. Discussion on 6LoWPAN and ROHC

In this subsection, we take a closer look for the most recent compression approaches, namely ROHC and 6LoWPAN.

ROHC is stateful compression method while 6LoWPAN is a stateless compression method. Thus, their advantages and disadvantages are like those of stateful and stateless methods discussed in Section 15.2. The stateful approach allows higher compression rate than stateless and is more adaptive for header changes. As a trade-off, stateful method is more complex and the compression state needs to be negotiated and agreed in advance.

ROHC supports both IPv4 and IPv6 compression with several protocols over them, e.g., TCP, UDP, UDP-Lite, ESP and RTP. 6LowPAN supports only IPv6 compression with UDP or TCP, and RPL and ICMP, so IPv4 is not supported at all.

Different header compression schemes are defined as profiles in ROHC, so there is no need to open the whole standard when a new protocol needs header compression support from it. In case of 6LowPAN, adding support for a new protocol would mean opening the whole standard again. To avoid this, Generic Header Compression for 6LoWPANs (6LoWPAN-GHC) [25] standard has been designed for allowing new compression schemes without opening the whole standard again.

Another big difference between ROHC and 6LowPAN is that ROHC is clearly meant only for header compression, whereas 6LowPAN also includes dedicated solutions concerning fragmentation, neighbor discovery, under mesh addressing, stateless address configuration, and management information base [26, 27].

The limitation of 6LowPAN is that it is designed only for 802.15.4 devices. If other devices communicating over, e.g., Bluetooth or UWB, like to have support of 6LowPAN, then a new standard is needed, e.g., [28] for IPv6 over Bluetooth Low Energy.

15.4. IoT Compression Solution with ROHC

Our IoT packet compression solution utilizes ROHC and CoAP. In our earlier study [5], we extended ROHC by presenting CoAP compression profile for ROHC. The profile requires basically two things: identifying and grouping the packets into "flows" so that the packet-to-packet redundancy can be exploited in the compression, and understanding the change patterns of the various header fields. Grouping packets into flows is usually based on source and destination IP addresses, transport protocol type, port numbers, and potentially some application layer identifiers, such as the synchronization source (SSRC) in RTP protocol. Header field change patterns can be classified to changing or static ones. Static ones remain constant throughout the lifetime of the packet flow and some of them can be used as identifiers for the flow. It is enough to communicate the value of the static field only once in the beginning of the flow. The changing fields require more sophisticated methods based on their expected change patterns.

In the CoAP compression profile for ROHC, we used source and destination IP addresses, UDP port numbers, and CoAP Type and Code fields for grouping the packets into flows. Message ID field was identified as the only changing field in the CoAP header. Thus, it is enough to communicate the static CoAP header parts only once in the beginning and after that it is enough to communicate only the Message ID field. In our current implementation, we do not exploit the change pattern of the Message ID field, but in the future, we could enhance the compression ratio by using it. Otherwise, the UDP and IP header parts of the packet were compressed as in the existing UDP/IP profile. Our CoAP profile was implemented for the open source ROHC library [29]. A version 1.6.1 of the ROHC library was used as the basis for the implementation.

In the following subsections, we present our test bed, testing scenario and performance evaluation results for three different radio technologies: WLAN, LTE and XBEE.

15.4.1. Test Bed and Testing Scenario

The test bed consists of a laptop and Model B+ Raspberry Pi (RPi) computers. A transmission link between RPi and laptop was made with XBEE, LTE and WLAN radios. VTT's CNL laboratory's Willab network was used to make it possible to connect the laptop to the WLAN or RPi to the LTE base station. In the XBEE test, the test link was made directly between RPi and laptop. The power consumption was monitored from RPi's main power supply by using a current probe and an oscilloscope. During the test, the network traffic and power consumption data was logged.

The XBee test bed is shown in Fig. 15.6. The XBee radio shield was connected to RPi's GPIO port. To make it possible to transfer IP packets through XBee network, the XBee-Tunnel-Daemon [30] was used. It is a daemon that creates a virtual network interface allowing transmission of UDP/IP packets over the XBee module.

In the LTE radio scenario, the radio stick was connected to RPi's USB port (see Fig. 15.7). The LTE base station was connected to the Willab network. The WLAN test bed is described in Fig. 15.8. In that case, the server was connected to Willab using WLAN, while the RPi used Ethernet.

In all cases, an UDP-tunnel was established between the laptop and the RPi by using the tunnel application provided by the ROHC library for easy testing of the library. It is to be noted that the tunneling was used only for practical testing reasons since we do not have a real ROHC implementation integrated in the protocol stack in the both ends of the link. The laptop served as a CoAP server running the example CoAP server software provided by the C-based libcoap-4.0.3 library [31]. The RPi acted as a CoAP client requesting periodically the 'time' resource from the server using libcoap's example CoAP client software. Thus, it created a GET request flow to the server. The server responded to the requests by sending acknowledgements piggy-packing the CoAP RESPONSEs. The exact packet formats are presented in the next section in parallel with the results for the packet size.

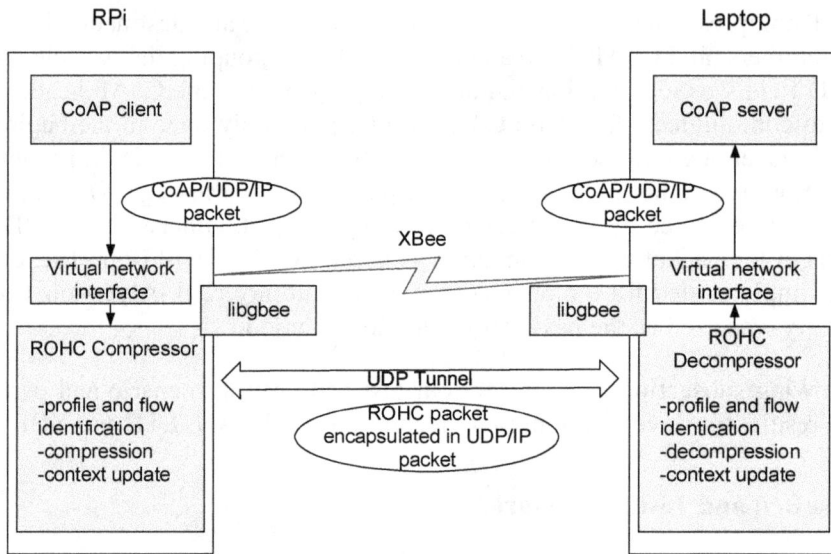

Fig. 15.6. XBee test bed.

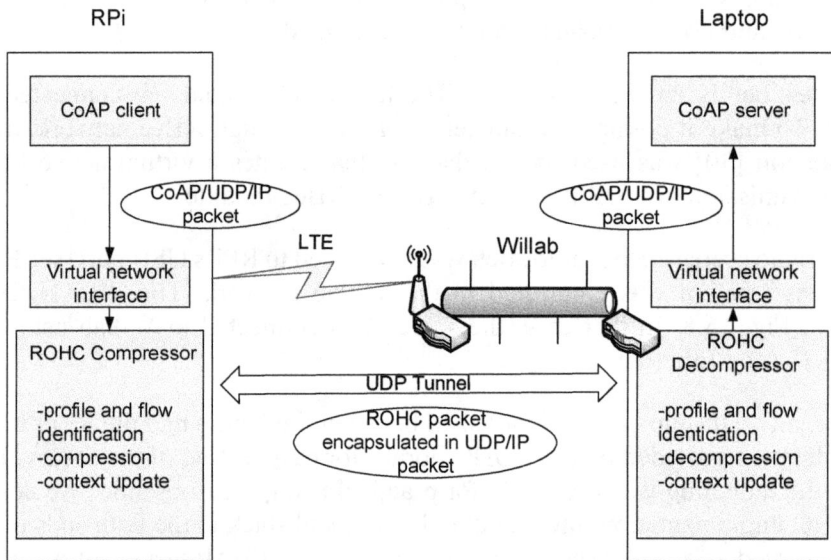

Fig. 15.7. LTE test bed.

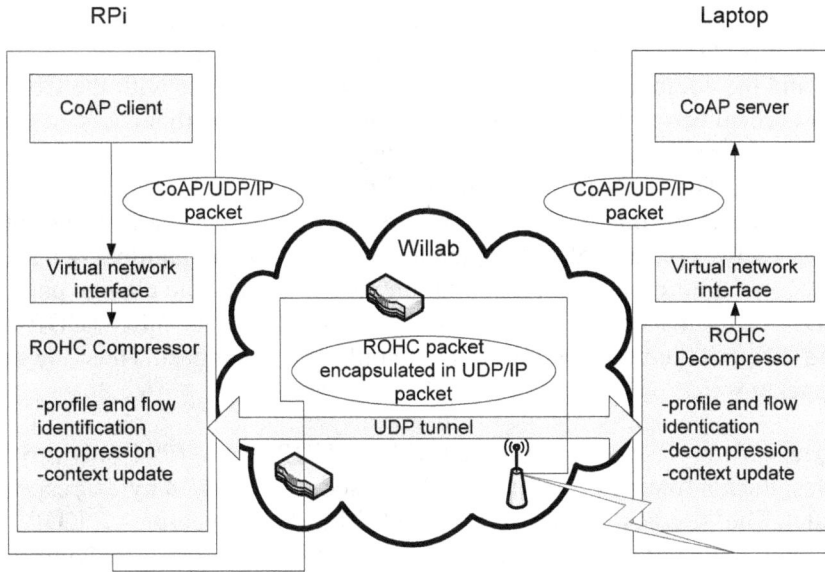

Fig. 15.8. WLAN test bed [5].

15.4.2. Header Compression Results

In this section, we first study the performance results for the compression ratio, i.e., the packet size. After that, we take a look at delays and energy consumption in three different networks: WLAN, LTE and XBEE.

15.4.2.1. Packet Size

Compression ratio, i.e., the packet size, does not depend on the used radio technology. In all test cases, the created CoAP GET request message consisted of the following CoAP header fields:
- version = 1
- type = 0 (confirmable)
- tkl = 0 (no token)
- code = 1 (GET request)
- message ID = 58468 (changing number)
- option Uri-Path = 'time' in TLV format =0xb474696d65; Type=11, Length=4, Value='time'
- no payload

In the beginning, the compressor starts in the ROHC Initialization and Refresh (IR) state and transmits full headers. The uncompressed packet size was 37 bytes consisting of IPv4 header of 20 bytes, UDP header of 8 bytes, and CoAP header of 9 bytes. One extra byte was needed for the CID in ROHC, so the compressed packet size was in the beginning 38 bytes.

However, after couple of packets when the context was created and the compressor changed into First Order (FO) state, only the dynamical (changing) header parts were transmitted and the compressed packet size decreased to 15 bytes with the CoAP profile. The only dynamical part of the CoAP header is the Message ID that takes 2 bytes. So the 9-byte CoAP header was compressed to only 2 bytes, i.e., the savings in the CoAP header part were 77.8 %. The rest of the compressed packet was used by the UDP and IPv4 headers. The savings for the whole packet were 59.5 %. When the compressor changes into Second Order (SO) state, also the change patterns of the dynamical header parts are taken into account. Our current implementation did not exploit the change pattern for the Message ID, but further compression was still available in the other (UDP and IPv4) headers, and the packet could be compressed to only 5 bytes at best. Thus, the savings for the whole packet were 86.5 %.

The same packet flow was also compressed with the UDP/IP and IPv4-only profiles. With these profiles, the compressed packet sizes decreased to 22 and 28 bytes, respectively, in the FO state. The savings were 40.5 % and 24.3 %, respectively. UDP/IP profile compresses only the UDP and IP header parts, while the IP-only profile compresses only the IP header part. These profiles do nothing for the CoAP header part. In SO state, the IPv4 compression profile compressed the packet to 18 bytes at best, so the savings were only 51.4 %.

The server's acknowledgement message piggy-packed also the response, and its length was 51 bytes uncompressed. IPv4 header took 20 bytes, UDP header 8 bytes, CoAP header 8 bytes, and CoAP payload 15 bytes. The CoAP header part consisted of the following header fields:

- version = 1
- type = 2 (ACK)
- tkl = 0 (no token)
- code = 69 (2.05 Content)
- message ID = 58468 (changing number)
- options Content-Format (Length = 0) and Max-Age (length 1 and value 1) in TLV format = 0xc02101;
- Payload Start Marker = 0xFF
- Payload = 'Oct 14 19:45:32'

For the ack/response packet, the header parts could be compressed as much as in the request messages, but the overall savings were smaller because the payload part could not be compressed. Some of the messages also contained ROHC feedback data. With the CoAP profile, the compressed packet size (including the payload of 15 bytes) was 53 or 58 bytes in IR state (depending on the amount of feedback data), 31 or 22 bytes in FO state, and 21 bytes in SO state. So the savings in the total packet size were 58.8 % at best in SO state. With IPv4-only profile, the packet size could be compressed to 33 bytes at best, meaning 35.3 % savings in total packet size.

We also studied the packet sizes of the same CoAP request and ack/response messages when using IPv6 protocol. IPv6 protocol header takes 40 bytes, i.e., 20 bytes more than

IPv4. Thus, the uncompressed CoAP request packet size was now 57 bytes and the corresponding ack/response packet 71 bytes. In the IR state, the compressed packet sizes (with ROHC header and feedback information) were 61 or 68 bytes for the CoAP request packet and 75 bytes for the ack/response packet. In the next state, when only dynamical header parts were transmitted, the compressed packet sizes decreased to 12 and 27 bytes, respectively. Finally, in the SO state, the compressed packet sizes were only 5 and 20 bytes, respectively, with the developed CoAP profile. As percentages, these mean total savings of 91.2 % and 71.8 % in the request and ack/response packet sizes, respectively. With the UDP profile, the SO state packet sizes were 12 and 26 bytes, respectively, meaning 78.9 % and 63.3 % savings. With IPv6-only profile, the SO state packet sizes were 18 and 32 bytes, respectively, meaning 68.4 % and 54.9 % savings. Compared to the IPv4 packet sizes, IPv6 results in larger packets in the beginning of the compression, but in the end, the packets can be compressed to about the same sizes (or even smaller) as with IPv4. Thus, the percentile savings in the packet sizes are actually bigger with IPv6 than with IPv4.

So, even if the CoAP header is designed to have only a small overhead, header compression can still make it even smaller. During the transmission, these smaller packets have lower probability to have bit errors, and hence the packet loss will be smaller than for packets with the original size [5]. Thus, packet loss and response time decrease, which together provide better performance. This will realize especially in wireless and low bandwidth links, where bit errors and packet loss are common. The header compression is beneficial when the payload is small compared to the header part. With larger payloads, the benefit of header compression gets smaller and smaller.

15.4.2.2. Delay and Energy Consumption

Let's take a more careful look how compression affects to processing delay and radio transmission time in the case of a low bandwidth XBEE link. The whole process consists of packet creating and compression, transmission, and receiving and decompression. The compression and decompression processes create delay compared to uncompressed packet processing, as can be seen in Fig. 15.9.

The delays are 1.5 ms and 3.0 ms in compression and decompression phases, respectively. On the other hand, because of the smaller packets, compression will speed up radio transmission by about 1.1 ms, see Fig. 15.10. When the delays are calculated together, the total delay remains at 3.4 ms.

Using Figs. 15.9 and 15.10, we can calculate the energy consumption by multiplying the time and the average power consumption during the packet creating & compression, transmission and receiving & decompression processes. Table 15.1 presents the total energy consumption in the case of CoAP/UDP/IPv4 packets utilising ROHC's CoAP profile in SO, i.e., full compression state. Even if the time saved in the transmission was shorter than the time spent for compression and decompression processing, the total energy efficiency improved by 14.2 μJ/packet (0.173 %), when the packets were compressed. This is because during the transmission, the power consumption is on a much higher level than during packet processing. The trade-off for saving bandwidth and for lower energy consumption was the extra delay of 3.4 ms.

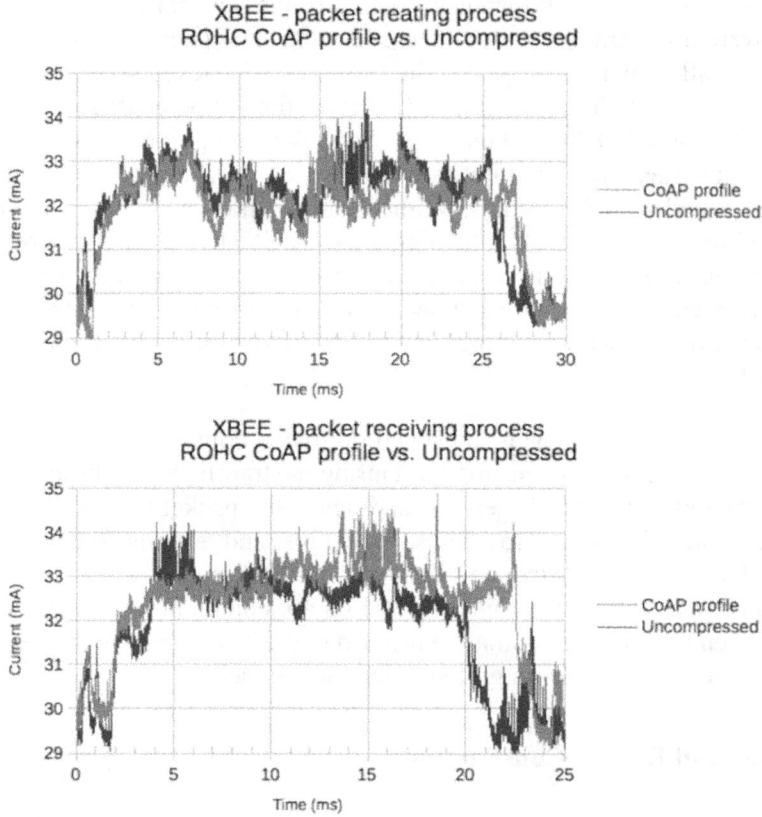

Fig. 15.9. CoAP/UDP/IPv4 packet creating and receiving processes in RPi using XBEE radio.

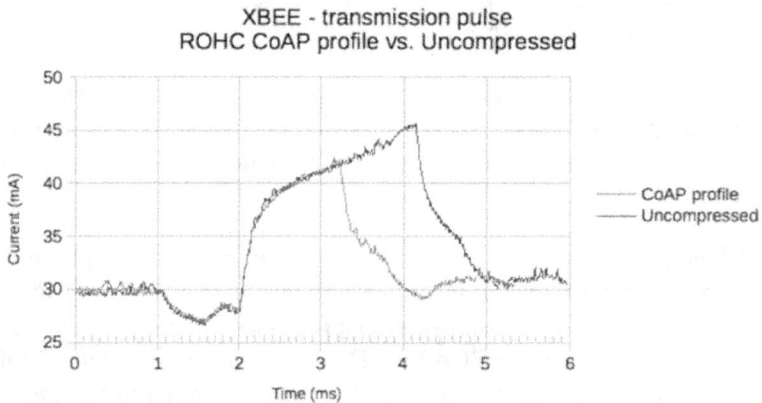

Fig. 15.10. XBEE transmission pulse when using ROHC with CoAP and Uncompressed profiles.

Table 15.1. Energy consumption in XBEE using ROHC's CoAP/UDP/IPv4 compression profile.

Packet processing	Uncomp. packet [ms]	Comp. packet [ms]	Time gap [ms]	Energy gap (μJ)
Creating	26.5	28	-1.5	-14.7
Receiving	21	24	-3	-40.0
Transmission	3.1	2.0	1.1	68.9
TOTAL			-3.4	14.2

As depicted in Table 15.2, transmission in LTE and WLAN is more than 200 time faster than in XBEE transmission. Thus, we could not detect the transmission peaks in our measurements, only delays of compression and decompression processing could be found out, see Fig. 15.11 and Fig. 15.12.

Table 15.2. Transmission times with different radios for compressed and uncompressed packet.

Radio	Transmission speed (Mbit/s)	Transmission time (μs)	
		Uncomp. packet (736 bits)	Comp. packet (440 bits)
XBEE	0.25	2944	1184
WLAN 802.11b	54	14	8
LTE	100	7	4

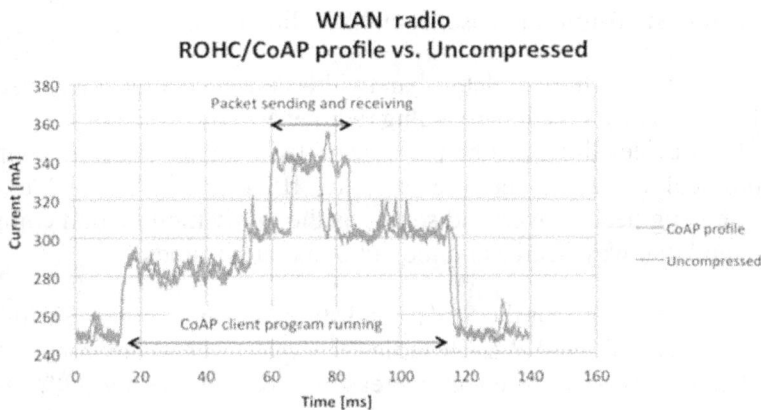

Fig. 15.11. ROHC's CoAP/UDP/IPv4 compression profile vs. uncompressed profile in WLAN [5].

LTE radio

ROHC/CoAP profile vs. Uncompressed

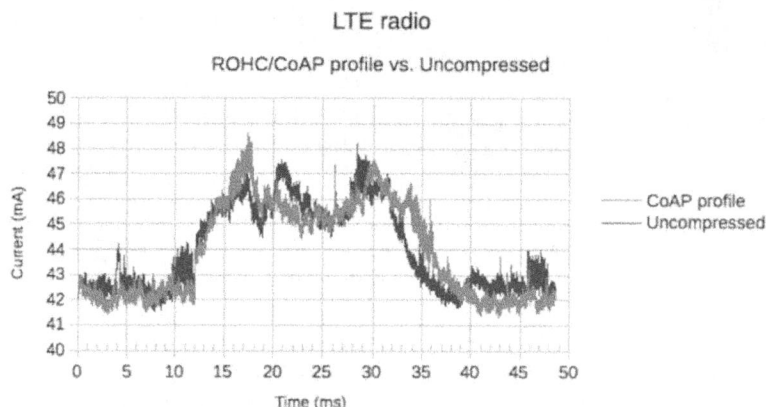

Fig. 15.12. ROHC's CoAP/UDP/IPv4 compression profile vs. uncompressed profile in LTE.

So, in the case of fast speed radios like LTE and WLAN, direct energy savings during transmission will be neglible and the cost of extra compression/decompression processing will rule and the total process will consume more energy than without compression. For example, in case of WLAN, the energy consumption increased by 2.5 % [5]. However, in lossy links, packet loss will decrease when the packet size decreases [32], i.e., when the packet is compressed. This will enhance transmission by avoiding retransmissions, and that way energy could be saved. However, the potential savings depend on how many packet losses actually can be avoided.

15.5. Conclusions

Due to the extensive growth of Internet of Things (IoT), the number of wireless devices connected to the Internet is forecasted to grow to 26 billion units installed in 2020 representing an almost 30-fold increase from 0.9 billion in 2009 [1].

In wireless IoT networks, the available bandwidth and energy is often restricted. That resource can be effectively saved by reducing radio operation: transmission and reception. In the case of IoT devices like sensors, the transmitted data may be only couple of bytes, whereas the protocol headers of the packet (MAC, IP, TCP/UDP, CoAP, etc.) are many tens of bytes. This big header overhead serves as the motivation behind compressing the packet headers and that way decreasing the radio operation times.

We presented a CoAP profile for the ROHC and studied its performance with a low speed XBEE and high speed LTE and WLAN radios in a real test bed environment. We found that the CoAP compression profile can decrease the packet size by 90 % or more. The smaller packet size speeds up the radio operation during transmission and reception. This reduces the energy consumption during the radio transmission. Smaller packet will also reduce packet loss in lossy links, which can further improve the energy efficiency. The trade-off for smaller packet size is the delay and extra processing power needed for the compression and decompression.

310

In the case of XBEE, the energy saving during radio transmission was bigger than energy consumption during compression/decompression processes. The total energy savings were 14.2 µJ/packet (0.173 %). The compression and decompression processes themselves consume roughly ten times less energy than the whole packet creating and receiving processes. However, the delay caused by the compression and decompression was bigger than the saved time during the radio transmission, so the trade-off for smaller packet and energy savings was the extra delay of 3.4 ms.

In the case of high speed radios (LTE and WLAN), the savings in transmission time and energy were neglible compared to the extra processing needed for compression and decompression. Thus, there were no energy savings during individual packet transmissions. In fact, the energy consumption increased by 2.5 % in case of WLAN. However, the smaller packet size will reduce packet loss in lossy links and this way it could be possible to save energy, enhance throughput and decrease delay.

The compression and decompression processes cause quite long delays. They can possible be decreased by having a more optimal software implementation. Reducing the delay caused by the compression/decompression processing automatically improves also the energy efficiency.

ROHC has good compression gain and it is used in current mobile networks, which make it a promising candidate for IoT header compression solution. The disadvantage is the stateful operation and more complexity compared to 6LowPAN. One possible future work item could be to merge 6LowPAN functionalities to ROHC to have some stateless compression at the Initialization and Refresh (IR) state, and perhaps to extend the compression profiles to cover also MAC headers.

Acknowledgements

This work was supported by VTT Technical Research Centre of Finland Ltd.

References

[1]. Gartner Says the Internet of Things Installed Base Will Grow to 26 Billion Units By 2020, (http://www.gartner.com/newsroom/id/2636073).
[2]. P. Koskela, M. Valta and T. Frantti, Energy Efficient MAC for Wireless Sensor Networks, *Sensors & Transducers*, Vol. 12, Issue 10, October 2010, pp. 133–143.
[3]. Z. Shelby, K. Hartke and C. Bormann, The Constrained Application Protocol (CoAP), *Internet Engineering Task Force (IETF)*, Request for Comments: 7252, Category: Standards Track, 2014, pp. 1-112.
[4]. P. Koskela and M. Majanen, Robust Header Compression for Constrained Application Protocol, *Internet of Things - Finland*, No. 1, 2014, pp. 36-39, (http://www.internetofthings.fi/extras/IoTMagazine2014.pdf).
[5]. M. Majanen, P. Koskela and M. Valta, Constrained Application Protocol Profile for Robust Header Compression Framework, in *Proceedings of the 5th International Conference on*

Smart Grids, Green Communications and IT Energy-aware Technologies (ENERGY'15), Rome, Italy, May 24 – 29, 2015, pp. 47-53.

[6]. P. Koskela, M. Majanen, M. Valta, Packet Header Compression for the Internet of Things, *Sensors & Transducers,* Vol. 196, Issue 1, January 2016, pp. 43-51.

[7]. Razzaque M. A., Bleakley C. and Dobson S., Compression in wireless sensor networks: A survey and comparative evaluation, *ACM Transaction on Sensor Networks*, 10, 1, Article 5, 2013, 44 pages.

[8]. Shivare M. R., Maravi Y. P. S. and Sharma S., Analysis of Header Compression Techniques for Networks: A Review, *International Journal of Computer Applications*, Vol. 80, No. 5, 2013, pp. 13-20.

[9]. T. Srisooksai, K. Keamarungsi, P. Lamsrichan and Araki K., Practical data compression in wireless sensor networks, *Journal of Network and Computer Applications*, Vol. 35, No. 1, 2012, 37-59.

[10]. L. Yu, J. Li, S. Cheng, S. Xiong and H. Shen, Secure Continuous Aggregation in Wireless Sensor Networks, *IEEE Transactions on Parallel and Distributed Systems*, Vol. 25, No. 3, 2014, pp. 762-774.

[11]. S. Roy, M. Conti, S. Setia and S. Jajodia, Secure Data Aggregation in Wireless Sensor Networks: Filtering out the Attacker's Impact, *IEEE Transactions on Information Forensics and Security*, Vol. 9, No. 4, 2014, pp. 681 - 694.

[12]. P. Ostovari, J. Wu and A. Khreishah, Network coding techniques for wireless and sensor networks, in The Art of Wireless Sensor Networks, H. M. Ammari (Ed.), *Springer,* 2013, pp. 1-35.

[13]. B-N. Cheng, J. Zuena, J. Wheeler, S. Moore and B. Hung, MANET IP Header Compression, in *Proceedings of the IEEE Military Communications Conference (MILCOM'13),* 2013, pp. 494-503.

[14]. Effnet AB WHITE PAPER Library, An introduction to IP header compression, 2004, (http://www.effnet.com/sites/effnet/pdf/uk/Whitepaper_Header_Compression.pdf).

[15]. V. Jacobson, Compressing TCP/IP Headers, *Internet Engineering Task Force (IETF)*, Request for Comments: 1144, Category: Standards Track, 1990, pp. 1-46.

[16]. M. Degermark, B. Nordgren, S. Pink, IP Header Compression, *Internet Engineering Task Force (IETF),* Request for Comments: 2507, Category: Standards Track, 1999, pp. 1-47.

[17]. S. Casner and V. Jacobson, Compressing IP/UDP/RTP Headers for Low-Speed Serial Links, *Internet Engineering Task Force (IETF)*, Request for Comments: 2508, Category: Standards Track, 1999, pp. 1-24.

[18]. C. Bormann, C. Burmeister, M. Degermark, H. Fukushima, H. Hannu, L-E. Jonsson, R. Hakenberg, T. Koren, K. Le, Z. Liu, A. Martensson, A. Miyazaki, K. Svanbro, T. Wiebke, T. Yoshimura and H. Zheng, RObust Header Compression (ROHC): Framework and four profiles: RTP, UDP, ESP, and uncompressed, *Internet Engineering Task Force (IETF),* Request for Comments: 3095, Category: Standards Track, 2001. pp. 1-168.

[19]. G. Montenegro, N. Kushalnagar, J. Hui and D. Culler, Transmission of IPv6 Packets over IEEE 802.15.4 Networks, *Internet Engineering Task Force (IETF)*, Request for Comments: 4944, Category: Standards Track, 2007, pp. 1-30.

[20]. M. Degermark, H. Hannu, L. Jonsson, K. Svanbro, Evaluation of CRTP performance over. cellular radio links, *IEEE Personal Communications,* Vol. 7, No. 4, August 2000, pp. 20-25.

[21]. Tian Ye, K Xu, N. Ansari, TCP in wireless environments: problems and solutions, in *IEEE Communications Magazine,* Vol. 43, No. 3, March 2005, pp. 27-32.

[22]. Pelletier G. and Sandlund K., RObust Header Compression Version 2 (ROHCv2): Profiles for RTP, UDP, IP, ESP and UDP-Lite, *Internet Engineering Task Force (IETF)*, Network Working Group, Category: Standards Track, Request for Comments: 5225, 2008, pp. 1-124.

[23]. K. Sandlund, G. Pelletier and L-E. Jonsson, The RObust Header Compression (ROHC) Framework, *Internet Engineering Task Force (IETF)*, Category: Standards Track, Request for Comments: 5795, 2010, pp. 1-41.

[24]. G. Pelletier, K. Sandlund, L-E. Jonsson, and M. West, RObust Header Compression (ROHC, pp. A Profile for TCP/IP (ROHC-TCP), *Internet Engineering Task Force (IETF)*, Category: Standards Track, Request for Comments: 6846, 2013, pp. 1-96.

[25]. C. Bormann, 6LoWPAN-GHC: Generic Header Compression for IPv6 over Low-Power Wireless Personal Area Networks (6LoWPANs), *Internet Engineering Task Force (IETF)*, Request for Comments: 7400, 2014, pp. 1-24.

[26]. Z. Shelby, S. Chakrabarti, E. Nordmark, C. Bormann, Neighbor Discovery Optimization for IPv6 over Low-Power Wireless Personal Area Networks (6LoWPANs), *Internet Engineering Task Force (IETF)*, Request for Comments: 6775, 2012, pp. 1-55.

[27]. J. Schoenwaelder, A. Sehgal, T. Tsou, C. Zhou, Definition of Managed Objects for IPv6 over Low-Power Wireless Personal Area Networks (6LoWPANs), *Internet Engineering Task Force (IETF)*, Request for Comments: 7388, 2014, pp. 1-27.

[28]. J. Nieminen, T. Savolainen, M. Isomaki, B. Patil, Z. Shelby, C. Gomez, IPv6 over BLUETOOTH(R) Low Energy, *Internet Engineering Task Force (IETF)*, Request for Comments: 7668, Oct 2015, pp. 1-21.

[29]. ROHC library, (https://rohc-lib.org/).

[30]. libgbee, (http://sourceforge.net/projects/libgbee/).

[31]. libcoap: C-Implementation of CoAP, (https://libcoap.net/).

[32]. N. Golmie, Coexistence in Wireless Networks: Challenges and System-Level Solutions in the unlicensed band, *Cambridge University Press*, 2006, pp. 40-41.

Chapter 16
Multifunction Sensing System for Wireless Monitoring of Chronic Wounds in Healthcare

Alex Hariz

16.1. Introduction

The treatment of chronic wounds such as venous leg ulcers and diabetic foot ulcers are becoming increasingly costly for healthcare systems around the world [1, 2]. A recent estimate shows the economic cost of wound-care activities in the world is distributed as 15-20 % materials, 30-35 % nursing time and more than 50 % as hospitalization time [3]. In 2012, approximately 7 million people suffered from chronic wounds in the USA, and the cost for their treatment was estimated at almost $25 billion annually [4].

The most effective and economical treatment of wounds is to cover them with a suitable dressing or bandage in order to protect damaged skin from external infections such as those caused by microorganism attacks [5]. For certain chronic wounds such as venous leg ulcers, appropriate compression bandages are applied to increase the healing rate [6, 7]. These bandages may be retention (low pressure), light support (medium pressure), or compression (high pressure) bandages. The method of applying compression bandages on the affected limb is very important as the efficacy and maintenance of sub-bandage pressure depends on it [8]. Compression bandages can produce a pressure up to 60 mmHg at the ankle (extra high pressure), while the recommended high pressure value is 40 mmHg at the ankle [6, 8]. Depending on the applied pressure range and the type of bandage used, the sub-bandage pressure may vary significantly during the physical movement of the patient, thus affecting the healing rate [9]. In addition to compression bandages, healing rates may also be increased by managing moisture produced by the wound (exudate) through moisture-retentive dressings such as Anasept® (hydrogel) and Hydrocolloids [10, 11] for wounds with low exudate and dressings such as Allevyn® (foam) or Melgisorb® (calcium alginate) for wounds with moderate to high exudate. In addition to moisture levels, the temperature and pH under the dressing may change as a result of an infection [12, 13]. Unfortunately, these parameters associated with the dressings are not

Alex Hariz
School of Engineering, University of South Australia, Adelaide, Australia

315

currently monitored in clinical practice. There is an opportunity for advanced sensor technologies to contribute to improved wound monitoring and diagnostics.

In this chapter, we demonstrate a flexible wireless telemetric system for continuous sensing and monitoring of the wound environment, which is proposed in our published review article [14]. Preliminary results indicate that the system is capable of measuring and transmitting real-time information on temperature, moisture, and sub-bandage pressure from under the bandage or within the wound dressing at programmable transmission intervals [15]. The selection of sensors and their calibration processes have been discussed in our journal article [16]. The sensing system is fabricated on a flexible printed circuit material, while the sensors are micro-sized and flexible, thus making the system minimally invasive to wounds and the human body. The receiver is portable with the capability to receive data accurately within a distance of 4-5 meters. The system has been tested on a human volunteer using various compression bandages and moisture-retentive dressings. The results from these trials confirm the clinical utility of this system in a wound environment.

16.2. Methods and Materials

In applications involving chronic wound monitoring, the diagnostic device is required to perform reliably in a delicate environment involving human skin and wound fluid. In addition to satisfy the essential criteria of flexibility, protection from wound chemicals, and bio-compatibility, the device needs to fulfil certain performance requirements as well, which sets the foundation for minimum measurement resolutions. For meaningful temperature measurements, the device must be able to detect changes in temperature of less $\pm$ 0.5 °C. The cases with pressure and moisture are different. Although, the aim of compression bandages and stockings is to maintain a constant sub-bandage pressure at certain positions on limb, however, it may be anticipated that a $\pm$ 5 mmHg variation in bandage pressure would not have a significant impact on wound healing. Similarly, a $\pm$ 5 % RH resolution could be expected for moisture measurements. The wireless device for this application does not need to transmit continuously, as the wound conditions do not change rapidly. A complete packet of information transmitted twice an hour would be sufficient. The proposed sensing system satisfies all of the above mentioned medical and performance requirements as explained in the following sections.

16.2.1. Selection and Calibration of Sensors

For wound monitoring applications, the sensors and their assemblies need to be biocompatible and minimally invasive to the human body, as the sensors would be placed within a wound dressing or compression bandage over a human limb [17]. Having metallic inflexible structures, the sensors and circuit components would create discomfort to the patients. It has been revealed through online surveys that the majority of available sensors do not qualify for this particular application because of their large size, invasive structure, complex principle of measurement, and the need for additional on-board circuit components for operation.

The sensors needed to be carefully selected, calibrated and characterized for the intended environment in order to capture reliable information. Any error in the sensor's measurement would spread through the whole system and in some scenarios might get amplified. This would create ambiguities and false diagnosis by clinicians and health practitioners. For wound-site temperature monitoring, we chose the LM94021B (Texas Instruments, USA) temperature sensor for its small size and reliable performance. This ultra-low power sensor typically consumes just 9 μA current at a rated 5 V supply voltage. With a size of 2.15 mm × 2.40 mm × 1.1 mm (L × W × H) and a nominal accuracy of ±1.5 °C in the temperature range 20-40 °C [18], the sensor is quite suitable for our wound monitoring system.

For moisture sensing, Honeywell HIH4030 piezoelectric and Multicomp's HCZ-D5 piezoresistive moisture sensors were calibrated, characterized, and used with a prototype wireless sensing system. The piezo-resistive sensor (HCZ-D5) was found the most suitable sensor for this application because of its small size (10 mm × 5 mm × 0.5 mm). The sensor was calibrated and characterized using a dedicated experimental setup. An interface circuit was also designed to properly operate the moisture sensor.

For sub-bandage pressure measurement, we used the Interlink Electronics' FSR406 piezoresistive pressure sensor with a square sensing area of 38 mm × 38 mm. This sensor is non-invasive, flexible and is only 0.5 mm in thickness. The pressure sensor was calibrated using a clinical-grade pressure meter HPM-KH-01 for validation of pressure measurements up to 40 mmHg which is regarded as the desired value for high sub-bandage pressure [8]. An interface circuit was also designed to properly operate the pressure sensor. A commercial compression bandage system (CobanTM 2) was used to create pressure over the sensor placed on a mannequin leg.

16.2.2. Flexible Wireless Sensing System

A number of System-on-Chip (SoC) devices are commercially available with telemetry functions. Most of these devices use standard wireless transmission protocols, such as WiFi®, Bluetooth®, Bluetooth Low Energy®, and ZigBee®. Some examples of such SoCs are CC2530/31 (ZigBee), CC2540/41 (Bluetooth Low Energy), CC2560/64 (Bluetooth), CC3000/3100/3200 (WiFi), EM358x (ZigBee), 88MZ100 (ZigBee), ATMega128RFA1 (ZigBee) etc. In our proposed system, we have chosen ZigBee® for its simplicity, reasonable range, and low power operations. For this purpose, we have used Atmel's ATMega128RFA1 RF transceiver in our sensing system [19]. The transceiver is of small size (9 mm × 9 mm × 1 mm) and operates at 2.45GHz ISM (industrial, scientific, medical) frequency band using IEEE 802.15.4 ZigBee® protocol. It has a -100 dBm sensitivity and a 3.5 dBm programmable output power. It also contains a programmable serial interface and a 10-bit analog-to-digital (ADC) converter, thus minimizing the hardware overhead.

The information captured by the sensors was first converted into digital format and then stored into the TX frame buffer register. The first byte of TX frame buffer register was written with frame length information followed by the sensors' data. Before transmitting, the transmitter was passed through a set of pre-defined states i.e. RESET → TRX_OFF → PLL_ON → TX_ON. After transmitting one packet of information, the transmitter was turned off (TRX_OFF) to save battery energy.

The net weight of the sensing system is 1.938 g without sensors, 6.709 g with all sensors attached, and 9.740 g with sensors and an alkaline battery having dimensions 14 mm × 10 mm (length × diameter). The nominal and maximum current consumptions of the sensing system were measured as 13.58 mA and 17.4 mA, respectively. Hence, the peak power consumption of the sensing system is 57.4 mW at 3.3 V supply voltage. As most of the power is consumed in the RF circuits, reducing the RF switching frequency could save a considerable battery power. In a chronic wound monitoring application, even a very low data transmission rate (e.g. one packet of information per quarter an hour) would be sufficient because changes in wound parameters are very slow. A high data transmission rate would only mean a high switching rate of RF components resulting in unnecessary loss of battery power.

16.2.3. Interfacing the Sensors with the Radio Transmitter

The product of the ADC conversion is stored in two 8-bit registers; ADCH for high byte and ADCL for low byte. The maximum reference voltage for the ADC was 1.8 V, which was internally generated and stabilized. All the sensors in our system work at a 5.0 V supply for all input values exceeding its reference voltage and would not be able to detect any input signal above 1.8 V. This problem was resolved by reducing the ADC input voltage range such that the maximum input voltage of any sensor was less than the ADC reference voltage. For this purpose, a simple voltage divider circuit was designed using two surface mount resistors R_1 and R_2 for each sensor (Fig. 16.1 (b)).

16.2.4. Information Display Module

The transmitted data in ZigBee® 802.15.4 protocol was received by the handheld receiver, processed by the T6963C LCD controller, and displayed on a 10 cm × 8 cm LCD screen [20]. The LCD controller was programmed through the receiver module (ATMega128RFA1) by sending the appropriate commands with required timings as recommended in the datasheet of T6963C [21]. The first transmitted byte consists of frame length information followed by the temperature, moisture and sub-bandage pressure data, respectively. Upon reception, the sensor's data was stored in a 128-byte frame buffer register. This data was then sent to the LCD controller (T6963C) through a parallel interface for processing and display. The LCD controller was also programmed to display the device ID (TX device), frequency channel (2405-2480 MHz in 5 MHz steps), and received signal strength (i.e. RSSI) in dBm.

The TX device ID may be used as a unique identifier for the patient under observation. In a hospital environment where a number of patients would be using these devices, unique frequency channels may be allocated to all devices to avoid any interference between devices or corruption of data. The RSSI level could be indicative of how trustworthy the information of measured parameters is. Information associated with much lower RSSI levels (e.g. -90 dBm) may be discarded on the grounds that the TX device might be too far away from the RX or there might be some obstacle between them, thus inhibiting a reliable communication link.

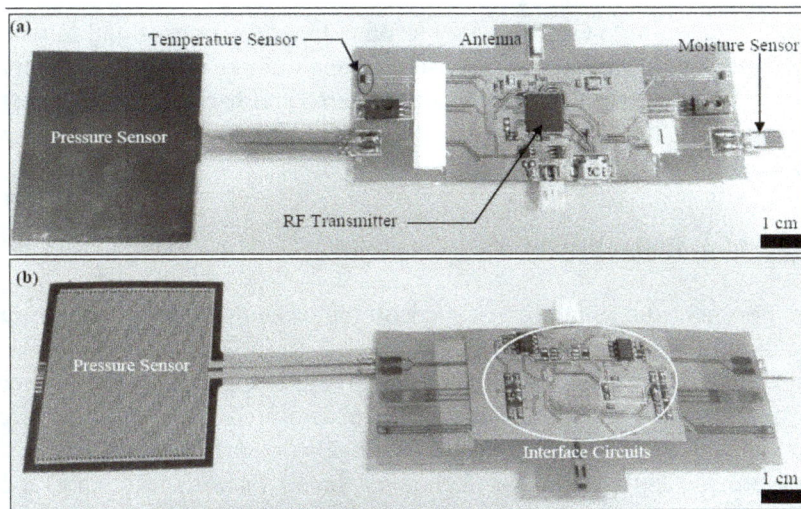

Fig. 16.1. (a) Photo of the wireless sensing system fabricated on a 0.15 mm thickness 2-layer flexible printed circuit board. The diagram shows various parts of the sensing system, (b) Rear side of the sensing system showing the voltage divider network used as interface circuits between the sensors and the ADC.

16.3. Experiments and Results

The developed wireless wound sensing system was first tested in a room environment using commercial compression bandages and a mannequin leg. The compression bandages used for the experiments consisted of AMS Bi-Flex® elastic bandage, Hartmann Lastodur Light® long-stretch bandage, 3M CobanTM 2 two-layer, and Hartmann Lastolan® short-stretch bandages. Adjunct components used in the application of compression bandage systems such as Soffban®undercast padding and Idealcrepe® crepe retention bandage were also used. Using the experimental setup, a number of experiments were performed using the mannequin leg in ambient environment. The sensing system was placed flat on the central portion of the mannequin to avoid any damage to the components from bending. After the application of dressing, distilled water was sprayed over the bandage in the proximity of the moisture sensor. The measurements were taken with an interval of 5 minutes by the handheld receiver placed 3 m apart in line of sight to the wireless transmitter. The experiment was performed at 25 °C. The average values of received energy level, temperature, and sub-bandage pressure were calculated as -79.92 dBm, 23.21 °C and 15.14 mmHg, respectively. The average errors in temperature, moisture, and sub-bandage pressure measurements were calculated as 0.93 °C, 3.0 % RH, and 1.50 mmHg, respectively. However, the error in sub-bandage pressure measurements is expected to increase on the human leg because of flexible morphology and muscle movements. The errors in moisture and temperature measurements, however, are immune to these factors.

The manufactured wound sensing and monitoring system was extensively tested for its performance on a healthy volunteer. The objectives of these trials were to validate the performance of the sensing system under various clinical scenarios and to determine the optimum placement of the system and sensors on the human body. These scenarios were realized using different compression-bandage systems and various possible postures, both believed to influence the real-time measurements. These trials and their results are discussed in detail in the following sections.

16.3.1. Establishing Reliability of Measurements

In the first experiment the reliability of measurement results was tested, particularly the sub-bandage pressure measurements as they are prone to variations due to body movements. The sensing system was covered with a transparent adhesive silicone gel on both sides. The trial was performed using a 4-layer compression system Profore® (Smith & Nephew©) as well as a 2-layer inelastic bandage system CobanTM 2 (3M©) for sub-bandage pressure measurements. The bandages in all experiments were applied by a Wound Management Nurse Practitioner adept in managing chronic wounds and bandaging techniques. The sub-bandage pressure and moisture measurements were performed in two separate experiments. During the sub-bandage pressure measurements, the moisture sensor was placed directly on exposed skin with the sensor facing the skin while the pressure sensor was placed over the flat area of the exposed skin, directly above the medial malleolus (ankle), with the sensor facing the skin. Initial measurements were taken in each experiment prior to applying bandages in order to find and nullify any offset readings. The bandages were applied in accordance with the manufacturer's instructions, achieving 40 mmHg starting pressure. The measurement results are plotted in Fig. 16.2 for a 4-layer type, at one minute interval using various common postures. Variations in sub-bandage pressure readings are measured by calculating the standard deviation (SD), shown as a red horizontal line in graphs with sub-bandage pressure measurements.

As the moisture and the temperature measurements are independent of posture changes, different setups were used from those used for sub-bandage pressure measurement. For moisture measurements, a moisture-retentive foam dressing AllevynTM Adhesive (Smith & Nephew) was used. A small slit was made to insert micro-volume extension set tubing in one corner, and another slit was made to insert the moisture sensor in the opposite corner of the dressing. Fluid was then injected in the tubing every five minutes until the sensor was soaked and started providing moisture values. Subsequently, the dressing was placed upside down so that the fluid moved away from the sensor until it was completely depleted of moisture. Moisture results are plotted in Fig. 16.3 (a).

For temperature measurements, the sensing system was placed over the lower leg area with the temperature sensor facing the skin, and then various temperature readings were recorded continuously for more than 15 minutes. The experiment was performed at 23 °C room temperature, and with 40 % humidity. The temperature measurements are plotted in Fig. 16.3(b). The average value of the measured skin temperature was 35 ± 1 °C while the standard deviation was 0.93 °C for skin temperature measurements.

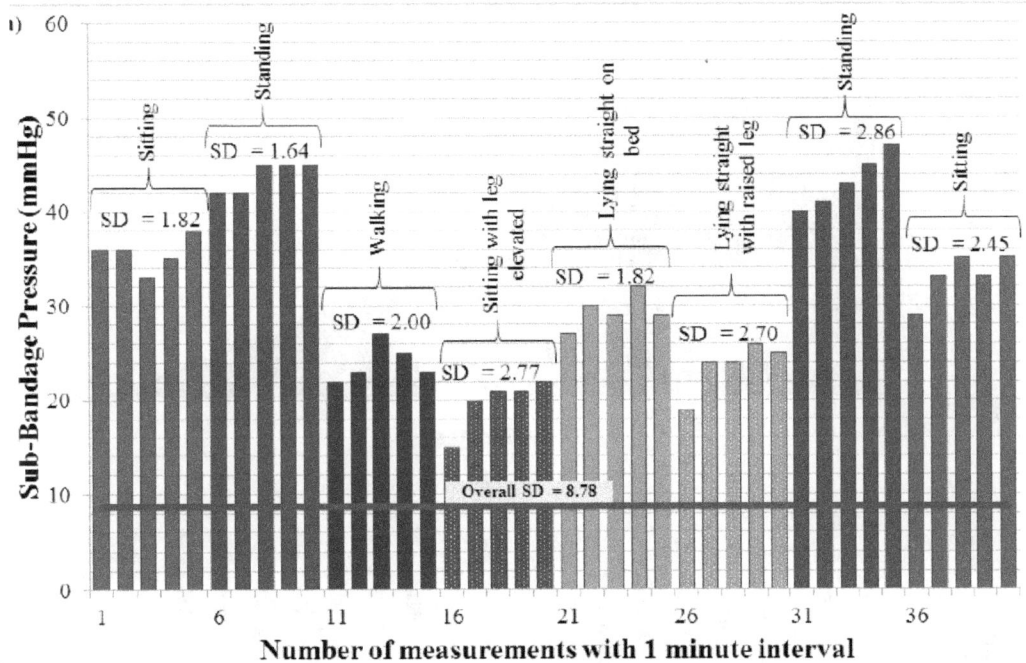

Fig. 16.2. Graphical plots of pressure measurements during the first trial using 4-layer compression bandage system. The measurements are recorded in various routine movements and postures.

The sub-bandage pressure measurement results (Fig. 16.2) for both bandages have shown a reliable and repeatable performance of the sensing system under required sub-bandage pressure. In each posture, five consecutive pressure readings were taken over a period of 5 minutes. For the 4-layer bandage system, the maximum deviation (SD value) observed was 2.86 mmHg during standing, while the same for the 2-layer bandage system was 3.35 mmHg during walking. The moisture level measurement results (Fig. 16.3 (a)) have also shown a reliable system performance. A steep rise in moisture level was observed as the injected fluid reached the proximity of the sensor. Next, a gradual and consistent drop in the measured moisture level was observed as the sensor was placed in an upward position to allow backward flow of the fluid. In about six hours, the moisture level dropped to zero when the sensor was completely depleted of the fluid.

Thus, the first trial concludes that the measurement of temperature, moisture, and sub-bandage pressure with the developed sensing system is reliable, repeatable and consistent with the changes in experimental conditions.

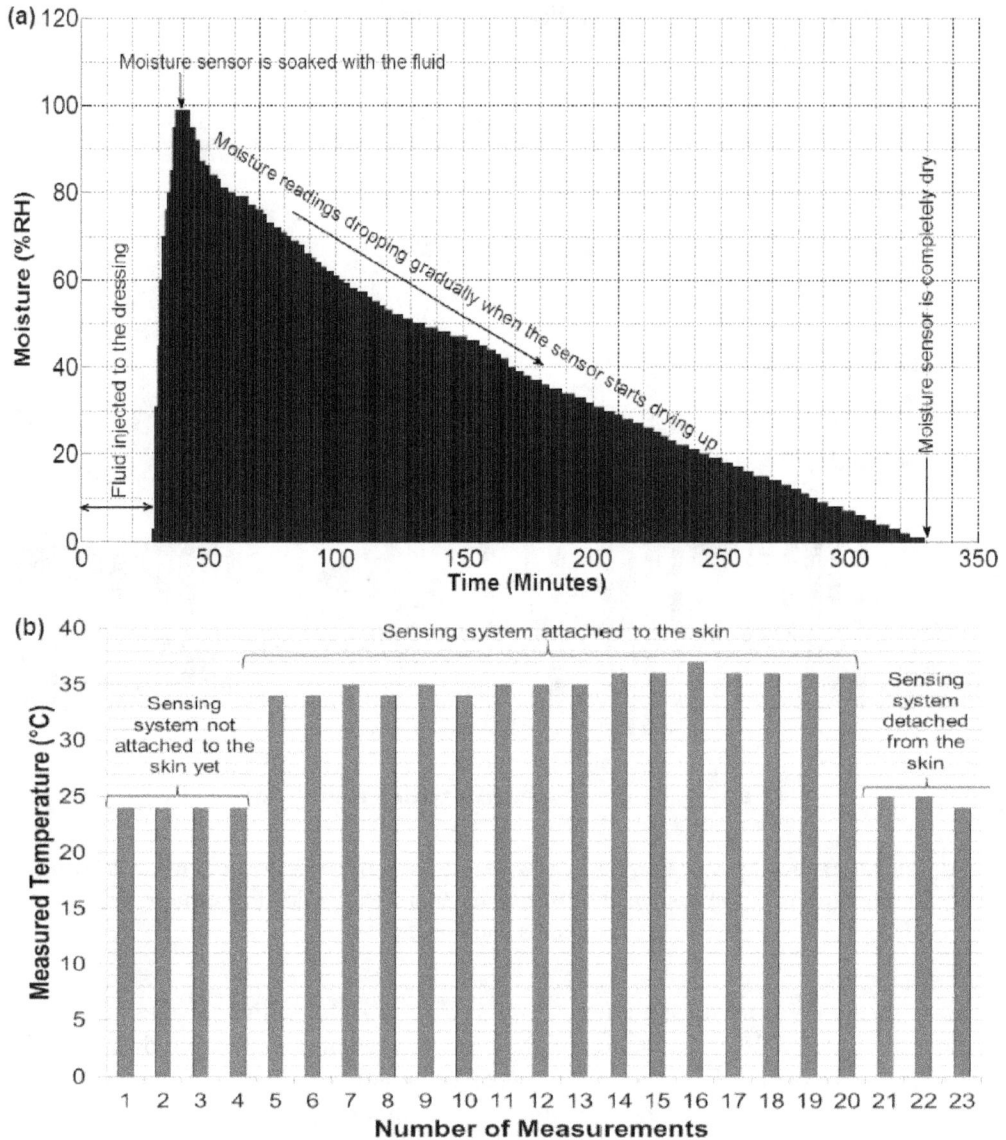

Fig. 16.3. (a) Graphical plot of moisture values measured using moisture-retentive dressing. Fluid was injected into the dressing through micro-volume extension tubing. The graph indicates a natural rise and fall of moisture level over time (b) Graph showing temperature measurements using the flexible sensing system. The graph shows almost constant readings of the room and the skin temperatures.

16.3.2. Measurements at Ankle Level with a 4-layer Bandage

In this trial, a 4-layer bandage system known to apply 40 mmHg approximately at ankle was used. Bandages were applied as per manufacturer's instructions. A small slit was made in the outermost cohesive bandage (layer 4) to allow battery connection. AllevynTM

Adhesive (Smith & Nephew) 12.5 cm × 12.5 cm dressing was applied to the lower calf and a small cut was made in the back of the dressing. A micro-volume extension set tubing was attached to the dressing and the slit sealed with film tape to allow injection of fluid (as real wound fluid was not available soy sauce diluted with water was used). The fluid was injected under the dressing to mimic wound exudate and soy sauce was chosen so that the spread of fluid was visible during experiments. The sensing system was then attached to the lower leg. The pressure sensor was placed proximal to the medial malleolus between exposed skin and bandages. The moisture sensor was placed at the bottom corner of the AllevynTM dressing between the exposed skin and the dressing. The results of second trial are plotted in Fig. 16.4.

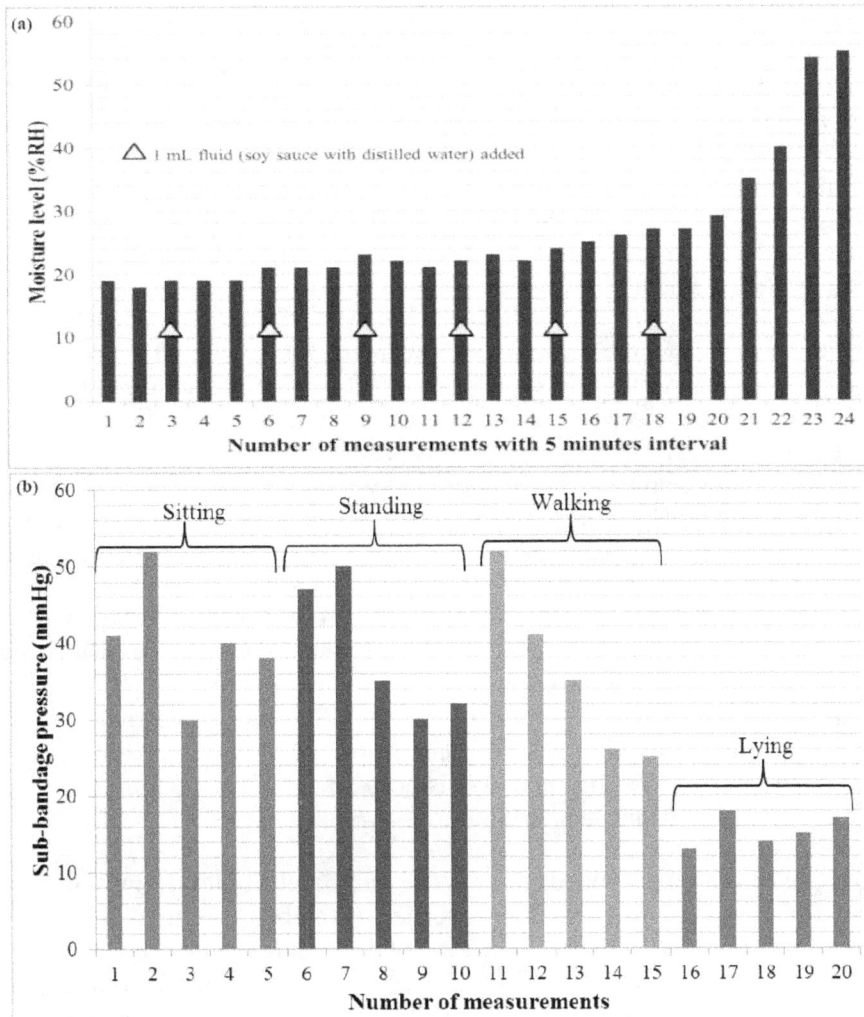

Fig. 16.4. Graphical plots of the measured values of (a) moisture for second trial, (b) sub-bandage pressure for second trial in various postures.

The average value of temperature during this trial was 33 ± 1 °C. The moisture values were increasing gradually as more fluid was injected to the dressing until the sensor was soaked with the fluid, the point from where the readings started rising up (readings 21-24 in Fig. 16.4 (a)). The sub-bandage pressure values (Fig. 16.4 (b)) were dependent on posture, being higher during walking and standing and lower during sitting or lying. It can also be observed from the graph that the pressure readings were consistently dropping over time. This phenomenon can be attributed to the loosening of bandage layers with movement.

16.4. Interpretation

In the first trial, the measurements for pressure, moisture, and temperature were taken separately using distinct experimental setups. The measurements were reliable and consistent with the applied conditions as can be observed in all the figures. The sub-bandage pressure values are expected to fluctuate with the movement, and the graphs in Figs. 16.2 and 16.4 have verified this phenomenon. The overall fluctuation of sub-bandage pressure values measured as standard deviation was 8.69 mmHg for the 4-layer and 7.27 mmHg for the 2-layer bandage system. The postures were changed in a cyclic fashion (i.e. starting from 'sitting' and ending in 'sitting') to isolate the source of fluctuations in measurements.

The average sub-bandage pressure values recorded during the first and last 'sitting' postures for the 4-layer were 35.6 mmHg and 33.0 mmHg respectively for the 4-layer bandage, and were 42.0 mmHg and 43.2 mmHg respectively for the 2-layer bandage. Similarly, the average sub-bandage pressure values during both 'standing' postures were recorded as 43.8 mmHg and 43.2 mmHg respectively for the 4-layer bandage, and were 45.8 mmHg and 44.0 mmHg respectively for the 2-layer bandage. The twin average values during respective postures for each bandage system were pretty close to each other.

These measurements proved that the major source of fluctuations was not the sensing system but the movement of the subject. The variations in measured moisture level over time confirmed that the measured values were consistent with the dry, wet and intermediate conditions of the moisture sensor. Similarly, the body temperature measurement results were also reliable and stable measuring an average value of 35 ± 1 °C. Hence, the reliability and consistency of readings performed on a healthy human volunteer were established during the first trial.

In all trials, variations in sub-bandage pressure can be attributed to muscle movement, blood flow direction, and the type and properties of the bandage used. The temperature and moisture measurements did not manifest any dependency on these factors. However, the moisture readings may be strongly affected by the combined effect of the gravity pull and the location of moisture sensor with respect to fluid entry point. In a clinical scenario, a consistent rise in measured temperature could mean infection at wound site. A gradual decrease in moisture level might indicate the start of healing and the opposite might mean the worsening of healing state. A consistently increasing sub-bandage pressure could be a sign of infection or excessive swelling in the limb, and a gradual decrease in sub-bandage pressure might mean loosening of the bandage layers.

Using this device, a clinician would be able to visualize the state of healing of a chronic wound remotely without disturbing the wound, a phenomenon which was never possible in chronic wound management before. However, the device needs further miniaturization and performance improvements, such as enhancing measurement resolutions, in order to have a significant impact on chronic wound monitoring. Although, the moisture and temperature sensors could be used for any wound, however, the pressure sensing capability is useful only for venous leg ulcers. The device works with an external battery which needs to be present at all times. The sensing system and sensors are re-usable on a single patient after proper sterilization. The cost incurred by the system will be compensated by reduced frequency of dressing changes, reduced nursing time, and reduced use of hospital resources [22].

16.5. Conclusions

Traditional tools and methods have proven insufficient to effectively monitor the chronic wound healing progress. Inaccuracies in chronic wound measurements are responsible for a significant loss in healthcare budgets e.g. a 30 % loss in the US healthcare budget [2]. Engaging an integrated measurement approach (i.e. measuring other critical parameters [12] in addition to wound dimensions) is believed to be more effective to provide stronger evidence of healing than using only one kind of measurement. Eventhough the wound area and volume calculations over time, and 3D surveillance techniques are useful for this purpose, the proposed sensing system would be a valuable addition to these approaches to mitigate the losses incurred by human errors in chronic wound measurements.

In this chapter, we have presented a novel wireless chronic monitoring system which is flexible, biocompatible, and reliable in performance. The sensing system has been fabricated on a flexible circuit material, enabling it to adapt to human limb contours. Low-power profile of the sensing device enables it to be operated continuously under a compression bandage over a longer time period (e.g. 3-4 days) without any disturbance to the bandaged wound. Battery life-time can be further enhanced (e.g. 2-3 weeks) by reducing the frequency of wound data transmission. An effective data transmission range of 4-5 m would enable clinicians and nurses to visualize the current wound state from a distance in a hospital environment.

Experimental results on a healthy human volunteer ascertained that the sensing system was capable of accurately measuring the instantaneous changes in sub-bandage pressure, moisture, and temperature under compression bandages and dressings. Gradual changes in measured wound parameters could reveal the status of the healing progress e.g. a consistent rise in sub-bandage pressure might be an indication of infection or swelling. However, the device needs to be tested on patients with chronic wounds to analyze its performance in real environment. The device needs to be tested further even after successful clinical trials to determine its clinical and financial impacts on chronic wound management. The suitability and efficacy of the sensors could be determined during these clinical trials. Nonetheless, the light weight, reliable performance on human limb, flexible and non-invasive structure, low-power consumption and wireless connectivity of the sensing system make it a strong candidate for use in continuous wound sensing and monitoring applications. Future works will incorporate miniaturization of the telemetric sensing system, and its communication with smart phones and display of information on screens using a smart phone application. The system would also incorporate WiFi or 3G communication technologies to share the measured wound parameters to the clinician remotely. This would enable clinicians to gather information on the state of a chronic wound while the patient stays at home.

Acknowledgments

We are thankful to the Wound Management Innovation Cooperative Research Centre (CRC) Australia for funding this work. The content of this chapter is based on research carried out by Nasir Mehmood for his PhD thesis under the author's supervision. We also acknowledge significant contributions from Nico Voelker of the Future Industries Institute, and from Sue Tempelton of the Royal District Nursing Service.

References

[1]. J. Cadogan, et al, Identification, diagnosis and treatment of wound infection, *Nursing Standard*, Vol. 26, 2011, pp. 44-8.
[2]. N. Mark, and M. Christine, Evidence-Based Wound Surveillance-A 3-Dimensional Approach To Measuring, Imaging and Documenting Wounds, *Aranz Medical*, 2014.
[3]. The Economic Cost of Wounds, 2014, Available: http://www.smith-nephew.com/about-us/what-we-do/advanced-wound-management/economic-cost-of-wounds/
[4]. Higher wound care costs are driving treatment research, *McKnight's Long-Term Care News*, 2012, http://www.mcknights.com/higher-wound-care-costs-are-driving-treatment-research/article/244578/
[5]. T. Abdelrahman, and H. Newton, Wound dressings: principles and practice, *Surgery (Oxford)*, Vol. 29, 2011, pp. 491-495.
[6]. H. Partsch et al., Classification of compression bandages: practical aspects, Dermatologic surgery, *Official Publication for American Society for Dermatologic Surgery*, Vol. 34, 2008, pp. 600-609.
[7]. A. Fletcher et al., A systematic review of compression treatment for venous leg ulcers, *British Medical Journal*, Vol. 315, 1997, p. 576.
[8]. A. Hopkins, How to apply effective multilayer compression bandaging, *Wound Essentials* (Wounds UK), Vol. 1, 2006.
[9]. B. Kumar et al., Analysis of sub-bandage pressure of compression bandages during exercise, *Journal of Tissue Viability*, Vol. 21, 2012, pp. 115-124.
[10]. L. G. Ovington, Advances in wound dressings, *Clinics in Dermatology*, Vol. 25, 2007, pp. 33-38.
[11]. G. Lagana, and E. H. Anderson, Moisture Dressings: The New Standard in Wound Care, *The Journal for Nurse Practitioners*, Vol. 6, 2010, pp. 366-370.
[12]. K. Harding, Diagnostics and Wounds, A consensus document, *World Union of Wound Healing Societies* (WUWHS), 2007.
[13]. T. R. Dargaville et al., Sensors and imaging for wound healing: A review, *Biosensors and Bioelectronics*, Vol. 41, 2013, pp. 30-42.

[14]. N. Mehmood et al., Applications of modern sensors and wireless technology in effective wound management, *Journal of Biomedical Materials Research Part B: Applied Biomaterials*, 2013, pp 1552-4973.

[15]. N. Mehmood et al., An Innovative Sensing Technology for Chronic Wound Monitoring, in *Proceedings of the Australian Biomedical Engineering Conference,* Canberra, Australia, 20-22 Aug. 2014.

[16]. N. Mehmood et al., Calibration of sensors for reliable radio telemetry in a flexible wound monitoring device, *Sensing and Bio-Sensing Research*, Vol. 2, 2014, pp. 23-30.

[17]. M. Ochoa et al., Flexible Sensors for Chronic Wound Management, *IEEE Reviews in Biomedical Engineering,* Vol. 7, 2014, pp. 73-86.

[18]. Online datasheet Texas Instruments temperature sensor LM94021, 2013, Available: http://www.ti.com/lit/ds/symlink/lm94021.pdf

[19]. Online datasheet of RF transceiver ATMega128RFA1, 2013, Available: http://www.atmel.com/devices/ATMEGA128RFA1.aspx

[20]. Online datasheet of DS-G160128STBWW LCD module, 2013, Available: https://www.sparkfun.com/datasheets/LCD/DS-G160128STBWW.pdf

[21]. Online datasheet of Toshiba T6963C dot matrix LCD controller, 2013, Available: http://www.mikroe.com/downloads/get/1910/t6963c_spec.pdf

[22]. A. Hariz, N. Mehmood, and N. Voelker, Sub-bandage sensing system for remote monitoring of chronic wounds in healthcare, *Proc. SPIE Micro+Nano Materials, Devices, and Systems,* 2015, Vol. 9668, p. 966853.

Chapter 17
Direction of Arrival Using Smart Antenna Array

Ihedrane Mohammed Amine, Bri Seddik

17.1. Introduction

Smart antenna has been widely used in many applications such as radar, sonar and wireless communication systems. Considerable research efforts have been made to estimate the direction of arrival (DOA) and various array signal process techniques for DOA estimation have been proposed. In particular, the DOA estimation for uniform circular arrays (UCAs) has been developed in these scenarios, which desired all-azimuth angle coverage. By the virtue of their geometry, UCAs are able to provide 360° of coverage in azimuth plane. Moreover, they are known to be is isotropic. That is, they can estimate the DOA of incident signal with uniform resolution in the azimuth plane. In addition, direction patterns synthesized with UCAs can be electronically rotated in the plane of the array without significant change of beam shape. As well, search and rescue always require the location of electromagnetic beacon sources. All these applications perhaps can be counted as the main reason for the recent increased interest in determining the direction of arrival (DOA) of radio signals in wireless systems. In fact, estimating the direction of arrival of several radio signals impinging on an array of sensors is required in a variety of other applications as well, including radar, sonar, and seismology. Another technology that has become equally glamorous is smart antenna technology [2, 3]. In smart antenna technology, a DOA estimation algorithm is usually incorporated to develop systems that provide accurate location information for wireless services [4].

A smart antenna, for this chapter discussion, is a system that combines multiple antenna elements with a signal processing capability to optimize its radiation and/or reception pattern automatically in response to the system's signal environment, and can be used to enhance the coverage through range extension and to increase system capacity [2, 3]. Smart antennas can also be used to spatially separate signals, allowing different subscribers to share the same spectral resources, provided that they are spatially separable at the base station. This spatial division multiple access (SDMA) method allows multiple users to operate in the same cell and on the same frequency/time slot provided by utilizing the adaptive beamforming techniques of the smart antennas. Since this approach allows more users to be supported within a limited spectrum allocation, compared with conventional antennas, SDMA can lead to improved capacity.

The smart antenna technology can be divided into three major categories depending on their choice in transmit strategy:

• *Switched lobe (SL):* This is the simplest technique and comprises only a basic switching function between predefined beams of an array. When a signal is received, the setting that gives the best performance, usually in terms of received power, is chosen for the system to operate with.

• *Dynamically with phased array (PA)*: This technique allows continuous tracking of signal sources by including a direction-of-arrival (DOA) finding algorithm in the system; as a result, the transmission from the array can be controlled intelligently based on the DOA information of the array.

• *Adaptive array (AA)*: In this case, a DOA algorithm for determining the directions of interference sources (e.g., other users) is also incorporated in addition to finding the DOA of the desired source. The beam pattern can then be adjusted to null out the interferers while maximizing the transmit power at the desired source. The importance of DOA estimation for smart antenna can be understood by studying the architecture of smart antenna as described in the following section.

17.2. Smart Antenna Architecture

Typical smart antenna architectures for a base station can be divided into following functional blocks, as presented in Figs. 17.1 and 17.2.

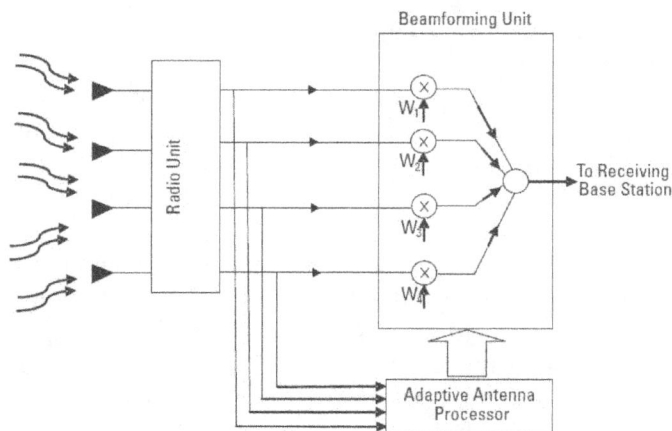

Fig. 17.1. Smart antenna receivers.

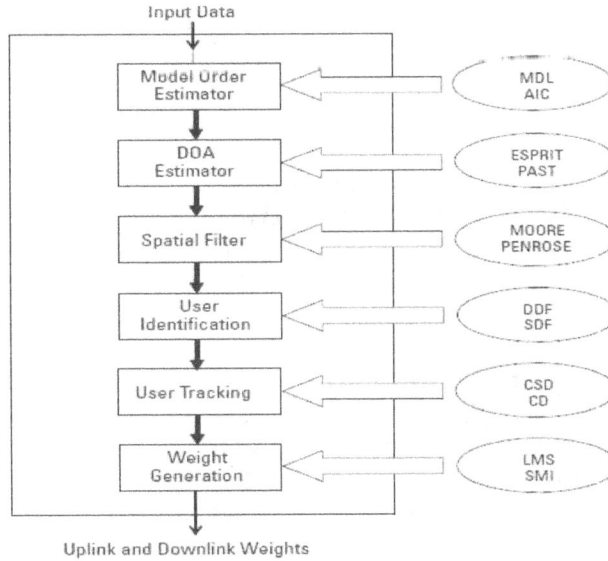

Fig. 17.2. Adaptative antenna processor.

• *Radio unit*: This unit mainly consists of: (1) antenna arrays that intercept radio frequency (RF) signals from the air, (2) downconversion chains that remove the carrier(s) of the RF signals received by the antenna array, and (3) analog-to-digital converters that convert the no-carrier signals to the corresponding digital signals for further processing. Antenna arrays can be one, two, or even three-dimensional, depending on the dimension of the space one wants to access. The radiation pattern of the array depends on the element type, the relative positions, and the excitation (amplitude and phase) to each element [5].

• *Beamforming* unit: The beamforming unit is responsible for forming and steering the beam in the desired direction. In it, the weighting of the received (or transmitted) signals is applied. Basically, the data signals X_k, k = 0, …, M − 1 received by an N element array are directly multiplied by a set of weights to form a beam at a desired angle. In other words, by multiplying the data signals with appropriate sets of weights, it is possible to forming in a signal peak at the output of a beamformer.

• *Adaptive antenna processor*: The function of the adaptive processor unit is to determine the complex weights for the beamforming unit. The weights can be optimized from two main types of criteria: maximization of the data signal from the desired source (e.g., switched lobe or phased array) or maximization of the signal-to-interference ratio (SIR) by suppressing the signal from the interference sources (adaptive array). In theory with N antenna elements, one can "null out" M − 1 interference sources, but due to multipath propagation, this number will be normally lower. The method for calculating the weights will differ depending on the types of optimization criteria. When a switched lobe is used, the receiver will test all the predefined weights and choose the one that gives the strongest received data signal. However, if the phased array or adaptive array approach is used, which consists of directing a maximum gain towards the strongest signal component, the directions of arrival (DOAs) of the signals are first estimated and the weights are then calculated in accordance with the desired steering angle.

The adaptive antenna processor consists of several computation processes:

• *Model order estimator*: From the input data X_k, k = 0, ..., $M - 1$ received by the antenna elements, the number of wavefronts impinging on the array is estimated using model order estimation algorithms, such as *AIC* or *MDL*. The knowledge of number of the signals impinging on the array is crucial to DOA estimation algorithms; hence, these algorithms are run prior to DOA estimation algorithms.

• *DOA estimator*: This forms the vital stage of the adaptive antenna processor where algorithms like MUSIC or ESPRIT are used for estimating the direction of arrival of all the signals impinging on the array. This stage gives DOAs of all the relevant signals of the user sources and other interference sources. To make the process faster, instead of estimating the signal space every time, subspace tracking algorithms like dPAST and PAST (Projection Approximation Subspace Tracking) are used to recursively track the signal subspace. Usually, the signal subspace is only slowly time-varying. It is therefore more efficient to track those changes than to perform full subspace estimation. The DOAs can then be estimated faster from these signal subspaces.

• *Spatial filter:* After the DOAs of all the signals impinging on the array are obtained, the signals are filtered by reconstructing the signals for each of the DOAs estimated. Estimating the signals from the estimated DOAs is usually called signal reconstruction or signal copy. With the knowledge of DOAs, the corresponding steering vectors a and eventually the estimated steering matrix A are constructed.

• *User identification:* Once the signals are separated with respect to their distinct DOAs, the desired user corresponding to these DOAs needs to be identified. By comparing the received midambles (training sequences) with the desired user mid-amble, the number of bit errors within the training sequence can be calculated. A spatially resolved wavefront and thus the corresponding DOA are attributed to a user, when the number of bit errors is smaller than a threshold. In this way not only a single user path but also all paths that correspond to the intended user can be identified. The DOA of the user path with the strongest instantaneous power is then detected. As a training sequence detector, standard sequence estimators like delayed decision feedback (DDF) and soft decision feedback (SDF) are applied [6].

17.3. DOA Algorithms Using Uniform Linear Array

17.3.1. Mathematical Model for MUSIC Algorithm

Multiple Signal Classification (MUSIC) method [7] is widely used in signal processing applications for DOA [8]. In estimation, it is applied to only narrow band signal sources, i.e., frequencies of interest are narrowband [9]. Consider M number of narrow band signal sources arriving from different angles $\theta_i = 1, 2...M$, impinging on a uniform linear array of N equispaced array elements (where $N > M$), as presented in Fig. 17.3.

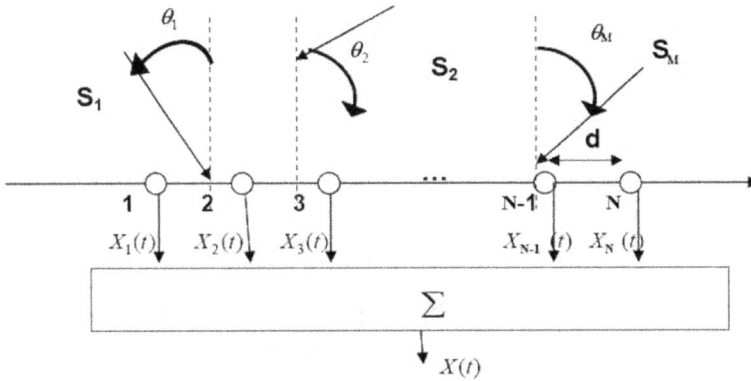

Fig. 17.3. N linear element array with M signals.

At different instances of time t, t = 1, 2 ... K, where K is the number of snapshots, the array output will consist of a signal, along with noise components [7].

We choose a signal source S(t) impinging on the array with an angle θ. If the received signal at the first element is x1(t) = s(t), then the delay at element i is:

$$\Delta i = \frac{(i-1)d \sin \theta}{c} \tag{17.1}$$

The received signal at sensor *i* is:

$$xi(t) = e^{-j\omega\Delta i}S1(t) = e^{-j\omega\Delta i} S1(t) = e^{-\frac{j\omega(i-1)d \sin\theta}{c}}S1(t) \tag{17.2}$$

The received signal at N elements due to a single source is:

$$X(t) = \left[1, e^{-\frac{j\omega d \sin\theta}{c}}, e^{-\frac{j\omega 2d \sin\theta}{c}}, ..., e^{-\frac{j\omega(N-1)d \sin\theta}{c}}\right]S(t) = a(\theta)S(t) \tag{17.3}$$

If there are M sources, the signals received at the array are given by

$$X = AS + W \tag{17.4}$$

$$A = [a(\theta 1), a(\theta 2), ... a(\theta M)] \tag{17.5}$$

$$S = [s1(t), s2(t), ... sM(t)]^T \tag{17.6}$$

The correlation matrix of received vectors can be written as:

$$R = E[XX^H] = E[ASS^H A^H] + E[WW^H] = AVA^H + \sigma^2, \tag{17.7}$$

where V is the covariance matrix of signal vector (S), which is a full rank matrix of order M×M, given by

333

$$V = E[SS^H] = \begin{bmatrix} E[|S_1|^2] & \cdots & 0 \\ 0 & E[|S_2|^2] & 0 \\ 0 & 0 & E[|S_M|^2] \end{bmatrix} \quad (17.8)$$

Where the statistical expectation is denoted by E [ ], and R_S is a signal covariance matrix of order (N×N), with rank M given by

$$R_S = \begin{bmatrix} E[|S_1|^2] & \cdots & \cdots & 0 & \cdots & 0 \\ 0 & E[|S_2|^2] & \cdots & 0 & \cdots & 0 \\ \vdots & \ddots & \cdots & \vdots & \ddots & 0 \\ 0 & 0 & \cdots & E[|S_M|^2] & \ddots & 0 \\ 0 & 0 & \cdots & 0 & \cdots & 0 \end{bmatrix} \quad (17.9)$$

So, R_S, has N-M eigenvectors, corresponding to zero eigenvalues. We know that steering vector a(θ1), which is in the signal subspace, is orthogonal to noise subspace; let Q_n be such an eigenvector.

$$R_S Q_n = AVA^H Q_n = 0 \quad (17.10)$$

$$Q_n{}^H AVA^H Q_n = 0$$

Since V is a positive definite matrix

$$AVA^H Q_n = 0 \quad (17.11)$$

$$a^H(\theta_i) Q_n = 0 \quad (17.12)$$

This implies that signal steering vectors are orthogonal to eigenvectors corresponding to noise subspace. So, the MUSIC algorithm searches through all angles, and plots the spatial spectrum.

$$P_{MUSIC}(\theta) = \frac{1}{a^H(\theta_i) Q_n} \quad (17.13)$$

Assume the number of signals, M, is known. Given the data set X (k), k = 1, 2. . . K, the MUSIC algorithm proceeds as per the following steps shown in Fig. 17.4.

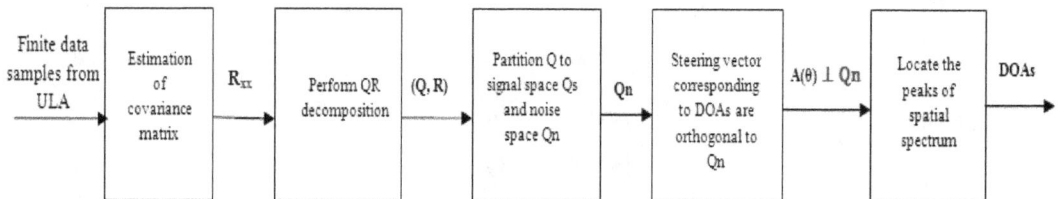

Fig. 17.4. Steps for MUSIC algorithm.

334

17.3.2. Mathematical Model for ESPRIT Algorithm

ESPRIT's acronym stands for Estimation of Signal Parameter via Rotational Invariance Technique. This algorithm is more robust with respect to array imperfections than MUSIC [12, 13]. Computation complexity and storage requirements are lower than MUSIC, as it does not involve extensive searching throughout all possible steering vectors. However, it explores the rotational invariance property in the signal subspace created by two subarrays derived from the original array with a translation invariance structure [14].

It is based on the array elements placed in identical displacement forming matched pairs, with N array elements, resulting in m = N/2 array pairs called "doublets" [12] presented in Fig. 17.5.

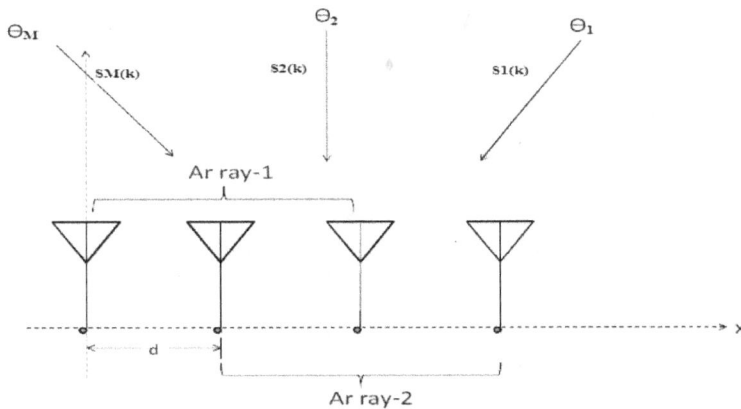

Fig. 17.5. Uniform linear antenna arrays with M incident signals.

Computation of signal subspace for the two subarrays, Sub_array-1 and Sub_array-2, are displaced by distance d. The signals induced on each of the arrays are given by

$$x_1(k) = A_1 * s(k) + n_1(k) \tag{17.14}$$

$$x_2(k) = A_1 * \Lambda * s(k) + n_2(k), \tag{17.15}$$

where $\Lambda = \text{diag}[e^{jkd\sin(\theta_1)}, e^{jkd\sin(\theta_2)}, ..., e^{jkd\sin(\theta_D)}]$ (D×D) diagonal unitary matrix with phase shifts between doublets for DOA. Creating the signal subspace for the two subarrays results in two matrices, V1 and V2. Since the arrays are translationally related, the subspaces of the eigenvectors are related by a unique non-singular transformation matrixφ, such that [6]

$$V_1 \varphi = V_2 \tag{17.16}$$

There must also exist a unique non-singular transformation matrix T, such as

$$V_1 = AT \text{ and } V_2 = A\Lambda T \tag{17.17}$$

And, finally, we can derive

$$T\varphi T^{-1} = \Lambda \tag{17.18}$$

Thus, the eigenvalues of φ must be equal to the diagonal elements of Λ, such that

$\lambda 1 = e^{jkdsin(\theta_1)}, \lambda 2 = e^{jkdsin(\theta_2)}, \ldots\ldots, \lambda M = e^{jkdsin(\theta_M)}$, once the eigenvalues of $\varphi, \lambda 1, \lambda 2 \ldots \lambda_M$ are calculated, we can estimate the angles of arrivals as

$$\theta_i = \sin^{-1}\left(\frac{\arg(\lambda_i)}{kd}\right). \tag{17.19}$$

Clearly the ESPRIT eliminates the search procedure and produces the DOA estimation directly in terms of the eigenvalues without many computational and storage requirements. This Eigen structure method has an excellent accuracy and resolution in many experimental and theoretical studies.

17.3.3. Mathematical Model for Maximum-Likelihood Algorithm

This method depends on spatial spectrum [15]. DOAs are obtained as locations of peaks in the spectrum. The concept of localisation is simple, but offers modest or poor performance in terms of resolution [16]. One of the main advantages of these techniques is that it can be used in situations where we lack information about properties of the signal [17]. The estimate is derived by finding the steering vector A, which minimizes the beam energy AVA^H subject to the constraint $EA^H = 1$.

$$F = AVA^H + \alpha(EA^H - 1) \tag{17.20}$$

When the gradients of A and A^H are evaluated, they are found to be complex conjugates of each other. Setting one of them to zero results in the solution

$$A = -\alpha V^{-1}/2 \tag{17.21}$$

The quantity α is determined from the constraint $EA^H = 1$. Hence

$$A = R^{-1}E(E^H V^{-1}E)^{-1} \tag{17.22}$$

Thus, the power spectrum in the beam is given by

$$P(\theta) = AVA^H = (E^H V^{-1}E)^{-1} \tag{17.23}$$

As expected, the peaks of $P(\theta)$ correspond to the direction of arrival of the given signal. Hence, the following algorithm steps:
- Collect the data samples X;
- Estimate the correlation matrix R;
- Estimate the number of signals;
- Evaluate P (θ).

17.3.4. Matrix Pencil Method

The Matrix Pencil formulations use the real Matrices to estimate the DOA of multiple signals simultaneously impinging on the ULA [18, 19]. The vector x (n) is the set of voltages measured at the feed point of antenna element of the ULA. Therefore, x (t) can be modeled by a sum of complex exponentials. The observed voltage is given by:

$$y(t) = x(t) + n(t) = \sum_{i=1}^{N} R_i e^{s_i t} + n(t) \qquad (17.24)$$

where y (t) = observed voltages at a specific instance t;
n(t) = noise associated with the observation;
x(t) = actual noise free signal.

Therefore, one can write the sampled signal as,

$$y(P) = \sum_{i=1}^{N} R_i z_i^P + n(p), \text{ for } p = 0, 1, \dots, N - 1, \qquad (17.25)$$

where

$$Z_i = e^{j\frac{2\pi}{\lambda} d \sin(\theta)}, \text{ for } i = 1, 2, \dots, N \qquad (17.26)$$

In this chapter, it has been assumed that the damping factor α_i is equal to zero. The objective is to find the best estimation for θ. Let us consider the matrix Y which is obtained directly from x (p). Y is a Hankel matrix, and each column of Y is a windowed part of original data vector, {x (0) x (1) x (2) … x (N -1)}.

$$\begin{bmatrix} x(0) & x(1) & \dots & x(L-1) \\ x(1) & x(2) & \dots & x(L) \\ \vdots & \vdots & \ddots & \vdots \\ x(N-L) & x(N-L+1) & \dots & x(N-L) \end{bmatrix}_{(N-L+1) \times (L)} \qquad (17.27)$$

The parameter L is called the pencil parameter. L is chosen between N/3 and N/2 for efficient noise filtering [12, 13]. The variance of the estimated values of R_i and Z_i will be minimal if the values of L are chosen in this range [14]. From the matrix Y, we can define two sub-matrixes, say

$$Y_a = \begin{bmatrix} x(0) & x(1) & \dots & x(L-1) \\ x(1) & x(2) & \dots & x(L) \\ \vdots & \vdots & \ddots & \vdots \\ x(N-L-1) & x(N-L) & \dots & x(N-2) \end{bmatrix}_{(N-L) \times (L)} \qquad (17.28)$$

$$Y_b = \begin{bmatrix} x(1) & x(1) & \dots & x(L-1) \\ x(2) & x(2) & \dots & x(L) \\ \vdots & \vdots & \ddots & \vdots \\ x(N-L) & x(N-L+1) & \dots & x(N-1) \end{bmatrix}_{(N-L) \times (L)} \qquad (17.29)$$

We can also write

$$Y_a = Z_a R\, Z_b \tag{17.30}$$

$$Y_b = Z_a\, R_0\, Z_0 Z_b \tag{17.31}$$

$$Z_a = \begin{bmatrix} 1 & 1 & \cdots & 1 \\ Z_1 & Z_2 & \cdots & Z_M \\ \vdots & \vdots & \ddots & \vdots \\ Z_1^{(N-L-1)} & Z_2^{(N-L-1)} & \cdots & Z_M^{(N-L-1)} \end{bmatrix} \tag{17.32}$$

$$(N-L) \times (L)$$

$$Z_b = \begin{bmatrix} 1 & Z_1 & \cdots & Z_1^{(L-1)} \\ 1 & Z_2 & \cdots & Z_2^{(L-1)} \\ \vdots & \vdots & \ddots & \vdots \\ 1 & Z_M & \cdots & Z_M^{(L-1)} \end{bmatrix} \tag{17.33}$$

$$M \times L$$

$$Z_0 = \mathrm{diag}\,[Z_1, Z_2, \ldots, Z_M] \tag{17.34}$$

$$R_0 = \mathrm{diag}\,[R_1, R_2, \ldots, R_M] \tag{17.35}$$

Now, let us consider the Matrix Pencil

$$Y_b - \lambda Y_a = Z_a\text{-}R_0\,[Z_0 - \lambda I]Z_b, \tag{17.36}$$

where I is the (M×M) identity matrix, one can resolve that the rank of $Y_b - \lambda Y_a$ will be M, provided that M≤ L≤ N-M [15-16-17]. However, if $\lambda = Z_i$, i=1, 2, …, M the ith row of $[Z_0 - \lambda I]$ is zero, then the rank of this matrix is (M×1). Therefore, the parameters Z_i can be found as the generalized eigenvalues of the matrix pair $\{Y_a^{+}Y_b - \lambda I\}$ where Y_a+ is the Moore-Penrose pseudo inverse of Y_a, which i s defined as

$$Y_a^{+} = \{Y_a^{H}\, Y_a\}^{-1}\, Y_a^{H} \tag{17.37}$$

The DOA is obtained from

$$\theta_i = \sin^{-1}\left(\frac{\mathrm{Im}\,(\log Z_i)}{\pi d}\right), \tag{17.38}$$

where Z_i is defined in eq. (26)

17.4. DOA Algorithms Using Uniform Circular Array

17.4.1. System Model and 2-d Music Algorithm

We assume that there are N uniform circular array, M narrow band far field signals from different incident direction [10, 11]. The radius of the circular array is denoted as r and wavelength of narrow band is denoted as λ. The incident angle of the signals is illustrate in Fig. 17.6.

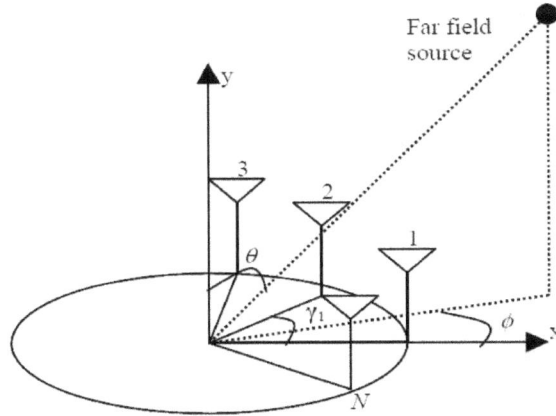

Fig. 17.6. Uniform circular array geometry.

The received array signal can

$$X(t) = AS(t) + N(t), \tag{17.39}$$

where $A=[a\,(\theta_1,\varphi_1),\ldots,a\,(\theta_M,\varphi_M)]$ is a matrix of the M steering vectors, which represents the possible value set of DOA and

$$a(\theta_1,\varphi_1)\left[e^{j2\pi\frac{r}{\lambda}\sin\Phi_m\cos\theta_m},\ldots,e^{j2\pi\frac{r}{\lambda}\sin\Phi_m\cos(\theta_m-\frac{2\pi(N-1)}{N})}\right]$$

and $S=[S1(t),\ldots,SM(t)]^T$ is a signal source vector of size ($M\times1$).

The correlation matrix of received vectors can be written as

$$R = E[XX^H] \;=\; E[ASS^H A^H] + E[WW^H] = \; AVA^H + \sigma^2, \tag{17.40}$$

where V is the covariance matrix of signal vector (S), which is a full rank matrix of order $M\times M$, given by

$$V = E[SS^H] \;=\; \begin{bmatrix} E[|S_1|^2] & \cdots & 0 \\ 0 & E[|S_2|^2] & 0 \\ 0 & 0 & E[|S_M|^2] \end{bmatrix}, \tag{17.41}$$

where the statistical expectation is denoted by E [ ], and R_S is a signal covariance matrix of order ($N\times N$), with rank M given by

$$R_s \;=\; \begin{bmatrix} E[|S_1|^2] & \cdots & \cdots & 0 & \cdots & 0 \\ 0 & E[|S_2|^2] & \cdots & 0 & \cdots & 0 \\ \vdots & \ddots & \cdots & \vdots & \ddots & 0 \\ 0 & 0 & \cdots & E[|S_M|^2] & \ddots & 0 \\ 0 & 0 & \cdots & 0 & \cdots & 0 \end{bmatrix} \tag{17.42}$$

339

So R_S, has N-M eigenvectors corresponding to zero eigen values. We know that steering vector $a(\theta_1, \varphi_1)$, which is in the signal subspace is orthogonal to noise subspace let Q_n be such an eigenvector.

$$R_S Q_n = AVA^H Q_n = 0 \tag{17.43}$$

Since V is a positive definite matrix:

$$a^H(\theta_i, \varphi_i)Q_n = 0 \tag{17.44}$$

This implies that signal steering vectors are orthogonal to eigen vector corresponding to noise subspace. So the MUSIC algorithm searches through all angles and plots the spatial spectrum:

$$P_{MUSIC}(\theta, \varphi) = \frac{1}{(a^H(\theta,\varphi)Q_n Q_n{}^H a(\theta,\varphi))} \tag{17.45}$$

17.4.2. The Proposed Algorithm

In the modified algorithm, we will reconstruct the data matrix:

$$Y = T X^* \tag{17.46}$$

The '*' represents complex conjugate, T is an N order inverse identity matrix, which is called transition matrix. The covariance matrix of the data Y is

$$R_Y = T R_X{}^* T \tag{17.47}$$

We introduce a new array covariance matrix, which is the sum of R_Y and R_x

$$R = R_Y + R_X = AR_sA + T[AR_sA]^* T + 2\sigma^2 I \tag{17.48}$$

According to matrix theory, if q is an eigenvector corresponding to a zero eigen value of matrix AR_SA, then q must also be an eigenvector correspond to the zero eigen value of matrix $T[AR_SA]*T$. We observe that matrix R_x, R_Y and R have the same noise subspace. By performing eigen value decomposition with R, we get its eigen values and its eigen vectors. According to the estimated number of signal sources, the noise subspace among the eigen vectors can be distinguished. With the new noise subspace, we can construct MUSIC spatial spectrum:

$$P_{MUSIC}(\theta, \varphi) = \frac{1}{(A(\theta,\varphi)^H q_n q_n{}^H A(\theta,\varphi))} \tag{17.49}$$

17.4.3. 2D- Matrix Pencil

We consider a circular array of isotropic antennas evenly spaced dx and dy respectively along the axes O_X and O_Y. Where we assume dx = dy = d. The network receives signals Ms with the angles of incidenc (θ_q, φ_q), which are respectively φq, θq and the directions of arrival in elevation and in azimuth. The information on the arrival direction is contained in the eigenvalues of the two transformation matrices that bind respectively subnets1 and 2 in the X direction and subnets 3 & 4 in the Y-direction [20, 21]. The values of αx and αy are written in the following form:

$$\alpha_{xi} = \exp\left(j \left(\frac{2\pi \Delta}{\lambda_0} \right) \sin\theta i \cos\Phi i \right) \tag{17.50}$$

$$\alpha_{yi} = \exp\left(j \left(\frac{2\pi \Delta}{\lambda_0} \right) \sin\theta i \sin\Phi i \right) \tag{17.51}$$

The elevation and azimuth are expressed by the following equation:

$$\theta i = \text{Arcsin} \left[\left(\frac{-j\lambda_0}{2\pi\Delta} \right) \sqrt{(\text{Ln } \alpha_{xi})^2 + (\text{Ln } \alpha_{yi})^2} \right] \tag{17.52}$$

$$\Phi i = \text{Arctg} \left(\frac{\text{Ln}\alpha_{xi}}{\text{Ln } \alpha_{yi}} \right) \tag{17.53}$$

where i= 1, 2, ... , Ms

17.5. Results and Discussion

A comparative study [22-24] has been made between MUSIC, ESPRIT and MLE algorithms for DOA estimation, using the MATLAB software tool. We analyzed the performance of these algorithms by varying a number of parameters relating to antenna arrays, such as the number of array elements N, spacing between the array elements d, and the number of snapshots taken at any time. In this simulation, we have considered the M number of stationary signal sources impinging on the number of uniform linear array elements, which are equispaced with a separation of $\lambda/2$. We also considered the randomly generated symbols for each of the signal sources with equal magnitudes. The noise is assumed to be additive white Gaussian having unit variance. Simulations have been done for three signals arriving from different angles $\theta_1 = -30°$, $\theta_2 = 30°$, and $\theta_3 = 60°$, and our algorithm spatially searched through angles from $-90°$ to $90°$.

The simulation results of MUSIC, ESPRIT, and MLE algorithms on 3 signals coming from different angles (-30, 30, 60), indicate in Tables 17.1 – 17.6, we resolve clearly that, if array size increases from 4 to 12 elements, the peak spectrum becomes sharp. The resolution capacity increases also if the number of snapshots increases (from 128 to 1024). The 3 signals are clearly identified, we observe also that if the spacing between the antenna array changes from 0.25 λ to 0.75 λ, we get better resolution of estimated peaks, but we also observe some peaks in the case of d = 0.75 λ due to grating lobes.

A comparison can be made between the 3 methods in terms of errors; the tables indicate that MUSIC presents less errors then ESPRIT and MLE. The tables illustrate that, for different numbers of array, values of snapshot, and distances between the elements of array, MUSIC presents a maximal error of 11 % and a minimal error of 0.16 %, compared with ESPRIT with a maximal and minimal error of 33.3 and 0.33 %, respectively, and MLE with a maximal error of 29.66 % and a minimal error of 0.33 %. We have proved that the MUSIC algorithm provided great resolution and accuracy. In previous studies

[22], the authors indicate that the spectrum does not contain side lobes, but they omitted that if the distance between elements of antenna exceed 0.6 λ the spectrum contain side lobes. The computation complexity and storage requirements for ESPRIT are lower than MUSIC, as it does not involve extensive searching throughout all possible steering, as was presented in other work [23].

Table 17.1. DOA estimation (k = 1024, d = 0.5λ).

N	$\theta_{in}(°)$	$\theta_{MUSIC}°$	Δ_{MUSIC}	$\theta_{ESPRIT}°$	Δ_{ESPRIT}	$\theta_{MLE}°$	Δ_{MLE}
	-30	-30.2	-0.2	-29.8	+0.2	-29.5	+0.5
4	30	30.2	0.2	29.6	-0.4	31.6	+1.6
	60	59.9	-0.1	59.4	-0.8	55.4	-4.6
	-30	-30.1	-0.1	-29.9	+0.1	-29.9	+0.1
6	30	30	0	30.1	0.1	30	0
	60	59.8	-0.2	60.1	-0.2	60.4	+0.4
	-30	-29.9	+0.1	-29.8	+0.2	-30.1	-0.1
8	30	30	0	30.2	0.2	29.9	0.1
	60	60	0	60	0	60	0
	-30	-30	0	-30	0	-30	0
10	30	30	0	29.8	-0.2	30.2	+0.2
	60	60	0	59.9	-0.1	60	0
	-30	-30	0	-29.8	+0.2	-30	0
12	30	30	0	30.1	0.1	29.9	-0.1
	60	60	0	59.9	-0.1	60	0

Table 17.2. DOA estimation (k = 128, d = 0.5λ).

N	$\theta_{in}(°)$	$\theta_{MUSIC}°$	Δ_{MUSIC}	$\theta_{ESPRIT}°$	Δ_{ESPRIT}	$\theta_{MLE}°$	Δ_{MLE}
	-30	-30.6	-0.6	-30.8	-0.8	-30.4	-0.4
4	30	29.4	-0.6	31.5	+1.5	32.6	+2.6
	60	58.6	-1.4	55.1	-5.9	56	-4
	-30	-29.9	+0.1	-30.9	-0.9	-30.1	-0.1
6	30	30	0	28.8	-1.2	30.1	+0.1
	60	59.9	-0.1	61	+1	60.1	+0.1
	-30	-30	0	-30.1	-0.1	-30.2	-0.2
8	30	30.2	0.2	30.1	+0.1	30	0
	60	59.6	-0.4	59.7	-0.3	60.1	0.1
	-30	-30	0	-29.9	+0.1	-29.9	+0.1
10	30	30.1	0.1	30	0	30.1	+0.1
	60	59.9	-0.1	60.5	+0.5	60	0
	-30	-30	0	-30.2	-0.2	-30.1	-0.1
12	30	30	0	30	0	30	0
	60	59.9	-0.1	60.6	+0.6	59.9	-0.1

Table 17.3. DOA estimation (k = 1024, d = 0.25λ).

N	$\theta_{in}(°)$	$\theta_{MUSIC}°$	Δ_{MUSIC}	$\theta_{ESPRIT}°$	Δ_{ESPRIT}	$\theta_{MLE}°$	Δ_{MLE}
	-30	-30.6	-0.6	-31.9	-1.9	-29.5	+0.5
4	30	31	1	40.2	+10.2	32.6	+2.6
	60	64.4	+4.4	55.1	-5.1	42.6	-12.6
	-30	-29.8	+0.2	-29.6	+0.4	-30.3	-0.3
6	30	30	0	30.4	+1.4	30.1	+0.1
	60	60.2	+0.2	61	+1	42.6	-12.6
	-30	-30.1	-0.1	-30.2	-0.2	-30.1	-0.1
8	30	30	0	30.1	+0.1	31.3	1.3
	60	59.9	-0.1	60.1	+0.1	56.7	-4.3
	-30	-30	0	-29.9	+0.1	-30	0
10	30	30	0	29.9	-0.1	30.4	+0.4
	60	59.9	-0.1	59.8	-0.2	59.4	-0.6
	-30	-30	0	-30	0	-30.2	-0.2
12	30	30	0	30.1	0.1	29.9	-0.1
	60	59.9	-0.1	60.5	+0.5	60.2	+0.2

Table 17.4. DOA estimation (k = 128, d = 0.25λ).

N	$\theta_{in}(°)$	$\theta_{MUSIC}°$	Δ_{MUSIC}	$\theta_{ESPRIT}°$	Δ_{ESPRIT}	$\theta_{MLE}°$	Δ_{MLE}
	-30	-29.9	+0.1	-28.9	+1.1	-31.6	-1.6
4	30	33.7	3.3	26.5	-4.5	22.6	-7.4
	60	60.4	+0.4	45.1	-15.9	44.5	-16.5
	-30	-30.4	-0.4	-31.9	-1.9	-30.2	-0.2
6	30	29.9	-0.1	32.8	2.8	30.1	+0.1
	60	59.1	-0.9	57.5	-3.5	42.2	-17.8
	-30	-29.9	0.1	-29.5	+0.5	-28.6	+1.4
8	30	30.1	0.1	30.7	+0.7	30.4	+0.4
	60	59.9	-0.1	61	+.1	60.1	+0.1
	-30	-29.7	0.3	-30.5	-0.5	-29.9	+0.1
10	30	27.7	-2.3	30.3	-30.3	30.6	+0.6
	60	59.9	-0.1	60.5	+0.5	58.9	-1.1
	-30	-30	0	-30.8	-0.8	-29.7	+0.3
12	30	30	0	29.3	0.7	30.1	+0.1
	60	59.9	-0.1	59.2	-0.8	60.1	+0.1

The comparisons between MUSIC method investigate in this work and the proposed method in [24] indicates that the performance of MLE degrades by changing the parameters: number of antenna, samples and distance between elements, moreover the results of the MLE and MUSIC indicate that the MLE algorithm presents more errors at the level of angles compared to the MUSIC result.

To validate our studies, a comparison was made with experimental results. Table 17.7 illustrates the comparison between the MUSIC, ESPRIT, and MLE methods. The authors in [25] indicate that, to decrease the SNR and avoid undesirable lobes, the number of antennas should be increased and the distance between the elements should be limited to $\lambda/2$. However, they have omitted two essential factors: the signal power and the spectrum of angles that are linked to the number of snapshots. If we increase them, the power increases and the peaks become sharp. The angles (-2 3) are properly determined with precision and with an important magnitude = 49.9 dB compared to [26], as illustrate in Fig. 17.7.

Table 17.5. DOA estimation (k = 1024, d = 0.75λ).

N	$\theta_{in}(°)$	$\theta_{MUSIC}(°)$	Δ_{MUSIC}	$\theta_{ESPRIT}(°)$	Δ_{ESPRIT}	$\theta_{MLE}(°)$	Δ_{MLE}
	-30	-29.8	+0.2	-26.5	+3.5	-28.8	+1.2
4	30	30	0	30.1	+0.1	30	0
	60	56.7	-3.3	62.6	+2.6	58.4	-1.6
	-30	-28.4	-1.6	-29.9	+0.1	-29	1
6	30	30	0	30.1	2.8	29.9	-0.1
	60	59.1	-0.9	56.7	-3.5	57.7	-2.3
	-30	-29.9	+0.1	-27.8	+0.5	-29	+1
8	30	30	0	29.9	+0.7	30.1	+0.1
	60	56.6	-3.4	60.1	+0.1	58.2	-1.8
	-30	-29.8	0.2	-27.7	+2.5	-28.9	+1.1
10	30	30	0	30	0	30.2	+0.2
	60	59.6	-0.4	60.3	+0.3	58.3	-1.7
	-30	-30	0	-29.9	+0.1	-28.8	+1.2
12	30	30	0	30	0	30	+0
	60	59.8	-0.2	56.6	-3.6	58.5	-1.5

Table 17.6. DOA estimation (k = 1024, d = 0.75λ).

N	$\theta_{in}(°)$	$\theta_{MUSIC}(°)$	Δ_{MUSIC}	$\theta_{ESPRIT}(°)$	Δ_{ESPRIT}	$\theta_{MLE}(°)$	Δ_{MLE}
4	-30	-28.7	+1.3	-28.7	+1.3	-28.9	+1.1
	30	29.6	-0.4	30.3	+0.3	30.6	0.6
	60	58.5	-1.5	58.6	-1.4	58.4	-1.6
6	-30	-29.1	+0.9	-24.6	+5.4	-28.8	+1.2
	30	30.1	0.1	30	0	29.9	-0.1
	60	57.9	-2.1	56.5	-3.5	58.1	-1.9
8	-30	-29.5	+0.5	-28.2	+1.8	-29	+1
	30	29.9	-0.1	30.1	+0.1	30.1	+0.1
	60	57.2	-2.8	59.5	-0.5	58.1	-1.9
10	-30	-29.3	+0.7	-29.8	+0.2	-28.9	+1.1
	30	30	0	30	0	30.1	+0.1
	60	57.6	-2.4	56.9	-3.1	58.3	-1.7
12	-30	-29.7	+0.3	-29.9	+0.1	-28.7	+1.3
	30	30	0	30.1	0.1	30.1	+0.1
	60	58.8	-1.2	60.4	0.4	58.3	-1.7

Table 17.7. DOA comparison with d= 0.5 λ

N	θ_{in} (°)	θ_{MUSIC} [25]	θ_{MUSIC}	θ_{ESPRIT} [25]	θ_{ESPRIT}	θ_{MLE} [25]	θ_{MLE}
2	-10	18	-11.4	-51.8	-15.9	-16.75	-17.6
	-30	-49	-28.8	-44.8	-32.3	-48.75	-40.8
	50	36	53.6	75.42	49.6	35.5	42.9
4	-10	3	-9.3	-27.1	-13.2	-6.5	-5.3
	-30	-7	-35.1	-34.45	-39.3	-28.25	-20.7
	50	58	49.9	46.9	52.8	57.2	56.6

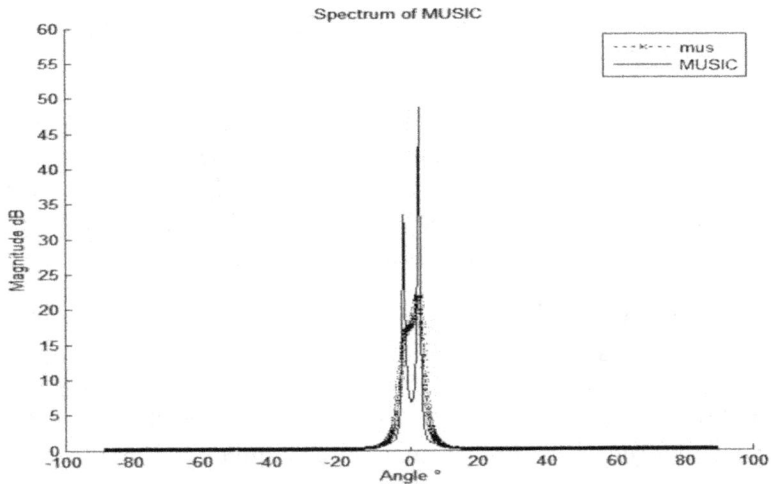

Fig. 17.7. Spectrum of MUSIC and proposed method.

Table 17.8 groups the results of a comparison between the MUSIC algorithm [27] and the proposed MUSIC algorithm. This table indicates that, for [27], the angles 50 and 60° present an error of 10 and 3.33 %, contrary to the proposed algorithm, which assures a minimum error of 0.8 % for 50° and 0.6 % for 60°. Therefore, we note that the proposed algorithm MUSIC is more robust and precise in detection of the angles.

Table 17.8. MUSIC comparison.

θ_{in} (°)	θ_{out} [27]	% Error	θ_{out}	% Error
50	55	10	50.4	0.8
60	62	3.33	59.6	0.6

In the following step, the simulation results of the proposed MUSIC algorithm investigated at this research work was compared with the experimented one [26 - 28] using Uniform Circular array (UCA). The comparison was made in the same condition: UCA with five antennas, radius r = 124 mm. the radiation source is pulse signal and the distance between the radiating antenna and the direction finder receiving antenna approximately is 8 m. The carrier frequency is 6 GHz. And according to the estimated received signal, the receiving data SNR is above 20 dB according to [28], a 4×4 planar antenna array with 0.5 λ element spacing for [26] and UCA with 8 antenna element, 2 source and noise = 12 dB, with BPSK modulation, the search step of MUSIC is 0.1° and the noise intensity is -12 dB according to [27].

In the simulations illustrate in Figs. 17.8 and 17.9, we note that the proposed can resolve clearly the azimuth elevation (133.6°, 137.8°) and (78.6°, 82.4°) respectively and the peaks are sharp, while the Music only fond one peak around there. To confirm the first simulation result, another simulation presented in Fig. 17.9, using unequal power signal arriving at azimuth and elevation are (128.4, 116) and (78, 84). The Fig. 17.9 confirms that the proposed Music algorithm can resolve clearly the angles(θ, φ) and the peaks become sharp.

From the simulation results indicate in Figs. 17.8 and 17.9, both of the two algorithms can get a correct estimation of the direction angle of independent signals. Because of reconstructing the data covariance matrix in the modified algorithm, which is equivalent to utilize the information of the data one more time, the peak of spectrum becomes sharper and the precision is higher compared to the results indicated in [28].

Fig. 17.10 indicates that angles have the same values but the proposed method presents a good magnitude for angles (99.48, 50.13); (64.88, 15.1), the magnitude is 40.15 dB, 38.84 dB respectively. We resolved that the proposed method can detect, estimate the DOA clearly and present an efficient magnitude with +0.24 % for angles (99.48 50.13) and +0.02 % for angles (64.88 15.1) compared to result given in [26].

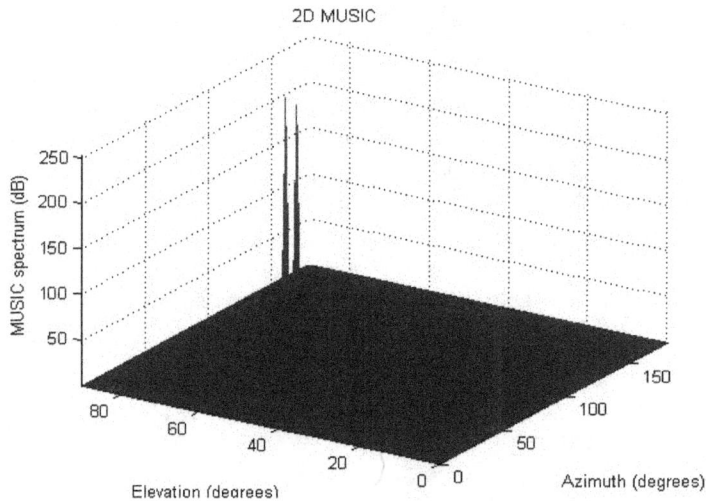

Fig. 17.8. Spectrum of proposed MUSIC method for azimuth and elevation
(133.6, 137.8); (78.6, 82.4).

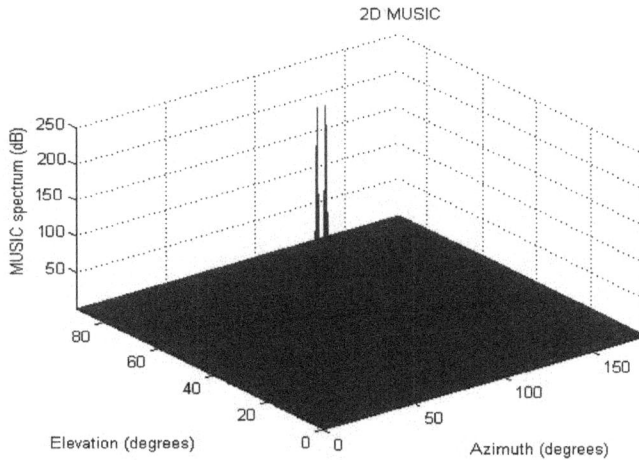

Fig. 17.9. Spectrum of proposed MUSIC method for azimuth and elevation
(128.4, 116); (78, 84).

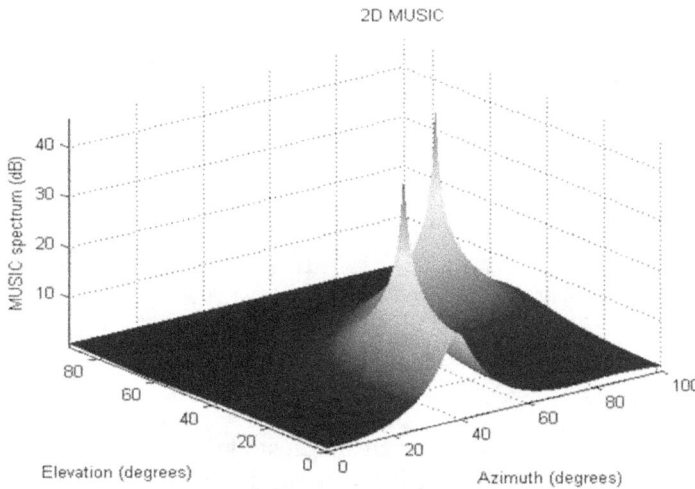

Fig. 17.10. Spectrum of proposed MUSIC method for azimuth and elevation
(128.4, 116); (78, 84).

Table 17.9 illustrates the result of the comparison between proposed method and MUSIC method indicate at [28]. It can be seen that the proposed method estimates 3 DOA more accurately while the experimented one cannot detect angles when the Number of signals exceeds 2. The proposed method gives a less error margin to estimate DOA.

Table 17.10 indicates that even if the values of number radius change, the proposed method give a higher precision and value of peaks, contrary to experimented one [27]. It cannot detect all angles even if number of signals increases. Results indicate that proposed method MUSIC based on UCA does not have a problem of aperture vagueness.

Table 17.9. Comparative result for different signals.

Signal		θ_{in}	θ_{out}	$\Delta\theta_{out}$	φ_{in}	φ_{out}	$\Delta\varphi_{out}$
1	[28]	78 84	78.5 82.5	+0.0064 0	128.4 116.0	129.5 117	+0.0082 +0.0086
	Pm	78 84	78 84	0 0	128.4 116.0	128.1 116.2	-0.0023 +0.0017
2	[28]	77.0 85.8	76.0 86.5	-0.0065 0	128.2 120	129.5 121.5	77.0 85.8
	Pm	77.0 85.8	77.0 85.8	0 0	128.2 120	128.2 120.1	00 +0.0008
3	[28]	78.6 82.4	0 0	-1 -1	133.6 137.8	0 0	-1 -1
	Pm	78.6 82.4	79.0 85.4	+0.005 +0.036	133.6 137.8	134 137	+0.0029 -0.0028

Pm: is the proposed method.

Table 17.10. Comparative result for different radius.

Radius	Pm	[27]	Error (deg) [27]	Error (deg) Pm
0.1 λ	57.100000	57.100002	+0.10002	+0.10002
0.5 λ	57.100004	57.100004	+1.00004	+1.00004
0.8 λ	56.900000	56.900002	-1.90002	-1
1.0 λ	57.000000	57.000004	+0.00004	0

The computer simulation results are given to illustrate the performance of the Matrix Pencil method. The noisy signal model is formulated from (10). n(k) id treated as a zero mean Gaussian whit noise with variance σ^2. The distance between any two elements of the smart antenna is half wavelength. y(k) is the voltage induced at each of the antenna elements, for k=0, 1, …,N-1. In order to demonstrate the numerical properties, a comparative study was made between the Matrix Pencil and pencil method indicates in [29 - 32].

Table 17.11 illustrates the comparison between Matrix Pencil investigate in this research work using UCA and Matrix Pencil indicate at [30]. Through to comparison The authors in [30] indicate that the matrix pencil can accuracy estimate the direction of arrival of signal on smart antenna only with one snapshot but the pencil method studied in this work gives more precision than [30] because we have chosen a better value of parameter of pencil 'L' to increases the precision and the uniform circular arrays have better performance than the other geometry.

Table 17.11. Comparative results for Matrix Pencil with various elements.

Number of elements		θ_{in}	θ_{pencil}	$\Delta\theta_{pencil}$
7	Pencil [30]	30 60	37.5900 61.8400	0.253 0.030
	This Work	30 60	30.0017 60.0928	0.226 0.025
8	Pencil [30]	30 60	30.1800 61.6400	0.006 0.027
	This work	30 60	30.0477 60.0267	0.003 0.018
10	Pencil [30]	30 0.2	30.3100 0.2100	0.010 0.05
	This Work	30 0.2	30.0004 0.2010	0.006 0.005
14	Pencil [30]	30 60	30.1700 59.4900	0.005 0.0085
	This Work	30 60	29.9975 60.0009	0.0026 -0.0033

In Table 17.12, the MP given in this work gives a good precision compared to the proposed one indicate at [29]. The authors in [29] didn't give the number of snapshots using in this case contrary to the second case when they use 100 element with 5 faulty elements, they are used one snapshots to demonstrate the efficiency of their proposed method, but in our case we have chosen the same number of snapshots in order to improve the robustness of our method with only one snapshots even if the number of elements changed or one of them stops working.

We compare the LMS investigate at this work and LMS indicate at [31-32]. From Fig. 17.11 we observe that the final weighted array which has a sharp peak at the desired direction of 0° and a null at the interfering direction of -60°. In Fig. 17.12 it is observed that the array output acquires and tracks the desired signal after 20 iterations contrary to the results indicated in [31-32]. If the signal characteristics are rapidly changing, the LMS algorithm may not allow tracking of the desired signal in a satisfactory manner.

It is describes the algorithmic changing the weighting in each iteration from Fig. 17.13, it shows the graph of signal versus number of iteration. In which array output acquire and track after 20 iterations. If the characteristic of the signal rapidly changing, the LMS algorithm may not allow tracking of the desired signal in a satisfactory manner. Finally, we simulated the MSE error in each iteration shows in Fig. 17.14, and shows the relationship between mean square error and number of iteration. In which MSE is decreases with iteration and after 20 iterations it become converged.

Table 17.12. Comparative results for Matrix Pencil for 100 and 95 elements.

Number of elements		θ_{in}	θ_{out}	RMSE
100	Pencil [29]	0	0.0004	0.0019
		5	5.0001	
		10	9.9948	
		15	15.0014	
		20	20.0045	
		30	29.9984	
	This Work	0	0.0000	0.0006
		5	5.0001	
		10	9.9991	
		15	15.0001	
		20	20.0000	
		30	30.0001	
95	Pencil [29]	0	0.0067	0.0023
		5	5.0024	
		10	10.0014	
		15	14.9976	
		20	19.9965	
		30	29.9991	
	This Work	0	0.0005	0.0018
		5	5.0006	
		10	10.0008	
		15	15.0009	
		20	20.0000	
		30	29.9999	

*RMSE is the Root Mean square Error is defined as: RMSE=$\sqrt{E(\theta_{Act}-\theta_{Mes})^2}$, where E is the mean value, θ_{Act} is the actual DOA and θ_{Mes} is the estimated values.

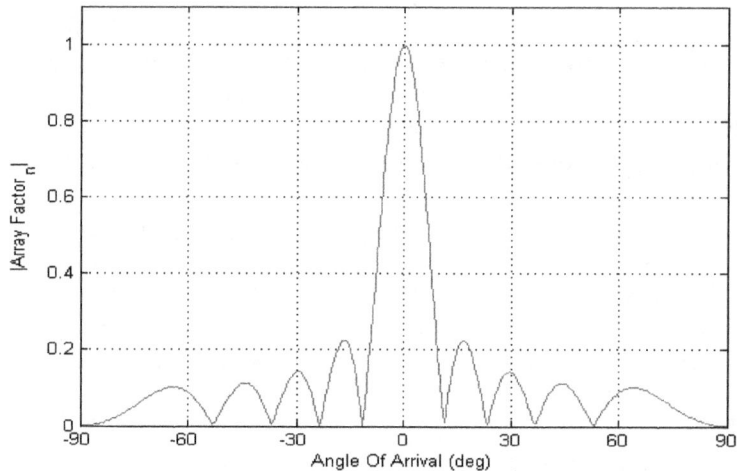

Fig. 17.11. The array factor using LMS algorithm.

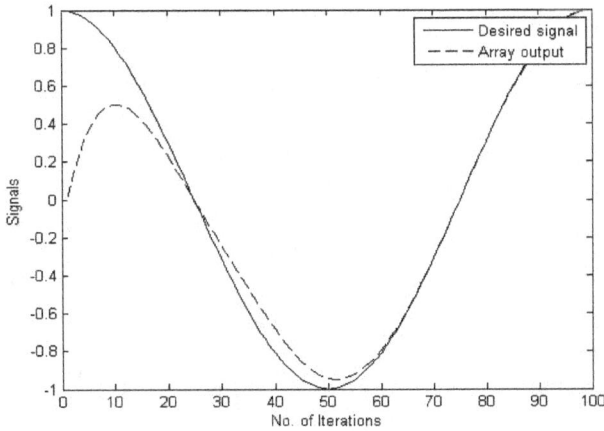

Fig. 17.12. Acquisition and tracking of desired signal using LMS algorithm.

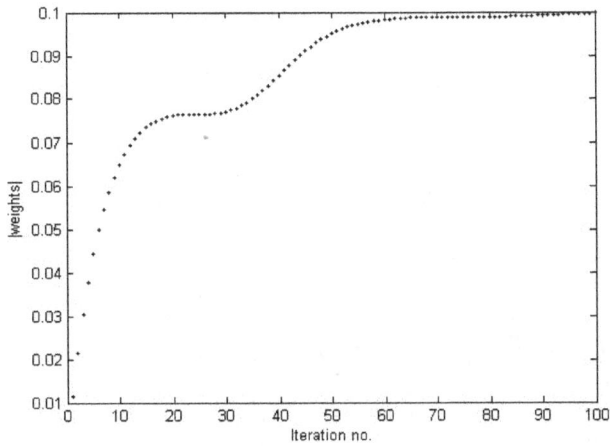

Fig. 17.13. Magnitude of array weights using LMS algorithm.

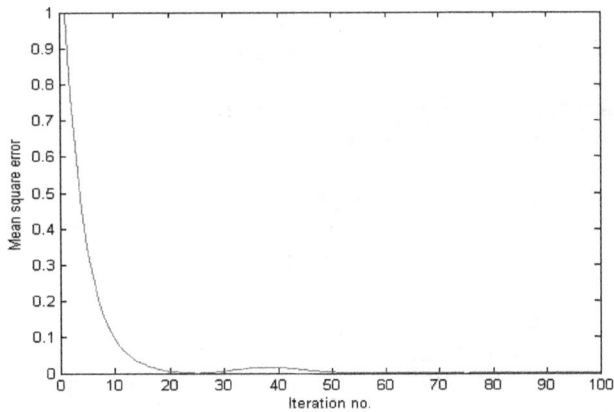

Fig. 17.14. Magnitude of array weights using LMS algorithm.

17.6. Conclusion

The modern high resolution methods based on the concept of subspace are among the most efficient for estimating the directions of arrival (DOA) of signals using array antennas. In this study presented an analysis of randomly spaced smart antennas. First, randomly spaced linear and planar signal models were studied. After understanding the non-uniform model, several DOAs and Pencil algorithms were theoretically analyzed; such as the Esprit, MLE and MUSIC algorithms for direction of arrival, and the LMS. These algorithms were implemented by using MATLAB .The first part of the experiments analyzes the DOA algorithms by considering two different scenarios: with using uniform linear array for 'Music, Esprit and MLE' and uniform circular array for Matrix Pencil, MLS and MUSIC. Both algorithms yield a good results for detecting DOAs; however, the Pencil algorithm and MLS algorithm yields more accurately.

Acknowledgements

This work is supported by the University Moulay Ismail, Meknes –Morocco.

References

[1]. Federal Communications Commission, http://www.fcc.gov/e911/
[2]. Liberti, Jr., J. C. and T. S. Rappaport, Smart Antennas for Wireless Communications: IS-95 and Third Generation CDMA Applications, *Prentice-Hall,* Upper Saddle River, NJ, 1999.
[3]. Sarkar, T. K. Smart Antennas, *IEEE Press/Wiley-Interscience*, New York, 2003.
[4]. Kuchar, A., et al., *A Robust DOA-Based Smart Antenna Processor for GSM Base Stations,* in *Proceedings of the IEEE International Conference on Communications,* Vol. 1, June 6–10, 1999, pp. 11–16.
[5]. Balanis, C. A., Antenna Theory: Analysis and Design, 3rd ed., *Wiley,* New York, 2005.
[6]. R. O. Schidt, Multiple emitter location and signal parameter estimation, *IEEE Tran. Ant. Pro.,* Vol. 34, 3, 1986, pp. 276-80.
[7]. A Paulraj, R Roy and T Kailath. A subspace rotation approach to signal parameter estimation. *Proc. IEEE,* Vol. 74, 1986, pp. 1044-1046.
[8]. L. Osman, I. Sfar and A. Gharsallah, Comparative study of high-resolution direction-of-arrival estimation algorithms for array antenna system, *Int. J. Res. Rev. Wir. Comm.,* Vol. 2, 2012, pp. 72-77.
[9]. R Roy and T Kailath, ESPRIT-estimation of signal parameters via rotational invariance techniques, *IEEE Trans. Acou. Spe. Proc.*, Vol. 37, 1989, pp. 984-995.
[10]. M. Jalali, M. N. Moghaddasi and A. Habibzadeh, Comparing accuracy for ML, MUSIC, ROOT-MUSIC and spatially smoothed algorithms for 2 users, in *Proceedings of the IEEE Conference Mediterranean on Microwave Symposium,* Tangiers, Morocco, 2009, pp. 1-5.
[11]. K. Huang, J. Sha , W. Shi and Z. Wang, An Efficient FPGA Implementation for 2-D MUSIC Algorithm, *Circ. Syst. Sign. Proc,* Vol. 35, 2016, pp. 1795-1805.
[12]. G. Mao, B. Fridan and B. Anderson, Wireless sensor network localization techniques, *Comput. Net,* Vol. 51, 2007, pp. 2529-2553.
[13]. P. Yang, F. Yang and Z. P. Nie, DOA estimation with sub-array divided technique and interpolated esprit algorithm on a cylindrical conformal array antenna, *J. Pro. Ele. Res,* Vol. 103, 2010, pp. 201-216.

[14]. F. Vincent, O. Besson and E. Chaumette, Approximate unconditional maximum likelihood direction of arrival estimation for two closely spaced targets, *IEEE Sig. Pro. Let,* Vol. 22, 2015, pp. 86-89.

[15]. J. Xin, N. Zheng and A. Sano, Simple and efficient nonparametric method for estimating the number of signals without eigen decomposition, *IEEE Tran. Sig. Pro,* Vol. 55, 2007, pp. 1405-1420.

[16]. P. Stoica and A. Nehorai, Music, maximum likelihood, and Cramer-Rao bound, *IEEE Trans. Aco. Spe. Sig. Pro,* Vol. 37, 1990, pp. 720-741.

[17]. Y. Khamou, S. Safi and M. Frikel, Comparative study between several direction of arrival estimation methods, *J. Tel. Inf. Tech,* Vol. 1, 2014, pp. 41-48.

[18]. N. Yilmazer, A. Seckin, and T. Sarkar, Multiple snapshot direct data domain approach and ESPRIT method for direction of arrival estimation, *Digital Signal Processing - A Review Journal,* Vol. 18, 2008, pp. 561-567.

[19]. A. El Fadl, S. Bri, M. Habibi, Uniform rectangular smart antenna based on Matrix Pencil, *Journal of Basic and Applied Scientific Research,* December 2011, pp. 3322-3329.

[20]. C. R. Dongarsane, A. N. Jadhav and S. M. Hirikude, Performance analysis of ESPRIT algorithm for smart antenna system, *Int. J. Comput. Comm. Tech,* Vol. 2, 2011, pp. 51-54.

[21]. C. Joe, C. Ralph, E. Hudson and K. Yao, Maximum-likelihood source localization and unknown sensor location estimation for wideband signals in the near field, *IEEE Tran. Sig. Pro,* Vol. 50, 2002, pp. 1843-1854.

[22]. F. A. de Leon and J. J. S. Marciano, Application of MUSIC, ESPIRIT and SAGE algorithms for narrowband signal detection and localization, in *Proceedings of the IEEE Region 10 Conference TENCON,* Hong Kong, China, 2006, pp. 1-4.

[23]. W. J. Si, X. Y. Lan and Y. Zou. Novel high-resolution DOA estimation using subspace projection method, *J. Chi. Uni. Pos. Tel,* Vol. 19, 2012, pp. 110-116.

[24]. B. Li, Y. X. Zou and Y. S. Zhu. Direction estimation under compressive sensing framework: A review and experimental results, in *Proceedings of the IEEE International Conference on Information and Automation (ICIA),* Shenzhen, China, 2011, pp. 63-68.

[25]. F. A. de Leon and J. J. S. Marciano, Application of MUSIC, ESPIRIT and SAGE algorithms for narrowband signal detection and localization, in *Proceedings of the IEEE Region 10 Conference TENCON,* Hong Kong, China, 2006, pp. 1-4.

[26]. T. Varum, J. N. Matos and P. Pinho, Direction of Arrival Estimation Analysis Using a 2D Antenna Array, *Proce. Techn,* Vol. 17, 2013, pp. 667-624.

[27]. B. Sun., MUSIC Based on Uniform Circular Array and Its Direction Finding Efficiency, *Int. J. Sign. Proce. Syst,* Vol. 1, 2013, pp. 273-277.

[28]. W. J. Si, X. Y. Lan and Y. Zou, Novel high-resolution DOA estimation using subspace projection method, *J. Chi. Uni. Pos. Tel,* Vol. 19, 2012, pp. 110-116,.

[29]. Yerriswamy T. and S. N. Jagadeesha, Fault Tolerant MatrixPencil Method for Direction of Arrival Estimation, *Sig Ima Proc: An Inter J,* Vol. 2, 2011, pp. 55-67.

[30]. A. ELFadl, S. Bri, M. Habibi, Multipath Elimination using Matrix Pencil for Smart Antenna with Uniform Linear Array, *Eur. J. of Scie. Rese,* Vol. 87, 2012, pp. 397-405.

[31]. D. M. Motiur Rahaman, Md. Moswer Hossain, Md. Masud Rana, Least Mean Square (LMS) for Smart Antenna, *Univ. J. Comm. Netw,* Vol. 1, 2013, pp. 16-21.

[32]. S. Singh, Er. M. Kaur, A LMS and NLMS Algorithm Analysis for Smart Antenna, *Inter. J. Adva. Rese. Comp. Scie. Soft. Engi,* Vol. 5, 2015, pp. 380-384.

Chapter 18
UAV Control Based on Optimized Neural Network

Roderval Marcelino, Vilson Gruber, Luan Casagrande, Renan Cunha, Yuri Crotti, Sarah de Rezende Guerra

18.1. Introduction

In the last years, there was a rising interest in relation to the utilization of unmanned aerial vehicle (UAV) mainly because of the technological advances that occurred in essential equipment such as computers, sensors, batteries, electric motors and speed electronic control (ESC). This allowed the development of new UAVs with less weight, size, cost, and with more endurance of flight. In addition, with the growing interest of the scientific community, the UAV is becoming more accurate, thus reducing the risk of crashes and collisions.

As a result, mainly by lower cost and risk that it creates, the UAV has become a reliable alternative for jobs that were previously done by manned aircraft. Both in the military and in the civil sector, this equipment is increasingly present in various tasks, including defense and security, mapping, image acquisition, among others.

Considering this growing use of UAVs, a demand for flight control systems to act as an automatic pilot is increasing significantly. Different methodologies are being tested for this purpose, including Neural Network [1], PID (Proportional Integral Derivative) controller [2], sliding mode control [3], Neuro-Fuzzy system [4], Fuzzy system [5], among others. Currently the most widely used controller on the market for this kind of application is the PID [4]. According to [4], the justification for this massive preference is the fact that its implementation is relatively simple and cheap.

However, the fixed wing UAV category has the possibility to move in six degrees of freedom (DoF). The dynamics of an airplane in this category consists of a non-linear model with a high complexity degree. Considering this problem, a PID system would require recalibration every flight in which any of the basic flight characteristics change.

Roderval Marcelino
Universidade Federal de Santa Catarina (UFSC), Brazil

As a solution for the control, other methods based on artificial intelligence are being proposed, such as the use of control systems with Fuzzy, Artificial Neural Network (ANN), among others [2]. Such methods offer new possibilities for modeling the problem in a non-linear manner without calibration in each flight.

This work proposes a new solution using an ANN multilayer perceptron optimized through the use of genetic algorithm (GA) to control two degrees of liberty (pitch and roll) of a fixed wing UAV. Section 18.2 will introduce UAV, its classification, and degrees of liberty. Section 18.3 will introduce ANN, GA and the proposed solution for optimization. Section 18.4 will describe the simulation system. Section 18.5 will describe the optimization system. Section 18.6 will describe the results and Section 18.7 will conclude this chapter and present future work.

18.2. UAV

18.2.1. Classification

UAV can be denominated as: remotely piloted aircraft (RPA), unmanned aerial vehicle system (UAVS) or Drones. These vehicles are built with systems which operate as either remotely controlled, semi-autonomously, or autonomously [6].

There are a huge variety of UAVs and systems to classify this equipment, but generally it is classified by dimension [4], functionality [7], sustainability [1], among others. This chapter will follow the system of classification based on the dimension of the system [8]. In this classification system, UAVs are divided between small, medium and large. Each category is defined by some specific features.

A. Small UAV
   a) Maximum take-off weight up to 150 kg;
   b) Flown within the visual line-of-sight of the operator up to 500 feet above the ground.

B. Medium UAV
   a) Maximum take-off weight between 150 kg and 600 kg;
   b) Typically operated below 18000 feet above sea level;
   c) Range of less than 800 km;
   d) Endurance of only a few hours.

C. Large UAV
   a) Maximum take-off weight above 600Kg;
   b) Flight range is not defined;
   c) Long autonomy.

The small UAV class is experiencing tremendous growth mainly due to the low cost, ease of use and less difficulty of integration in the national airspace [8]. On the other hand, medium UAVs are generally operated by militaries and large UAVs are only operated by militaries, and as a consequence, the use of airplane in these categories is controlled. In this work is being proposed the use of a small UAV to simulate the control system based on ANN optimized.

18.2.2. Control Surfaces

As defined above, an airplane has six DoL. These six degrees can be divided into two different movements. The first one, angular movement, is composed by pitch, roll, and yaw. The second one, directional movement, is composed by the X, Y, and Z movement. All these movements are being demonstrated in the Fig. 18.1.

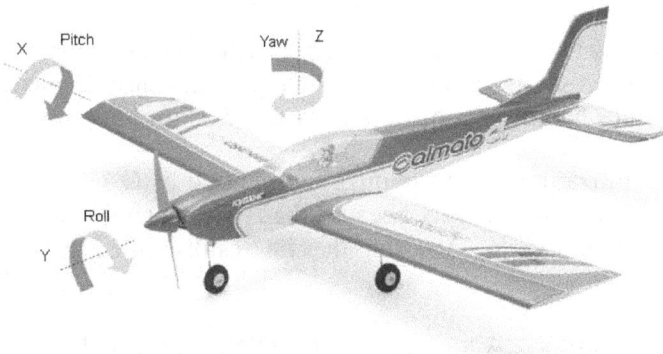

Fig. 18.1. Angular movement and directional movement.

The aircraft used in this work consists of all the traditional surfaces, and as a consequence, it uses four actuators to control:
- Motor: Controls the velocity of the airplane;
- Elevator: Controls the pitch angle;
- Aileron: Controls mainly the roll angle;
- Rudder: Controls mainly the yaw angle.

The control system will realize the lateral and longitudinal control. In the lateral control, the optimized ANN will act directly on the aileron, then adjusting the roll angle to zero. In the longitudinal control, the system will act using the elevator and the motor, causing the pitch angle to approach zero.

18.3. Genetic Algorithm and Artificial Neural Network

Artificial Neural Network (ANN) simulates organizations that have the capacity to be adapted to their training environment [9]. Basically, ANN simulates a human brain, because each node is connected to other nodes simulating the connection between neurons. Furthermore, each node has a specific skill and each connection between the nodes has a specific weight called synaptic weigh. This synaptic weight is an important parameter to define the importance of the connection between the nodes.

On the other hand, Genetic Algorithm (GA) is a biological metaphor for the evolution of species, as described by Charles Darwin. All the individuals, that have their own genome, will be evaluated and the best ones will survive the elimination process. The worst individuals will be eliminated and new ones will be created through the crossover and mutation techniques. As output, it doesn't guarantee the best solution possible for the problem, but this algorithm has the capacity to find an acceptable and optimized solution [9].

It is known that the topology of an ANN influences in solving the problem, because it can directly affect the accuracy of the result. Other important aspects of an ANN such as generalization, learning rate, connectivity and tolerance to network problems are strongly linked to the topology choice [10].

The most common way to find the best topology is using the heuristic methods. However, as the problem in discussion is considerably complex, heuristic methods can lead to inaccurate results, thus hampering the operation of the control system.

Consequently, the utilization of an optimization method as GA becomes an interesting solution to the problem in discussion, because the ANN structure can be assessed by the performance of the network, thus ensuring that the result will be the best possible in the search space.

There are different ways for GA to interact with ANN, such as [11]:

- ANN is used to support GA: the network creates a primary population, while AG optimizes the population.

- GA is used to optimize ANN: In this case, GA can contribute selecting topology, parameters, learning rate, weights and/or bias.

In [10 - 15], some of these interactions were proposed in different problems, and in all these works the optimization was successfully obtained.

Considering the benefits that GA can create to ANN and the complexity of the problem, a solution based on both algorithms was proposed. In this solution, GA contributes directly to the ANN, finding its best topology and the main characteristics of the network through the performance analysis using the method mean squared error (MSE).

18.4. Optimization System Descript

The structure of the system is divided between the generations of individuals representing the structures of ANN, fitness analysis using MSE, sort the individuals, crossover and mutation.

The first step is to generate the individuals. In this system, besides randomize the structure of the ANN, the transfer function, the backpropagation function and the learning rate was randomized too. These characteristics were chosen to ensure that the main variants of the ANN are optimized.

To reduce the computational cost, the variation in the number of hidden layers in the structure of the ANN ranged from 1 to 10 layers, where each internal layer can be defined between 1 and 10 neurons. Each topology has a number of randomly generated hidden layers and each layer has a number of neurons also generated randomly.

Each layer in the structure has one transfer function that is randomly defined too. For this problem, it was defined as an option the linear transfer function, log-sigmoid transfer function (18.1) and the hyperbolic tangent sigmoid transfer function (18.2).

$$x(n) = \frac{1}{(1+e^{-n})},$$
(18.1)

$$x(n) = \frac{2}{(1+e^{-2*n})},$$
(18.2)

The learning rate is defined randomly between the maximum and minimum value too. In addition, each structure will have a backpropagation function defined randomly in a list of 10 possibilities. Considering the variety of topologies that will be created through the code, different backpropagation algorithms should be analyzed. This is necessary because each function with its own specificity can improve a range of structures. The algorithms included in this problem are described below:

- Levenberg-Marquardt optimization;
- Bayesian Regularization Backpropagation;
- BFGS quasi-Newton method;
- Resilient Backpropagation;
- Scaled conjugate gradient backpropagation;
- Conjugate gradient backpropagation with Powell-Beale restarts;
- Conjugate gradient backpropagation with Fletcher-Reeves updates;
- Conjugate gradient backpropagation with Polak-Ribiére updates;
- One-step secant backpropagation;
- Gradient descent backpropagation.

After generating all the necessary data for each individual, these data are organized in a chromosome structure. Each chromosome is a layer of the ANN. The first chromosome is the input layer and stores all the weights between the input neurons and the neurons in the first hidden layer. Next, the chromosomes that determine the structures of the hidden layer are responsible to store all the weights between the current hidden layer and the next hidden layer, the bias from the previous layer, and the transfer function of the current layer. The last layer represents the output of the ANN and this chromosome stores the bias value from the last layer, the transfer function, learning rate of the structure, and the backpropagation function used. A representation of the structure is presented in Fig. 18.2.

Layer I	Layer 1	...	Layer n	Layer O

n = 1, 2, 3, 4

Layer I - Input

Weights

Layer n - Hidden

Weights

Bias

Transfer Function

Layer O - Output

Bias

Transfer Function

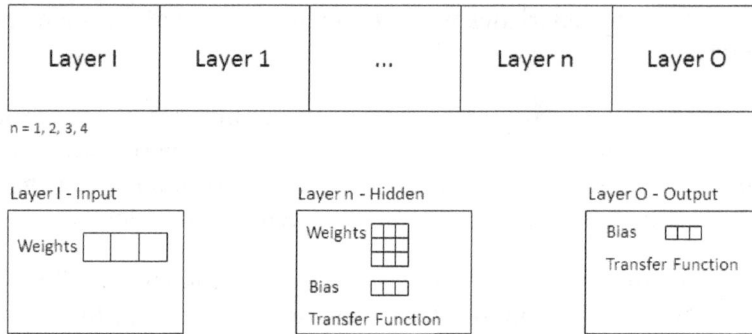

Fig. 18.2. General Structure of an Individuous.

After the generation of individuals, all the artificial neural networks are created and finally each ANN is evaluated using MSQ. After, these ANN are sorted using the performance from the last step. This order is necessary to start the crossover and mutation process.

In the crossover, it was defined that the algorithm would select a group composed of 50 % of the group of individuals. This group would be composed of the individuals who have the worst performances in the last step. In addition, it was defined that the algorithm would select two of the best individuals to generate a new individual. The crossover point in each individual was defined exactly in the middle.

After dividing each individual in half, a new individual is generated using the parameters from each half. However, considering the fact that the connection between the parts will be broken, it must generate new random weights and bias for the connection between the two halves.

In the mutation process, it was defined a mutation rate of 40 % for the individuals. After being selected, the individual has an equal chance of mutation in the learning rate, backpropagation function and transfer function.

18.5. Simulation

The model that describes a flight controller based on neural networks requires consecutive adjustments of few parameters certain parameters until it achieves satisfactory results for a generic network that support all flight situations. Considering this difficulty, it was chosen to design the controller through computer simulations.

The simulation is composed of a number of tools, as is listed below:

* XPlane 10

The software XPlane 10 is a flight simulator certified by the FAA (Federal Aviation Administration – US) and it has extremely realistic physical aspects that provides a high degree of accuracy. An important feature of this simulator is that it has a robust and complete mechanism to interact with other software through the communication protocol UDP (User Datagram Protocol).

- Advanced Data Communication Software

This software was developed by the authors of this present work, and its function is to serve as a channel of communication with the simulator. This was necessary to send commands to the aircraft, to read sensor data from the aircraft, among other information.

- Ardupilot

The ardupilot is a set of tools that makes a complete system of UAV flight control. In this work it was used just the logical part of the Ardupilot, i.e. only the flight control software. It was connected directly to the X-Plane simulator 10, so it was possible to trace routes, set flight plan, set a flight mode, among others.

It was defined to use the SITL (Software-in-the-loop) methodology to run the simulation. This methodology allows simulation of a system without executing the software in a hardware, in other words, everything occurs virtually through the software. In [15], the authors point out that the software-in-the-loop combines the flexibility and low cost of the simulator with the accuracy of the hardware emulator. This technique can be used in the modeling and testing phase.

In this work, the communication software and the Ardupilot were executed using SITL. The Control Software (CS) was used in two distinct steps, as described below:

- 1st phase: Acquisition of training data

In this phase, the airplane was manually positioned in a specific situation in such way that the airplane has an angle of roll and pitch greater than zero in the simulator. After, the Ardupilot controller was triggered to stabilize the airplane. During this process, the communication software received and stored the information send from the X-Plane (pitch and roll angle, motor, aileron, rudder, and elevator). With this data, it was possible to create the training set necessary to the ANN.

- 2nd phase: ANN Execution

At this stage, the optimal topology for ANN has already been defined through the optimization system. Thus, the communication software was executed in a loop with a frequency of 10 Hz. In each loop, the CS captured the angles of roll and pitch, passing these to the ANN as input and then obtaining the output. Then, this output was used to control the aircraft through the actuators. In this phase, the performance of the controller system was evaluated through the simulator.

18.6. Results

Considering that the ANN's training process is based on a model, it is necessary a set of data that is commonly referred to training data. This data set lists a series of input values with their respective outputs.

As mentioned in previous sections, this data was obtained using the communication software, Ardupilot, and X-Plane 10. The final training set was composed of 2944 lines describing different flight situations.

With the training set, the optimization code was triggered to find an ideal ANN for the simulation. The number of generations was the stop criterion defined for the GA. In this experiment, it was defined 10 generations with 500 individuals in each generation. The optimization experiment used a mutation rate of 40 % and crossover rate of 50 %.

This configuration outputs a minimum performance of 2.04×10^{-07} for the last generation, as shown in the Table 1. The individual with the best performance had a maximum error of 0.0055, representing a maximum angular error of 0.99 degrees. In other words, the maximum error in the control system is imperceptible to the problem.

Table 1. Top 10 individuals.

	Learning Rate	Backpropagation Function	Hidden Layers	Performance
1	0.9404	Levenberg-Marquardt	5	2.042×10^{-07}
2	0.2294	Bayesian Regularization	5	3.224×10^{-07}
3	0.3454	Bayesian Regularization	6	4.070×10^{-07}
4	0.2319	Bayesian Regularization	5	7.172×10^{-07}
5	0.2417	Bayesian Regularization	5	7.993×10^{-07}
6	0.9771	Bayesian Regularization	6	9.240×10^{-07}
7	0.2417	Bayesian Regularization	6	1.438×10^{-06}
8	0.3050	Levenberg-Marquardt	6	1.786×10^{-06}
9	0.6934	Gradient Descendent	6	3.861×10^{-06}
10	0.4130	Levenberg-Marquardt	4	4.061×10^{-06}

Considering the top 10 individuals for the last generation, it is possible to conclude that 5 of the 10 individuals have 6 hidden layers and other 4 individuals have 5 hidden layers. Moreover, it is clear that the backpropagation function more present among the best individuals is the Bayesian Regularization backpropagation function. This feature was present in 7 of the top 10 individuals.

On the other hand, the learning rate is random among the best individuals, since it varies between 0.2294 and 0.9771. In addition, the number of neurons per layer also varies considerably.

The best topology is shown in Fig. 18.3, where "E" represents an input, "O" an output, "H" a hidden layer. The first value after "H" represents the level of the layer. The second value represents the neuron identification in the level. The numbers concatenated to the letters "E" and "S" also represents the neuron identification in the level.

The learning rate for this structure was defined as 0.9404 and the backpropagation function as Levenberg-Marquardt optimization.

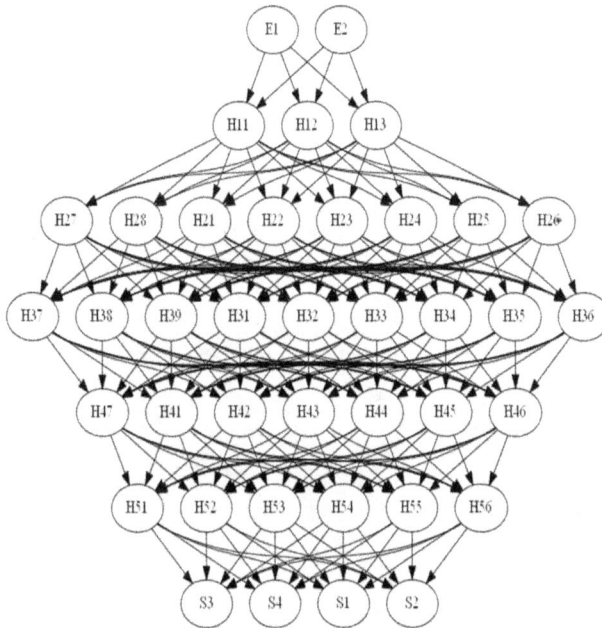

Fig. 18.3. Final Structure used in the Simulation.

To analyze the quality of the optimization software, it was used statistical data, such as the best performance per generation and medium performance per generation. The Fig. 18.4 represents the minimum performance per generation.

It is possible to conclude that the GA produced a considerable improvement between the first generation and the octave generation, decreasing from 1.25×10^{-4} to 2.04×10^{-7}, thus demonstrating the efficiency of the proposed problem. However, after the octave generation, the system has stabilized at an optimized network.

Moreover, another important statistical data to be analyzed to evaluate the performance of the GA is the average performance per generation. Fig. 18.5 shows this statistical index.

It is visible that there was a considerable improvement in the individuals until the seventh generation. After this generation, the average performance began to stabilize. Also, it is possible to conclude that there is still room for improvement in individuals, because the average error is high compared if compared with the minimum error.

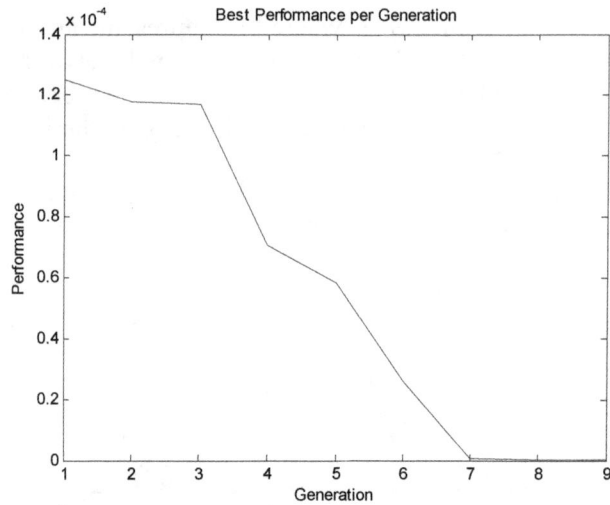

Fig. 18.4. Best Performance per Generation.

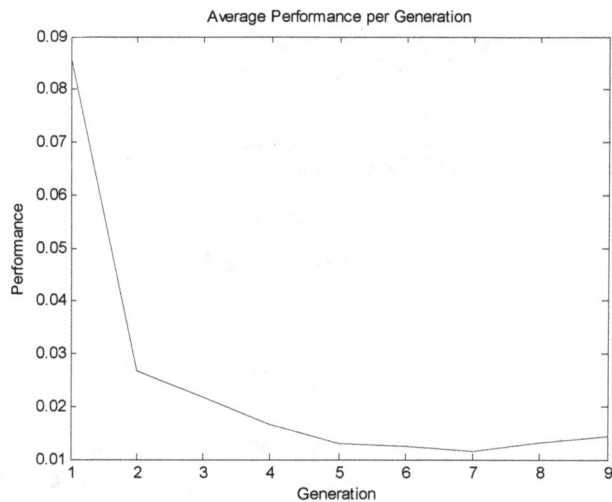

Fig. 18.5. Average Performance per Generation.

After the optimization step, the ANN was tested in the simulator to make sure that it was really doing what was purposed. These tests were made using the X-Plane 10 simulator and the communication software.

The aircraft was positioned in the simulator with approximately 28° of pitch angle and 55° of roll angle, and then was triggered the stabilization system proposed. Fig. 18.6 and Fig. 18.7 show the performance of the system. It is important to cite that the data capture frequency was defined in 10 Hz, and as a consequence, in every second 10 frames were captured.

364

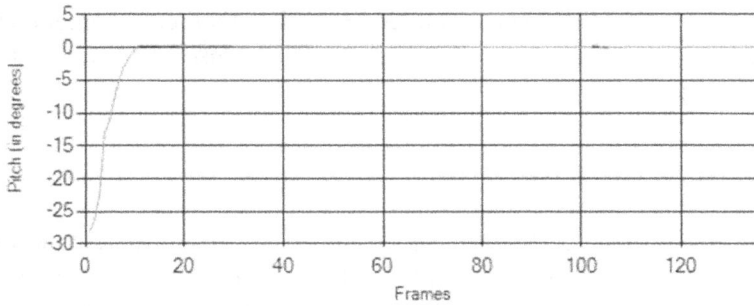

Fig. 18.6. Recovery of the pitch angle.

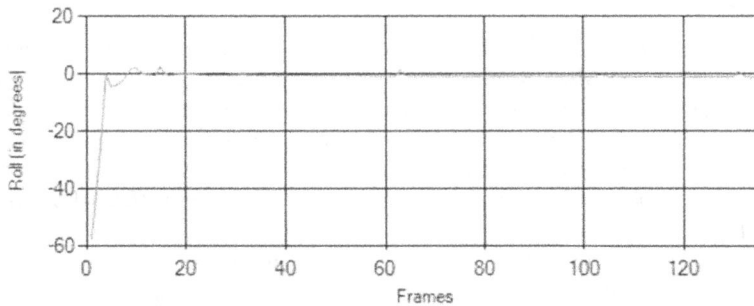

Fig. 18.7. Recovery of the roll angle.

Since the purpose of the control system was to act in the aircraft so that it could fly with roll angle and pitch angle around 0 degrees, it is clear that the system performance was satisfactory. The settling time, which is the time that the system takes to reach the final state, was approximately 2 seconds.

As can be seen in Fig. 18.6, approximately around the frame 105, there was a small disturbance in the pitch angle of the aircraft. This disturbance may be due to the simulation processing problems, such as loss of packets sent to the simulator. It is important to note that even though a variation generated by external factors, the stabilizer acted aiming to correct this discrepancy. The same phenomenon can be observed in Fig. 18.7, approximately around the frame 65.

18.7. Conclusion

PID has been the first choice UAV control systems. However, our work shows an artificial neural network multilayer perceptron optimized for a stabilization system of a fixed wing aircraft. Simulation tests have shown that the ANN is another viable option for this kind of controller since it achieved the purpose of this work. In addition, the optimization algorithm has shown that, due to the difficulty of choosing the best parameters and topology for an ANN, is essential to define the best structure for the proposed problem.

The same work can be modified and extended to the development of other controls for the aircraft. In addition, the optimization algorithm can be used to find optimized neural networks to other problems.

References

[1]. V. Puttige, Neural network based adaptive control for autonomous flight of fixed wing unmanned aerial vehicles, Doctor of Philosophy Thesis, *School of Aerospace, Civil and Mechanical Engineering, University of New South Wales,* Australia, 2008.

[2]. F. A. Warsi et al., Yaw, Pitch and Roll controller design for fixed-wing UAV under uncertainty and perturbed condition, in *Proceedings of the IEEE 10th International Colloquium on Signal Processing & its Applications (CSPA),* Kuala Lumpur, 2014, pp. 151-156.

[3]. A. Melbous, Y. Tami, A. Guessoum, and M. Hadjsadok, UAV Controller Design and Analysis using Sliding Mode Control, *Laboratory of Electric System and Remote Control, Electronic Department University of Saad Dahleb Blida,* Algeria.

[4]. I. de Medeiros Esper and P. F. Ferreira Rosa, Heading Controller for a Fixed Wing UAV with Reduced Control Surfaces Based on ANFIS, in *Proceedings of the IEEE International Conference on Dependable Systems and Networks Workshops (DSN-W)*, Rio de Janeiro, 2015, pp. 118-123.

[5]. R. C. Hoover, M. P. Schoen, and D. S. Naidu, Fusion of hard and soft control for Uninhabited Aerial Vehicles, Measurement and Control Engineering Research Center, *Idaho State University,* Pocatello, USA.

[6]. H. Eisenbeiss, UAV photogrammetry, *Institute of Geodesy and Photoframmetry, ETH,* Zurich, 2009.

[7]. S. G. Grupta, M. M. Ghonge, and P. Jawandhiya, Review of unmanned aircraft system (UAS), *Int. J. Adv. Res. Comput. Eng. Technol.,* Vol. 2, No. 4, 2013, pp. 1646-1658.

[8]. United Nations, Study on Armed Unmanned Aerial Vehicles, *United Nations Publication,* New York, 2015.

[9]. H. Chihi and N. Arous, Recurrent Neural Network learning by adaptive genetic operators: Case study: Phonemes recognition, Sciences of Electronics, in *Proceedings of the 6th International Conference on Technologies of Information and Telecommunications (SETIT),* Sousse, 2012, pp. 832-834.

[10]. D. Dasgupta and D. R. McGregor, Designing application-specific neural networks using the structured genetic algorithm, in *Proceedings of the International Workshop on Combinations of Genetic Algorithms and Neural Networks (COGANN'92)*, Baltimore, MD, 1992, pp. 87-96.

[11]. V. S. Dykin, V. Y. Musatov, A. S. Varezhnikov, A. A. Bolshakov and V. V. Sysoev, Application of genetic algorithm to configure artificial neural network for processing a vector multisensor array signal, in *Proceedings of the International Siberian Conference on Control and Communications (SIBCON),* Omsk, 2015, pp. 1-4.

[12]. W. Yu, J. Chuanwen, W. Chengmin, and S. Yufei, Loss calculation method for distribution network based on measured data, in *Proceedings of the 5th WSEAS/IASME International Conference on Systems Theory and Scientific Computation (ISTASC'05),* Stevens Point, Wisconsin, USA, 2005, pp. 90-93.

[13]. L. D. Fu, K. S. Chen, J. S. Yu and L. C. Zeng, The Fault Diagnosis for Electro-Hydraulic Servo Valve Based on the Improved Genetic Neural Network Algorithm, in *Proceedings of the International Conference on Machine Learning and Cybernetics*, Dalian, China, 2006, pp. 2995-2999.

[14]. Z. G. Çam, S. Çimen and T. Yıldırım, Learning parameter optimization of Multi-Layer Perceptron using Artificial Bee Colony, Genetic Algorithm and Particle Swarm Optimization, in *Proceedings of the IEEE 13th International Symposium on Applied Machine Intelligence and Informatics (SAMI)*, Herl'any, 2015, pp. 329-332.

[15]. S. Demers, P. Gopalakrishnan and L. Kant, A Generic Solution to Software-in-the-Loop, in *Proceedings of the IEEE Military Communications Conference (MILCOM'07)*, Orlando, FL, USA, 2007, pp. 1-6.

Chapter 19
Electrical Power and Energy Measurement Under Non-sinusoidal Condition

Soumyajit Goswami, Arghya Sarkar, Samarjit Sengupta

19.1. Introduction

In an ideal power system, voltage and current waveforms are pure sinusoids. However, in practice under different circumstances, voltage and current waveform become distorted in nature and causes harmonics and/or interharmonics. The integral multiple of alternating current (ac) system fundamental frequency is defined as harmonics of voltage or current signals. On the other hand, interharmonics is nonintegral multiple of ac system fundamental frequency. Traditional harmonics may cause negative effects such as signal interference, overvoltage, data loss, equipment malfunction, equipment heating, and damage. On the other hand, power system interharmonics also creates lot of complications such as thermal effects, low-frequency oscillation of mechanical system, light and cathode-ray-tube flicker, interference of control and protection signals, high frequency overload of passive parallel filter, telecommunication interference, acoustic disturbance, saturation of current transformer, sub-synchronous oscillations, voltage fluctuations, malfunctioning of remote control system, erroneous firing of thyristor apparatus, and the loss of useful life of induction motors. As a result, the measurement of different power components such as active, reactive and apparent power are error prone. It has been established that there are direct relationship between electrical power and energy measurement. In this chapter, instead of considering both power and energy, different power components measurement under non-sinusoidal condition have been discussed.

19.2. Need of Power and Energy Measurement

Power component measurements are important for many purposes such as designing of power system equipment, setting tariffs and designing compensation devices for improving the quality of the electric energy. Under sinusoidal conditions, the conventional instruments provide excellent accuracy in power quantities estimation. But unfortunately,

Soumyajit Goswami
IBM India Private Limited, Saltlake, Sector V, Kolkata-700091, WestBengal, India

in modern electrical energy systems, voltages and especially currents become less sinusoidal and periodical due to the large number of domestic and industrial non-linear loads. The nonlinear characteristics of semiconductor devices as well as the operational function of most power electronic circuits cause distorted current and voltage waveforms in the supply system. These loads are commonly referred to as "power electronics loads", "power system polluters" or "distorting sources" in the relevant literature. These nonlinear loads primarily generate harmonic currents, which upon passing through the system impedances produce voltage harmonics which distort the system voltage waveform. A typical distorted waveform along with the fundamental and harmonic component has been shown in Fig. 19.1.

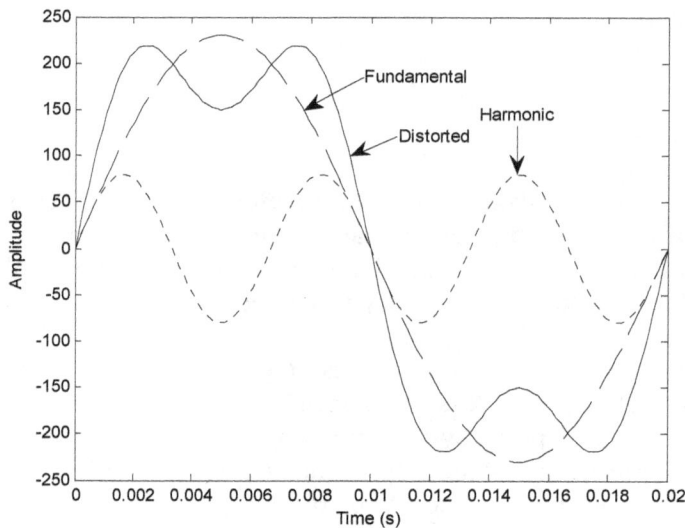

Fig. 19.1. Distorted waveform.

The standard measuring equipment does not comply with any valid definition in this non-sinusoidal situation, and may, therefore, exhibit large errors. Harmonics can cause signal interference, over voltages and circuit breaker failure, as well as equipment heating, malfunction and damage. The consequences of harmonics in power system are:

- Failure of capacitor banks due to dielectric breakdown or reactive power overload;
- Interference with ripple control and power line carrier systems, causing failure of systems which accomplish remote switching, load control and metering;
- Excessive losses resulting in heating of induction and synchronous machines;
- Over-voltages and excessive currents on the system from resonance to harmonic voltages or currents in the network;
- Dielectric breakdown of insulated cables resulting from harmonic over-voltages in the system;
- Inductive interface with telecommunication systems;
- Errors in meter readings;

370

- Signal interference and relay malfunction, particularly in solid state and microprocessor-controlled systems;
- Interference with large motor controllers and power plant excitation systems;
- Mechanical oscillations of induction and synchronous machines;
- Unstable operation of firing circuits based on zero crossing detection or latching.
- Data stored in the computer may be lost up to ten times.

As society in general and economy in particular become more and more dependent on sophisticated information technologies, digital circuits and computer controlled processes, the demand of uninterrupted and conditioned power has grown accordingly. To ensure it, complex operation and control action is required, in which precise measurement of electrical parameters plays important roles.

19.2.1. Importance of Power Quantities Measurements

The following are the key reasons for rising importance of power components measurements.

19.2.1.1. Power System Operation, Control and the Design of Mitigation Equipment

The flows of active and reactive power in a transmission network are influenced by different control actions. Active power control is closely related to frequency control and reactive power control is closely related to voltage control. As constancy of frequency and voltage are important factors in determining the quality of power supply, the control of active and reactive power is vital to the satisfactory performance of power systems. In modern power system, automatic generation control (AGC) and automatic economic load dispatch have been performed for load-frequency control (LFC) whereas static var compensators are utilized to maintain proper voltage levels. All control actions and the design of controller and compensating devices are standing on accurate measurement of active and reactive power and also their direction of flow [1–3]. In order to determine the line and generator capacity, the apparent power knowledge is also necessary [1–3].

19.2.1.2. Tariff Assessment

During recent years, market oriented restructuring of the electricity supply industries has led to consumers' choice, promote competition, improve the quality and maximize plant efficiency. Reliability will be a key issue in this deregulated utility industry. As deregulation takes over the industry, there is a tendency to deteriorate the level of service and investment in the system in terms of harmonics, transients, sags, and many others. Hence, regulators will want to prevent this by developing and standardizing some indices to facilitate the characterizing of power quality levels on the system [4–5]. Consumers are required to limit the disturbances they cause on the utility grid, and may incur penalties when other consumers are adversely affected. Power contracts are based on energy uses as well as maintaining acceptable power factor, imbalance, flicker and harmonic levels.

The monitoring equipment requirements will depend on the specific power quantity variations that must be characterized to evaluate system performance.

19.2.1.3. Evaluation of Electric Energy Quality

Now-a-days, the end use devices and equipment have become more sensitive to power system disturbances. No matter the kind of power disturbances encountered, all pose problems not only to industrial facilities, relaying on electrical devices and electronic components for process control, but also for everything that uses electronic equipment, including common house hold devices at televisions or microwaves. In a competitive market (under deregulated environment) while customers have the option to purchase electric energy from a variety of supply or retail companies, they are more demanding with regard to information on the voltage quality so that electronic components or systems are not seriously disrupted or interrupted. Hence, proper power quality measuring instrument being a concern in today's technical world for both, users and power providers alike [6–8].

19.3. Classical Methods of Power and Energy Measurement

There are many commonly used electrical power components and frequency measuring methods, like the methods using electromechanical indicating instruments or induction type watt-hour meters, which could call as classical methods. There is no doubt that the future is for automatic and computer supported measuring systems. But classic instruments are still present in our lives.

The classic power components and frequency measuring instruments are well described in [11–13] and many other textbooks. Therefore, the discussion is kept short and only the basic ideas are presented.

19.3.1. Active Power Measuring Instruments

The instrument most commonly used for active power measurement is the electrodynamometer wattmeter. It is essentially an inherent combination of an ammeter and a voltmeter and, therefore, consists of two coils known as current coil and pressure coil. The current coil is inserted in series with the line carrying current to be measured and the pressure coil in series with a high non-inductive resistance connected across the load or supply terminals. The operating torque is produced due to interaction of fluxes on account of currents in current and pressure coils, and reading is proportional to current flowing through its current coil, potential difference across potential coil and cosine of the phase angle between voltage and current i.e. active power of the system. For three phase power measurement, "two wattmeter principle" is generally employed. The accuracy of the electrodynamometer type wattmeter is mainly influenced by inductance and capacitance of the pressure coil and the eddy current induction in the metal parts of the meters.

19.3.2. Active Energy Measuring Instruments

The most traditional and widely used ac energy meter (watt-hour meter) is the induction type meter. It operates by counting the revolutions of an aluminum disc which is made to rotate at a speed proportional to the active power consumption. The number of revolutions is thus proportional to the energy usage. The metallic disc is acted upon by two coils. One coil is connected in such a way that it produces a magnetic flux in proportion to the voltage and the other produces a magnetic flux in proportion to the current. The field of the voltage coil is delayed by 90 degrees using a lag coil. This produces eddy currents in the disc and the effect is such that a force is exerted on the disc in proportion to the product of the instantaneous current and voltage. A permanent magnet exerts an opposing force proportional to the speed of rotation of the disc. The equilibrium between these two opposing forces results in the disc rotating at a speed proportional to the power being used. The disc drives a register mechanism which integrates the speed of the disc over time by counting revolutions, much like the odometer in a car, in order to render a measurement of the total energy used over a period of time. The aluminum disc is supported by a spindle which has a worm gear which drives the register. The register is a series of dials which record the amount of energy used.

In an induction type meter, creep is a phenomenon that can adversely affect accuracy. It occurs when the meter disc rotates continuously with potential applied and the load terminals are open circuited.

19.3.3. Reactive Power Measuring Instruments

In an electric circuit, reactive power can be measured by a varmeter (volt ampere reactive meter). This is an electrodynamic wattmeter in which pressure coil circuit is a large inductive reactance or phase shifting transformer (instead of the series resistance in wattmeter) so that the pressure coil current is in quadrature with the applied voltage. As a result an extra 90° phase shift has been introduced between voltage and current signals, and the instrument reads reactive power drawn by the circuit. Since, varmeter is basically an electrodynamic wattmeter, all the advantage and disadvantage of the electrodynamic wattmeter are inherently present within this. Moreover, this instrument does not read correctly if the system frequency differs from the calibration time frequency.

19.3.4. Power Factor Measuring Instruments

Power factor meters, like wattmeters, have two circuits namely current circuit carrying a circuit current (or a definite fraction of it) whose power factor is to be measured and pressure circuit usually split up into two parallel circuits-one non-inductive and other inductive. The deflection of the instrument depends upon the phase difference between the load current and the currents in the two branches of the pressure circuit i.e. upon the power factor of the circuit.

There are two types of power factor meters in common use (1) dynamometer type and (2) moving iron type, both electrodynamic type. The dynamometer type instrument is more accurate than the moving iron type. But its' scale arc is limited and single phase meters are highly sensitive to system frequency change. On the other hand, moving iron power factor meters are simple, robust in construction, cheap and its' scale extends over 360°. However, errors are introduced in these meters owing to losses in iron parts. Moreover, the calibration of these instruments is appreciably affected by variations in power frequency, voltage and waveform.

19.4. Problems Associated with Classical Methods

Conventional measuring instruments, as described in the previous section, initially calibrated on purely sinusoidal signal and subsequently used on a distorted electricity supply, can be prone to error. There are several advantages of traditional electromechanical instruments like simplicity, reliability, and low price. The most important advantage is that the majority of such instruments can work without any additional power supply. On the other hand, there are several drawbacks associated with electromechanical analog indicating instruments. They do not provide electrical output signal, thus there is a need for operator's activity during the measurement. Moreover, such instruments generally use moving mechanical parts, which are sensitive to shocks, aging or wearing out. Regrettably, it can be stated that most of the electromechanical analog instruments are rather of poor quality. In most cases these instruments are not able to measure with uncertainty better than 0.5 %.

The presence of harmonics will overload magnetic shunts, and greatly influence the frequency-sensitive elements like aluminum disc, quadrature and anti-friction loops. Since, the response of this meter to frequencies outside the design parameter is inefficient, the meter is subjected to two error causing effects [14]:

- The magnetic flux driving the rotating disk inside the meter decreases in magnitude proportionally with the order of the harmonic, and hence the meter slows down at higher frequencies when constant power is drawn.

- An erroneous registration of nonexistent "phantom energy" occurs when voltage and current harmonics in quadrature are applied to the meter.

All these effects will have an influence on the electricity bill, which will show an amount that will become more uncertain as the distortion increases.

It is not yet possible to determine exact accuracy of the classic reactive energy meters under non-sinusoidal conditions, as for active energy meters, since there is still no agreement on the definition of reactive energy in the presence of harmonics. The accuracy of the different types of analog power-factor meters (used for low frequencies) under non-sinusoidal conditions generates large errors. The classic frequency meters are also subjected to large error at the presence of harmonics.

19.5. Measurement at Non-Sinusoidal Condition

As discussed in the previous section, all the classic instruments lack in the required accuracy in distorted power system. In order to design proper measuring devices under non-sinusoidal conditions, the following requirements should be dealt carefully:

- Traditional definitions of power quantities with pure sinusoidal waveforms used to work well and were universally accepted, can no longer be a reliable measure in actual power systems. Within a context of practical usefulness, electric power quantities are required to cover all aspects of power quality and to allow determining the direction of the energy transfer and the disturbances correctly.

- Power components and frequency monitoring involves the capturing and processing of voltage and current signals at various points of the power system. The signals to be captured are normally of high voltage and current levels, and thus require large transformation ratios before they can be processed by the instruments. The performances of these transducers should be satisfactory in transforming current and voltage signals containing harmonics.

Different standards have been prepared from time to time by the Institute of Electrical and Electronics Engineers (IEEE) and the International Electro-technical Commission (IEC) which provide guidelines for power quality usages, practices and instrumentation. However, in real-time measurements following two standards are most relevant. Different power components definitions proposed in the IEEE Standard 1459-2010 are discussed.

19.5.1. International Electro-technical Commission (IEC) 61000 4-7

IEC 6100 4-7 [16] standard is the most important document of the IEC 61000 series, covering the subject of testing and measurement techniques. It is a general guide on discrete Fourier transform (DFT) based harmonic and inter-harmonic measurements and instrumentation for power systems and equipment connected thereto. It is "applicable to instrumentation intended for measuring spectral components in the frequency range up to 9 kHz which are superimposed on the fundamental of the power supply systems at 50 Hz and 60 Hz. The standard defines the instrumentation to be used for emissions testing for individual pieces of equipment as well as for the overall measurement of harmonic/ interharmonic voltages and currents in supply systems.

19.5.2. Institute of Electrical and Electronics Engineers (IEEE) 1459-2010

This sub-section provides power components' definitions contained in the IEEE Standard 1459–2010 [15] for distorted power systems.

At steady-state conditions, the non-sinusoidal instantaneous voltage, v, and current, i, of fundamental angular frequency ω_F may be represented by Fourier series of the form

$$v = \sqrt{2}V_1 \sin(\omega_F t - \alpha_1) + \sqrt{2}\sum_{h\neq1} V_h \sin(h\omega_F t - \alpha_h) = v_1 + v_H \tag{19.1}$$

$$i = \sqrt{2}I_1 \sin(\omega_F t - \beta_1) + \sqrt{2}\sum_{h\neq1} I_h \sin(h\omega_F t - \beta_h) = i_1 + i_H, \tag{19.2}$$

where v_1, i_1 represent the power system frequency components, and v_H, i_H represent the harmonic components. V_1 and I_1 are the fundamental rms values and α_1, β_1 are the fundamental phase angles of voltage and current signals, respectively. V_h and I_h indicate the rms values and α_h and β_h represent the phase angles of hth harmonic components of the voltage and current waveforms, respectively.

The direct voltage and the direct current terms V_0 and I_0, obtained for $h = 0$, must be included in v_H and i_H. They correspond to a hypothetical $\alpha_0 = \beta_0 = -45°$; $\sin(-\alpha_0) = = \sin(-\beta_0) = \sin 45° = 1/\sqrt{2}$. Significant dc components are rarely present in ac power systems; however, traces of dc are not uncommon.

19.5.2.1. RMS Calculations

The squared rms values of the non-sinusoidal voltage and current signals are

$$V^2 = \frac{1}{k_c T}\int_\tau^{\tau+k_c T} v^2 dt = V_1^2 + \sum_{h\neq1}V_h^2 = V_1^2 + V_H^2 \tag{19.3}$$

$$I^2 = \frac{1}{k_c T}\int_\tau^{\tau+k_c T} i^2 dt = I_1^2 + \sum_{h\neq1}I_h^2 = I_1^2 + I_H^2, \tag{19.4}$$

where T is the time period and k_c indicates the number of cycles utilized for estimation.

19.5.2.2. Active Power

The total active power P is defined as the average value of the instantaneous power $p = vi$ and has been expressed as

$$P = \frac{1}{k_c T}\int_\tau^{\tau+k_c T} vidt = \frac{1}{k_c T}\int_\tau^{\tau+k_c T} pdt = P_1 + P_H, \tag{19.5}$$

where P_1 is the fundamental active power, defined as

$$P_1 = \frac{1}{k_c T}\int_\tau^{\tau+k_c T} v_1 i_1 dt = V_1 I_1 \cos(\alpha_1 - \beta_1) \tag{19.6}$$

and P_H is the harmonic active power, expressed as

$$P_H = \sum_{h\neq1} V_h I_h \cos(\alpha_h - \beta_h) = P - P_1 \tag{19.7}$$

376

19.5.2.3. Reactive Power

The fundamental reactive power Q_1 is

$$Q_1 = V_1 I_1 \sin(\alpha_1 - \beta_1) \tag{19.8}$$

The harmonic reactive power or Budeanu's reactive power Q_{BU} is given by

$$Q_{BU} = Q_1 + Q_{BH} = Q_1 + \sum_{h \neq 1} V_h I_h \sin(\alpha_h - \beta_h) \tag{19.9}$$

19.5.2.4. Apparent Power

Fundamental apparent power S_1 and its components P_1 and Q_1 are the actual quantities that help define the rate of flow of the electromagnetic field energy associated with the 60/50 Hz voltage and current. This is a product of high interest for both the utility and the end-user. The fundamental apparent power is defined as

$$S_1 = V_1 I_1 = \sqrt{P_1^2 + Q_1^2} \tag{19.10}$$

The total apparent power becomes

$$S = VI \tag{19.11}$$

The separation of the rms current and voltage into fundamental and harmonic terms resolves the total apparent power in the following manner:

$$S^2 = (VI)^2 = \left(V_1^2 + V_H^2\right)\left(I_1^2 + I_H^2\right) = \left(V_1 I_1\right)^2 + \left(V_1 I_H\right)^2 + \left(V_H I_1\right)^2 + \left(V_H I_H\right)^2$$
$$= \left(S_1^2 + S_N^2\right), \tag{19.12}$$

where square of non-fundamental apparent power S_N is

$$S_N^2 = S^2 - S_1^2 = \left(V_1 I_H\right)^2 + \left(V_H I_1\right)^2 + \left(V_H I_H\right)^2 = D_I^2 + D_V^2 + S_H^2, \tag{19.13}$$

where $D_I = V_1 I_H$ is the current distortion power in var unit, $D_V = V_H I_1$ is the voltage distortion power in var unit and $S_H = V_H I_H$ is the harmonic apparent power.

19.5.2.5. Nonactive Power

This power lumps together both fundamental and non-fundamental nonactive components. In the past, this power was called fictitious power. The nonactive power N_{na} is defined as

$$N_{na} = \sqrt{S^2 - P^2} \qquad (19.14)$$

19.5.2.6. Power Factor

The fundamental power factor or displacement power factor has been defined as

$$PF_1 = \cos\theta_1 = \frac{P_1}{S_1} \qquad (19.15)$$

The true power factor has been defined as

$$PF = \frac{P}{S} \qquad (19.16)$$

These definitions are summarized in Table 19.1.

Table 19.1. Summary and Grouping of the Definitions of Power Quantities in Single-Phase Systems with Non-Sinusoidal Waveforms.

Quantity or indicator	Combined	Fundamental Powers (50/60 Hz)	Non-fundamental Powers (Non-50/60 Hz)	
Apparent	S (VA)	S_1 (VA)	S_N (VA)	S_H (VA)
Active	P (W)	P_1 (W)	P_H (W)	
Nonactive	N_{na} (var)	Q_1 (var)	D_I D_V (var)	D_H
Line utilization	$PF = P/S$	$PF_1 = P_1/S_1$	—	
Harmonic pollution	—	—	S_N/S_1	

19.6. Modern Technologies for Power and Energy Measurement at Non-Sinusoidal Condition

Several new techniques have been explored, aimed at measuring electrical power components and frequency under non-sinusoidal conditions. These have been motivated by the existing problems associated with the use of classic measuring instruments at the presence of harmonics. The major trends in measurement techniques have been presented here.

19.6.1. Power Components Measurement

From literature survey, it has been observed that there are some methods or algorithms which can estimate all the power components whereas some other techniques also exist which are particularly used for measurement of specific power components. Noteworthy algorithms which are applicable to estimate all the power components are as follows.

19.6.1.1. Fourier Transform based Assessment

The oldest and conventional approach used in power quantity assessment under non-sinusoidal condition is Fourier transform or its discrete time implementation algorithm fast Fourier transform (FFT) [17]. It permits mapping of signal from time domain to frequency domain by decomposing a signal into several frequency components. However in applying FFT, the phenomena of aliasing, leakage and picketfence effects may lead to inaccurate estimation.

To incorporate time-frequency information, short time Fourier Transform (STFT) based algorithms are introduced which provides a better time resolution as well as equidistant frequency resolution [18]. But, limitation of fixed window-width is inconvenient for the non periodic signal processing.

19.6.1.2. Wavelet Transform based Assessment

The Fourier transform approach can provide amplitude-frequency spectrum while losing time-related information. Moreover, it carries a heavier computational burden. To overcome these limitations, the wavelet transform has been used to measure voltage and current root mean square (RMS) value [19], active power [19] and reactive power [20]. In [21], all power components definitions, introduced in the IEEE Standard 1459–2000, are reformulated using the discrete wavelet transform (DWT). Using the DWT preserves the information concerning time and frequency and also reduces the computational time and effort by dividing the frequency spectrum into bands or levels. But, slow attenuation of quadrature mirror filter banks (QMF) and the overlay of the pass-bands of different levels, strongly influence the frequency decomposition of the DWT and introduce significant error. Moreover, it is often difficult to extract the fundamental or any other single harmonic component of the signal using this method. In order to get detail spectral resolution wavelet packet based algorithms are introduced [22], at the cost of further reduced accuracy and greater computational load.

19.6.1.3. Adaptive Linear Neuron (ADALINE) based Assessment

An adaptive neural network approach for the estimation of the harmonic distortions and power components in power networks have been presented in [23–26]. The neural estimator is based on the use of linear adaptive neural elements called ADALINE. The method presents a very promising approach for fast estimation of harmonics of the distorted power system signals. But, ADALINE is generally employed for known frequency system and unable to synchronize itself with the system's frequency deviation, thus, producing considerable error during off-nominal frequency conditions.

19.6.1.4. Newton Type Algorithm

A two-stage Newton-Type Algorithm for the measurement of power components is presented in [27–28]. To estimate their spectra and fundamental frequency, in the first stage, the current and voltage signals are processed, whereas in the second stage, the power components are calculated based on the results obtained in the first stage. The methods consider the frequency as an unknown parameter and simultaneously estimate it with the input signal spectrum, thus, the algorithm becomes frequency insensitive and the problem becomes non-linear. However, application of this approach is constrained due to its massive computational load.

19.6.1.5. Time Domain Technique

In [29–30], a time-domain strategy is presented for the evaluation of fundamental positive, negative, and zero-sequence components of voltages and currents for single-phase and three-phase applications, both three wire and four wire system. Thus, this strategy allows one to measure all of the power quantities according to IEEE Standard 1459-2000 without using any time-to-frequency transformation. Furthermore, active and reactive harmonic power flow can be measured directly without the evaluation of amplitudes and phase angles of currents. But, this approach is also suffered by large computational complexities.

19.6.1.6. DAQ-based Sampling Wattmeter

Data acquisition based sampling wattmeter (DAQ-SW) has been presented in [31] for the measurement of IEEE Std. 1459-2010 power quantities in non-sinusoidal conditions. The instrument makes use of two commercial DAQ boards, a non-inductive current shunt, a personal computer and a commercial software for data processing.

19.7. Conclusions

In this chapter, an overview of several classical methods applicable for electrical power measurement in the presence of harmonics and/or interharmonics have been discussed. At the same time, problems associated with classical methods and to overcome those, different modern technologies with their pros and cons are presented for power components measurement. The methods include: FFT, DFT, DAQ-SW, EPLL, NTA, ADALINE, S-ADALINE etc. We can conclude this chapter as the selection of the most suitable method depends on the requirements and on the available resources.

References

[1]. P. M. Anderson, and A. A. Fouad, Power System Control and Stability, *Iowa State University Press*, Ames, IA, 1977.

[2]. K. R. Padiyar, Power System Dynamics—Stability and Control, 2nd ed. Hyderabad, *BS Publications,* India, 2002.

[3]. L. L. Grigsby, Power System Stability and Control, *CRC Press, Taylor & Francis,* Boca Raton, 2007.

[4]. Voltage Characteristics of the Electricity Supplied by Public Distribution Systems, EN 50160, *CENELEC*, Nov. 1999.

[5]. Electromagnetic Compatibility (EMC) – Part 3: Limits, *IEC Standards 61000-3*, IEC.

[6]. J. Arrillaga, N. R. Watson, and S. Chen, Power System Quality Assessment, *John Wiley & Sons,* UK, 2000.

[7]. R. C. Dugan, M. F. McGranaghan, S. Santosa, and H. W. Beaty, Electric Power System Quality, 2nd ed., *McGraw-Hill,* NY, 2003.

[8]. IEEE Working Group on Power System Harmonic, Power system harmonics: an overview, *IEEE Transactions on Power Apparatus and Systems*, Vol. 102, No. 8, August 1983, pp. 2455–2460.

[9]. V. Eckhardt, P. Hippe, and G. Hosemann, Dynamic measurement of frequency and frequency oscillations in multiphase power systems, *IEEE Transactions on Power Delivery,* Vol. 4, No. 1, January 1989, pp. 95–102.

[10]. Z. Salcic, Z. Li, U. D. Annakkage, and N. Pahalawaththa, A comparison of frequency measurement methods for underfrequency load shedding, *Electric Power System Research,* Vol. 45, No. 3, June 1998, pp. 209–219.

[11]. S. Tumanski, Principles of Electrical Measurement, CRC Press, *Taylor & Francis, Boca Raton,* 2006.

[12]. W. Golding, and F. C. Widdis, Electrical Measurements and Measuring Instruments, 5th ed., *Wheeler*, Allahabad, India, 1991.

[13]. K. Sawhney, A Course In Electrical and Electronic Measurements and Instrumentation, 2nd ed., *Dhanapath Rai & Co.,* New Delhi, 2001.

[14]. J. Driesen, T. V. Craenenbroeck, and D. V. Dommelen, The registration of harmonic power by analog and digital power meters, *IEEE Transactions on Instrumentation and Measurement,* Vol. 47, No. 1, February 1998, pp. 195–198.

[15]. IEEE Standard 1459-2010, IEEE standard definitions for the measurement of electric power quantities under sinusoidal, non sinusoidal, balanced or unbalanced conditions, *IEEE Standard*, March 2010.

[16]. Electromagnetic Compatibility (EMC) – Part 4: Testing and Measurement Techniques – Section 7: General Guide on Harmonics and Interharmonics Measurement and Instrumentation for Power Supply Systems and Equipment Connected Thereto, IEC 61000-4-7, 2002.

[17]. C. Gherasim, J. V. Keybus, J. Driesen, and R. Belmans, DSP implementation of power measurements according to the IEEE trial-use standard 1459, *IEEE Transactions on Instrumentation and Measurement,* Vol. 53, No. 4, August 2004, pp. 1086–1092.

[18]. S. Zhang, Z. X. Geng, and Y. Z. Ge, The algorithm of interpolating windowed FFT for harmonic analysis of electric power system, *IEEE Transactions on Power Delivery,* Vol. 16, No. 2, April 2001, pp. 160–164.

[19]. W.-K. Yoon, and M. J. Devaney, Power measurement based on the wavelet transform, *IEEE Transactions on Instrumentation and Measurement,* Vol. 47, No. 5, October 1998, pp. 1205–1210.

[20]. W.-K. Yoon, and M. J. Devaney, Reactive power measurement using the wavelet transform, *IEEE Transactions on Instrumentation and Measurement,* Vol. 49, No. 2, April 2000, pp. 246–252.

[21]. W. G. Morsi, and M. E. El-Hawary, Reformulating power components definitions contained in the IEEE Standard 1459–2000 using discrete wavelet transform, *IEEE Transactions on Power Delivery,* Vol. 22, No. 3, July 2007, pp. 1910–1916.

[22]. Vatansever, and A. Ozdemir, Power parameters calculations based on wavelet packet transform, *International Journal of Electrical Power and Energy System,* Vol. 31, No. 10, November / December 2009, pp. 596–603.

[23]. P. K. Dash, D. P. Swain, A. C. Liew, and S. Rahman, An adaptive linear combiner for on-line tracking of power system harmonics, *IEEE Transactions on Power Systems,* Vol. 11, No. 4, November 1996, pp. 1730–1735.

[24]. P. K. Dash, S. K Panda, A. C. Liew, B. Mishra, and R. K. Jena, A new approach to monitoring electric power quality, *Electric Power System Research,* Vol. 46, No. 1, July 1998, pp. 11–20.

[25]. Sarkar and S. Sengupta, On-line tracking of single phase reactive power in non-sinusoidal conditions using S-ADALINE networks., *Measurement,* Vol. 42, No. 4, May 2009, pp. 559-569.

[26]. Sarkar, S. Roy Choudhury and S. Sengupta, A self-synchronized ADALINE network for on-line tracking of power system harmonics, *Measurement,* Vol. 44, May 2011, pp. 784 – 790.

[27]. V. V. Terzija, and J. C. Mikulovic, Digital metering of active and reactive power in non-sinusoidal conditions using Newton type algorithm, in *Proceedings of the Instrumentation and Measurement Technology Conference (IMTC'97),* Vol. 1, May 1997, pp. 314–319.

[28]. V. V. Terzija, V. Stanojevi´c, M. Popov, and L. Sluis, Digital metering of power components according to IEEE standard 1459-2000 using the Newton-type algorithm, *IEEE Transactions on Instrumentation and Measurement,* Vol. 56, No. 6, December 2007, pp. 2717–2724.

[29]. M. Aiello, A. Cataliotti, V. Cosentino, and S. Nuccio, A novel time domain method to locate dominant harmonic sources, in *Proceedings of the XII IMEKO International Symposium on Electrical Measurements and Instrumentation,* Zagreb, Croatia, September 2002, pp. 1–6.

[30]. Cataliotti, and V. Cosentino, A time-domain strategy for the measurement of IEEE standard 1459-2000 power quantities in nonsinusoidal three-phase and single-phase systems, *IEEE Transactions on Power Delivery,* Vol. 23, No. 4, October 2004, pp. 2113–2123.

[31]. Cataliotti, Antonio, et al., A DAQ-based sampling wattmeter for IEEE Std. 1459-2010 powers measurements. Uncertainty evaluation in nonsinusoidal conditions, *Measurement,* Vol. 61, February 2015, pp. 27-38.

Chapter 20
Fine Curvature Measurements through Curvature Energy and their Gauging and Sensoring in the Space

Francisco Bulnes, Isaías Martínez, Omar Zamudio

20.1. Introduction

In the curvature measurements studies an important concept that will be relevant and determinant in the generalization of measure methods and new technology prototypes to measure curvature as field observable, or as microscopic deforming of the space-time associated to the gauge fields that enter in action with the quantum gluing of the matter and the constructing of the electric charge of the particles, is the curvature energy, which is obtained as a change of energy perceived by a curvature sensor designed through integral transforms and electronic cycles used in the fine measurement of the curvature to design of our transduction g-cell sensor that consignee these electrical data in curvature measurements that conform a spectra of curvature that is the energy representation space of the physical curved space (see the Fig. 20.1).

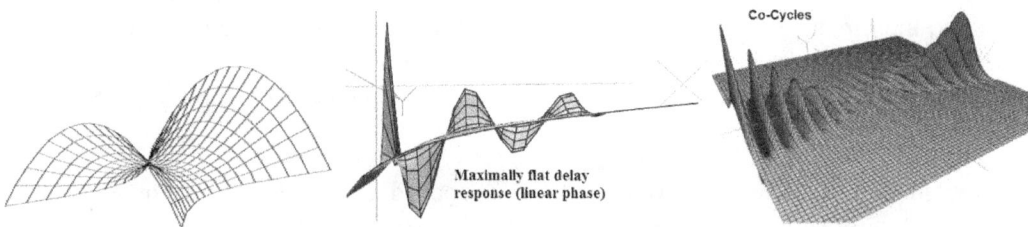

Fig. 20.1. Hyperbolic paraboloid represented through an energy representation space in co-cycles (spectral space of the hyperbolic paraboloid) to measurement curvature through their curvature energy.

Francisco Bulnes
Research Department in Mathematics and Engineering, GI-TESCHA, México

The perception of the curvature in the space has that to be given in terms of an electronic characteristic that has a relation with the geometrical enthrone where the direction change speed of the space due to the existence of curvature and which is established and perceived through of change of sensor field and their flux on the surface where is realized the measure.

The units of the energy curvature are $Volts/meter^{n+1}$, with $n = 1,2,3$, in the case of the 1-, 2- and 3-dimensional space, and the generalizing to n-dimensional spaces. In the case of the usual curvature on a curved trajectory in the space, the corresponding dimensional analysis says:

$$\kappa = \frac{\|\mathbf{v} \times \mathbf{a}\|}{\|\mathbf{v}\|^2} = \frac{\|\text{velocity} \times \text{acceleration}\|}{\|\text{velocity}\|^3} \left(= \frac{(\text{meter}/\text{sec}) \times (\text{meter}/\text{sec}^2)}{(\text{meter}/\text{sec})^3} \right)\left(= \frac{1}{\text{meter}} \right) (= m^{-1}),$$

Adjusting instruments to gauge signals (see the Fig. 20.1) and electronic systems, verifying the analytical expression of some components of the Hilbert inequality [1][1]:

$$[V]^2 \int_C hk^2 ds \geq (\int_C h^2 - k)^2 ds \geq \frac{1}{2} AV^2 \int_C k(\theta)d\theta, \qquad (20.1)$$

The energy representation spaces of the "curvature energy" are bounded in these inequalities. The central term of these inequalities include a kernel that can to characterize a transformation of the space curvature in curvature energy. Indeed, that kernel is given by spherizer operator [2, 3], which will be implicit in the energy quantity sensed to each curved region of the space.

This application must be connected to the creation of a new integral transform (perhaps included in the integral transforms of the curvature sensor) that can be called "curvature transform" which can to create curvature signals that are proper of their intrinsic spectral space and the direct recovering of objective functions of the real space where is realized the measurement of curvature. Of fact, this "curvature transform" is a geometrical integral transform of the physical space in a spectral space [4] with their corresponding recovering of physical data.

Possibly in this study we can find that the proposed transform is a divisor of some correlation function created when is used a filter that lets see only an aspect of the spectral space when this is product of a curved space. The values of the integrals are the change of energy due to the speed of direction change of the g-cell censoring on curved space.

---

[1] Here V is the applied potential energy of curvature, A is the area of the surface and h, their mean curvature and the last integral correspond to the curvature energy employed to measure the roundness in their displacement to along of a principal direction.

20.2. Cycles of Space Curvature and Co-cycles of Curvature Energy

The energy curvature obtained by the energy spectra is realized using the value of the integrals of a field interacting on the geometric pattern along their surface doing it on signals of finite energy where these are Gaussian pulses (the cycles), and that they will code the information of curvature in a spectral space $\mathcal{L}(H(\Omega^2(M)))$, (curvature energy) with M a 2-dimensional space or surface [3]; through of the signals given in the frequency (co-cycles) established in our detector device that detects and measures curvature (see the Fig. 20.2).

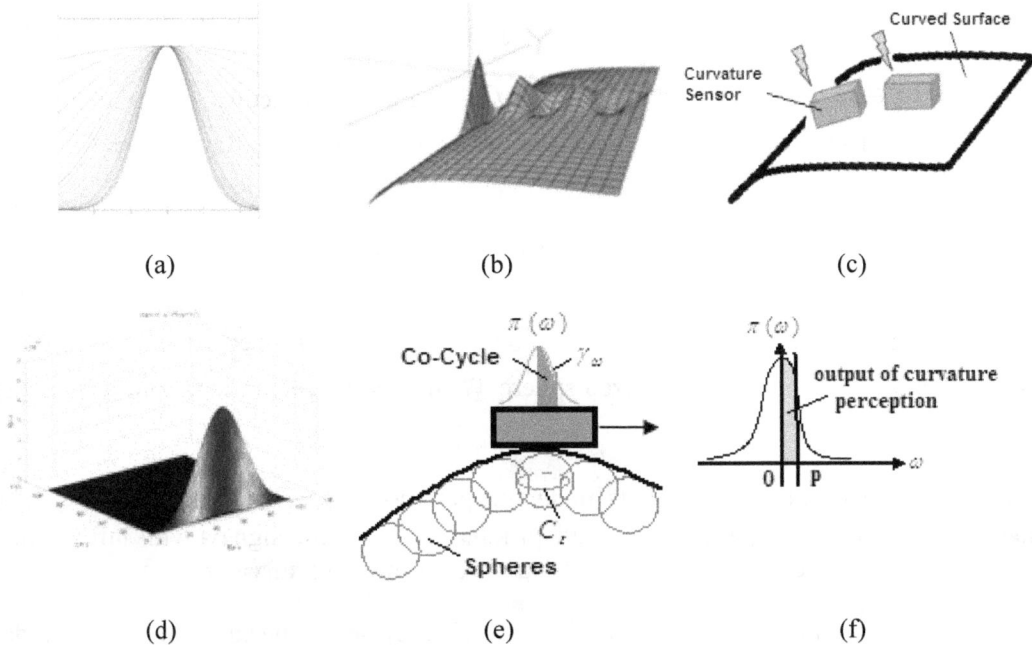

(a) (b) (c)

(d) (e) (f)

Fig. 20.2. (a) Cycles used in the electronical measure of space curvature. (b) Curvature sensor in curved surface M. Curvature energy surface as the surface πM. (d) Cocycles of curvature energy. (e) Net energy given trough their spherizer[1]. (f) Co-cycle as output of curvature energy.

If we consider our space M, as a homogeneous space[2] we can consider the value of the integrals in closed cosets of the corresponding homogeneous space G/H, where H, is a

---

[1] The curvature sensor by curvature energy has an operator defined by the spherizer:
$$\Theta_E = d(\gamma)Hom_K(M,S^2)\int\Omega,$$

Then sensor have the constant Gaussian factor $\Theta_E = 2(-1)^2 \times 1 \times (4\pi R^2) = 8\pi R^2$, to sphere of radious R.

[2] Let $M = G/H$. Let G, be a topological group and H, a closed subgroup of G. A homogeneous space G/H, is a lateral coset space gH, $\forall g \in G$, such that to any left translation T_ρ^*, of G, $\forall h \in G$, and $\rho \in H$, $T_\rho^* g = h, \forall h \in G.$ Concisely $G/H = \{gH | T_\rho^* g = h, \forall \rho \in H, g,h \in G\}.$

closed subgroup of G (that is to say, their elements are operators whose translation is closed in G) in the corresponding space M, that is a differentiable manifold[1].

We consider the integrable invariant of the projections of M, in said closed coset, that is to say

$$p_g : M \to gM, \tag{20.2}$$

Then their integral on G/H, is

$$\int_{G/H} f(x)dx = \int_{G/H} f(gx)dx_g, \tag{20.3}$$

$\forall x_g \in gH,\ x \in M, g \in G/H,$ and $f \in C_c^\infty(M),$ where in particular said integral on generalized spheres take the form of a Radon integral (that is to say, the space M, is divided in cycles):

$$\mathcal{R}(f(x)) = \int_\Xi f(x)d\sigma = \int_{\mu\Xi} f(x)d\sigma_g, \tag{20.4}$$

where Ξ, is the space M, but give or divided in cycles or generalized spheres [3, 5]. Here $d\sigma_g$ is the measure on the generalized sphere $g\Xi$, that is the generalized sphere Ξ, under proper movements of M.

We want cycles that are invariants under proper movements of M, and that can be *measured electronically*, and likewise adapt them as electronic signals with information of the property of the space that we want measure, in this case curvature.

Considering the studies established in [1, 4], we elect cycles of the space invariants under translations and rotations of the space, these cycles of M, are Gaussian pulses satisfaying:

$$\mathcal{R}(\pi(p,x)) = \int_\Pi \pi(p,x)d\xi = \int_{\mu\Pi} \pi(p,x)d\xi_g, \tag{20.5}$$

where Π is the space M, give or divided in Gaussian pulses which are in this case, the generalized spheres. Here $d\xi_g$ is the measure on the generalized sphere $g\Pi$, that is the generalized sphere Π, under proper movements of M. In particular, if we use isometries

---

[1] It's a topological space that locally is Euclidean, which can be "charted", that is to say, exist a biyection of the space in a nth − coupling of real numbers, which is differentiable.

as the Fourier transform[1] to manipulate energy pulses as $\pi(p,z)$, then their co-cycles also are invariant under proper movements of M. Then the space curvature can be write as

$$\kappa(p,x) = \mathfrak{hess}\int_{M} f(x)d\mu(\pi(p,x)) = \mathfrak{hess}\int_{\pi M} f(x)d\mu_{\pi}(x) \qquad (20.6)$$

Considering the relation between Fourier transform with Radon transform we have finally (see Fig. 20.2 (b)):

$$\kappa(\omega) = \int_{M} \kappa(p,x)d\mu(\tilde{\pi}(\omega)) = \int_{\tilde{\pi} M} \kappa(p,x)d\mu_{\tilde{\pi}}(\omega), \qquad (20.7)$$

where $d(\tilde{\pi}(\omega)) = \pi(\omega)d_{y}\mu$. To our very particular case (2-dimensional) our spectral Gaussian curvature will be:

$$\kappa(\omega_1,\omega_2) = \mathfrak{F}_2\{\mathfrak{hess}\mathcal{R}_{\sigma}(f(x,y))\}, \qquad (20.8)$$

where to a Gaussian pulse $\pi(x,y) = e^{-x^2-y^2}$, as cycle[2], the Radon transform is

$$\mathcal{R}_{\sigma}(f(x,y))\} = \int_{M} f(x,y)\delta(\pi(p,\varphi))dxdy = \int_{M} f(x,y)dS(\pi(p,\varphi)), \qquad (20.9)$$

with $dS(\pi(p,\varphi)) = \delta(\pi(p,\varphi))d(e^{-x^2-y^2})$, which is the corresponding Dirac measure of the pulse defining their metric. Then their curvature measure by Gaussian pulses will in the space:

$$\kappa(p,\varphi) = \mathfrak{hess}\mathcal{R}_{\sigma}(f(x,y)) = \mathfrak{hess}\hat{f}(p,\varphi))(h_1,h_2) = 1/2[h_1,h_2]\det\begin{pmatrix} \dfrac{\partial^2\hat{f}}{\partial p^2} & \dfrac{\partial^2\hat{f}}{\partial p\partial\varphi} \\ \dfrac{\partial^2\hat{f}}{\partial\varphi\partial p} & \dfrac{\partial^2\hat{f}}{\partial\varphi^2} \end{pmatrix}\begin{bmatrix} h_1 \\ h_2 \end{bmatrix}, \qquad (20.10)$$

Considering the tempered distribution [6, 7] $\hat{f}(p,\varphi)$, we have [4][3]

$$\kappa(\omega_1,\omega_2) = \int_{M} \kappa(p,\varphi)e^{-j(\omega_1 t_1 + \omega_2 t_2)}dpd\varphi, \qquad (20.11)$$

---

[1] The Fourier transform belongs to the space $\mathcal{L}^2(M)$, that is to say, represents traslations and rotations

[2] $f(x,y) = \begin{cases} e^{-x^2-y^2}, & \text{if } |x|,|y| < \dfrac{1}{2\pi} \\ 0, & \text{if } |x|,|y| \geq \dfrac{1}{2\pi} \end{cases}$

[3] $\kappa(\omega_1,\omega_2) = \int_{M} \mathfrak{hess}\hat{f}(p,\varphi)e^{-j(\omega_1 t_1 + \omega_2 t_2)}dpd\varphi.$

which is our spectra of curvature to a measure realized by our accelerometer in an instant t.

20.3. Transitory Analysis of Response and Bordering Conditions to Curvature Energy

The response (output) time of our accelerometer in the curvature sensoring is $\tau = 2ms \approx 0.2cs$. We have a parameter depending of R, and C, to that relates the electronic part with their geometrical representation given by the Gaussian pulses that are co-cycles of curvature energy to conform our spectral curvature of the measured space [1, 8]:

$$\kappa(\omega_1,\omega_2) = \int_M \kappa(p,\varphi)e^{-j(\omega_1 t_1 + \omega_2 t_2)}dpd\varphi,$$

But the co-cycles are of the form $\frac{2}{\sqrt{\pi}}\alpha e^{-h(\omega_1^2 - \omega_2^2)}$, then is necessary to determine the parameter $h(R,C)$, [9] which come from a transitory analysis of response of our accelerometer[1].

From the differential equation to inputs and outputs system (see solution (20)[2] in [4]) we have a transference function $H(s)$, and applying $\mathcal{L}^{-1}$, we have the dynamic solution of the system $\forall t \geq t_0$,

$$h(t) = V_{in}e^{-\left(\frac{1}{RC}\right)t}, \tag{20.12}$$

---

[1] Remember that, geometrically the relations between voltage outputs and displacement of the accelerometer in the space are given in the design of sensor. Likewise if we consider a one-directional displacement, for example in the ξ – axis, then we have a differential equation of type: $\frac{dV(\xi)}{dt} + \xi V(\xi) = V_E$, whose solutions in function of displacement are $V(\xi) = \frac{2}{\sqrt{\pi}}\int_0^\xi e^{-t^2}dt$. Then the variation of this voltage respect to the displacement is $\frac{d}{d\xi}V(\xi) = \frac{2}{\sqrt{\pi}}e^{-\xi^2}$, that is to say, the space is understanding through Gaussian pulses. Enters of the Hessian matrix must be curvature spectra of these pulses

[2] $I(t) = \lambda e^{\frac{1}{RC}t}$.

388

But considering that there exist a reference voltage $V_{\text{Ref}} = \beta\theta,$ [1] that affects the obtaining of voltage of the capacitance C, we have that $V_{out} = V_{in} - RCV = V_{in} - V_c$, where H, is given by the system:

$$V_{in} \to \boxed{\overset{H(s)}{}} \to V_C, \tag{20.13}$$

where

$$h(t) = V_{out} = V_{in} - V_C = V(1 - e^{-\left(\frac{\beta}{RC}\right)t}), \tag{20.14}$$

which constitutes the electronic characterizing of the Gaussian pulse (co-cycle) having the 95 % free of noise when is considered in pulse $\pi(\omega_1, \omega_2) = \dfrac{2}{\sqrt{\pi}}\alpha e^{-h(\omega_1^2 - \omega_2^2)}$, that is to say the Gaussian pulse is the more optimus in the electronic measure.

Indeed, we prove the consistence of the electronic parameter equation (20.14), and their existence in the co-cycles of curvature energy. Using the principles and definitions given in the Table 20.1, we have the equation[2]

$$Vdt = RC\frac{dV_C}{dt} + V_C \tag{20.15}$$

Table 20.1. Characteristics of the electrical behavior.

Functional Characteristics of the Electrical Behavior		
Electrical Element	**Descriptive Equation**	**Accumulated Energy/Dissipated Power**
CAPACITOR	$V = \dfrac{1}{C}\int idt$ $i = C\dfrac{dv}{dt}$	$\mathcal{E} = \dfrac{1}{2}CV^2$
RESISTANCE	$V = Ri$ $i = \dfrac{V}{R}$	$P = \dfrac{1}{R}V^2$

---

[1] $\theta = Arc\tan^{-1}\left(\dfrac{y}{x}\right).$

[2] We obtain the circuit equation by the Kirchoff law: $V_{in}dt - V_R - V_C = 0$, where $V_{in}dt = iR + V_C$. Consideration that the current in a circuit in serie is equal and only the fall of voltage is different, we have the differential equation $V_{in} = RC\dfrac{dv}{dt} + V_C$. Now we demostrate that if we change V_{in}/V_{ref}, affects to V_C, as in the accelerometer, that is to say: $\dfrac{dV_C}{dt} = \dfrac{1}{R}(Vdt - V_Cdt)$, where integrating we have:

$\int\dfrac{dV_C}{V(t)} = \int\dfrac{1}{RC}(Vdt - V_Cdt) = \ln|V_C(t)| = \dfrac{1}{RC}t$, or equivalently $V_C(t) = \lambda e^{-\beta t/RC}$

Considering the equation of the geometrical analogy of (15), we have

$$a_1 \frac{d\theta}{dt} + a_0 \theta_0 = b_0 \theta_1, \tag{20.16}$$

where we can to consider $a_1 = RC, a_0 = 1,$ and $b_0 = 1.$ Then a solution of (16) is:

$$\theta_0 = \left(\frac{b_0}{a_0} \right) \theta_1 (1 - \exp\{-(a_0 t / a_1)\}), \tag{20.17}$$

having the boundering conditions, if $t = 0,$ we have that $\exp\{-(a_0 t / a_1)\} = 1.$ For other side, if $t = \infty,$ $\exp\{-(a_0 t / a_1)\} = 0.$ Then $\theta_0 = U + V,$ where U is the transitory response and V is the forced response. The stable modus (fix point) is $\theta_0 = \left(\frac{b_0}{a_0} \right) \theta_1,$ where the transference function in stable modus is $G_{ss} = \frac{a_0}{a_1} = \frac{b_0}{b_0}.$ Then in their electric characterization we have $G_{ss} = \frac{V_C}{e^{-\frac{t}{RC}}},$ where (17) takes the form

$$\theta_0 = G_{ss} \theta_1 (1 - \exp\{-(a_0 t / a_1)\}), \tag{20.18}$$

with a net gain $G_{ss} \theta_1.$ If the time is $t = (a_1 / a_0),$ then the exponential term has the value $e^{-t} = 0.2,$ which is mentioned at the beginning of this section. Then we can assign $\beta = V_{ref} / \theta,$ that affects the differential equation to the obtaining of the voltage of $C,$ that is to say, the voltage $V_C.$ Likewise, $\theta = \tan^{-1}(x / y),$ which is the angle that takes the accelerometer (see the Fig. 20.3), having $\Omega = V(1 - \exp\{-(\beta t / RC_*)\}),$ where Ω is a constant that relates the electronic part with their geometrical representation and C_* is the variable capacitance.

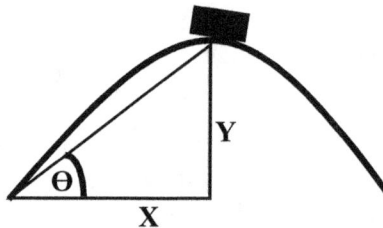

Fig. 20.3. Angle to capacitance considered by $V_C.$ The accelerometer takes said angle establishing a geometricalparameter in their transitory analysis of voltage with $\Omega = V_C.$

20.4. Dimensional Analysis and Metrology

As has been mentioned, the units of curvature energy are $Volts/meter^{n+1}$, which are justified under the design of our curvature sensor that involves a detector of accelerometer type designed under *electrostatic tension* produced by *the variation of the electrostatic field* E, (subjacent in their design) for unit of area over surface under change of direction experimented by the our sensor when this realizes change of position (Fig. 20.4).

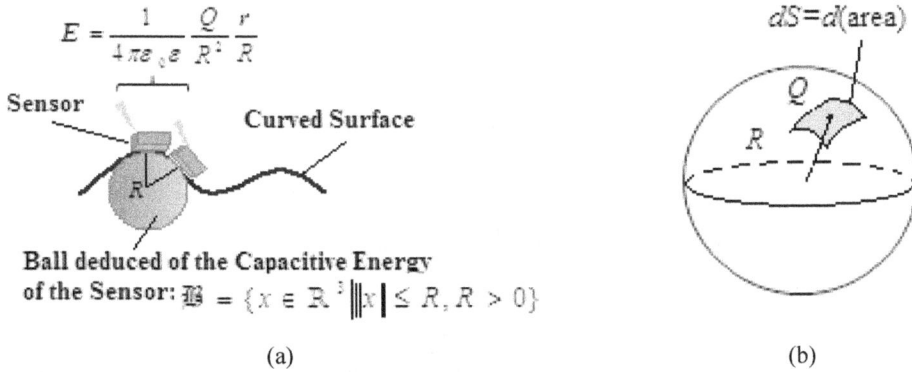

$$E = \frac{1}{4\pi\varepsilon_0\varepsilon}\frac{Q}{R^2}\frac{r}{R}$$

Sensor

Curved Surface

Ball deduced of the Capacitive Energy of the Sensor: $\mathbb{B} = \{x \in \mathbb{R}^3 \mid \|x\| \le R, R > 0\}$

(a)

$dS = d(area)$

Q

R

(b)

Fig. 20.4. Variation of the electrostatic field.

Theorem (F. Bulnes). 4. 1. The curvature energy is given by the capacitive energy available in the curvature of the space and registered in the variable capacitor is the $\mathbb{B}$, ball charge energy $AQ^2/2\Theta_E C$, where $\Theta_E{}^1$ is the spherizer operator [3, 2]. Their curvature is E/V.

Proof. We consider the case to 2-dimensional surface. We consider the electric force that has a curved surface whose curvature has a sphere of radious R, adequated (see the Fig. 20.4 (b)). Then the electric field due to their capacitance when is situated the sensor in the point where is measuring is:

$$E \bullet K_T = \frac{1}{\varepsilon\varepsilon_0}\frac{CV}{R^2}\frac{R}{R}, \qquad (20.19)$$

For other way, the designed capacitance of the sensor obeys to

---

[1] Operator that permits adequate spheres. This operator is deduced through the Gauss-Bonnet theorem in geometry. Here K_T, is the total Gussian curvature of a compact and orientable geometrical surface that is the topological invariant. The surfaces that we measure are of this class.

$$C(D,A) = \frac{A}{D}\varepsilon_0\varepsilon, \tag{20.20}$$

But $CV = Q$ is the charge due to energy sphere of radious R, that can be set pointed to the curved surface. But by the spherizer we have that $\boldsymbol{E} \bullet \boldsymbol{K}_T = \boldsymbol{E} \bullet \mathcal{O}_E / \chi(S^2) = \boldsymbol{E} \bullet 4\pi$, and considering that the point of sensing is the point of charge application Q then we can consider from (19) and (20) that $D \approx R$, where

$$k = \frac{E}{V} \tag{20.21}$$

is the sectional curvature. Indeed, by the dimensional analysis

$$\frac{\|\boldsymbol{E}\|}{V} = \frac{\left(\dfrac{Newton}{Coulumb}\right)}{\left(\dfrac{Kg \times meter^2}{Ampere \times \sec^3}\right)} = \frac{Newton \times Ampere \times \sec^3}{Coulumb \times Kg \times meter^2} =, \frac{\dfrac{Kg \times m}{\sec^2} \bullet \dfrac{Coulumb}{\sec} \bullet \sec^3}{Coulumb \times kg \times meter^2}$$

$$= \frac{kg \times meter \times coulumb}{Coulumb \times kg \times meter^2} = \frac{1}{meter} = \frac{1}{R},$$

Our sensor device is designed to do vary the capacitance in a distance range at most R, (due to that our sensor has a variable capacitor whose radious of energy ball vary in $0 \leq x \leq R$ (see the Fig. 20.4 (a)), in an area A, of a surface portion where is applicable the sphere of radious R, considering that $K_T(S^3) = 8\pi^3\chi(S^3)$, [1] where S^3 is the ball or 3-sphere of radious R. Then capacitive energy available in the curved space which is available in the ball $\mathfrak{B}$ is:

$$\mathcal{E} = \int_0^Q V dQ = \frac{1}{K_T} \int_0^Q E dQ, \tag{20.22}$$

But the corresponding design of our sensor obeys to the capacitive perception of curvature given by the electrostatic force $F = \dfrac{1}{2}\dfrac{AV^2}{D^2}$, that causes the deflection in the g – cell component in accelerometer. Remember that this electrostatic force is designed in our sensor by the basic isotopic component of Gaussian factor to lectures of curvature defined as $\alpha(-1)^2 \times 1 \times (4\pi(1)^2)$, where $(-1)^2$ is the basic charge given in function of the milimetric potential V, the factor $4\pi(1)^2$ is the sphere surface of unit $S^2(1)$, α is the degree of the spherical map used in the *transduction* of the physical model to measure, which comes

---

[1] The generalized Gauss-Bonnet theorem is considered.

like a factor of electromagnetic adjustment such that signed by the equation $E \bullet K_T$, of the sensor on curved surface, and 1 is the positive charge generated inside the sensor (see the Fig. 20.5). Then the integral of energy takes the forma

$$\mathcal{E} = \frac{1}{4\pi} \frac{A}{2R^2} \int_0^Q \frac{V^2}{Q} dQ = \frac{AQ^2}{16\pi R^2 C}, \tag{20.23}$$

But $K_n \Theta_E = K_T \chi(S^2)$, [3], then (23) takes finally the form $\mathcal{E} = AQ^2 / 2\Theta_E R^2 C$. Indeed realizing the dimensional analysis we have:

$$\frac{AQ^2}{2\Theta_E R^2 C} = \left(\frac{meter^2 \times Coulumb^2}{meter^2 \times \dfrac{\dfrac{Coulumb}{1}}{\dfrac{Kg \times meter^2}{Ampere \times sec^3}}} \right) = \left(\frac{meter^2 \times coulumb^2}{meter^2 \times \dfrac{Coulumb \times Ampere \times sec^3}{Kg \times meter^2}} \right)$$

$$= \left(\frac{meter^2 \times coulumb^2 \times kg \times meter^2}{meter^2 \times coulumb \times \dfrac{coulumb}{sec} \times sec^3} \right) = \left(\frac{coulumb \times kg \times meter^2}{coulumb \times sec^2} \right)$$

$$= \frac{kg \times meter^2}{sec^2} = Joule,$$

which are energy units.

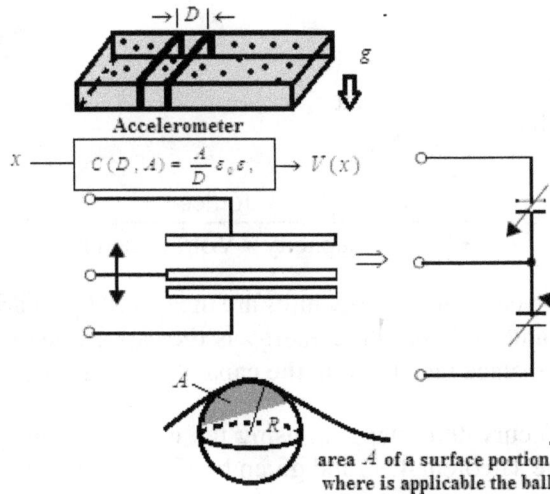

Fig. 20.5. Capacitive energy available in the curved space.

We consider the following result that is direct consequence of the before theorem.

Corollary (F. Bulnes, I. Martínez) 4. 1. The curvature energy measured in *Volts / meter*^{3}, over a curved surface is the energy obtained by the electric force on the electrical charge produced by the curved surface for unit of area, that is to say

$$\mathbf{E} \bullet d^{-3} (= \frac{Nw}{C} \times meter^{-2}), \tag{20.24}$$

Proof. By the theorem 4. 1, and the inequality (1) [1], and using the electrical-geometrical consideration of the before section 20.3, we have:

$$Volts / meter^3 \left(= \frac{1kg \times meter^2}{Ampere \times seg^3} \frac{1}{meter^3} \right) \left(= \frac{1kg \times meter^2}{seg^2} \frac{meter}{Ampere \times seg} \frac{1}{meter^3} \right)$$

$$\left(= 1 Joule \times \frac{1}{Coulumb} \frac{1}{meter^3} \right) \left(= 1 Nw \times meter \frac{1}{Coulumb} \frac{1}{meter^3} \right)$$

$$\left(= \frac{1Nw}{C} \times meter^{-2} \right).$$

Now is necessary to prove that is our *curvature energy* given by the theorem and gauged by the electronic instruments. Also we can to define our curvature energy as the density of electrical energy of a curved region of the space $\left(= 1 Joule \times \frac{1}{Coulumb} \frac{1}{meter^3} \right)$.

But before is necessary consider the accumulated capacitive energy available, that is to say, the energy available inversely to the electrical charge in a complex in $\mathbb{R}^3$, because our units involves volume units $(= 1 / meter^3)$ (that is characterized in a ball), we use the equation of energy to a capacitor given in the Table 20.1, considering the demonstration of the corollary 4. 1:

$$\frac{Capacitive\ Energy}{Charge} \times meter^{-3} =$$

$$= \frac{\mathcal{E}}{QVol} = \frac{\frac{1}{2}CV^2}{CV \bullet Vol} \left(= \frac{Joules}{Capacity \times Volts \times meter^3} \right) \left(= \frac{Volts}{meter^{-3}} \right) \tag{20.25}$$

that is to say, indeed the curvature energy units are measured by a variable capacitor inside our curvature sensor, and said curvature energy is the capacitive energy available by the proper curvature of the space registered in the capacitor as has been mentioned.

Then we can to give a curvature mapping using the curvature energy through of the net energy estimated in the Theorem 4. 1, and given by (23) on our test platform with curved surfaces as is showed in the Fig. 20.6 (a). Then the energy as the function $\mathcal{E}(C, \mathcal{O}_E)$, can be illustrated in the Fig. 20.6 (b), to little portion of the test surface.

measured non-symmetrical incurved surface region

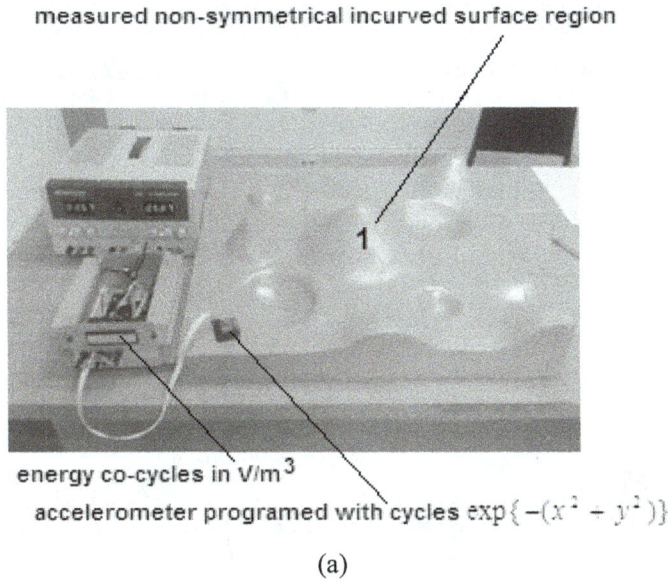

energy co-cycles in V/m^3

accelerometer programed with cycles $\exp\{-(x^2 + y^2)\}$

(a)

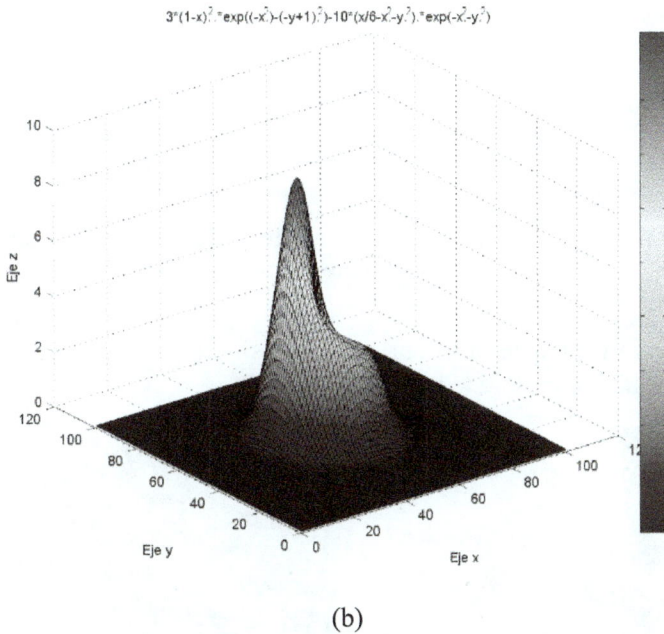

$3^*(1-x)^2 {}^* \exp((-x^2)-(-y+1)^2)-10^*(x/6-x^3-y^5) {}^* \exp(-x^2-y^2)$

(b)

Fig. 20.6 (a, b). (a) Region 1 of the test surface; (b) Energy as a function $\mathcal{E}(C, \mathcal{O}_E)$.

To major incurved region the invest energy by our variable capacitance is major and here appears more energy intensity (red color). In the case of the flat region the intensity of this energy is null (obscure blue color (Fig. 20 (b))). In the case of negative curvature, that is to say $K_G < 0$, the charge is given by $(-1)^3$, to a milimetric potential V, per volume unit, because the sensor has displacement in Z – axis but the milimetric potential used is

395

minimal and gives negative potential $-V$, then the available energy is completely potential and from a point of electrical view appears as obscure blue in the Fig. 20.6 (d). However, their capacitance in variable and creates a new Gaussian pulses with a little variation of the accelerometer.

The registered variations outputs in the spectra correspond to that the measured surface is non-symmetrical (see the Fig. 20.6 (c)), and .the Gaussian pulses are adjusted to the curvature perception of the curvature sensor on the surface and their incurved elements (Fig. 20.6 (d)).

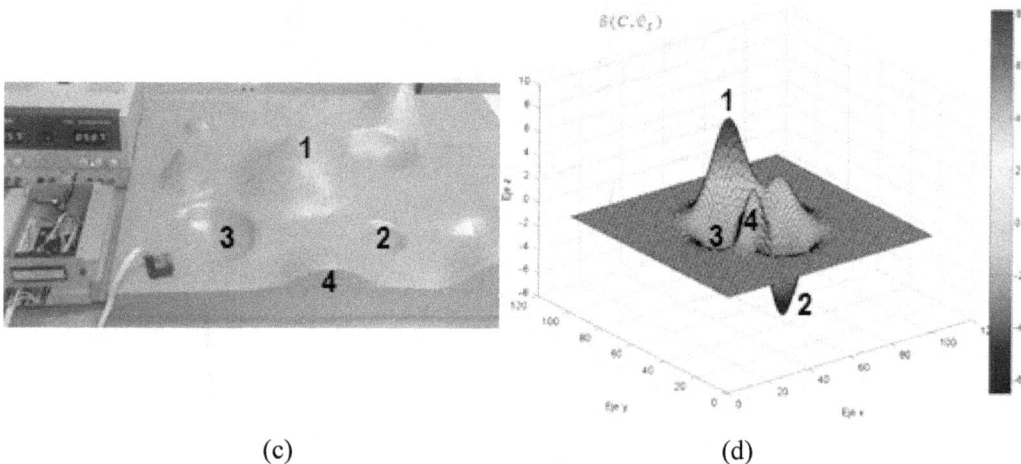

(c)   (d)

Fig. 20.6 (c, d). (c) The Gaussian pulses are adjusted in the vicinity of the incurved elements **1, 2, 3 y 4,**. (d) Their curvature spectra by co-cycles.

We observe the following spectra of curvature considering the same modelling realized before to a hyperbolic paraboloid (Fig. 20.7).

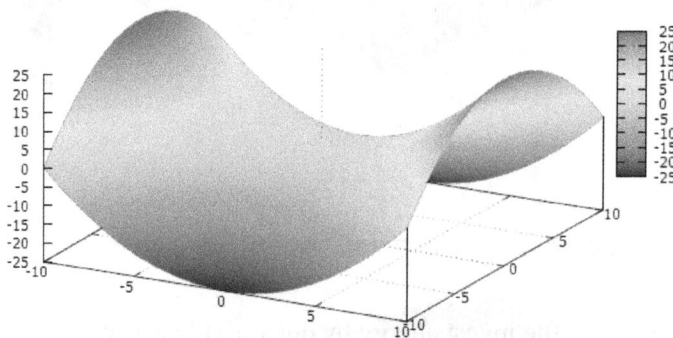

Fig. 20.7. Newly the red color shows the region of maximum curvature. The yellow color shows an average curvature and the light blue a flat spaces. Finally the strong blue is the corresponding to the negative curvature (concave regions).

20.5. Instrumentation and Laboratory Measurements

The experiments will consider the relation between our gauge field given by the electrical field of the proper sensor device that helps to detect and measure the curvature through the variation of their capacitance (sensing part of sensor) and the field of reference system of the space created by the spatial positioning and their coordinates correlation with the implicit sphere in the net energy that involves the spherizer operator as has been demonstrated in the before section.

Likewise, we consider the equation $E = \kappa V$, of the *Theorem 4. 1,* where we are considering a field of reference system as the generated by the field of reference system of a sphere Σ (kernel of spherizer or kernel of curvature transform[1]) given in the programming of the sensor, thus we have in the space $\mathbb{R}^3$, to electrical field E, that [10][2]

$$E = \begin{pmatrix} E_x \\ E_y \\ E_z \end{pmatrix} = \begin{pmatrix} \omega_{11} & \omega_{12} & \omega_{13} \\ \omega_{21} & \omega_{22} & \omega_{23} \\ \omega_{31} & \omega_{32} & \omega_{33} \end{pmatrix} \begin{pmatrix} U_1 \\ U_2 \\ U_2 \end{pmatrix} = \begin{pmatrix} 0 & \omega_{12} & \omega_{13} \\ 0 & 0 & \omega_{23} \\ 0 & 0 & 0 \end{pmatrix} \begin{pmatrix} U_1 \\ U_2 \\ U_3 \end{pmatrix},$$ (20.26)

where $U_i = \nabla_i V$ is the corresponding gradient of the potential V. In the case of the sphere the gradient has the normal direction and satisfies with the unitary tangent vector that $U_i \mathbf{v} = 0$, (see the Fig. 20.8a)). For other side, each connection form $\omega_{ij} = \nabla_v U_i U_j (p)$, $\forall p \in \mathbb{R}^3$.

From (20.26) and using as dual reference system[3] the spherical coordinates (Figs. 20.8 (b), (c) and (d)) we have:

$$\omega_{12} = \sin\varphi d\theta, \quad \omega_{13} = -\cos\varphi d\theta, \quad \omega_{23} = -d\varphi,$$ (20.27)

where realizing several measurement on the superior spherical cap surface (black points in the spherical surface Σ [4] in the Figs. 20.8 (b), (d), we have the following Table 20.2.

These measurements were realized in a foam ball as is showed in the laboratory photograph (Fig. 20.8 (d)). We observe the uniformity of the values presenting certain periodicity in all electrical outputs of the curvature sensor to different position on the superior spherical cap surface. The outputs values are represented by the little histograms behind the high histograms (Fig. 20.9). The histograms in the first plane (high histograms)

---

[1] This kernel comes given by the spherical mapping or Gauss mapping $G : M \to \Sigma$, in the spherizer operator [1, 3] which appear in the invested energy by the sensor.

[2] Here the equation $E = \kappa V$, is generalized to $3D$, as the vector equation $E^i = \Omega_{ij} V^i, i = 1,2,3$.

[3] $z_1 = r\cos\varphi d\theta, \quad z_2 = rd\varphi$.

[4] Remember that we consider to spherical mapping $G : M \to \Sigma$, that in an orientable surface $M \subset \mathbb{R}^3$, we have the total Gaussian curvature through the integral of the form $KdM = G*(d\Sigma)$.

show the uniform or constant increase of the measurements taken in the octants I, II, III, and IV (Fig. 20.8 (d)).

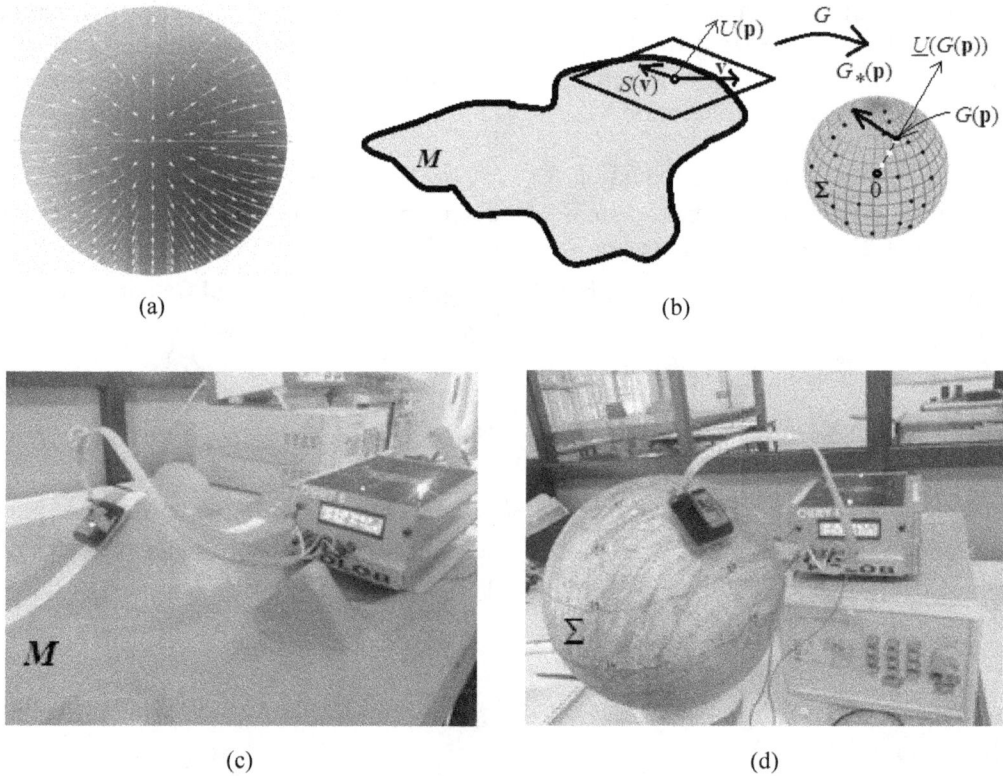

(a)

(b)

(c)

(d)

Fig. 20.8. Experiments that show the relation between the electrical outputs measurements obtained by the curvature sensor and the field of reference systems given by the shape operator.

Table 20.2. Electrical measurements realized by the sensor directly on superior spherical cup surface (octants I, II, III, and IV).

Measurement of the curvature of the sphere with bending sensor					
Coordinates in the sphere			Voltages at the accelerometer exes		
X	Y	Z	U_1	U_2	U_3
1	1	1	0.01	-0.5	-0.1
2	2	2	0.01	-0.9	-0.4
3	3	3	0.01	-1	-0.7
4	4	4	-0.1	-1.2	-1.4
5	5	5	0.11	-0.4	-0.1
6	6	6	0.11	-0.8	-0.3
7	7	7	0.11	-1	-0.7
8	8	8	-0.2	-1.2	-1.3
9	9	9	0.31	-0.8	-0.4
10	10	10	0.21	-0.8	-0.3
11	11	11	0.02	-1.1	-0.8

Table 20.2 (Continued). Electrical measurements realized by the sensor directly on superior spherical cup surface (octants I, II, III, and IV).

Measurement of the curvature of the sphere with bending sensor					
Coordinates in the sphere			Voltages at the accelerometer exes		
X	Y	Z	U_1	U_2	U_3
12	12	12	0.01	-1.2	-1.3
13	13	13	-0.1	-0.5	-0.1
14	14	14	0.11	-0.9	-0.4
15	15	15	0.11	-1.2	-0.9
16	16	16	-0.1	-1.3	-1.4
17	17	17	0.16	-0.4	-0.1
18	18	18	0.15	-0.8	-0.3
19	19	19	0.01	-1.1	-0.8
20	20	20	-0.1	-1.3	-1.4
21	21	21	0.21	-0.4	-0.1
22	22	22	0.12	-0.9	-0.3
23	23	23	-0.1	-1.1	-0.8
24	24	24	-0.1	-1.2	-1.4
25	25	25	0.11	-0.4	-0.1
26	26	26	0.01	-0.8	-0.3
27	27	27	0.01	-0.1	-0.8
28	28	28	0.01	-1.2	-1.3
Reference Point (Rp)					
X	Y	Z	Xv	Yv	Zv
0	0	0	0	0	0

Corollary. 5. 1. (F. Bulnes, I. Martínez, O. Zamudio, C.). The electrical boundering conditions to our curvature sensor applying (26) through their shape operator (see the Fig. 20.2 (e)) and 20.9 (b)) are:

$$E_{n_1} - E_{n_2} = P, \qquad R(E_{\tau_1} - E_{\tau_2}) = V, \qquad (20.28)$$

Proof. The gradient on the sphere Σ, comes given by $U = \dfrac{1}{R}\Sigma x_i U_i$. Then their shape operator is:

$$\nabla_v V = \frac{1}{R}\Sigma v[x_i]V(p) = \frac{v}{R}, \qquad (20.29)$$

But for electromagnetism of the sensor device field $E = -\nabla_v V = -\dfrac{v}{R}$, where the electrical field has the same direction of the tangent vector (because is the field of the device that is displaced on surface), thus $E \bullet v = E_\tau$ is the tangential component. Then to two different position of the sensor we have the two tangential components $E_{\tau_1} = E \bullet v_p$, and $E_{\tau_1} = E \bullet v_q$, $\forall p,q \in \Sigma$, then in the two different positions of the sensor is had that

Voltage Outputs in the superior Spherical Cap Xv
$=U_1$, Yv $= U_2$, and Zv$= U_3$

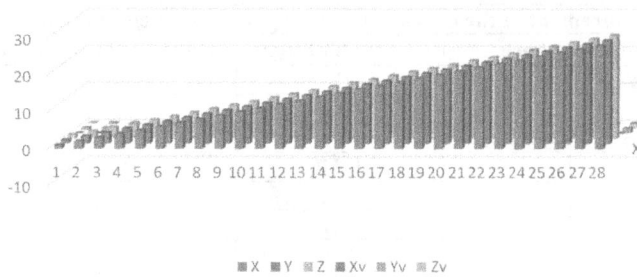

■ X ■ Y ▨ Z ■ Xv ■ Yv ▨ Zv

(a)

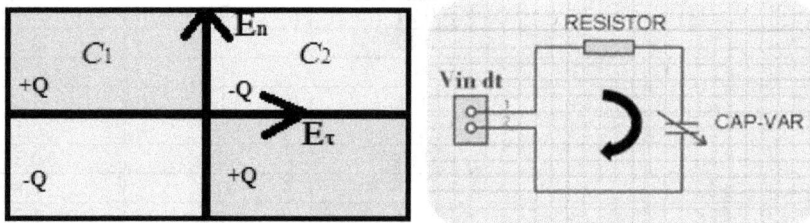

(b)

Fig. 20.9. (a) Voltage Outputs distribution on superior spherical cap surface. (b) Scheme of tangential and normal boundering conditions. The normal component of electric field varies due to the change of charges inside the accelerometer (Polarization of the accelerometer). The tangent component varies due to the invested voltage along displacement distance which depend of the R, taken. Here C_1, and C_2, are two capacitance values.

$$(E_{\tau_1} - E_{\tau_2}) \propto E_\tau = V\kappa = V\frac{1}{R} = \frac{V}{R}, {}^1 \qquad (20.30)$$

The gauging of units is correct because the dimensional analysis gives

$$\frac{V}{R}\left(\frac{1Kg \times meter^2}{Ampere \times \sec^3}\right) = \left(\frac{1Kg \times meter^2}{\frac{\frac{1}{Coulumb}}{\sec} \times \sec^3}\right) = \left(\frac{\frac{1Kg \times meter^2}{meter}}{\frac{Coulumb \times \sec^2}{1}}\right)\left(\frac{\frac{1Kg \times meter}{\sec^2}}{Coulumb}\right)$$

$$\left(=\frac{1Nw}{Coulumb}\right)(=E),$$

---

[1] There is a value that does proportional to the voltage used for the device respect to the tangential electric field component. This value is the curvature value.

In the case of the normal components as the normal vector varies due to the polarization P, due to the change of charges for each capacitance as was showed in the Fig. 20.9 (b), then in Σ, we have $E_{n_1} - E_{n_2} = P$. Their gauging of units is obvious.

What happens in the signal analysis context? The implicit spherizer action inside the device establish a behavior of electric field that reflex coordinates correlation with the implicit sphere in the net energy (20.23) that involves the spherizer operator. This has a contextualization more wide, for example the application of the Gaussian pulses and their harmonic analysis behavior (see the Fig. 20.10), when the action is realized by the spherizer. Curiously, this behavior is similar to the spherical mode of the electric field on charged sphere (Fig. 20.11 (c)).

Spherizer Actions in each Spatial Component

——X ——Y ——Z

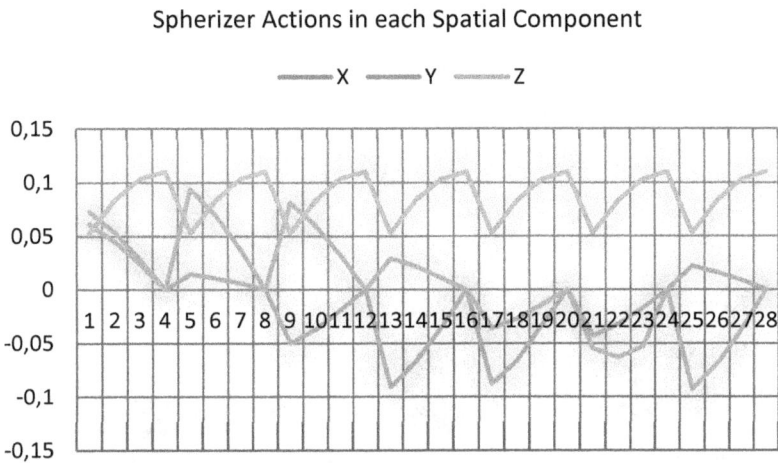

Fig. 20.10. The Spherizer actions $\mho_E$, (curvature transform) in each spatial component of electrical field of the curvature sensor. The graph shows $\Im_{\mho_E} - E$, where E is the electric field [11].

Then this demonstrate the relation between the variation of their capacitance (sensoring part of sensor) and the field of reference system of the space created by the spatial positioning and their coordinates correlation with the implicit sphere in the net energy that involves the spherizer operator.

20.6. Conclusions

The curvature as a geometrical property of the space is the geometrical invariant more important that characterizes the shape of space. But also in the physical sense the curvature is a field observable where the field has a direct inherence on the space creating their scenery that obeys a geometrical characteristics inherent of the proper space.

Spherical Harmonics in Curvature Sensor

(a)

(b)

(c)

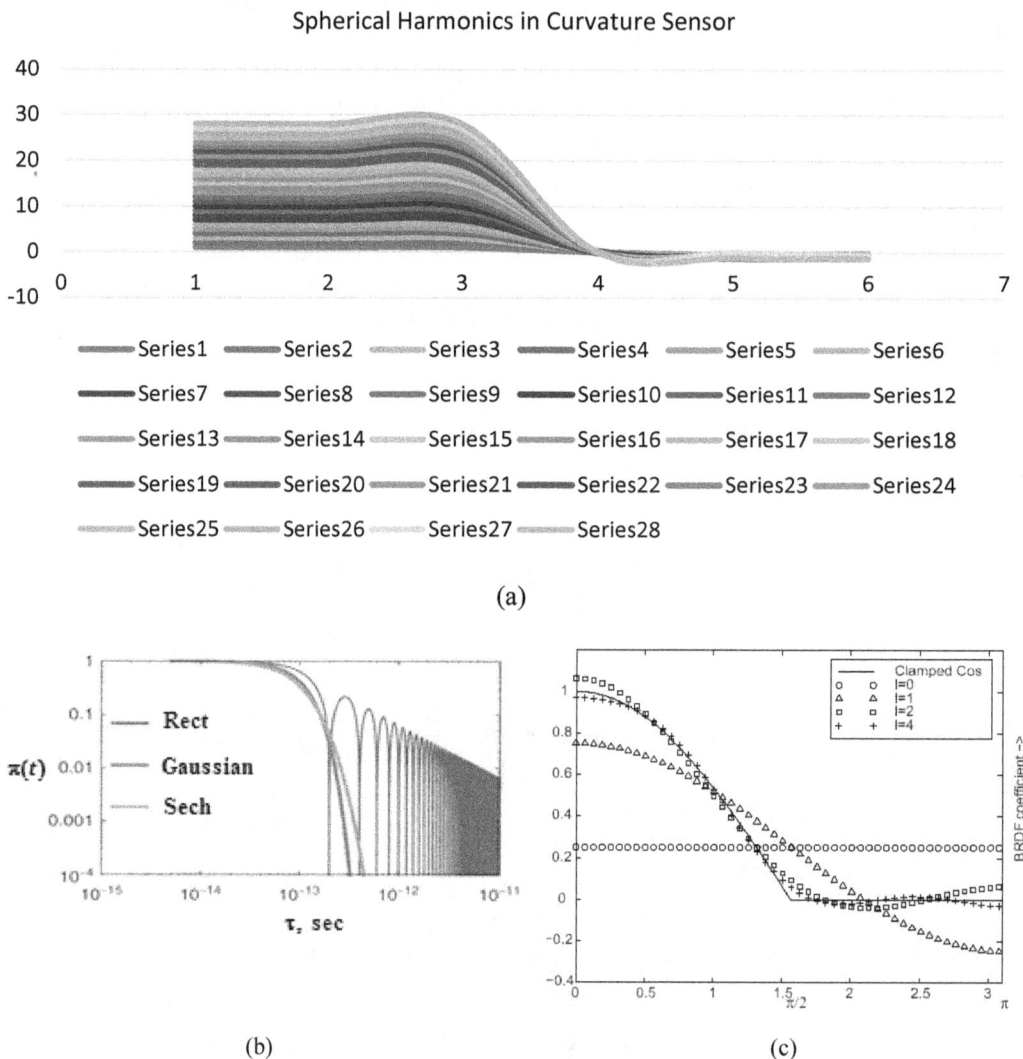

Fig. 20.11. Spherical harmonic modes [11]. (a) Spherical harmonics in curvature sensor. (b) Gaussian energy pulse used in each time $\tau = 2ms \approx 0.2cs$, established in the section 20.3. (c) Successive approximations to clamped cosine by adding more spherical harmonics.

In this chapter the goal is demonstrate the consistence of the units of curvature energy, which precisely generalizes the way to measure and detect curvature not only as geometrical invariant of the space or physical observable of the space, but also as a property that measure the deforming action of a field on the space producing energy product of this deformation creating likewise the concept of curvature energy which can be manipulated through finite energy signals that carry the information of curvature through the capacitance variation. The units in MKS system of curvature energy are Volts/m^3, whose net energy invested for unit of surface is the given by an energy sphere whose energy is $\mathcal{E}(\mathcal{O}_E, C)(=\text{Joules})$. Likewise the electromagnetic field of the sensor device has as component the given in the two directions, the normal and tangent vectors to the curved surface which satisfy the boundering conditions established in (20.28), where $C(D, A) = \varepsilon \varepsilon_0 A / D$, is the variable capacitance of the accelerometer. Finally, we can establish that curvature energy is the spectral curvature $\kappa(\omega_1, \omega_2)$, whose cycles are determined by (20.11) and whose correlation function between this curvature and the spherizer is given by the application of these co-cycles whose outputs in our curvature sensor is given in volts. These measurements will be useful to be applied in advanced measurement studies in curvature and torsion of the space, for example in the detection of the quantum gravity, where also will be measured in terms of curvature energy [12, 13].

Acknowledgements

Our acknowledgement to Edgar Daniel Sánchez Balderas, TESCHA Director, and Evaristo Vázquez, Financial Department Director. Also we are grateful with René Rivera Roldán, Chief of TESCHA Electronic Engineering Division for the available and facilities of the laboratory equipment. We greatful to Cuauhtemoc for the modelling realized in the Fig. 20.6 (b), and the measurements obtained to demonstrate the electrical gauging of the curvature sensor.

References

[1]. F. Bulnes, I. Martínez, A. Mendoza, M. Landa, Design and Development of an Electronic Sensor to Detect and Measure Curvature of Spaces Using Curvature Energy, *Journal of Sensor Technology,* 2, 2012, pp. 116-126.

[2]. F. Bulnes, E. Hernandez and J. Maya, Design of Measurement and Detection Devices of Curvature through of the Synergic Integral Operators of the Mechanics on Light Waves, in *Proceedings of the International Mechanics Engineering Conference and Exposition (ASME-IMECE'09),* Orlando Florida, USA, 16 November 2009, pp. 91-102.

[3]. F. Bulnes, Research on Curvature of Homogeneous Spaces, *TESCHA,* Mexico, 2010, pp. 44-66. http://www.gimathematics.org

[4]. F. Bulnes, I. Martínez, O. Zamudio, G. Negrete, Electronic Sensor Prototype to Detect and Measure Curvature Through Their Curvature Energy, *Science Journal of Circuits, Systems and Signal Processing*, Vol. 4, No. 5, 2015, pp. 41-54.

[5]. F. Bulnes, Curvature Spectrum to 2-Dimensional Flat and Hyperbolic Spaces through Integral Transforms, *Journal of Mathematics,* Vol. 1, No. 1, pp. 17-24.

[6]. W. Rudin, Functional Analysis, *McGraw-Hill, Inc.,* USA, 1973.

[7]. J. E. Marsden, A. J. Tromba, Vectorial Calculus, *Addison Wesley,* 1991.

[8]. V. Kaajakari, Closed Form Expressions for RF MEMS Switch Actuation and Release Time, *Electronics Letters,* Vol. 43, No. 3, 2009, pp. 149-150.

[9]. A. T. Alastalo, V. Kaajakari, Intermodulation in Capacitively Coupled Microelectromechanical Filters, *IEEE Electron Device Letters,* Vol. 26, 2005, pp. 189-191.

[10]. K. Kobayashi, K. Nomizu, Foundations of Differential Geometry, *Wiley and Sons,* New York, 1969.

[11]. O. Holmgen, K. Kokkonen, T. Veijola, T. Mattila, V. Kaajakari, A. Oja, J. V. Knnuttila, M. Kaivola, Analysis of Vibration Modes in a Micromechanical Square Plate Resonator, *Journal of Micromechanics and Microengineering,* Vol. 19, No. 1, 2009, Article ID: 015028.

[12]. F. Bulnes, Electromagnetic Gauges and Maxwell Lagrangians Applied to the Determination of Curvature in the Space-Time and their Applications, *Journal of Electromagnetic Analysis and Applications,* Vol. 4 No. 6, 2012, pp. 252-266.

[13]. F. Bulnes, Quantum Gravity Sensor by Curvature Energy: their Encoding and Computational Models, in *Proceedings of the IEEE Science and Information Conference (SAI'14)*, London, UK, 2014, pp. 855-861.

Chapter 21
Artificial Intelligence Based Medical Computer Vision Image Registration System and Algorithm

Jong-Ha Lee

21.1. Introduction

Emphysema is an obstructive lung disease affecting millions of people, especially smokers around the world. It is characterized by a loss of elasticity, which results in an early airway closure during exhalation. The normal anatomy of the lung is altered in such a way that gas exchange becomes poor. According to the National Center for Health Statistics, the number of non-institutionalized adults who have ever been diagnosed with emphysema was 4.1 million in 2006 [1]. In the United States, emphysema also contributes to more than 100,000 deaths each year, and costs more than 2.5 billion dollars in annual health care expenses [2]. National emphysema treatment trial shows that lung volume reduction surgery (LVRS) improves the quality of life by improving the exercise capacity [3]. The same study also concludes that for patients with both predominantly upper-lobe emphysema and low base-line exercise capacity, LVRS reduces mortality. A more recent study shows the improvements in pulmonary function and exercise ability after LVRS [4-5]. There are also studies that LVRS improves neuropsychological function and sleep quality [6]. Thus, LVRS benefits the patients with severe emphysema, and the improvements of LVRS equipment and procedure should also benefit the patients.

Computed tomography (CT) scan is the choice of imaging modality for patients with suspected emphysema [Screaton and Koh 2004]. Out of different possible CT images such as high resolution computed tomography (HRCT) and spiral CT, the spiral CT images printed with pixels below 960 HU (Housfield Unit), called density-masked images, gave the best estimate of the degree of emphysema in patients undergoing evaluation for LVRS [Cederlund et al. 2002]. CT scans are relatively fast—it takes less than 30 seconds to obtain the lung CT scan—and cost effective compared to MRI. Thus, CT scan images are extensively used to determine the severity of emphysema. CT images are used in diagnosis

Jong-Ha Lee
Keimyung University, School of Medicine, Dept. of Biomedical Engineering, Daegu, South Korea

of emphysema, but the data is not used in the operating room by the surgeons because the shape of the open lung is not identical to the shape of the lung when the CT scan was performed. Thus surgeons depend on their experiences of the tactical and visual senses to determine the line of resection. We develop an image-to-physical space overlay method that will aid surgeons in deciding the line of resection during the LVRS.

The protocol requires intraoperative geometric data to measure and compensate for tissue deformation in the organ. In this paper, we use laser scanner to accomplish these tasks intraoperatively. The laser scanners are capable of generating texture point clouds describing the surface geometry and intensity pattern. Once we obtain the image of the lung from the laser scanner with the associated physical coordinate system, then we need to correspond the image with the CT image. Thus we need a method to register the non-rigid images. In this chapter robust point matching (RPM) has been considered. RPM algorithm uses continuous relaxations of correspondence variables and non-rigid mapping [Chui *et al.* (2000)]. A recent comparative study of various non-rigid registration methods suggests that RPM are appropriate when the set of control point correspondences is fewer than a thousand and variation in spacing between the control points is not large [Zagorchev 2006], which is the case in our application. In this chapter, TPS affine transformation and non-affine deformation matrixes between CT data and laser scanner data are computed by RPM. Two matrices overlay CT image to laser scanner image and surgeons identifies their region of interest intraoperatively. The preliminary version of our chapter has been presented in [Lee *et al.* 2008].

21.2. Image Overlay Method and System Design

21.2.1. Image Overlay Method

a) A patient is scanned by a CT scan and a 3D image is created. We determine the lung pressure of the patient.

b) The CT scan data is segmented to the 3D torso surface image and lung images.

c) In the operating room, prior to draping, the patient is scanned by a laser range scanner.

d) Register the coordinate space of the patient's torso (physical space) and the scanned laser image of the torso (scanned image space).

e) Then 3D torso CT scan image is matched with laser torso image. This is to save time when the lung images are matched.

f) Drape the patient and get ready for open lung surgery.

g) After one of the lungs is visible, we inflate the lung to the predetermined pressure from step "a". Then we scan the lung using the laser range scanner and coordinate measuring machine (CMM).

h) Register the coordinate of the patient's lung (physical space) and the coordinate of the laser scanned image (scanned image space).

i) CT scan image is matched to the patient lung surface data obtained by the laser scanner using visible control points such as lung boundary, fissures, and carina.

j) Overlay and display the resection boundary on the laser scanned lung image in the computer screen.

The theoretical and algorithmic developments are mostly given in steps "d", "e", "h", and "i". These steps are the registration process to relate the physical space, laser scanner image space, with the CT image space. Steps "d" and "h" are similar, and here we register the scanned image space with the physical space. Steps "e" and "i" are similar and here we use non-rigid registration of the scanned image with the CT image. The steps "d" and "e" are performed *a priori* as an initial alignment.

In the following subsections we describe the overall design of the Image Overlay System, preoperative image acquisition method, intraoperative image acquisition method, and coordinate transformation. The critical part of the Image Overlay System is non-rigid image registration. The next section is dedicated to the design, implementation, and test of a non-rigid image registration method.

21.2.2. Image Overlay System

The system is divided into the Preoperative Subsystem and Intraoperative Subsystem, and within each subsystems there are hardware unit and software unit. Furthermore, within the software unit there are multiple modules. The outline of the whole system is given in Fig. 21.1 Siemens CT scanner software, VB10 and Syngo Multimodality Workplace (MMWP) VE20A SL08P62, is the first module in the Preoperative Subsystem. This module will capture and display the three dimensional image of the lung. Using these images as the input, Emphysema Boundary module will determine the degrees of emphysema based on the Housfield unit (HU), and determine the optimal resection boundary. The final module in the Preoperative Subsystem is a MATLAB Interface Module. We designed software to import the dicom files generated by the CT Scanner module into MATLAB.

On the Intraoperative Subsystem side, we have four modules: Laser scanner software module, Point Matching module, Radial basis function (RBF) module, and Display module. The laser scanner module will be the laser scanner company's software. This will enable us to see and manipulate the laser scanned image. Also, this module have a file conversion part to convert 3D stereolithography (STL) or ASCII files to other MATLAB readable file format. The Point Matching module will be based on the RPM algorithm. The RBF module will be used to fit a smooth surface to the laser scanned images using RBF. The final software module, the Display module, will visualize the patient's lung on the computer screen and overlay the emphysema boundary on the screen.

Preoperative Subsystem

Software Unit		Hardware Unit	
CT Scanner S/W Module	Image Interface Module	Siemens CT Scanner	IBM Blade Server

Intraoperative Subsystem

Software Unit			
Laser Scanner S/W Module	Image Interface Module	Point Matching Module	RBF Module

Hardware Unit				
Optix Laser Scanner	CMM (Faro Arm)	CMM Laser Interface Hardware	Cart	Laptop

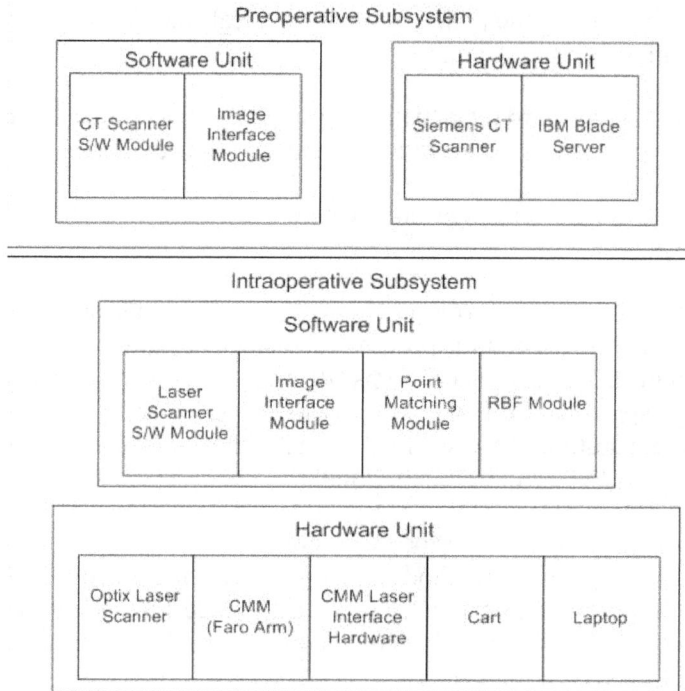

Fig. 21.1. Image overlay system hardware and software design.

21.2.3. Preoperative Image Acquisition

Preoperative images from various different modalities are usable with our approach. In this paper, we used 3D CT images from a Siemens Somatom Sensation CT scanner in Temple University Hospital. The CT produced 64 multiple 2D slices per rotation and the resolution was 0.24 mm ultra high isotropic resolution. For our lung phantom, the highest rotation time was 0.33 seconds. The CT data were manipulated through Siemens Syngo acquisition workplace and stored in a computer. When the data were available in the server, we accessed clinical data through Syngo multimodality workplace and rendered 3D CT images through CT clinical engines and Syngo Expert-1. Every slice was transmitted and stored in the Digital Imaging and Communication in Medicine (DICOM) file format. The size of DICOM files obtained from the CT is directly proportional to the number of obtained 2D slices. Later, we imported these 2D slices image into our software package. Then we rendered the 3D image and determined the outline and the region of interest for the pig's lung.

21.2.4. Intraoperative Image Acquisition

Different modalities such as CT, MRI, or SPECT can be used with our approach to obtain intraoperative images. Here, we used a laser scanner (Optix 400M, 3D Digital Corporation, Sandy Hook, CT) to obtain intraoperative images. The laser scanning is a non-contact, non-invasive, fast, safe, and relatively simple method. The laser scanner is capable of generating point clouds with a resolution of 0.175 mm at a distance of 30 cm and 0.375 mm at a distance of 65 cm. The maximum point density was 1,000 points per line, up to 1,000 lines, and the field of view was 30 degrees. If the object is glossy or light absorbing such as black in color, then the scanned image may be noisy or the sensor may not detect the laser light reflection. We have tested the use of a laser scanner with a pig's lung which is simulated as a smoker's lung. Even though, the simulated smoker's lung made from a pig's lung was dark in color with smooth, light, porous, spongy texture, the laser scanner was able to successfully scan it.

We also have experimented with the point density, resolution, and scanning time. The results are summarized in Table 21.1. The preliminary result shows that the best resolution of X and Y was 0.25 mm with 90 seconds scanning time and the worst resolution was 0.40 mm of X and 0.75 mm of Y with 20 seconds scanning time. The organ such as a lung is not stationary for 90 seconds. However, surgeons can inflate one lung at a time and hold it at a constant pressure for 90 seconds [9]. Then we can scan the lung during this period. The laser scanner that we use is a class II laser, which will not cause ocular damage as long as human does not directly look into the laser for an extended period of time.

Table 21.1. Swine lung scanning test with Optix 400M laser scanner.

Lines	Points	Resolution, x (mm)	Resolution, y (mm)	Time (seconds)
1000	1000	0.25	0.25	90
500	1000	0.40	0.25	75
300	1000	0.75	0.25	50
500	255	0.40	0.75	20

After acquiring data, Radial Basis Function (RBF) was used to fit a smooth surface to the laser scanned images. In the OR, there is a need to relate the laser scanner space with a fixed physical coordinate system as the laser scanner moves to obtain three dimensional images of the organ. To relate the laser scanned images to the physical coordinate, we designed the system to use a Coordinate Measurement Machine (CMM). After the initial alignment, the CMM can track the location and orientation of the laser scanner and the scanner images are related to the physical coordinates.

21.2.5. Non-rigid Registration

Once the preoperative and intraoperative images are obtained, we need a method to relate these two images accurately. One of the most popular registration methods is Iterative Closest Point (ICP) algorithm. While ICP algorithm is very simple, fast, and guarantees to converge to the local minimum, it is not very robust. The algorithm is easily degenerated by noise and large deformation. In this paper, another algorithm, robust point matching (RPM) has been considered and modified [Chui et al. (2000)]. Moreover, we would like to determine region of interest (ROI) such as the location of the tumor or the line of resection within the intraoperative images. In order to perform these tasks, we augmented a RPM method to estimate ROI.

Consider we have a CT image and a laser scanner image as our preoperative and intraoperative images, respectively. We will consider the line of resection as the ROI of an organ. CT image control points consist of two point sets depending on whether correspondence points exist or not. Let $A = \{a_1, a_2, ..., a_M\}$ be a set of control points selected along the contour line of an object in a CT image. CT image includes line of resection where laser scanner cannot detect. Let $B = \{b_1, b_2, ..., b_K\}$ be a set of control points on the line of resection of a CT image. These points do not have a corresponding points in C. The outline of the organ is also fully detected on a laser scanner image, and all of control points in this set have correspondences. It is in the case of 100 % field of vision. We use $C = \{c_1, c_2, ..., c_N\}$ to denote a set of control points in the laser scanner image.

Each landmark a_i is represented as a 2D homogeneous coordinate vector $(1, a_{ix}, a_{iy})$. In our application, one to one matching is desired, but in general, one to one matching is not achieved due to outliers. To handle this problem, two point sets A and C are augmented to $\hat{A} = \{a_1, a_2, ..., a_M, nil\}$ and $\hat{C} = \{c_1, c_2, ..., c_N, nil\}$ by introducing a dummy or *nil* point. From these sets we determine the matching correspondence matrix, H. For a point $m \in \hat{A}$ and $n \in \hat{C}$, if m is matched to n, then $h_{mn} = 1$, otherwise, $h_{mn} = 0$, where h_{mn} is an element in matrix H. Points in A with the corresponding points in C are matched one to one, some of the points are determined as outliers if the matching probability is below the pre-determined threshold. Outliers in A are matched to the nil point in C, and vice versa. Multiple control points may be matched to a nil point. We want to find the matching correspondence and transformation $H, f : \hat{A} \to \hat{C}$ between these two point sets, which minimizes the bending energy of the Thin Plate Spline (TPS) model. Then we will use this matching function, f, to transform B, which is without correspondence to $\hat{B}$, which is the resection line on the intraoperative laser scan image; $f : B \to \hat{B}$. RPM method formulates the energy function to match the control points as closely as possible while rejecting outliers. The softassign technique and deterministic annealing algorithm are used to search for the optimal solution. Below is the summary of RPM method of [Chui et al. (2000), Yang et al. (2006)], which we modified to include the matching of the points without correspondence points. The optimal correspondence $\hat{H}$ and transformation $\hat{f}$ is obtained as follows.

$$[\hat{H}, \hat{f}] = \arg \min_{H, f} E(H, f) \tag{21.1}$$

where

$$E(H,f)=E_g(H,f)+E_t(f)+E_d(H)+E_w(H) \tag{21.2}$$

subject to the following constraints.

$$\sum_{j=1}^{N+1} h_{ij} = 1 \ \text{ for } i \in \{1,...,M\}$$

$$\sum_{i=1}^{M+1} h_{ij} = 1 \ \text{ for } j \in \{1,...,N\} \tag{21.3}$$

with $h_{ij} \in [0,1]$ and it indicates the matching probability.

$E_g(H,f)$ is the geometric feature based energy term according to the Euclidean distance,

$$E_g(H,f)=\sum_{j=1}^{N}\sum_{i=1}^{M} h_{ij} \left\| c_j - f(a_i) \right\|^2$$
$$=\sum_{i=1}^{M} \left\| v_i - f(a_i) \right\|^2, \tag{21.4}$$

where

$$v_i = \sum_{j=1}^{N} h_{ij} c_j \tag{21.5}$$

The variable v_i can be regarded as a newly estimated position that corresponds to a_i.

$E_t(f)$ is the TPS smoothness energy term to generate a smooth spatial mapping. The constant λ regulates the smoothness.

$$E_t(f) = \lambda \iint \left[\left(\frac{\partial^2 f}{\partial x^2} \right)^2 + 2 \left(\frac{\partial^2 f}{\partial x \partial y} \right)^2 + \left(\frac{\partial^2 f}{\partial y^2} \right)^2 \right] dxdy. \tag{21.6}$$

$E_d(H)$ controls the fuzziness of softassign technique with the temperature T of deterministic annealing, $T \in [0,1]$

$$E_d(H) = T \sum_{j=1}^{N}\sum_{i=1}^{M} h_{ij} \log h_{ij}. \tag{21.7}$$

The final term $E_w(H)$ prevents the rejection of too many points as outliers.

$$E_w(H) = -\xi \sum_{j=1}^{N}\sum_{i=1}^{M} h_{ij}, \tag{21.8}$$

where ξ is just a constant.

RPM method involves a dual update process embedded within an annealing scheme. The first step is to update the correspondence:

$$h_{ij} = \frac{1}{T} \exp\left\{ -\frac{(c_i - f(a_i))^T (c_i - f(a_i))}{2T} \right\} \tag{21.9}$$

Then, with a fixed H, the mapping parameters of f are calculated. If a weighting parameter λ is fixed, a unique solution of f that minimizes TPS energy function exists,

$$E_{tps}(f) = E_g(H, f) + E_t(f) \tag{21.10}$$

A solution consists of two parameter matrices D and W

$$f(a_i, D, W) = a_i \cdot D + \phi(a_i) \cdot W \tag{21.11}$$

where D is a 3×3 matrix representing the affine transformation and W is a $M \times 3$ warping coefficient matrix representing the non-affine deformation. $\phi(a_i)$ is a $1 \times M$ vector for each landm a_i, where each entry $\phi_j(a_i) = \|a_j - a_i\|^2 \log\|a_j - a_i\|$, $j = 1, ..., M$. If we substitute the solution for f (21.12) into (21.11), the TPS energy function becomes,

$$E_{TPS}(f) = \|V - XD - \Phi W\|^2 + \lambda \text{trace}(W^T \Phi W), \tag{21.12}$$

where $\lambda \in [0,1]$. $\Phi = \{\phi(a_i): i = 1, ..., M\}$ is an $M \times M$ matrix formed from $\phi(a_i)$. To find the least-squares solutions for D and W, QR decomposition on A is applied to separate the affine and non-affine warping space [Wahba (1990)].

$$A = [Q_1 \mid Q_2] \binom{R}{0} \tag{21.13}$$

As the final optimal solutions for $\hat{D}$ and $\hat{W}$, we obtain

$$\hat{D} = R^{-1}(Q_1^T A - \Phi W), \tag{21.14}$$

$$\hat{W} = Q_2(Q_2^T \Phi Q_2 + \lambda I)^{-1} Q_2^T V. \tag{21.15}$$

The minimum value of the TPS energy function obtained at the optimum ($\hat{D}, \hat{W}$) is the bending energy. In RPM method, above two steps are iterated as the temperature is gradually reduced until the desired correspondence $\hat{H}$ and the transformation $\hat{f}$ are obtained. Because we know the bending energy in every determined annealing step, the line of resection landmark set B can be warped to the laser scanner image using the TPS warping field. This process will allow B transforming gradually without correspondence.

412

The resection boundary estimation method has several parameters. To setup these parameters, some practical techniques are utilized. For our convenience, we scale coordinates of x-axis and y-axis into unit square and set T_o as 0.05. In the beginning, the preoperative image's control points are placed as the center of the initial search boundary. Then the search boundary is set large enough to enclose all intraoperative control points. We set the initial weighting parameters λ as 1. We initialize the affine transformation matrix D to a zero matrix and non-affine deformation matrix W to an identity matrix. During the annealing procedure, T gradually decreases according to a linear annealing schedule, $T_{new} = T_{old}\tau$, τ is the annealing rate. We set τ as 0.94. The annealing rate of λ is T and decreases by $\lambda_{new} = \lambda_{old}T$. The updates occur alternatively and are repeated until they converge at each temperature value. The deterministic annealing is repeated until T reaches T_{final} and we set T_{final} as 0.005. The pseudo-code for the resection boundary estimation algorithm is summarized in Table 21.2.

Table 21.2. The Resection Boundary Estimation Algorithm Pseudo-code.

Input: CT image contour landmark set A,

CT image line of resection landmark set B,

Laser image landmark set C.

Initialization: $D \leftarrow 0$, $W \leftarrow 0$, $T \leftarrow To$, $\lambda \leftarrow \lambda o$

Do 1: Deterministic Annealing $\hat{A}$ and $\hat{C}$

Do 2: Alternating Update

Update $\hat{H}$ based on current $\hat{A}$ and $\hat{C}$

Update D and W based on current $\hat{H}$

Update $\hat{B}$ based on updated D and W

Until 2: D and W are converged

End Alternating Update

$T \leftarrow T\tau$, $\lambda \leftarrow \lambda T$

Until 1: $T < T\_final$

End Deterministic Annealing

Merge warped $\hat{A}$ and warped $\hat{B}$.

21.2.6. Registration Validation Metrics

The measurement of root mean square (RMS) distance between corresponding points is the most common performance metric of registration accuracy in rigid control points based registration [Fitzpatrick et al. (1998)]. A metric for non-rigid registration is measuring mean RMS distance error between control points on one image and its corresponding closest control points on the other image. This is called mean registration error (MRE) and is defined as follows.

$$\sqrt{\frac{1}{m}\sum_{i=1}^{m}[(\hat{a}_i - b_i)(\hat{a}_i - b_i)^T]},\tag{21.16}$$

where $\hat{a}_i$ is the warped control points, b_i is the truth control points, and m is the total number of control points. However this metric is inconsistent and sometimes leads to misleading results in the non-rigid registration problem [5]. Fig. 21.2 shows the failure of mean RMS distance error for the registration accuracy. In Fig. 21.2 (a), three control points (x) are matched to one landmark (o) simultaneously. In contrast, in Fig. 21.2 (b), control points (x) and control points (o) are matched perfectly. If we calculate mean RMS distance between corresponding points, however, both results have the same mean RMS distance error. This problem occurs more prevalently if there are many control points and for a non-rigid registration.

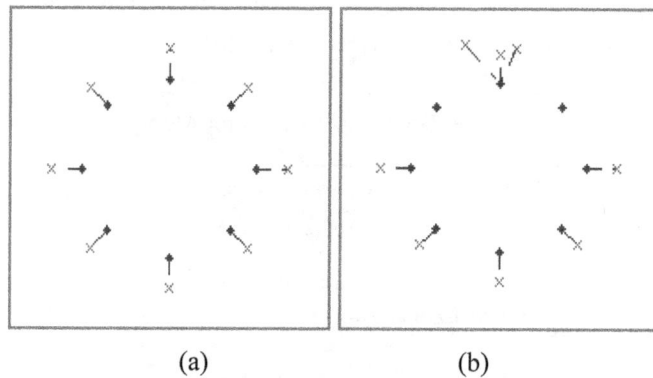

(a)   (b)

Fig. 21.2. An example of the shortcomings of mean RMS distance error for the registration accuracy in the non-rigid registration problem. Many to one registration result (a), and one to one registration result (b) have same mean RMS distance error.

In order to overcome this drawback of RMS distance and to have a more intuitive metric, we define a new metric called Non-Overlapping Ratio (NOR). See Fig. 21.3 for the visual depiction of the NOR error. Our main concern is the mismatched area after the registration. We assumes that control points have m vertices $(a_x^i, a_y^i), i=1,...,m$ and a polygon is made of line segments between m vertices. The last vertex (a_x^m, a_y^m) is assumed to be the same as the first, thus the polygon is closed. The NOR between polygon area A consisting of m vertices, and polygon area B consisting of n vertices is given by

$$\frac{R(A)+R(B)-2R(C)}{R(B)}\times100 \text{ (\%)},\tag{21.17}$$

where $R(\cdot)$ indicates the area made of vertices. $R(C)$ indicates the intersection area of $R(A)$ and $R(B)$. In this study, the MATLAB function polyarea is used to calculate the polygon area. Typically, in non-rigid image registration using control points, perfect matching is rarely achieved, thus we assume there is always some overlapping; i.e., there are no 0 % nor 100 % NOR. Then NOR gives more intuitive metric for the registration quantity. This metric can be generalized to 3D volume error.

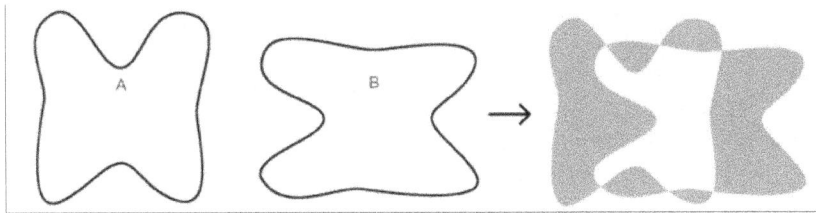

Fig. 21.3. An example of NOR. The dark area indicates non-overlapping region between A and B.

21.3. Result

In the first case, we used a balloon as a phantom. We used a digital camera as our first modality. Before taking a picture, we drew a line on the balloon's surface to indicate a truth line of resection. The black points on the Fig. 21.4 represent ROI area. We took the first digital image with an inflated balloon and took the second image with a deflated balloon to represent the deformation of organs. We chose 200 landmarks for each image, 170 landmarks were chosen along the outlier and 30 landmarks were chosen along the resection line. The resection line on the deflated balloon image was used as the "truth" value.

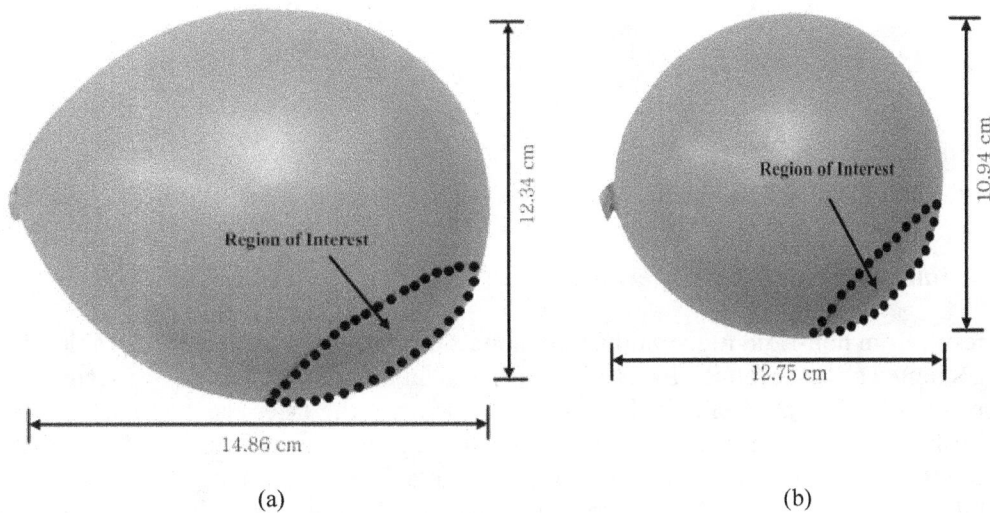

(a)   (b)

Fig. 21.4. (a) Inflated balloon digital image, and (b) deflated balloon digital image.

Fig. 21.5 (a) shows landmarks before registration. The landmarks represented by the cross "X" are from the first image and the circle "o" are the second image. Notice that there are significant differences between two outlines landmarks and two resection lines landmarks. Fig. 21.5 (b) shows the result after registration; we registered just using the outline landmarks. We notice that the outlines match relatively well, but on the left and right corner of the line of resection landmarks, there are some errors. After rescaling it back to the actual coordinate system, we calculate the corresponding area. The result shows that before registration, the difference area was 31.48 cm^2 for outline area and 2.41 cm^2 for ROI area. However, after registration, it becomes much smaller, and the difference area was 0.04 cm^2 for outline area and 0.21 cm^2 for ROI area. The corresponding NOR was 0.53 % for outline area and 0.74 % for ROI area. We note that outline area has smaller NOR than ROI area. We conjecture that this is because there are more landmarks in the outline area and they are registered using corresponding landmarks. We remind the readers that ROI landmarks were estimated without correspondence. We also note that we can improve the accuracy by carefully choosing more landmarks for both areas.

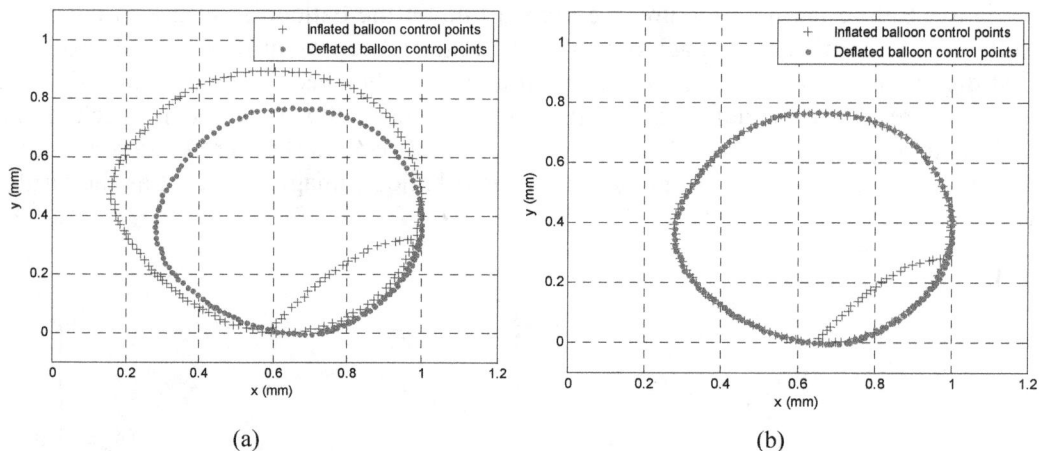

Fig. 21.5. (a) Balloon digital image landmarks (a) before registration and (b) after registration.

21.3.1. Image Registration Experiments

The result from non-rigid registration experiments on pig's lung are shown in Table 21.3. A pig's lung was first scanned by a CT scanner and an inflated pig's lung was scanned by a laser scanner. Fig. 21.6 shows the CT scan image and the laser scanned image of the lung. A total of 700 control points are used for the registration. We assume that our ROI is the left lung, right lung, and a plastic gasket attached on the lung. To increase the reliability of experiments, the phantom was scanned 10 times, and for each scan, the control points were extracted and the registration is performed.

416

Table 21.3. Registration results between CT image and laser scanner image. Three ROI were used as targets.

	Before Registration		After Registration	
	MRE (mm)	NOR (%)	MRE (mm)	NOR (%)
Left lung	15.42 ± 2.23	23.03 ± 8.46	2.32 ± 0.23	1.21 ± 0.28
Right lung	16.22 ± 2.36	24.78 ± 4.63	3.36 ± 0.26	1.91 ± 0.13
Trachea	14.78 ± 3.01	21.47 ± 5.63	1.58 ± 0.17	1.23 ± 0.23
Cardiac notch	14.25 ± 2.97	21.57 ± 7.45	1.32 ± 0.21	1.01 ± 0.35

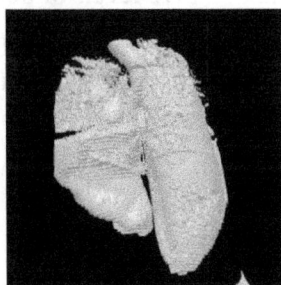

(a)

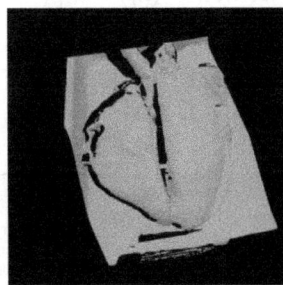

(b)

Fig. 21.6. CT image (a), and laser scanner image (b) of a phantom surface.

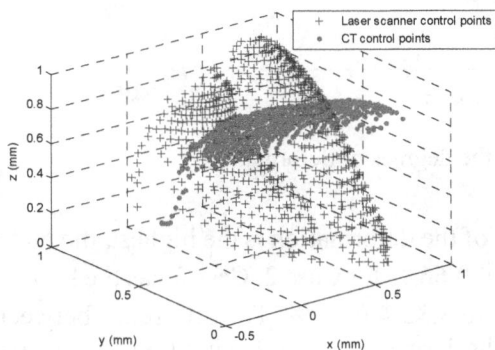

(a)

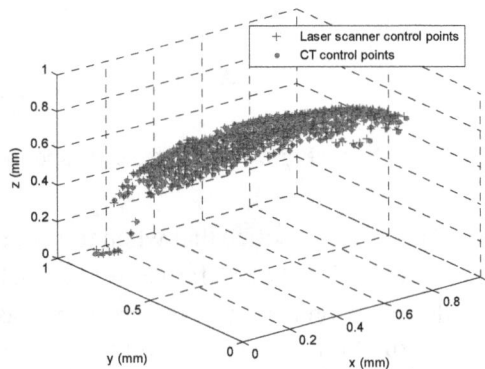

(b)

Fig. 21.7. Swine lung CT /laser scanner image landmarks (a) before registration, and (b) after registration.

21.3.2. Deformation Experiments

During the surgery, an organ's shape change. Therefore, the degree of organ's deformation and visible percentage of the organ during a surgery are two important factors. For this reason, two sets of data from three images were obtained. We designed a computer simulation to measure the robustness of our algorithm under different degree of deformation and organ visibility. In the deformation experiment, 700 control points on each set of images are used and 10 different scanned data sets are chosen for each degradation level. Because the CT scan and laser scanned images are taken at the different times, the registration accuracy will depend upon the degree of deformation between the two registering images. We assume that the laser scanned image is the deformed image and register it to the CT scan image. So, we quantify the degree of nonlinear deformation using Gaussian Radial Based Function (GRBF), which is generated by choosing a special form of the kernel, $\phi_j(x_i) = \exp(-\|x_j - x_i\|^2 / \varepsilon^2)$, $i, j = 1, ..., l$. Five levels of deformation: 0.01, 0.03, 0.05, 0.07, and 0.09 are used depending on the parameters ε of GRBF. These values correspond to the degrees of deformation, 1 %, 3 %, 5 %, 7 %, and 9 %. The results are shown in Fig. 21.8.

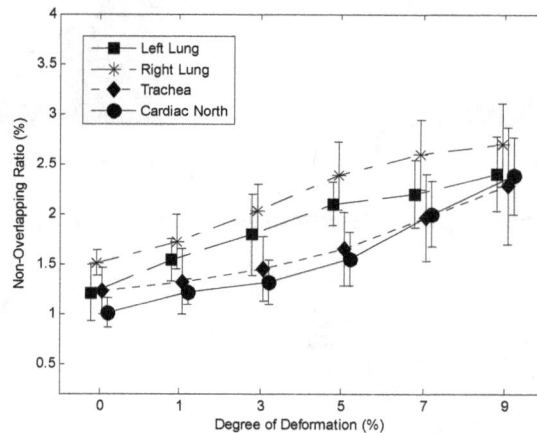

Fig. 21.8. Values correspond to the degrees of deformation.

From Fig. 21.8, we notice that when the degree of the deformation is the highest, the NOR increases to 2.23 ± 0.65 % for Case 1 and 3.19 ± 0.66 % for Case 2. Case 3 has the highest value deformation error and the error increases to 3.32 ± 0.65 %. The difference between 1 % deformation and 9 % deformation was the largest in case 2 with 1.33 % and the smallest in case 1 with 0.78 %. Case 3 was in between these two and the difference was 1.07 %.

21.3.3. Organ Visibility Experiments

During a surgery, often only a part of the organ is visible. Thus, we investigate the effect of our method if the organ is partially visible in this section. For the organ visibility test, blocking windows are used to the original image and the images are reduced accordingly. The control points are re-chosen based on the reduced image. Six different organ visibility percentages are considered: 100 %, 96 %, 93 %, 90 %, 87 % and 85 %. Different percentages of organ visibility were also simulated and the results are shown in Fig. 21.9. In Fig. 21.9, we noticed that NOR error increases as percentage of visibility decreases, which is consistent with the intuition. If the surgeon sees about 96 % of organ, NOR was 0.94 ± 0.30 %, 1.35 ± 0.22 %, and 1.66 ± 0.54 % for Cases 1, 2, and 3. However, if the organ visibility drops to 84 %, the error increased and NOR was 3.28 ± 0.33 % in Case 1 and 4.14 ± 0.52 % in Case 3. Case 2 had the error of 4.01 ± 0.49 %. The statistical test verifies that our approach is tolerant with various deformation and organ visibility.

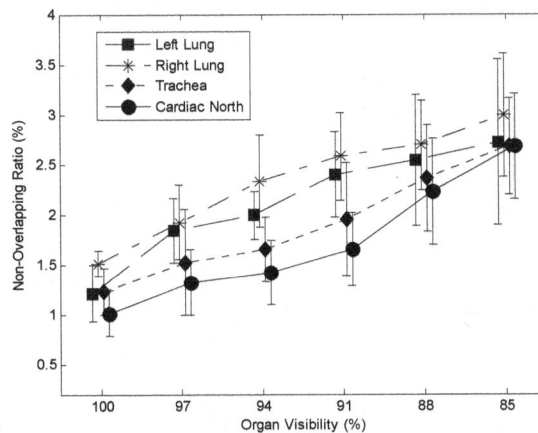

Fig. 21.9. Simulated results in Different percentages of organ visibility.

21.4. Discussion

With the multitude of medical diagnostic equipment and frequent use of them, there is a need to integrate data from different modalities. Despite the fact that a surgeon can determine the region of interest such as a line of resection from preoperative CT scan image, there is no good method to relate this information to the open organ in the operating room. The work presented in this chapter estimates the operative surgical line of resection in the OR using a preoperative image such as CT data and intraoperative image using laser scanner data. Once again we emphasize that the line of resection does not have a corresponding points on the intraoperative image. We, however, use a novel non-rigid image registration method to relate these two images including the line of resection. The registration and estimation methods are implemented through customized software.

The proposed method will minimize unintended removal of the healthy part of the organ. We hypothesize that this method will improve the precision and quality of the LVRS, consequently improving the morbidity and mortality of patients with severe emphysema. Furthermore, the image-to-physical space registration method can be used train new surgeons in determining the line of resection in severe emphysema surgery. Furthermore, this research will have a direct applicability in image-guided surgery of other non-rigid structures of the body such as the heart, liver, and the brain. Similar technology will be applicable to tracking of moving or deformable organs.

21.4.1. Non-rigid Image Registration

Our registration method uses control points on the images. Thus, it is an intrinsic method. Different approaches based on the extrinsic method are also available [Maintz et al. (1998)]. In extrinsic method, invasive markers are introduced. The use of intrinsic method is more suitable registration approach for the following reasons. Extrinsic methods depend on artificial fiducials such as invasive screw markers or skin markers. Patients should be preoperatively imaged with artificial markers attached to them. Intraoperatively, surgeons touch markers with the optical tracking probe and a system determines the transformation between those markers. The main drawback of this method is that fiducials must be attached or marked on the patient's body before the surgery. The patients need to be at OR earlier for the preparation, and markers make patients feel uncomfortable. Furthermore, the transformations is mostly designed for a rigid model. Consequently, this method introduces larger error on most non-rigid organ surgeries. Our intrinsic method, however, depends only on the patient's anatomical images. Registration is based on the control points or voxel properties of the images. The advantages of our method are relatively fast, simple, non-contact, and non-invasive.

Furthermore, a new non-rigid transformation method has been developed for points without correspondence. This is important because even though our intraoperative images do not have diagnostic capability, we are able to determine the region of interest from the preoperative image. This method can be used to identify tumor locations and lines of resection in the OR. In our study, the landmark based non-rigid registration was performed using RPM. The line of resection was estimated on the approximate TPS warping field. This warping filed was calculated through two novel techniques, soft-assign and deterministic annealing. Using this method the non-overlapping region (NOR) errors were determined; for the phantom case, the error was less than 1 % and for the lung case the error was less than 2 % non-overlapping region.

Although, this method has been applied to lungs, other organs may be used. Other organs such as breasts may work well with this method. For example, the location of the breast cancer cell predetermined by MRI may be related to the intraoperative image (laser scanned, digital image, etc.). Moreover, this method can be used for time series analysis. For example, the tumor growth may be determined if there are multiple images separated by time are available.

21.4.2. Registration Accuracy Metric

In order to determine how well the ROI are estimated, non-overlapping ratio (NOR) error was introduced and determined using the vertices consisting of control points on the contour. In fact, we found out that measuring RMS distance (i.e., mean registration error), the traditional index of the registration accuracy as the registration error metric, is not very accurate and sometimes leads to misleading results. This problem becomes worse, when the control points increase or the deformation degree increases. Thus in this work, NOR was chosen to compensate the drawbacks of MRE as the metric to measure registration error. The NOR gives more intuitive sense for the mismatched region. The accuracy of NOR is highly depended upon the number of control points, because it is calculated as a polygon which is comprised of control points.

The validity of our approach was demonstrated through the experimental results, with a mean NOR in the range of 1.40±0.65 % (Table 21.3). A limiting factor to the accuracy of the estimation is represented by the limited organ's degree of deformation and organ's visibility. Our approach was also validated under the different degree of deformation and organ visibility, and mean NOR was 3.32±0.65 % when the deformation was the highest and 4.14±0.52 % when the organ visibility was the smallest. Here we note that, the most accurate NOR would obtained by scanning the object under the perpendicular view angle. Thus, in this work, organs were scanned by the CT and laser scanner with 90 degrees from the ground.

A limiting factor to the accuracy of CT / laser scanner image registration in the open organ is represented by the different breathing conditions in CT and laser scanner. In fact, laser scanner image acquired the information during few seconds. If the patients inhale and exhale constantly during this period, this long acquisition time impairs the spatial resolution and in results it makes difficult to register images. The CT freezes the organ during CT sampling. The error can be introduced by these different breathing conditions. In this work, we assumed that the surgeons would inflate one lung at a time and hold it at a constant pressure for about 90 seconds.

Instead of using landmark based registration, surface registration can be an alternative way. [Hill et al. (2001)]. However, surface registration method has difficulty estimating the region of interest, which is inside an organ. In addition, the complexity and computational time is higher than our method. Our method is capable of determining control points on the surface region of interest as well as the region of interest embedded in the organ. Our method took about 1 minute for 100 landmark image registration with Pentium IV 3 GHz and 2 GB RAM.

It should be also noted that our experimental results are for of 2D registration of CT and laser scanner images. For a volume calculation, 3D registration is necessary. TPS-RPM is easily extended to 3D coordinates. Also the NOR has to be defined as the volume instead of area. These extensions are left as the future work.

It is worth noting that the topics of multi-modal registration and accuracy have been widely investigated in the literature. Most papers, however, deal with the registration between preoperative images to validate the growth of the tumor or determine ROI without consideration of high degree of non-rigid deformation and the limited organ visibility. As far as we know, the registration between preoperative image before surgery and intraoperative image of an open organ in the OR, and finding the warping field to estimate resection boundary in near real time are new methodologies. The experiments using different phantoms and modalities in this chapter show the suitability of our approach.

21.5. Conclusions

We proposed an estimation method to find the operative region of interest in the OR. For this, we used a preoperative CT image, a laser scanned intraoperative image, and Robust Point Matching based non-rigid image registration algorithm. To intuitively quantify the registration error, we proposed Non-Overlapping Ratio. Extensive experiments were performed to demonstrate the robustness of our approach. For the lung phantom, we obtained 0.72 % non-overlapping region error. Under the non-rigid deformation and the limited organ visibility (84 % of the organ visible), our approach performed less than 4.14 % NOR. It shows that the approach taken in this chapter is relatively fast and convenient with high accuracy for the OR environment. The proposed method for CT / laser scanner image integration will be a useful method to support surgeon's region of interest determination.

Acknowledgements

This work (Grants No. C0395986) was supported by Business for Cooperative R&D between Industry, Academy, and Research Institute funded Korea Small and Medium Business Administration in 2017.

References

[1]. Brendel B., Winter S., Rick A., Stockheim M. and Ermert H., Registration of 3D CT and ultrasound datasets of the spine using bone structures, *Comput. Aided Surg.*, 7, 2002, pp. 146–155.
[2]. Cash D. M., Sinha T. K., Chapman W. C., Terawaki H., Dawant B. M., Galloway R. L. and Miga M. I., Incorporation of a laser range scanner into image-guided liver surgery: surface acquisition, registration, and tracking, *Med. Phys.*, 30, 2003, pp. 1671-1682.
[3]. Chui H. and Rangaraian A., A new point matching algorithm for non-rigid registration Algorithm for Non-rigid Point Matching, *Comput. Vis. Image Und.*, 89, 2003, pp. 114-141.
[4]. Fei B., Lee Z., Duerk J. L. and Wilson D. L., Image Registration for Interventional MRI Guided Procedures: Interpolation Methods, Similarity Measurements, and Applications to the Prostate, *Lect. Notes Comput. Sc.*, 2717, 2003, pp. 321-329.
[5]. Fitzpatrick J. M., West J. B. and Maurer C. R., Predicting error in rigid-body point-based registration, *IEEE Trans. Med. Imag.*, 17, 1998, pp. 694-702.
[6]. Gwinn R., Cleary K. and Medlock M., Use of a portable CT scanner during resection of subcortical supratentorial astocytomas of Childhood, *J. Pediatr. Neuro.*, 32, 2000, pp. 37-43.

[7]. Harberland N., Ebmeier K., Hliscs R., Grunewald J. P. and Kalff R. L., Intraoperative CT in image - guided surgery of the spine, *J. Medica.*, 43, 1999, pp. 24-32.

[8]. Herring J. L., Dawant B. M., Maurer C. R., Muratore D. M., Galloway R. L., Fitzpatrick M., Surface-based registration of CT images to physical space for image-guided surgery of the spine: a sensitive study, *IEEE Trans. Med. Imag.*, 17, 1998, pp. 743-752.

[9]. Hill D. L. G., Batchelor P. G., Holden M. and Hawkes D. J., Medical image registration, *Phys. Med. Biol.*, 46, 2001, pp. 1-45.

[10]. Jinzhong Y., James P. W., Yiyong S., Rick S. B. and Chenyang X., Non-rigid image registration using geometric features and local salient region features, in *Proceedings of the IEEE Computer Society Conference on Computer Vision and Pattern Recognition*, 2006, pp. 825-832.

[11]. Kanan A. and Gasson B., Brain tumor resections guided by magnetic resonance imaging, *J. Associ. Peri. Reg. Nurs.*, 77, 2003, pp. 583-589.

[12]. Kaufman C. L., Jacobson L., Bachman B. and Kaufman L., Intraoperative ultrasound facilitates surgery for early breast cancer, *Ann. Surg. Oncol.*, 9, 2007, pp. 988-993.

[13]. Lee J. H., Won C. H. and Kong S. G., Estimation of Operative Line of Resection Using Preoperative Image and Nonrigid Registration, in *Proceedings of the Annual International Conference of the IEEE Engineering in Medicine and Biology Society*, 2008, pp. 3983-3986.

[14]. Martin R. C. G., Husheck S., Scoggins C. R. and Mcmasters K. M., Intraoperative magnetic resonance imaging for ablation of hepatic tumors, *Surg. Endosc.*, 20, 2006, pp. 1536-1542.

[15]. Miga M. I., Sinha T. K., Cash D. M., Galloway R. L. and Weil R. J., Cortical surface registration for image guided neurosurgery using laser range scanning, *IEEE Trans. Med. Imag.*, 22, 2003, pp. 975-985.

[16]. Schulder M. and Carmel P. W., Intraoperative magnetic resonance imaging: impact on brain tumor surgery, *Cancer Control*, 10, 2003, pp. 115-124.

[17]. Sohmura T., Nagao M., Sakai M., Wakabayashi K., Kojima T., Kinuta S., Nakamura T. and Takahashi J., High-resolution 3-D shape integration of dentition and face measured by new laser scanner, *IEEE Trans. Med. Imag.*, 23, 2004, pp. 633-638.

[18]. Sinha T. K., Miga M. I., Cash D. M. and Weil R. J., Intraoperative cortical surface characterization using laser range scanning: preliminary results, *Neurosurgery*, 59, 2006, pp. 368-377.

[19]. Wahba G., Spline models for observational data SBMS-NSF, *Seri. App. Math.*, 1990.

[20]. Wolverson M. K., Houttuin E., Heiberg E., Sundaram M. and Shields J. B., Comparison of computed tomography with high resolution real time ultrasound in the localization of the impalpable undescended testis, *J. Radiol.*, 146, 1983, pp. 133-136.

[21]. Zhong H., Peters T. and Siebers J. V., FEM-based evaluation of deformable image registration for radiation therapy, *Phys. Med. Biol.*, 52, 2007, p. 4721.

Chapter 22
SmartLab Magnetic: A Modern Student Laboratory on Magnetic Materials and Circuits

Javier Martinez-Roman, Angel Sapena-Baño, Manuel Pineda-Sanchez, Ruben Puche-Panadero

22.1. Introduction

Undergraduate laboratories should provide means to help the student visualize often complex concepts, to achieve a more lasting understanding and, thus, a more significant learning of the ideas presented in the lectures. As an example, magnetic circuits are often an introductory aspect of under-graduate electrical machines courses [1-5]. Basic learning objectives in these courses are that students should be able to understand how the magnetic circuits work, and calculate their main relationships, such as those between core dimensions, core flux and exciting current. Student must be aware that some lamination properties (such as permeability and total specific losses) are directly related to the electrical machine's fundamental performance figures such as efficiency, the no-load current, or the power factor. Traditionally, students have been able, to a very limited extent, to check these facts experimentally with no-load tests at different voltages on typical electrical machines, like transformers and induction machines, using simple measuring equipment like voltmeters and ammeters or more complex like oscilloscopes.

The student often faces challenges when integrating laboratory and lectures learning. With respect to electromagnetism and electrical machines these challenges are mainly related to the use of abstract concepts that are hard to visualize [6, 7], and to the lack of appropriate and easy to use measurement equipment and test rigs. In the case of electromagnetism, these difficulties are sometimes aggravated by the use of a mathematical approach when actually the students need to visualize the concepts to fully understand them. In recent decades the use of simulations to help visualize those abstract concepts [7], and to simulate the actual electrical machine performance during tests [8-11] has sometimes been adopted but then new drawbacks arise. The SmartLab

Javier Martinez-Roman
Universitat Politècnica de Valencia, Instituto de Ingeniería Energética, Valencia, Spain

approach, proposed here, blends the test of actual machines of classical experiments with the enhanced data processing and visualization offered by simulations, in an attempt to profit from their advantages while avoiding their weaknesses.

SmartLabs (SLs) complement the existing equipment under test of classical experiments with a sensors set, a Digital Acquisition (DAQ) board and a portable device App providing DAQ control, acquired data manipulation and handling, and the user interface. Similar approaches based on personal computers, often designated as Virtual Instruments (VI) using the widespread manufacturer terminology, has been shown to have many advantages over traditional instruments. Reference [12], as early as 1984, showed the suitability of VIs to help students get a 'real-time' handle on some electrical machine concepts. In [13], efficient data collection and manipulation by means of VIs is shown to help maintain student interest and reduce the time required to perform and evaluate experiments; similar advantages were reported in [14]. VIs' ability to provide user-friendly interaction with the experiment is also underlined in [15]. The modular and reusable nature of Vis, and therefore their suitability for integration into cost-effective systems is underlined in [16]. Additional benefits from having students collaborate in VI development is that these then reflect student interests and needs, as described, along with already mentioned advantages, in [17]. With respect to magnetic materials properties, some advantages are also cited in [18, 19] and [20].

SmartLab Magnetic (SLM) was the first SmartLab [21, 22] developed by the Installations, Systems and Electrical Equipment (iSEE) group (within the Institute for Energy Engineering of Universitat Politècnica de València) out of five already deployed and several more under development. SLM is based on the original magnetic circuits lab, which focused on changes in magnetic circuit structure and their effect on winding inductance and no-load current based on rms voltage and current measurements.

This chapter first briefly introduces, in Section 22.2, the main relationships between winding voltage, core flux, core dimensions and materials and exciting current in transformer-core type magnetic circuits, how these influence key electrical machine performance figures and how they relate to various aspects of the no-load current of a transformer. Section 22.3 describes the equipment under test, and the sensors set and DAQ card. Section 22.4 deals with the Android App User Interface (UI) design and the required data processing and handling that enhances visual recognition of the main relationships in Section 22.2 while providing a friendly user experience. Section 22.5 describes the tests to be performed and the data to be collected to arrive at the main results directly related to the lab's learning objectives. Finally, the student opinion on the relevance, usefulness and motivational effect of the laboratory is detailed in Section 22.6. The main conclusions of are presented in Section 22.7.

22.2. Introduction to Transformer-Core Type Magnetic Circuits

Transformer-core type magnetic circuits are characterized by a set of stretches linked in series (sometimes, as a result of symmetry two identical sets are arranged in parallel, as in the shell type core) along which the magnetic flux is quite approximately constant. As a result of this, the magnetic behavior of each stretch can be depicted by its reluctance:

$$F = \Re\Phi \; ; \; \Re = \frac{l}{\mu S} \tag{22.1}$$

For the whole magnetic circuit, as a series combination of several tracts or limbs, when excited by a single winding with N turns, results:

$$N \cdot I = mmf = \sum \Re_{tract} \Phi = \Re_{core} \Phi \tag{22.2}$$

Thus, longer magnetic circuit tracts or smaller available core cross-section result on a higher reluctance and a higher no load current for the same core flux. Also, as the core iron saturates its average permeability decreases and the higher reluctance means again more exciting current is needed.

The transformer being at no load, the winding sinusoidal voltage is almost equal to the induced electromotive force (emf) that, in turn, requires a sinusoidal core flux that must lag the winding voltage by ¼ period:

$$u(t) = \sqrt{2}U \cos(\omega t) = N \frac{d\Phi_{core}}{dt} \Rightarrow$$

$$\Rightarrow \Phi_{core} = \frac{\sqrt{2}U}{N\omega} \cos(\omega t - \pi/2) \tag{22.3}$$

The winding no-load current, necessary to provide the magnetic circuit magnetomotive force (mmf) and to support its losses, can then be split into its magnetizing and iron-loss components. The magnetizing current is in phase with flux pulsation but, due to core saturation, exhibits obvious peaks (in phase with core flux peaks which, in turn, coincide with the zero voltage instant due to the ¼ period lag) that results in a bell-shaped waveform. The iron-loss current is in phase with the winding voltage and, therefore, leads the magnetizing current by a ¼ period. The result of this phase shift between the two no-load current components appears as a left-peak-right asymmetry of the no-load current semi-periods (see for example [4], pp. 81-84 or [5], chapter 2.4.2.3).

From an electrical engineer's point of view, the core's magnetic materials main properties are the permeability and specific losses, because they are directly related to fundamental performance figures of the electrical machine. The iron core mmf accounts for between 30 % (in rotating electrical machines) to 90 % (in transformers) of the total required mmf, the remaining percentage being due to airgaps. In turn, the total mmf is directly related to the machine's no-load current and, through that, to its power factor and to the reactive power consumption. The ES specific losses are directly related to the core losses, amounting to about one third of the machine's total losses. The remaining losses are mainly Joule effect losses in the windings and friction losses in rotating machines. ES specific losses are closely proportional to core's induction squared and, therefore, for the same core flux, the smaller the available core cross-section the higher the specific losses.

22.3. Test Equipment

The test equipment that integrates the SLM has been derived from the original laboratory equipment used with classic instrumentation. It included a variable autotransformer to provide adjustable voltage at constant mains frequency, a three phase three leg transformer (220/380 V, 2 kVA), one voltmeter and one ammeter. The classic instrumentation was the only element enhanced for the SLM.

22.3.1. Magnetic Circuit Configurations

A three phase transformer is used to test different magnetic circuits with varying configurations (see Fig. 22.1 to Fig. 22.4) depending on the winding being fed or connected and thus give the student the opportunity to correlate changes in magnetic circuit configuration with the observed changes in winding inductance and overall magnetic circuit reluctance. Four tests where thus devised: Series-parallel I, Series-parallel II, Series and Open.

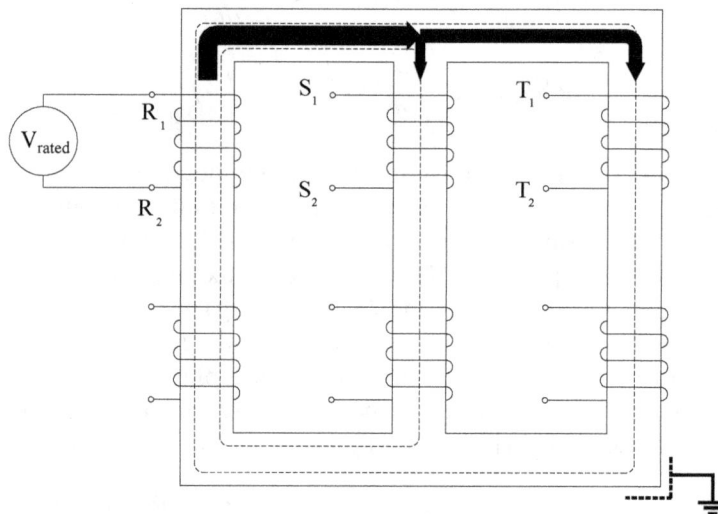

Fig. 22.1. Series-Parallel I magnetic circuit.

In the Series-parallel I configuration (Fig. 22.1) only the primary winding on the left leg is fed, at its rated voltage, keeping all other windings open. Thus, the flux required in the left leg goes through the left yoke to the junction between the central leg and the right yoke and leg, where it splits, mostly going through the shorter central limb.

428

In the Series-parallel II (Fig. 22.2) configuration only the primary winding on the central leg is fed, at its rated voltage, keeping all other windings open. Thus, the flux required in the central leg goes to the junction between the left and right legs where it splits approximately halves. In this configuration magnetic flux lines are, in average, shorter, and the full magnetic flux is confined in a single limb for a shorter tract. These two configuration changes result in a lower magnetic circuit reluctance.

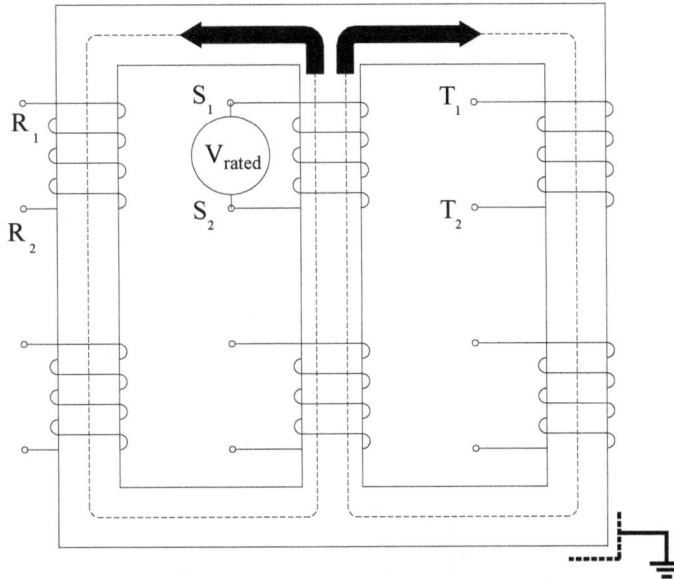

Fig. 22.2. Series-Parallel II Magnetic Circuit.

In the Series configuration (Fig. 22.3), again, only the primary winding on the central leg is fed, at its rated voltage, keeping all other windings open but the secondary winding on left leg, which is short-circuited. Thus, no voltage means almost no winding induced emf nor flux in the left leg and the flux required in the central leg goes to right leg almost fully. Now the magnetic flux lines are more or less the same length as those in Series-Parallel II magnetic circuit but the available iron cross-section has been halved along most of the magnetic circuit. Thus the magnetic circuit reluctance clearly increases.

Finally, in the Open Circuit configuration (Fig. 22.4), again, only the primary winding on the central leg is fed, at a fraction of its rated voltage, keeping all other windings open but the secondary winding on both the left and right legs, which are now short-circuited. Thus the flux required in the central leg must close through a mostly non-ferromagnetic path surrounding the winding. The deepest change in the magnetic circuit is the change of ferromagnetic to non-ferromagnetic material on most of its length and the magnetic circuit reluctance drastically increases.

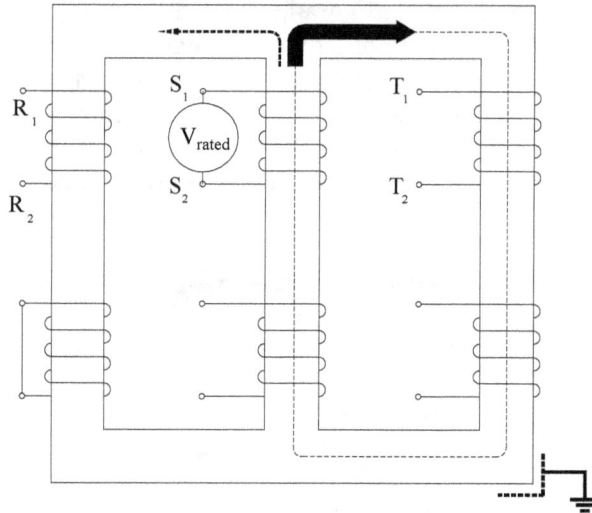

Fig. 22.3. Series Magnetic Circuit

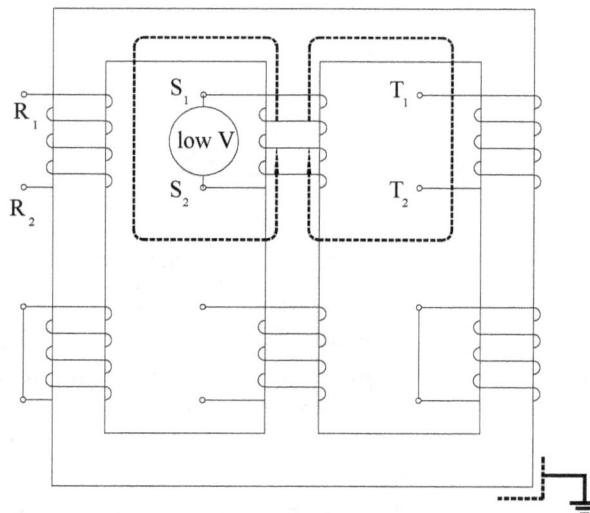

Fig. 22.4. Open Magnetic Circuit.

22.3.2. SLM Enhanced Instrumentation

The classic volt- and ammeter provided very interesting feedback on the magnetic circuit performance but that information can be clearly improved to enhance the student recognition of the basic phenomena taking place during the tests. Thus, instead of working with rms voltage and current sensors it was decided to substitute them by wide bandwidth voltage sensors and, additionally, to simplify the measurements, to include direct voltage measurement or the three primary windings.

The voltage sensor selected is a differential grounded voltage divider, which provides easily adjustable attenuation, very good bandwidth including DC voltages if required, and low common voltage suitable for a DAQ card differential analog input. Isolation is not required due to the use of low voltage mains and to the grounded divider configuration. The current sensor selected is a closed loop, compact, multirange Hall effect current transducer (LEM LTS 6 NP) which provides very good bandwidth, easy integration, 3 kV voltage isolation, and low common mode output while requiring single unipolar voltage supply.

Finally, as can be appreciated from Figs. 22.1 to 22.4, the flux distribution in the three core legs changes considerably for the four configurations being tested. This flux distribution can be directly correlated to the respectively fed/induced voltage in each of the primary windings, which makes it quite interesting to provide the user with voltage readings simultaneously for all primary windings.

The sensors described are complemented (Fig. 22.5) with a Measurement Computing BTH-1208LS DAQ card. This card is selected because it is one of the very few available with built in Bluetooth communication and Android support, thus enabling a direct integration with Android Apps and, especially, a simple and easily configurable link between the DAQ card and the mobile device.

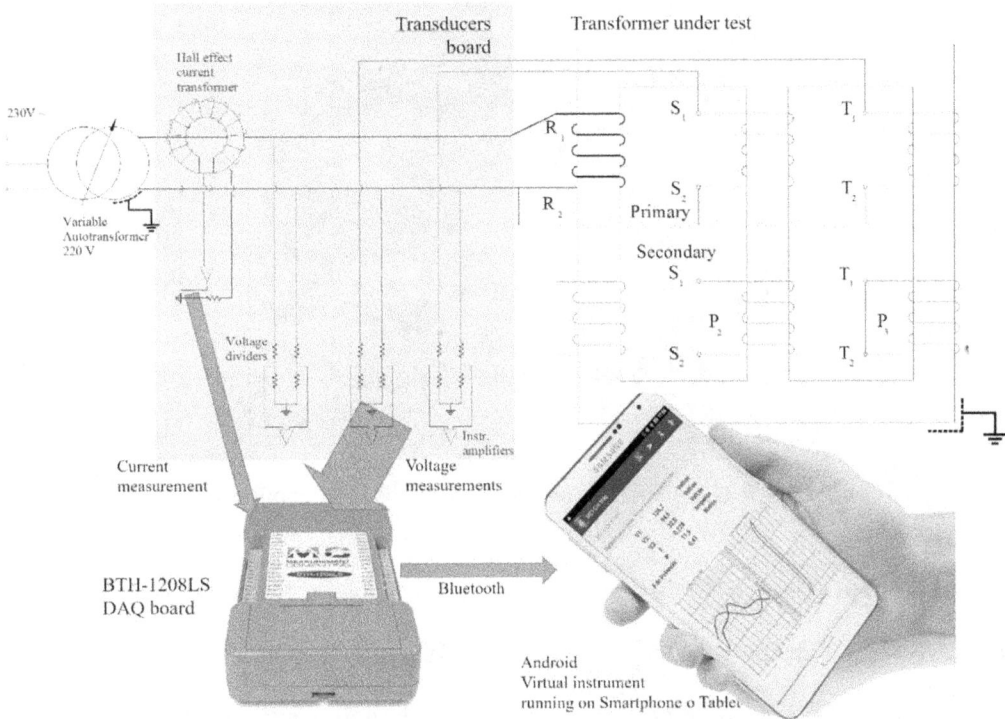

Fig. 22.5. SLM Instrumentation: Voltage and current sensors, DAQ Card and Virtual Instrument Android App.

The BTH-1208LS has 4/8 differential/single-ended analog inputs with a maximum shared sample rate of 47 kS/s, apart from two analog outputs, 8 digital input/outputs and a high-speed counter input, which can be used in different applications (e.g., reactive power compensation, discharge lamp lightning control or frequency converter drive operation, no described here [21-25]). The maximum sampling frequency of 47 kS/s is quite enough for this application as it requires four channels, one current and three voltages, with a suitable resolution to depict up to harmonic 19 at most. Thus an individual sampling frequency resulting in 5 samples per period for the highest harmonic, about 5 kS/s, is quite enough, which results in an aggregated sampling frequency of 25 kS/s, clearly below the DAQ card maximum sampling frequency.

The DAQ and sensors board are placed inside a rugged plastic case together with a short-circuit protection, the required power supply and the suitable mains and measurement connections (Fig. 22.6).

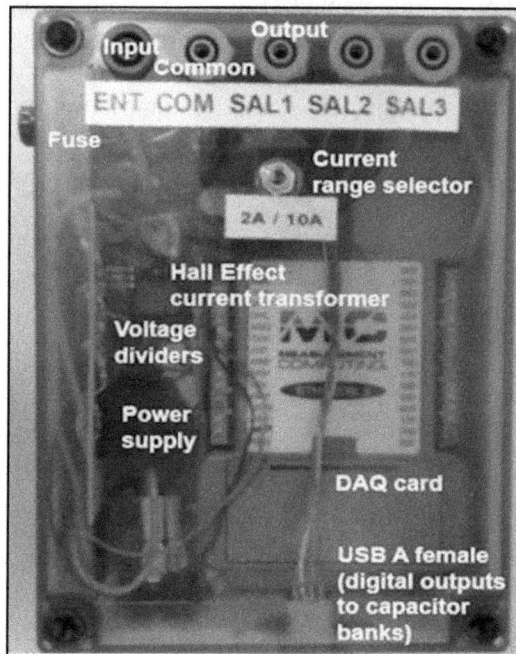

Fig. 22.6. DAQ and sensors casing with external connections.

22.4. SLM App User Interface Design

The SLM App UI design is based on these main objectives:

a)   Provide a clear reading of the test rms voltages and currents;
b)   Complement those readings with the fed voltage and current waveforms and
c)   Also with the comprehensive magnetic cycle of the magnetic circuit;
d)   Facilitate data gathering with in-built report generation and delivery.

To this end, the user interface is divided into three main areas: readings, graphs and App menu (Fig. 22.7).

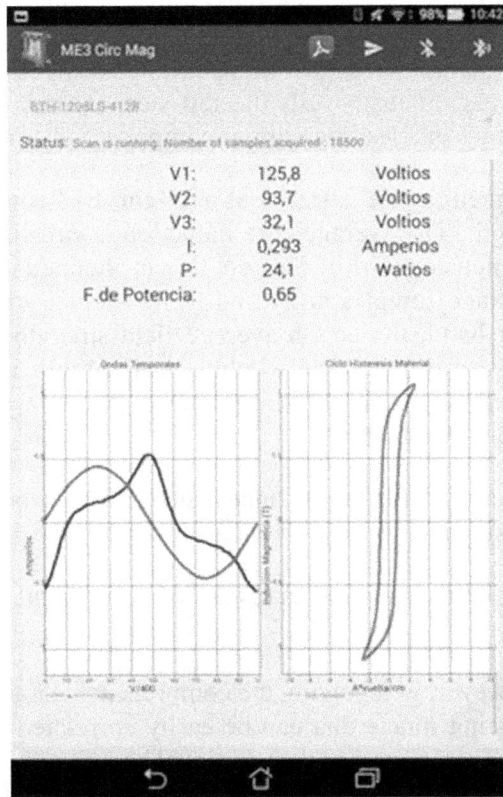

Fig. 22.7. SLM App user interface main areas: readings, graphs and App menu.

The readings area, apart from the rms voltages (three, one in each of the three primary windings) and current, includes also the power consumption and the power factor to provide the student notions on real magnetic circuits with actual iron losses and their importance on electrical machines operation and performance. The rms values are calculated as the square root of the current and voltage sample's squares average. The electric power is calculated as the average current and voltage sample's products and, with the apparent power, $V \cdot I$, is used to calculate de power factor.

The graphs area is shared between the current and voltage time-waveforms and the comprehensive magnetic cycle. The voltage and current time waveforms are a direct trace of the sampled voltage and current during one mains period. The voltage and current arrays are first rearranged to force a start with a voltage positive zero crossing. This simple operation helps readability as it behaves like a mains trigger locking the time-waveforms in the horizontal axis. The learning objectives especially related with the time-waveforms are the following ones:

a)   Mains connected electrical machines, and, specifically, transformers, operate with very closely sinusoidal voltage and flux, no matter how saturated the core is,

b)   Core saturation results in no-load current deformation with asymmetrical bell shaped waveforms that account for saturation and iron core losses, and,

c)   Current waveform asymmetry swells the left side of each half period around the current maximum, resulting in a leading current component, the iron loss current.

The comprehensive magnetic cycle is traced as a xy-curve of core average induction vs. core average field strength. The average core induction is calculated from the core flux using the known core dimensions, while the core flux is obtained through a quarter period shift of the winding voltage samples array and using known winding turns and mains frequency. On the other hand, the core's average field strength is calculated from the winding current samples array times the winding turns (mmf) and divided by the core mid-line length. The learning objectives especially related with the comprehensive magnetic cycle are the following ones:

a)   Changes in the magnetic circuit reluctance result in inversely related changes on the magnetic cycle average slope, and,

b)   Changes in the magnetic circuit losses result in direct changes on the magnetic cycle enclosed area.

The great advantage, however, in including the comprehensive magnetic cycle graph is to provide the student a lasting image that can be easily correlated with basic concepts of ferromagnetic materials. This image has the additional benefit of helping the user correlate simple changes in the magnetic circuit with directly recognizable features of the graph: for example, when the average flux lines length decreases from series-parallel I to series-parallel II, the user can directly associate that change to a slope increase in the magnetic cycle slope.

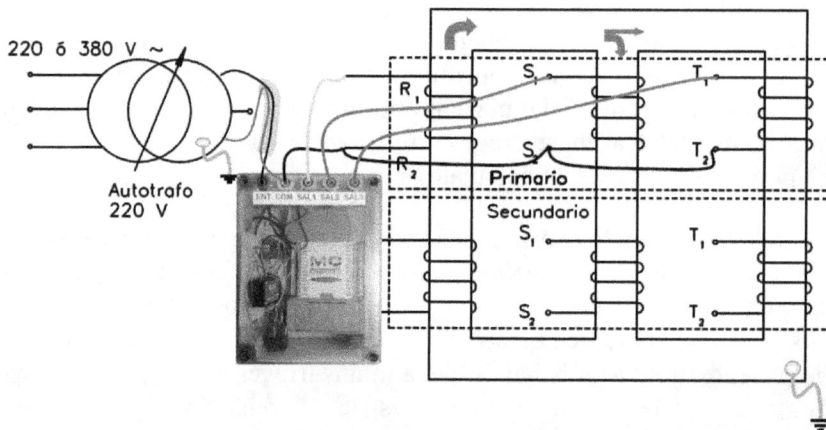

Fig. 22.8. Circuit Schematic for series-parallel I magnetic circuit in student laboratory notes.

Finally, with regard to the App menu, it includes buttons to generate a report in PDF format which includes both the readings and the graphs for the actual test being performed and also to send by email the reports generated during the laboratory session [26]. This actions are complemented with the Bluetooth connection and disconnection buttons required to establish or break the link with the DAQ card. The fact that the user does just need to press a button to generate a test report with the actual conditions and graphs and another to receive in his/her email inbox all the reports for the session experiments means that a significant amount of the session allotted time which was traditionally dedicated to data collection and processing can be now devoted to analyze test results, elaborate correlations, combine observations and prepare conclusions. This, by avoiding tedious tasks, adds also to the student motivation.

The SLM App is available for download in Google Play [27]. From there the students can download and install it in their Android mobile devices. One paired with the DAQ card by Bluetooth, they can use them to perform the measurements instead of with the tablet-PCs available in the Laboratory.

22.5. SLM Tests Guide and Main Results

The student laboratory work is organized in four consecutive tests that help him correlate different changes in the configuration of the magnetic circuit with the observed changes in the measured voltages and currents and, especially, with the shape (slope and area) of the comprehensive magnetic cycle. The four test to be performed are directly related with the four magnetic circuit configurations described in section 22.3.1.

The student work is guided with the help of easy to follow schematics (Fig. 22.9) of the electric circuits required in each test and a check list procedure to perform them.

Fig. 22.9. SLM page on Play Store (Google).

In these schematics, a simple color code is used to help distinguish the different connections and the changes to be made from one experiment to the next. Attention is also drawn to safety measures to avoid voltages or currents above those rated for the equipment under test or for the transducers.

The laboratory notes draw the attention to the changes to be observed between consecutive experiments. Thus, from the series-parallel I (Fig. 22.7) to the series-parallel II magnetic circuits (Fig. 22.10) the user can notice a reduction of the no load current and also of the magnetic losses, along with an increase in the comprehensive magnetic cycle average slope. These observations can be readily be associated to the reduction in average length of the magnetic flux lines together with an increase in the average available iron cross-section.

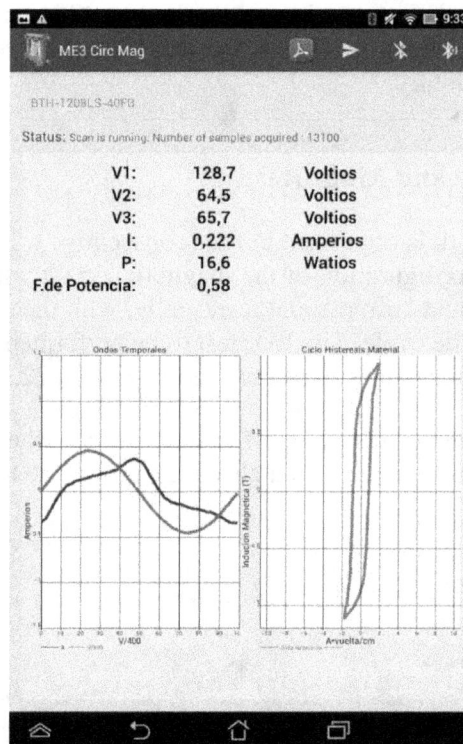

Fig. 22.10. Series-parallel II magnetic circuit test results.

Then, from the series-parallel II (Fig. 22.10) to the series magnetic circuit (Fig. 22.11) the user can notice an increase of the no load current and also of the magnetic losses, along with a reduction in the comprehensive magnetic cycle average slope. These observations can be readily be associated to the reduction in the average available iron cross-section.

Finally, from the series magnetic circuit to the open magnetic circuit the user can realize that current consumption is quite higher, even for a drastically reduced winding voltage, and a very reduced slope in the comprehensive magnetic cycle that has now a very small enclosed area (Fig. 22.11). These observations can be immediately connected to the high reluctance magnetic circuit in which magnetic flux lines have been drawn out of the iron core.

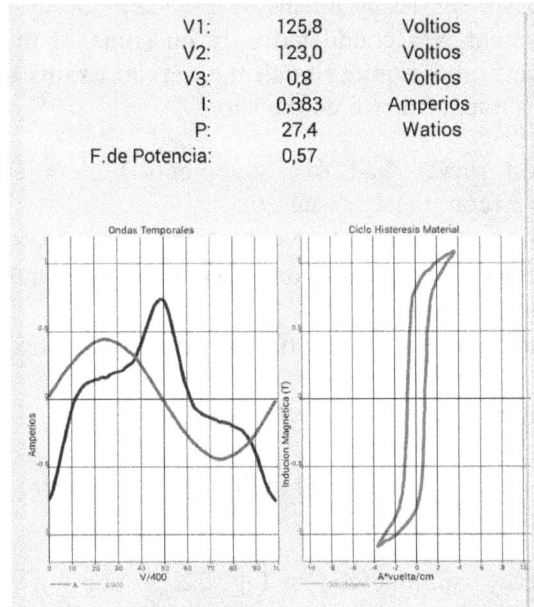

V1:	125,8	Voltios
V2:	123,0	Voltios
V3:	0,8	Voltios
I:	0,383	Amperios
P:	27,4	Watios
F.de Potencia:	0,57	

Fig. 22.11. Series magnetic circuit test results.

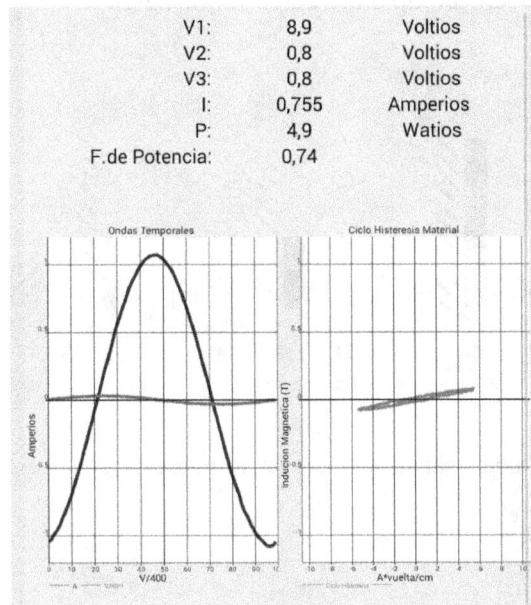

V1:	8,9	Voltios
V2:	0,8	Voltios
V3:	0,8	Voltios
I:	0,755	Amperios
P:	4,9	Watios
F.de Potencia:	0,74	

Fig. 22.12. Open magnetic circuit test results.

22.6. Evaluation

SLM was used during the Spring 2014 and 2015 semesters by UPV's GITI and GIE students, and had previously been tested with the students of earlier degree programs. SLM's effectiveness was assessed by means of a student survey administered after the laboratory. This assessment was conducted only on some of the laboratory sessions, because of the significant time required to sit the regular exams and then complete the survey and in total 127 students answered the survey.

The student satisfaction survey had five statements with a five-point scale, from 1 (completely agree) to 5 (completely disagree):
- S1. This laboratory was useful for your education on electrical machines;
- S2. This laboratory helped you to understand concepts and enabled you to apply them in the topic being studied;
- S3. This laboratory increased your motivation towards learning electrical machines;
- S4. This laboratory provided you with useful experience that might be of service in your future job;
- S5. The educational materials used in this laboratory were adequate for the tasks to be performed.

The results of this survey, summarized in Fig. 22.13, show a good level of student satisfaction with the laboratory, with responses falling mainly between "completely agree" and "agree". The best scores were for statements S1 (usefulness), S2 (aid to understanding and applicability) and S5 (adequacy of the materials), all with over 70 % of responses being between "completely agree" and "agree".

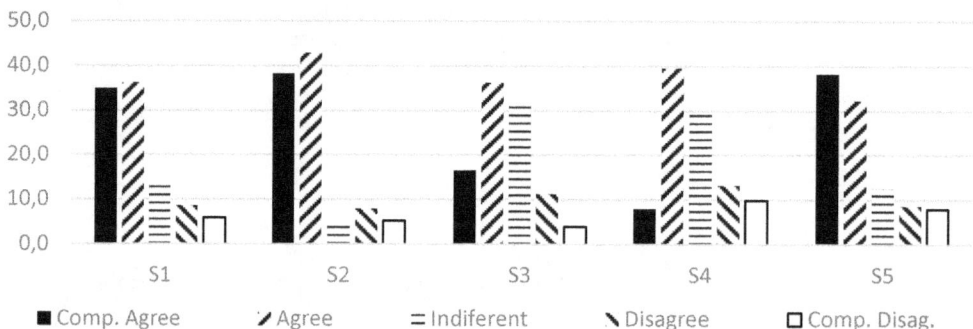

Fig. 22.13. Student satisfaction survey responses to statements S1 to S5.

22.7. Conclusions

Blending modern technologies like DAQ boards and mobile devices Apps with traditional electrical machines laboratory equipment has been shown to have many advantages. Modern test equipment and a set of laboratory tests on electrical machines' magnetic circuits configuration were developed to exploit these advantages, with the aim of helping students visualize complex concepts related with them, achieve a more lasting understanding and, thus, a more significant learning of the ideas presented in the magnetic circuit lectures.

The test equipment captures voltage and current waveforms during the no-load operation of a transformer fed in different ways to set-up four varied magnetic circuit configurations. It then processes and presents the actual measured voltage and current waveforms together with the calculated core flux waveform, rms values, power losses and the core magnetic cycle on the screen of an Android mobile device. In this way, through a guided set of tests, the students can directly link the main properties of the magnetic circuit (average flux lines length, iron cross-section, permeability and specific losses) with basic operation parameters of the transformer (mainly the no-load current and core losses) and, specially, with the time waveforms and the magnetic cycle peculiarities.

The combination of the guided exercises with the SLM test equipment developed results in high levels of student satisfaction with the laboratory work, and thus, improved student motivation showing that the main objectives set during the laboratory design were achieved.

References

[1]. S. A. Nasar and L. E. Unnewehr, Electromechanics and electric machines, 2nd ed., *Wiley*, 1983.
[2]. M. G. Say, Alternating current machines, 5th ed., *John Wiley & Sons*, 1983.
[3]. A. E. Fitzgerald, C. Kingsley, and S. D. Umans, Electric Machinery, 6th ed., *McGraw-Hill International Edition*, New York, 2003.
[4]. S. J. Chapman, Electric Machinery Fundamentals, 5th ed., *McGraw-Hill*, 2011.
[5]. L. Serrano-Iribarnegaray and J. Martínez-Román, Máquinas Eléctricas, 3r^d ed., *Editorial de la Universitat Politècnica de València*, 2014.
[6]. S. Bentz, Integration of basic electromagnetism and engineering technology, in *Proceedings of the Frontiers in Education Conference*, Vol. 2, November 1995, pp. 4a5.4–4a5.7.
[7]. V. Pulijala, A. Akula, and A. Syed, A web-based virtual laboratory for electromagnetic theory, in *Proceedings of the IEEE Fifth International Conference on Technology for Education (T4E)*, Dec 2013, pp. 13–18.
[8]. S. Ayasun and C. Nwankpa, Induction motor tests using Matlab/Simulink and their integration into undergraduate electric machinery courses, *IEEE Transactions on Education*, Vol. 48, No. 1, February 2005, pp. 37–46.
[9]. A. Bentounsi, H. Djeghloud, H. Benalla, T. Birem, and H. Amiar, Computer-aided teaching using Matlab/Simulink for enhancing an IM course with laboratory tests, *IEEE Transactions on Education*, Vol. 54, No. 3, August 2011, pp. 479–491.
[10]. M. Ojaghi, J. Faiz, M. Kazemi, and M. Rezaei, Performance analysis of saturated induction motors by virtual tests, *IEEE Transactions on Education*, Vol. 55, No. 3, 2012, pp. 370–377.
[11]. A. Syal, K. Gaurav, and T. Moger, Virtual laboratory platform for enhancing undergraduate level induction motor course using matlab/simulink, in *Proceedings of the IEEE International Conference on Engineering Education: Innovative Practices and Future Trends (AICERA)*, 2012, July 2012, pp. 1–6.

[12]. S. Gruber, A computer-interfaced electrical machines laboratory, *IEEE Transactions on Education,* Vol. 27, No. 2, 1984, pp. 73–79.

[13]. J. M. Williams, J. L. Cale, N. D. Benavides, J. D. Wooldridge, A. C. Koenig, J. L. Tichenor, and S. D. Pekarek, Versatile hardware and software tools for educating students in power electronics, *IEEE Transactions on Education,* Vol. 47, No. 4, 2004, pp. 436–445.

[14]. J. M. Jiménez-Martínez, F. Soto, E. D. Jodar, J. A. Villarejo, and J. Roca-Dorda, A new approach for teaching power electronics converter experiments, *IEEE Transactions on Education,* Vol. 48, No. 3, 2005, pp. 513–519.

[15]. F. Sellschopp and L. Arjona, An automated system for frequency response analysis with application to an undergraduate laboratory of electrical machines, *IEEE Transactions on Education,* Vol. 47, No. 1, 2004, pp. 57–64.

[16]. S. Durovic, Development of a simple interactive laboratory exercise for teaching the principles of velocity and position estimation, *International Journal of Electrical Engineering Education,* Vol. 50, No. 3, 2013, pp. 256–267.

[17]. W. Heath, O. Onel, P. Green, B. Lennox, Z. Gai, Z. He, and M. Rodriguez Liñan, Developing a student-focused undergraduate laboratory, *International Journal of Electrical Engineering Education,* Vol. 50, No. 3, 2013, pp. 268–278.

[18]. T. Sloane, Laboratories for an undergraduate course in power electronics, *IEEE Transactions on Education,* Vol. 38, No. 4, Nov 1995, pp. 365–369.

[19]. J. Van't Hof, J. Bain, R. M. White, and J.-G. Zhu, An undergraduate laboratory in magnetic recording fundamentals, *IEEE Transactions on Education,* Vol. 44, No. 3, August 2001, pp. 224–231.

[20]. J. Martinez-Roman, J. Perez-Cruz, M. Pineda-Sanchez, R. Puche-Panadero, J. Roger-Folch, M. Riera-Guasp, and A. Sapena-Bano, Electrical machines laminations magnetic properties: A virtual instrument laboratory, *IEEE Transactions on Education,* Vol. PP, No. 99, 2014, pp. 1–1.

[21]. J. Martinez-Roman, J. Fernandez-Molina, A. Sapena-Bano, M. Pineda-Sanchez, and R. Puche-Panadero, Magnetic materials and magnetic circuits smartlab, in *Proceedings of the XVII International Symposium on Electromagnetic Fields in Mechatronics, Electrical and Electronic Engineering (ISEF 2015),* 2015.

[22]. J. Martinez-Roman, A. Sapena-Bano, M. Pineda-Sanchez, and R. Puche-Panadero, SmartLab Magnetic: A Modern Paradigm for Student Laboratories, *Sensors & Transducers,* Vol. 197, Issue 2, February 2016, pp. 58-66.

[23]. J. Martinez Roman, R. Gomis-Cebolla, A. Sapena-Bano, and J. Perez-Cruz, Reactive power compensation smartlab, in *Proceedings of the XVII International Symposium on Electromagnetic Fields in Mechatronics, Electrical and Electronic Engineering (ISEF 2015),* 2015.

[24]. J. Martinez Roman, J. Perez-Cruz, M. Pineda-Sanchez, R. Puche-Panadero, and A. Sapena-Bano, Smartlabairgap: Rotating electrical machines airgap field laboratory, in *Proceedings of the XVII International Symposium on Electromagnetic Fields in Mechatronics, Electrical and Electronic Engineering (ISEF 2015),* 2015.

[25]. J. Martinez Roman, J. Bermudez-Campos, A. Sapena-Bano, J. Roger-Folch, and M. Riera-Guasp, Pumping station expert and control system on an android virtual instrument, in *Proceedings of the XVII International Symposium on Electromagnetic Fields in Mechatronics, Electrical and Electronic Engineering (ISEF 2015),* 2015.

[26]. iSEE (Instalations Systems and E. E. A. I. UPV, Smartlab magnetics test reports, 02 2016. [Online], Available: http://personales. upv. es/jmroman/SLM/test_reports.avi

[27]. Google play store: Smartlab-magnetics, [Online]. Available: https://play.google.com/store/apps/-details?id=com.smartlab.magnetic

Chapter 23
Recent Advances in Characterization of Sol-gel Based Materials for Sensor Applications

Viviane Dalmoro, Larissa Brentano Capeletti, João Henrique Zimnoch dos Santos

23.1. Introduction

The sol-gel technology provides a simple way for the production of chemical sensors with desirable properties through efficient incorporation of organic and inorganic compounds within a three dimensional network. Sol–gel materials synthesis includes the following steps: (i) sol (colloid suspension) preparation, (ii) gelation (sol-gel transition) and (iii) aging and drying of gel [1]. The formation of sol-gel material occurs via polymerization controlled through of hydrolysis and condensation reactions. To overcome the slow kinetics of these reactions commonly acids and bases has been employed as catalysts. The textures and morphological properties of sol-gel materials are affected by the catalysts employed during the production of the sol-gel materials. More specifically, the size and nature of the polycondensate structures in the sol, i.e., the length of linear chains, the degree of branching, or its density is determined by ratio of the rate constant of hydrolysis and condensation [1].

The sol–gel process is a multifaceted process that depends on many parameters, such as starting pH solution, the nature of the solvent, the water/precursor ratio, the presence of functionalized alkoxysilanes (containing amino, phenyl, methyl, vinyl, mercapto, and other organic groups) as well as on processing temperature and drying. Desired properties and morphology of sol-gel materials are achieved by tuning these chemical and physical parameters. For sensor application a frequently evaluated parameter is the introduction of organic groups directly bonded to sol-gel network, affecting the surface polarity, wettability, mechanical stability and porosity of resulting material. Additionally, this organofunctional groups may be anchored for sequential covalent modification of the

João Henrique Zimnoch dos Santos
Institute of Chemistry, Universidade Federal do Rio Grande do Sul
Av. Bento Gonçalves 9500 - CEP 91501-970, Porto Alegre, RS, Brazil

sensor or it can participate in the sensing process itself. The surface properties of hybrid sol–gel films can substantially influence their analytical performance [2]: Selectivity and sensitivity can be modulated by the ratio between hydrophobic/hydrophilic on the sensor surface [3].

Several remarkable advantages are provided by sol-gel materials (inorganic polymers) if compared to organic polymers such as: simpler fabrication process, higher chemical, photochemical and thermal stability, compatibility with various sensitive molecules, [4] as well as mechanical strength; optical transparency (down to 250 nm). Furthermore, synthesis may be conducted under mild conditions. The resulting material has highly cross-linked structure in a porous matrix, which enables the immobilization of organic or inorganic compounds according to the sensor final designing.

Several sensors have been developed by sol-gel approach: optical sensors and electrochemical sensor for detection of gases, atmospheric agents, neurotransmitters and drugs, just to mention a few. Physical chemical properties of sol-gel materials are influenced by the synthesis conditions and the characterization of these composts is a crucial step on the development of new sensors. Therefore, the interactions of element-receptor with the sol-gel network has to be clearly understood to allowing the tuning of the final performance of sensor (specially, concerning the improvement and specificity in the recognition properties) and to develop more simple, more rough, low cost, and with fast response (and, if possible with real time detection) devices.

Table 23.1 depicts the characterization techniques which have been employed for sol-gel based sensor recently published.

Table 23.1. Chemical and physical techniques employed for sol-gel based sensor for characterization published in 2015-2017.

Type of sensor	Element-receptor	Characterization techniques	Refs
NO_2 sensor	$Zr/CoTa_2O_6$	FE-SEM, FT-IR, XRD, XPS	[5]
O_2 sensor	Ca/ZrO_2	SEM, XRD	[6]
Ethanol sensor	$LaFeO_{3-\delta}$	SEM, TEM, XRD, XPS	[7]
Diethylstilbestrol sensor	AuNPs/MWCNTs-CS composites coupling with sol-gel molecularly imprinted polymer	CV, EIS, SEM	[8]
Hydrogen peroxide sensor	APTMS/MTMS/glutaraldehyde/graphite	EIS, FT-IR	[9]
Optical temperature sensor	Er^{3+}/Y^{3+}-codoped $Na_{0.5}Gd_{0.5}MO_4$	FE-SEM, TEM, XRD	[10]
Patulin sensor	APTES/TEOS/PVC	SEM	[11]
Acetone sensor	$LaFeO_3$	FE-SEM, XRD	[12]
Oxygen sensor	polydimethylsiloxane (PDMS)/MTMS-TEOS	FT-IR, UV	[13]

Table 23.1 (Continued). Chemical and physical techniques employed for sol-gel based sensor for characterization published in 2015-2017.

Type of sensor	Element-receptor	Characterization techniques	Refs
pH-responsive	ZrO_2 and HfO_2 functionalized porous silicon	SEM, FT-IR	[14]
Papaverine sensor	MTMS/TEOS/AuNPs/MWCNTs	CV, EIS	[15]
Dopamine sensor	silica matrix-poly(aniline boronic acid) hybrid	SEM, UV-vis, CV	[16]
HCl and NH_3 Sensing	TEOS-APTES	SEM, UV	[17]
Capsaicin sensor	graphene-doped sol-gel titania-Nafion composite	CV, EIS, SEM	[18]
Optical humidity sensing	yttria stabilized zirconia	SEM, TEM, UV–vis, XRD	[19]
Nalbuphine hydrochloride	TEOS	FT-IR, UV-vis, NMR	[20]

APTES (3-aminopropyl)triethoxysilane, AuNPs Gold nanoparticles, CV-Cyclic Voltammetry, EIS-Electrochemical Impedance Spectroscopy, FT-IR Fourier Transform Infrared Spectroscopy, MTMS methyltrimethoxysilane, MWCNTs multi-walled carbon nanotubes, NRM Nuclear Resonance Magnetic Spectroscopy, SEM- Scanning Electronic Microscopy, TEM Transmission Electron Microscopy, TEOS tetraethylorthosilicate, XRD X-ray Powder Diffraction, XPS X-ray Photoelectron Spectroscopy

According to Table 23.1, one can observe that the characterization of the sensors has been made with well-established and routine techniques. Most of them are concerned with structural (FT-IR, NMR, UV-vis, for instance) or morphological (SEM, TEM, for instance) characterization. Other (CV, EIS, for instance) seems to be employed to monitoring the sensor performance, which in turn may provide some insights of the sensor constitution. Thus, in the following sections, physical and chemical techniques used to characterize the sol-gel sensor will be discussed. More specifically this chapter is focused on structural (UV–vis, FT-IR and Raman spectroscopies, and electrochemical techniques), textural (BET and SAXS) and morphological (SEM, TEM and AFM) analysis.

23.2. Structural Analyses

23.2.1. UV-vis Spectroscopy

UV-vis spectroscopy enables the assessment of encapsulated species interactions with the sol-gel matrix. This is of paramount importance because the reactivity of the entrapped molecules depends mainly on their chemical nature and on the interactions between the compounds and the sol-gel matrices. Depending on these interactions, such as electrostatic and hydrogen bonds, it is possible to associate band wavelength shift with the interaction of chromospheres groups and the matrix. The electronic transitions n$\rightarrow \pi^*$ and $\pi \rightarrow \pi^*$ are detected in UV-vis spectroscopy. Two concepts are here involved: blue shift (hypsochromic shift) and red shift (bathocromic shift). The former corresponds to higher energy difference between the ground and excited state, due to a lowering of the energy of the ground state and an elevation of the energy of the excited state for $n \rightarrow \pi^*$ transitions [21]. More specifically, for sensor materials a hypsochromic shift for molecules bearing $n-\pi^*$ type transitions is detected with increase in the polarity of the medium because the indicator may interact with the silanol groups of silica surface [23]. In the case of red shift, the energy difference between the ground and excited state is decreased. This occurs due to a ground state energy increment and excited state energy reduced for $n \rightarrow \pi^*$ transitions [21]. The former is detected by a band shift towards lower wavenumbers, while the second one, to higher ones.

From the detection mode, it is worth mentioning that the solid nature of sensor demands to employ techniques other than absorbance/transmittance mode commonly employed in solution UV-vis measurements. No transmittance or even scattering properties under UV radiation by the solid devices demands the use of alternative modes of detection such as diffuse reflectance spectroscopy (DRS) or photoacoustic spectroscopy (PAS).

Frequently, the maximum band position of encapsulated molecule into sol-gel matrix can be influenced by several factors, such as steric effects, medium polarity, hydrogen bond and surface acidity. For example, Capeletti et al. using photoacoustic spectroscopy in the UV–vis region found that maximum band positions attributed to alizarin red were shifted depending on sol-gel route used to encapsulate the pH indicator within silica matrix [22]. For all the sol-gel routes (non-hydrolytic, acid catalyzed, basic catalyzed and without catalyst hydrolytic), a hypsochromic shift (blue shift) could be detected for alizarin red. This behavior can be explained by an enhancement on the polarity of the medium since the alizarin red molecules, bearing $n-\pi^*$ type transitions, may interact with the silanol groups on silica surface. In addition, pH sensors may electrostatically interact with the porous structure of silica and their activity towards the analyte may be reduced, resulting in the longer response times [2]. In this sense, the UV-vis analysis can be performed and correlated to the analytical response of sensor.

An apparent pKa, as opposed to a real thermodynamic pKa observed in solutions, is typical for encapsulated indicators because the protonation equilibrium is affected by heterogeneous microenvironment [23]. The absorption band of bromocresol purple encapsulated in silica nanoparticle associated to basic form (λ_{max} = 595 nm) has a blue shift of 5 nm compared to bromocresol purple in aqueous buffer [24]. An apparent *pKa* of 8.0 was determined for encapsulated indicator, changing with respect to *pKa* of bromocresol purple (*pKa* = 6.4). The authors explain that hydrophobic microenvironment inside the organically modified silica nanoparticles may obstruct the solvation of hydrophilic bromocresol purple, mainly due to the negatively charged basic form of pH indicator [24].

Depending on the organic groups present within the silica network, their interactions with element-receptor can induce band shift in UV spectra. For instance, bromocresol green immobilized in a sol–gel matrix shows a shift of the both absorption maxima (acid and basic form) with increasing (3-glycidyloxypropyl) trimethoxysilane (GPTMS) concentration [2]. An increase in the fraction of glycidoxypropyl group in the hybrid film results in hypsochromic shift of the absorption maximum of the protonated (neutral) form of the bromocresol green, and bathochromic shift of the deprotonated (anionic) form of the bromocresol green. These modifications in UV-vis spectra are consequences of the change in polarity of the microenvironment introduced by organofuncionalized sol–gel precursor. More specifically, it is arising from solvatochromism, which is resulting from different solvation effects on the ground and first excited states of the pH indicator molecule. A bathochromic shift of the absorption band is detected if the excited dye molecule is better stabilized by solvation than the molecule in its corresponding ground-state.

Besides the effect of the synthesis sol-gel route and the presence of organofunctionalized sol-gel precursors, the charge of surfactant may also affect its interactions with indicators containing chromophors groups leading to shifts in the electronic spectra. From interpretation of UV data, El-Nahhal et al. concluded that the presence of ethanediyl-1,2-bis(dimethyldodecylammonium chloride), a cationic surfactant, enhanced the acidity of silica silanols since the pKa value of 3.01 was detected for bromothymol blue into mesoporous silica containing the surfactant and in absence of surfactant the pKa was 4.01 [25]. This shift of pKa value to more acid region was explained due to the strong electrostatic attraction between the surfactant and anionic species of indicator, delaying the protonation of bromothymol blue by silanol groups. Similar results were reported in the literature [26], in which cetyl trimethyl ammonium bromide (CTAB), a cationic surfactant, interacts with the anionic form of investigated indicators (phenolphthalein, phenol red, bromophenol blue and cresol red) generating an ionic pairs. In the absorption spectra of the mixed free indicators two absorption bands were detected at 300 nm and 593 nm and the presence of two isosbestic points at 395 nm and 480 nm. This features point out the presence of to two equilibriums, the first one at pH 3–7 (between acid and neutral forms of indicators) and the second one at pH 7–12 (between neutral and base forms). The calculated value of pKa was 3.75, while for encapsulated indicators into silica–titania matrix containing CTAB the pKa was 9.3 and only one isosbestic point was detected at 490 nm. The change of pKa to more basic values was related to the anionic form of indicators, which involves weak Van der Walls interactions with silica–titania species in basic medium.

For efficient performance of sensors, the sol-gel material has to be produced with suitable porosity and permeability to allow the access of the analyte to the sensing centers and to prevent the leaching of the sensing molecules from the bulk of the sensor to the solution. In this sense, UV measurements has been carried out to monitoring the leaching of pH indicator of silica network to aquous solution. For example, Schyrr et al. [27] verified using UV assays that the leaching to phosphate buffered saline solution was dependent of the type of organofuncionalized silica precursor (methyltriethoxysilane (MTMS) or phenyltriethoxysilane (PTES) and ratio between TEOS/organofuncionalized silica precursor.

Another important feature in sensor development deals with the receptor element quantification in the sensor. For pH based sensors, in general, the pH indicator leached during the preparation of encapsulated systems or the supernatant liquid in the case of systems produced by adsorption are submitted to UV-vis analysis. Such measurements may provide indirectly the encapsulated amount of indicator. Nevertheless, the indicator has be extracted and solubilizated in a suitable solvent within the limits of linearity of Beer's law (ca. 10^{-3} to 10^{-6} M). Converserly, direct methods, especially Diffuse reflectance spectroscopy (DRS) in the UV-vis region, prevent the analyte loss or incomplete extraction, and constitute a potential relevant approach to quantify the amount of pH indicator encapsulated or grafted in the sensors. Thus, our group employed UV-vis DRS spectroscopy to quantify the acid–base indicators: Alizarin red, brilliant yellow and acridine encapsulated within silica based sensors produced from TEOS hydrolysis/condensation. Calibration curves for each system (indicator/silica produced by each sol–gel route) were built from a series of synthetic standards prepared using commercial silica to which different amounts of indicator were quantitatively added. It was obtained good linear correlations [28] for the three investigated systems. However, when the sensors were measured, inconsistent values were detected. In reflectance spectroscopy, the particle size and morphology may influence the optical signal. Thus, a second approach was used based on standard addition calibration method to eliminate the matrix effects. Increasing amounts of the pH indicator were added to the sensor prepared by acid catalyzed, base catalyzed and non-hydrolytic routes generating calibration curves with good linear correlation [28].

23.2.2. Fourier-Transform Infrared and Raman Vibrational Spectroscopies

Fourier Transform Infrared Spectroscopy (FTIR) and Raman Spectroscopy are powerful and complementary techniques to characterize either the immobilized/encapsulated compounds, as the materials network itself. In terms of the immobilization of compounds to serve as receptor elements, these vibrational techniques, especially FTIR, can provide information on the functional groups involved in the interaction between matrix and the encapsulated molecules. This type of information can be a key point in sensors development, since often it is necessary a specific site available for analyte detection. If this site is compromised with the matrix-receptor element interaction, the sensor functionality can be hindered loosing sensibility and/or increasing response time. FTIR can thoroughly provide information about the inorganic network of silica, while Raman can describe the organic portion of hybrid materials that are widely employed as sensors matrix in order to improve response time and avoid leaching [29-32].

In the case of sol-gel prepared silica, typical spectra (Fig. 23.1) use to present the following main bands: a broad band at high wavenumber region (3600–3000 cm^{-1}) corresponding to v(O–H) of hydroxyl groups (silanol) from silica network or adsorbed water on the surface; and the silica *fingerprint* region at 1250-700 cm^{-1} representing the network vibrations. The main bands between 1250 and 1000 cm^{-1} are associated with the sequential stretching Si-O mode from the Si-O-Si network. Si-O(H) stretching appears at slightly different positions in the IR (950 cm^{-1}) and Raman (ca. 980 cm^{-1}) spectra and symmetric mode of the band corresponding to the Si-O-Si vibrations is detected at ca. 795 cm^{-1}; Si-O^{-} rocking mode is observed at ca. 540 cm^{-1}; and siloxane ring breathing mode (with 3 or 4 SiO units) is located at ca. 490 cm^{-1} [29].

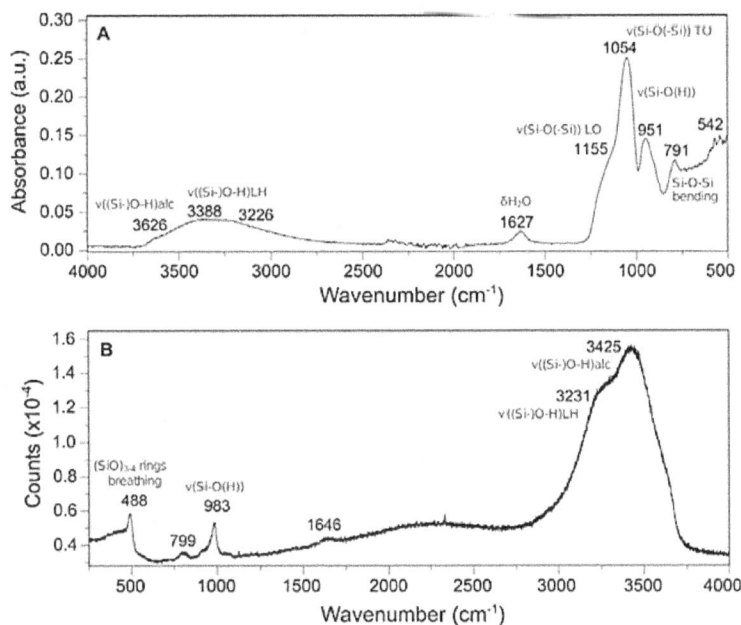

Fig. 23.1. (a) Attenuated total reflectance infrared and (b) Raman spectra of a silica sample prepared by sol-gel method [29].

In addition to the silica bands, sometimes the immobilized or encapsulated receptor element can also be detected, depending on the configuration of employed equipment and the compound concentration. Kriltz et al. employed this methodology to evaluate a fluorescent pH-sensitive dye named naphthalimide immobilized in sol-gel films prepared with tetraethoxysilane, 3-glycidoxypropyltrimethoxysilane and *N*-(3-(trimethoxysilyl)propyl)-ethylendiamine. The dye was covalently attached to the films by an ester activation of the dye with 2-succinimido-1,1,3,3-tetramethyluronium tetrafluoroborate. To determine the structure of the gel-dye system, FTIR spectra of films, both with and without the dye incorporation, were recorded and presented only small differences indicated by small shifts, caused by environmental effects. The primary amino group of *N*-(3-(trimethoxysilyl)propyl)-ethylendiamine formed an amide bond with release of protons when the dye is attached. The peak at 1578 cm^{-1} explicitly shows the presence of these amino groups, since all possible other peaks for these groups are overlapped with other bands. When the dye is added, both the peak at 1578 cm^{-1} and the other possible amino peaks are considerably reduced in intensity in the gel spectra. This fact indicates a bonding process of dye with amino groups, but there were also remaining primary amino groups. In comparison to the dye free gel the spectrum of the gel with dye exhibits small peaks at 1832, 1771, 1737, 1697 and 1650 cm^{-1}, which are considerable indications to the amide bonding of the dye to the gel film. In addition, the introduction of the dye into the gel results in carbonyl groups introduction to the system, detectable by C-O stretching vibrations at 1850–1650 cm^{-1} [33]. Mostly, silica network peaks dominates the vibrational spectra hindering the encapsulated compounds identification [34]. In addition, it is well documented that entrapped molecules fit themselves in the silica glass pores and did not make bonds with the network molecules. Therefore, spectra from bare and doped materials closely resemble each other [35]. Another possibility is to explore the interaction between the silica-based sensor and analyte, comparing spectra before and after contact with it. In some cases, this approach enables to elucidate binding sites and complexes formation involved in the detection process [36].

FTIR can be also employed to investigate the silica structure through the network vibrations mainly represented by (Si-O)-Si stretching mode appearing as the broad and dominating band in the region of 1000-1200 cm^{-1}. This band is represented by the longitudinal (LO) and transversal optical (TO) vibrational modes of four- and six-membered siloxane rings resulting in four different components. This LO/TO splitting is originated by long-range Coulomb interactions and the presence of the organic groups may interfere this phenomenon shifting the peak maximum wavelength of each component [37]. The shifts to lower wavenumbers are normally related to the network deformation in order to accommodate the organic groups within the inorganic silica matrix resulting in larger siloxane rings (six-membered) and greater Si-O-Si angles and longer Si-O bond lengths. LO and TO mode shifts occur mainly near the surface of the material, which can be better detected employing Attenuated Total Reflectance (ATR) detection mode for FTIR spectroscopy. The organic groups presence creates heterogeneous regions that may introduce local deformations in the network resulting in differences observed for Si-O bond lengths and Si-O-Si angles for hybrid materials [29]. In research performed by our group, the encapsulation of the indicators alizarin red, acridine and brilliant yellow using different sol-gel routes was investigated by this FTIR methodology. The different component modes contributions were obtained by deconvolution of the Si-O-Si stretching peak. The results for four-membered and six-membered siloxane rings showed an important correlation with the indicator content showing that higher percentage of the larger rings is also formed to accommodate entrapped organic molecules, similarly to the hybrid materials. These results may be important to evaluate interactions between the analyte and matrix and how it reflects in the response time as function of the silica network structure. Usually, matrices with high surface areas may facilitate analyte permeation through the network affording shorter response times, thereby improving the sensor performance. In this study, no relationship between the larger six-membered siloxane rings percentage and the sensor response time was found, indicating the analyte penetration through the silica matrix did not seem to be determined by the dimension of the rings. This observation may indicate that the analyte can access receptor elements through the spaces external to the siloxane rings and not through the siloxane rings themselves [38].

23.2.3. Electrochemical Techniques

A key point on the development of electrochemical sensors based in sol-gel materials is the mass transport properties inside pores of material which is directly related to electrochemical behavior. Moreover, the electrochemical response is also limited by charge-transfer properties at the electrode–film interface. The investigations of these properties can be performed by electrochemical techniques such as cyclic voltammetry (CV) and electrochemical impedance spectroscopy (EIS).

For electrochemical sensor the modification of electrode surface with silica films or silica films containing additives constitute an approach to improve the electrochemical signal of redox process. The first advantage of this process is the increase of active area and, consequently, increase of the current associated to electrochemical process. This phenomenon occurs due to the highest interfacial area or void fraction in sol-gel if compared to non-modified electrode surface [39]. To quantity the effect of surface modification on electrode, cyclic voltammetry has been used. For instance, Taei et al. verified, using $K_3Fe(CN)_6$ as an electrochemical probe, that electrode area for carbon paste electrode (CPE) modified with multi-walled carbon nanotubes (MWCNTs/CPE) and CPE electrode coated with MWCNTs modified with $ZnCrFeO_4$ magnetic nanoparticles synthesized via a sol–gel (MWCNTs/$ZnCrFeO_4$/CPE) are 4.9 and 6.2 times, respectively, higher than that for the CPE electrode prior to modification [40]. The specific role of additives into sol-gel matrix is linked to augmentation of conductivity. In this way, materials such as gold nanoparticles, carbon nanotubes, conductive polymer have been evaluated, which in turn of accelerated the electron transfer reaction at the electrode analyzed by EIS.

Cyclic voltammetry is a very interesting technique, allowing infers both mass transport and charge-transfer properties. Cyclic voltammetry can be performed in the presence of a variety of redox probes characterized by different charge and size: I^-, $Fe(CN)_6^{3-}$, ferrocenemethanol (FcMeOH), $Ru(bpy)_3^{2+}$, which are expected to cross the sol-gel film before being detected on the electrode surface. Thus it is possible to characterize the permeability of these films [41]. In their recent study, Gholivand et al. analyzed the Fe–Cu/TiO_2–CPE electrode by cyclic voltammetry and observed the same anodic and cathodic peaks with enhancement in peaks current and shift of the peak potential towards less positive potentials [42]. Moreover, they demonstrated that the modification not only favors the electron transfer kinetic, but also favors higher sensitivity for metformin monitoring. These results were explained by presence of copper oxide in the modifier, improving the possibility of adsorption of the metformin at the electrode surface. It should be remarked that apparent diffusion coefficient of redox species within the sol-gel matrix can be derived by voltammetric assays with sol-gel modified electrode. It was verified that not only average pore size, but also structure and pore interconnections, are likely to influence mass transport in sol-gel films [41].

Silica surface is covered by silanol groups, which may affect the charge of its surface depending on the pH. In acid solution, which pH is lower than the isoelectric point of silica, the surface is deprotonated and this negatively charged surface influence the electrochemical response. One consequence is that the positively-charged $Ru(bpy)_3^{2+}$ species can be adsorbed and/or accumulated by electrostatic forces on these surfaces before being electrochemically detected [41]. The second consequence is that diffusion of anionic compounds such as $Fe(CN)_6^{3-}$ becomes very limited, leading to low intensity signal, which can be justified by electrostatic repulsion between films and electrochemical species, imparting low current or not detection [41]. Thus, using cyclic voltammetry analysis associated to a technique for determination of surface charge of sol-gel based electrode it is possible to understand the role of surface charge on the detected electrochemical signal (current of redox process) and optimize the surface properties to improve the sensor performance.

Complementary information to CV assays can be obtained by Electrochemical Impedance Spectroscopy (EIS). This technique is a powerful tool for understanding the interfacial properties of surface modified electrodes and the electron-transfer resistance at the electrode surface. The impedance spectrum, more specifically the Nyquist plots, comprise: (i) a semicircle portion at high frequencies, corresponding to the electron transfer limiting process, which time constant concerns the electron charge transfer resistance (R_{ct}) and double layer capacitance, and (ii) a tail at low frequencies, resulting from the diffusion limiting step of the electrochemical process. For instance, impedance spectra were taken in a phosphate buffer solution (pH 9.0) containing 5.0×10^{-3} mol L^{-1} $[Fe(CN)_6]^{3-}/^{4-}$ of the unmodified and modified electrodes [43]. Well-defined semicircle was detected for the CPE electrode with electron transfer resistance equal to 10.5 kΩ and for modified electrode manganese titanate nanoceramics synthesized by the sol-gel method the charge transfer resistance was reduced to 3.9 kΩ. This behavior is explained by conductive layer formation on the surface electrode decreases the charge transfer resistance and accelerates the electron transfer at the surface of the modified electrode. Therefore, EIS analyses allows monitoring if the electrode modification favours the desired electrochemical process, i.e., decreases the R_{ct} and increase the electron transfer rate at the electrode surface.

An important application of electrochemical techniques is the monitoring of adsorption process on electrode surface. Li et al. al carried out cyclic voltammetry assays and correlates the current detected for dopamine oxidation with adsorbing capacity [16]. After set the data with Langmuir isotherm and it was found a Langmuirian binding process with association constant equal to 2.9×10^{-4} mol^{-1} L, indicating higher affinity of modified electrode towards dopamine in comparison to previous studies. More adsorptive molecules on the surface electrode could provide higher sensibility for sensor.

Another relevant application of electrochemical analysis consists on checking if the template molecule is extracted from molecularly imprinted sensors. These sensors may be produced based on covalent or non-covalent interactions between template molecule and functional sol-gel network progressively formed by condensation reactions and organized around the template. Subsequently, removal of the template can induce the formation of imprinted cavities that are complementary in size and shape to the target analyte (original template) [16], being applied as recognition elements in sensors. Several sol-gel processes for molecules imprinted sensor production have been reported. To overcome drawbacks such as low binding capacity, resulting in slow binding kinetics and long analysis time, organic modified silica precursor has been employed, allowing the improvement of the coordination of the molecule with the silica network. For example, the amino group of APTES (3-aminopropyl triethoxysilane) and the benzene ring of PTMS (phenyltrimethoxysilane) can provide recognition sites through hydrogen bonds and "π–π stacking" interaction with the template molecule (2-nonylphenol) [44]. Another drawback of molecularly imprinted sensor is a very weak electronic signal. The insertion of conducting polymers, gold nanoparticles and carbon nanotubes within silica result in a hybrid material with both electrical conductivity [8] and electrochemical activity and can offer new bond sites, which in turn may improve the imprinting efficiency of these sensors. EIS and CV analysis have been employed to verify the increase of conductivity of modified electrodes.

450

Cyclic voltammetry and Electrochemical Impedance Spectroscopy were carried out to characterize the imprinted sensor in the presence of electrochemical probe $[Fe(CN)_6]^{3-}/[Fe(CN)_6]^{4-}$. Generally, the silica film or a hybrid films composed by organic polymer and silica are coated on the surface electrode. When the template molecules remain incorporated within the matrix network, the permeation of ferricyanide from the solution to the electrode surface is hardship. Thus, the electron transfer was blocked. After template removal, the response obtained as current is higher than before molecule extraction of the template, evidencing that the small ions $[Fe(CN)_6]^{3-}/[Fe(CN)_6]^{4-}$ can penetrate in the silica network, access the electrode surface and suffer the electrochemical reaction. This could be explained since removal of the template leave some imprinted cavities for ferricyanide to arrive at the electrode surface more easily. After the imprinted molecular sensor has been incubated in template solution for certain time, and submitted to voltammetry analysis, the electrochemical current due to electron transfer of the redox probe ($[Fe(CN)_6]^{3-}/[Fe(CN)_6]^{4-}$) is reduced, revealing that some cavities were recombined with template molecules.

Zhang et al. observed a decrease of 99 % in the reduction peak current for the imprinted sensor compared to sol-gel multiwalled carbon nanotube (MWNT)–Nafion (NF) composite film immobilized on the glassy carbon electrode (MWNT–NF/GCE), because of difficulty for $K_3Fe(CN)_6$ to penetrate on sol–gel film and achieve electrode surface for electron transfer [44]. Nevertheless, a significant increasing (858 %) in the cathodic current was detected after removing 2-nonylphenol from the imprinted molecular sensor. After immersion in 60 mol L^{-1} 2-nonylphenol solution for 6 min, it occurs a decrease of 28 % in reduction peak current in comparison to the imprinted molecular sensor after extraction of 2-nonylphenol, suggesting that the target molecules (2-nonylphenol) could rebind up and to hinder the electrochemical probe to achieve at the electrode surface.

In this sense, EIS has been also employed to investigate the features of the surface modified electrode, in order to characterize the stepwise construction process of the sensor. In the EIS spectra, as mentioned before, the semicircle diameter at higher frequencies is associated to the charge-transfer resistance, and at lower frequencies the tail is related to mass transfer control process. After electropolymerization polymer composed from polypyrrole, sol-gel, gold nanoparticles, the resistance of the charge-transfer decreased compared to bare pencil graphite electrode (PGE), indicating that gold nanoparticles and polypyrrole layer had good electric conducting properties and could increase the electron transfer rate. Nevertheless, when the molecule template was present on modified film, the charge-transfer resistance was large. This suggests that the compact imprinted film acts as a non-conductive film, i.e, a barrier for the electron transfer. After removing the template, a small charge-transfer resistance is detected, because the formed cavities on sol-gel film could render easier the permeation of the electrochemical probe through the film [45].

The electrochemical techniques may be the great value in characterization of sol-gel pH sensor. As the production of porous silica can be conducted at low temperature, organic chromophores are incorporated without losing their electrochemical and chromophore properties. Capeletti et al. using electrochemical techniques concluded that the interactions of alizarin red, a pH indicator, with the silica network generated through its encapsulation via four sol–gel routes was dependent on the employed sol-gel route [22]. By differential pulse voltammetry (DPV) analysis, they found that the oxidation peak of encapsulated alizarin via acid route and hydrolytic sol–gel route without catalyst was dependent on the pH. This behavior pointed out a process that implicates two electrons and one proton, suggesting that alizarin red interacts with silica through the hydroxyl groups. For hydrolytic base-catalyzed route, the electrochemical behavior was different from those concerning the sensor prepared by acid route, these results were related to different of silica–dye interactions and/or formation of higher number of products during the chemical or electrochemical reactions. Based on the decreasing reduction peak intensity for alizarin in the non-hydrolytic route compared to acid route, it was suggested the consumption of alizarin red in its quinone form during the process. In sum, the nature of the encapsulating route may influence the behavior of the resulting sensor.

23.3. Textural Analyses

In the case of sensors, the solid phase materials should ideally present characteristics as large surface area, interconnected pores and an open framework. Nitrogen porosimetry and small angle X-ray scattering (SAXS) are techniques to evaluate these textural properties in terms of porosity and hierarchical organization, respectively. High porosity is an important feature for sensors, especially because the need of analyte molecules detection. An increase in the surface area also means larger active surfaces, improving recognition properties since more reactive groups are present as a result of the reactivity increase, which helps to reach high sensitivities. Silica gel materials and/or organically functionalized silica present the advantage of high surface areas ($S > 200$ m^2 g^{-1}) and also increased pore volume ($Vp > 0.2$ cm^3 g^{-1}). However, as a result of the nowadays effort to achieve better performance, even higher values can be obtained [46]. The evaluation of these parameters includes the usage of Brunauer-Emmet-Teller (BET) and Barrett-Joyner-Halenda (BJH) equations usually on nitrogen adsorption/desorption data to evaluate some materials physical properties as the specific surface area, pore volume, pore diameter and pore size distribution. In addition, the pore type can be determined from the profile of the hysteresis originated by the adsorption and desorption isotherms [36].

The porosity characteristics can be affected by several parameters of the sol-gel based materials preparation. The encapsulated materials use to immobilize inside the matrix pores reducing the total surface area. In the same way, the addition of organic groups via organosilane condensation can also reduce the porosity if these groups fill or block the silica pores. To evaluate this second hypothesis, Afshani et al. prepared a nanoporous silica-based optical sensor for detection of Fe^{3+}, Al^{3+} and CN^- in water. Using SBA-15 functionalized with aminopropyl groups, they could prove the uniform mesopores characteristic of such material from the H1 hysteresis type and also that pores remained open and accessible for further applications. Nevertheless, the SBA-15 channel walls functionalization with the organic groups has decreased surface area and pores diameter and volume [36]. Similar behavior was found by Qian et al., where SBA was functionalized with 4,4'-diaminodiphenyl sulfone to detect toluene [47]. On the other hand, the encapsulation of receptor elements can also influence the porosity properties as mentioned before. Dudas et al. report the preparation of hybrid silica-porphyrin materials that can be employed as pH fluorescence sensors. The matrix presents tailored pore sizes and they evaluated the influence of porphyrin addition over them, reporting that in all cases of the compound addition, surface area has decreased in comparison to the control samples without it highlighting the entrapment inside silica network. They also evaluated the silane precursors and the usage of an ionic liquid as additive in the textural properties. The larger pore sizes were obtained for samples prepared with TEOS and larger pore diameters were obtained in the ionic liquid presence, indicating a decrease of gel contraction process during synthesis in this case. When a mixture of different precursors was employed, the pore size was decreased and the authors attribute this behavior to the steric effect originated by the voluminous isobuthyl groups which limits the silica network increasing [48].

The textural properties can also be evaluated using SAXS, as mentioned before. This technique can provide information about the material fractal structure type indicating size and how the primary particles are organized. In terms of chemical sensors, a wide-open structure may impart faster mass transport, which is used to be the rate determining step [46]. SAXS is also a powerful tool to investigate periodic characteristics in ordered porous materials as the well-known and wide used SBAs. In this case, the curve presenting diffraction peaks corresponding to the three different spatial position planes indicates the hexagonal mesostructure with one-dimensional channels (Fig. 23.2).

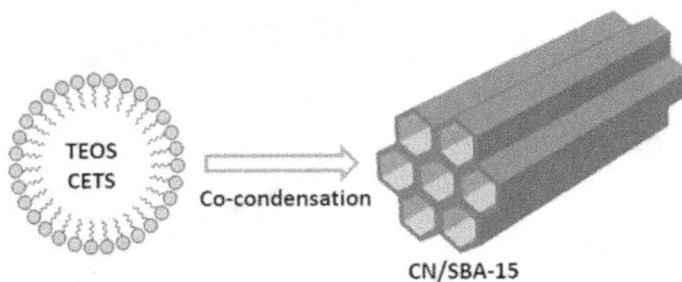

Fig. 23.2. SBA hexagonal structure with one-dimensional channels prepared from the co-condensation TEOS and 2-cyanoethyltriethoxysilane resulting in a cyanide functionalists SBA-15. Adapted from [47].

In addition, it is also possible to investigate this mesophase crystallinity evolution according to the material functionalization and/or immobilization of a receptor element [47]. Usually, a small peak intensity decreasing is observed as a result of pores and walls scattering contrast change and also due to the irregular organic groups covering the walls on the nanochannels [47, 49]. Yadavi et al. also observed that this behavior is an indication of SBA hexagonal structure maintenance, without collapsing pores along the surface modifying process. Using SAXS, they could also identify the walls thickness increase which is consistent with the organic moieties grafting in the silica porous walls [49]. When comparing SAXS and nitrogen porosimetry techniques, it is possible to obtain some results discrepancy. The nitrogen physisorption better describes the microporous characteristics and also mesoporous with less efficiency. However, it does not afford solid evidence for high mesoscopic empty pores. In fact, the functionalization process can hinder the nitrogen access to the mesopores, restricting thereby the mesoporosity measurement. At the same time, as the inorganic wall-to-wall distance is constant and the inorganic atoms have higher electronic contrast, these pores can be complementary determined using SAXS. [50].

For xerogel silica that presents non-ordered structure, this technique can also help to discover the network organization. Using the unified model equation to fit SAXS curves it is possible to obtain information about primary particles and aggregates size and also about the hierarchical organization regarding the fractal structure of the material [51, 52]. By using this model, the scattering curve can be adjusted using either one or more consecutive hierarchical levels. Each of these levels uses to be constituted by a shoulder-type Guinier region that gives details of the radius of gyration (Rg) and/or a linear Power-law decay which provides information about the material organization (α). Examples of SAXS curves for xerogel sensors prepared by sol-gel method and their respective unified fit modeling are shown in Fig. 23.3 [53].

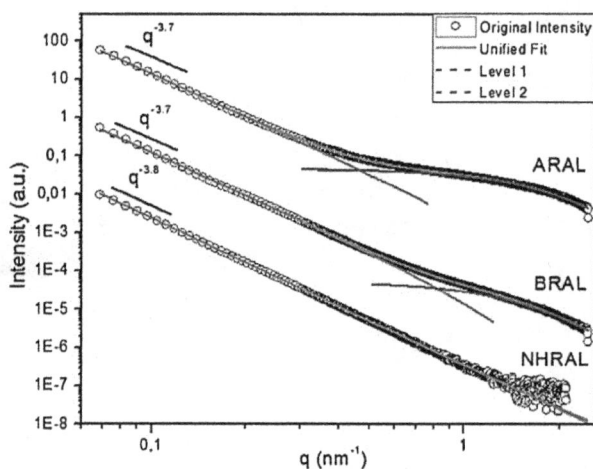

Fig. 23.3. SAXS plots of xerogel sensors prepared by different sol-gel routes and their respective unified fit modeling [53].

From the curves in Fig. 23.3 it is possible to observe the scattering profile differences originated by the use of different sol-gel routes. All the samples prepared by acidic catalyzed hydrolytic routes presented two levels at the measured q range, indicating the primary particles presence and its aggregation in a bigger structure, while non-hydrolytic route presented only one level. At low-q, the typical shoulder corresponding to Guinier region was not identified for any sample demonstrating the primary particles form aggregates with sizes larger than 100 nm. However, from this region it is possible to observe the power-law decay (green lines) with α slopes varying from 3.7 to 4.0. These values are characteristic of surface fractals that are represented by dense aggregates varying from completely rough surface ($\alpha = 3$) to smooth surfaces ($\alpha = 4$). For basic

454

catalyzed route, which usually result in spherical particles precipitation, it was possible to observe a roughness increase when the element receptor was added to the material. On the other hand, for acidic catalyzed and non-hydrolytic routes, which form gels that usually present α values between 1.8 and 2.5, surface fractals was also observed indicating that the drying process resulted in the formation of surfaces that dominate the scattering curve. This effect was especially prominent for non-hydrolytic route, where this scattering intensity may hinder the smaller structure detection.

At high-q (blue line) the Guinier region is well defined allowing to obtain the elementary silica particles Rg determination. In addition, a power-law decay followed the Guinier region, presenting α = 4, which indicates that all systems have smooth surfaces for this structural level. For the acidic route it was possible to observe an increase of the primary particles size with the encapsulated molecule (receptor element) molecular size [53]. SAXS curves can also provide surface/volume (S/V) ratio estimation and for this case it was observed an S/V decrease after encapsulation procedure. It is speculated that the ratio decrease is a result of the pores filling by the encapsulated molecules that usually lodge themselves inside the pores [38].

23.4. Morphological Analysis

Particles and fibers morphology and also films aspect are wide studied characteristics for sol-gel based sensors. Microscopy techniques as scanning electronic microscopy (SEM), transmission electronic microscopy (TEM) and atomic force microscopy (AFM) are extensive applied for this purpose. Characteristics like particle size and shape, aspect ratio, pores, aggregates and also surface topographies, roughness, and defects of thin films are the most studied ones for sensor applications.

In the case of films, the uniformity with improved optical quality is very important. This characteristic is mostly dependent on the sol stage stability and can be evaluated by microscopy techniques. Islam et al. prepared mesoporous silica-titania hybrid nanoparticles for pH sensing as optochemical sensor that were fully characterized by SEM, TEM, AFM and optical microscopies after deposition by spin coating in a quartz substrate. They found well distribution of titania particles in silica particles, however without the CTAB surfactant, agglomerates were observed as a result of the adsorbed water molecules and hydroxyls ions on hybrid films surface. When CTAB is added, the particles were found dispersed on the film, since the surfactant adsorption can increase the interaction forces between the particles, preventing the formation of film defects and cracks. After heat treatment for CTAB removal, observed large nanoparticles and pores might also help particles dispersion, which is desirable for sensing application. In addition, it is possible to observe nanometric pores that can contribute for a decrease in film refractive index affording better anti-reflectance properties. Using the microscopy techniques, it was also demonstrated that using appropriate aging procedure, films free of cracks could be obtained. TEM images also demonstrate this behavior, showing smooth surfaces and well-dispersed nanoparticles that after surfactant evaporation results pores and new agglomerates formation which is responsible for the surface roughness. Together, all these aspects would contribute for the optical characteristics and application of thin films [54].

Other materials morphologies are also employed as sensors and can be characterized by these techniques. In the case of nanofiber sensors, the aspect ratio (length/diameter) is important for the performance, since conductivity is directly related to the nanofibers diameter as demonstrated by Hussain et al [55] and therefore SEM characterization is mandatory. On the other hand, particulated materials as the mesoporous cellular foams (MCF) prepared by Xue et al. as an electrochemical sensor for capsaicin also need microscopic description. This material is similar to an aerogel: Three-dimensional, continuous and ultralarge-pore structure generated by the incorporation of swelling agents. These characteristics usually impinge much higher catalytic activity as a result of fast mass transfer kinetics and good accessibility, enabling to use the sensor also for large molecules. The SEM and TEM techniques could be performed to investigate these properties in terms of morphology and pores as shown in Fig. 23.4. The first one shows the MCFs morphology and their immobilization over carbon paste electrodes, while in TEM images it is possible to observe the honeycomb-like structure with interconnected and ultra-large pores, indicating MCFs as a promising material with high accessible active areas that can lead to increased analyte adsorption in the sample [56].

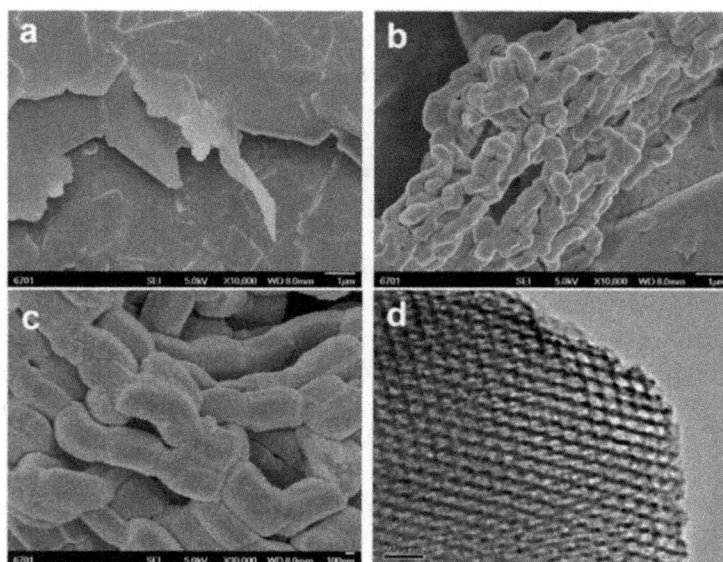

Fig. 23.4. SEM images of bare carbon paste electrode (a), mesoporous cellular foams/carbon paste electrode (b) mesoporous cellular foams (c) and TEM image of mesoporous cellular foams (d) [56].

23.5. Final Remarks

Sol-gel based materials are very flexible, versatile and can be used to suit different types of necessities depending on the application. The same trend is established for sensors and to achieve their best performance. The characterization techniques are key strategies in the development and optimization of such materials. From this point of view, it is important to seek for a set of complementary analytical techniques and their correlations which may provide direct clues or indirect sights of structural, textural and morphological characteristics of the resulting device. In terms of sensors, the structural analysis seems to be fundamental in order to keep a good compromise between absence of compounds leaching and their activity maintenance. In addition, the textural characterization highlights the importance of analyte access for fast sensor response, while the morphology characterization techniques helps to investigate the most appropriate design depending on the type of sensor. In sum, sol-gel method flexibility and this set of complementary techniques allow the preparation of highly tunable sensor materials with specific properties for the required purpose.

References

[1]. Schubert, U., Hüsing, N., Synthesis of inorganic materials, *Wiley*, Weinheinm, 2005.

[2]. Kassal, P., et al., Hybrid sol–gel thin films doped with a pH indicator: effect of organic modification on optical pH response and film surface hydrophilicity, *Journal of Sol-Gel Science and Technology*, 69, 3, 2014, pp. 586-595.

[3]. Ghazzal, M. N., et al., Tuning the selectivity and sensitivity of mesoporous dielectric multilayers by modifying the hydrophobic-hydrophilic balance of the silica layer, *Journal of Materials Chemistry*, 22, 42, 2012, pp. 22526-22532.

[4]. Barczak, M., C. McDonagh, and D. Wencel, Micro- and nanostructured sol-gel-based materials for optical chemical sensing (2005–2015), *Microchimica Acta*, 183, 7, 2016, pp. 2085-2109.

[5]. Liu, F., et al., High-temperature NO2 gas sensor based on stabilized zirconia and $CoTa_2O_6$ sensing electrode, *Sensors and Actuators B: Chemical*, 240, 2017, pp. 148-157.

[6]. Dhahri, R., et al., Enhanced performance of novel calcium/aluminum co-doped zinc oxide for CO2 sensors, *Sensors and Actuators B: Chemical*, 239, 2017, pp. 36-44.

[7]. Cao, E., et al., Effect of synthesis route on electrical and ethanol sensing characteristics for $LaFeO_3$-δ nanoparticles by citric sol-gel method, *Applied Surface Science*, 393, 2017, pp. 134-143.

[8]. Bai, J., et al., Ultrasensitive sensing of diethylstilbestrol based on AuNPs/MWCNTs-CS composites coupling with sol-gel molecularly imprinted polymer as a recognition element of an electrochemical sensor, *Sensors and Actuators B: Chemical*, 238, 2017, pp. 420-426.

[9]. Thenmozhi, K. and S. S. Narayanan, Horseradish peroxidase and toluidine blue covalently immobilized leak-free sol-gel composite biosensor for hydrogen peroxide, *Materials Science and Engineering: C*, 70, Part 1, 2017, pp. 223-230.

[10]. Du, P., et al., Citric-assisted sol-gel based Er3+/Yb3+-codoped Na0.5Gd0.5MoO₄: A novel highly-efficient infrared-to-visible upconversion material for optical temperature sensors and optical heaters, *Chemical Engineering Journal*, 306, 2016, pp. 840-848.

[11]. Fang, G., et al., Development and application of a quartz crystal microbalance sensor based on molecularly imprinted sol-gel polymer for rapid detection of patulin foods, *Sensors and Actuators B: Chemical*, 237, 2016, pp. 239-246.

[12]. Chen, Y., et al., Acetone sensing properties and mechanism of nano-$LaFeO_3$ thick-films, *Sensors and Actuators B: Chemical*, 235, 2016, pp. 56-66.

[13]. Xiong, Y., et al., A miniaturized oxygen sensor integrated on fiber surface based on evanescent-wave induced fluorescence quenching, *Journal of Luminescence*, 179, 2016, pp. 581-587.

[14]. Destino, J. F., et al., Robust pH-responsive group IV metal oxide functionalized porous silicon platforms, *Materials Letters*, 181, 2016, pp. 47-51.

[15]. Rezaei, B., H. Lotfi-Forushani, and A. A. Ensafi, Modified Au nanoparticles-imprinted sol–gel, multiwall carbon nanotubes pencil graphite electrode used as a sensor for ranitidine determination, *Materials Science and Engineering: C*, 37, 2014, pp. 113-119.

[16]. Li, J., et al., Electrochemical sensor for dopamine based on imprinted silica matrix-poly(aniline boronic acid) hybrid as recognition element, *Talanta*, 159, 2016, pp. 379-386.

[17]. Geltmeyer, J., et al., Colorimetric Sensors: Dye Modification of Nanofibrous Silicon Oxide Membranes for Colorimetric HCl and NH3 Sensing (Adv. Funct. Mater. 33/2016), *Advanced Functional Materials*, 26, 33, 2016, pp. 6136-6136.

[18]. Kim, D.-H. and W.-Y. Lee, Highly sensitive electrochemical capsaicin sensor based on graphene-titania-Nafion composite film, *Journal of Electroanalytical Chemistry*, 776, 2016, pp. 74-81.

[19]. Sikarwar, S., et al., Fabrication of nanostructured yttria stabilized zirconia multilayered films and their optical humidity sensing capabilities based on transmission, *Sensors and Actuators B: Chemical*, 232, 2016, pp. 283-291.

[20]. Elsayed, B. A., et al., Highly sensitive spectrofluorimetric analysis and Molecular Docking using benzocoumarin hydrazide derivative doped in sol-gel matrix as optical sensor, *Sensors and Actuators B: Chemical*, 232, 2016, pp. 642-652.

[21]. Drago, R. S., Physical Methods for Chemists, *Saunders College Pub*, 1992.

[22]. Capeletti, L. B., et al., Encapsulated alizarin red species: The role of the sol–gel route on the interaction with silica matrix, *Powder Technology*, 237, 2013, pp. 117-124.

[23]. Ye, F., M. M. Collinson, and D. A. Higgins, What can be learned from single molecule spectroscopy? Applications to sol-gel-derived silica materials, *Physical Chemistry Chemical Physics*, 11, 1, 2009, pp. 66-82.

[24]. Zhaobo, L., et al., pH-responsive silica nanoparticles for controllable $^1O_2$ generation, *Nanotechnology*, 21, 11, 2010, pp. 115102.

[25]. El-Nahhal, I. M., et al., Sol-gel encapsulation of bromothymol blue pH indicator in presence of Gemini 12-2-12 surfactant, *Journal of Sol-Gel Science and Technology*, 71, 1, 2014, pp. 16-23.

[26]. Islam, S., et al., Mesoporous SiO_2–TiO_2 nanocomposite for pH sensing, *Sensors and Actuators B: Chemical*, 221, 2015, pp. 993-1002.

[27]. Schyrr, B., et al., Development of a polymer optical fiber pH sensor for on-body monitoring application, *Sensors and Actuators B: Chemical*, 194, 2014, pp. 238-248.

[28]. Capeletti, L. B., J. H. Z. dos Santos, and E. Moncada, Quantification of indicator content in silica-based pH solid sensors by diffuse reflectance spectroscopy, *Analytical Methods*, 3, 10, 2011, pp. 2416-2420.

[29]. Capeletti, L. B., et al., Infrared and Raman spectroscopic characterization of some organic substituted hybrid silicas, *Spectrochimica Acta Part A: Molecular and Biomolecular Spectroscopy*, 133, 2014, pp. 619-625.

[30]. Capeletti, L. B. and Z. João Henrique, Fourier Transform Infrared and Raman Characterization of Silica-Based Materials, in Applications of Molecular Spectroscopy to Current Research in the Chemical and Biological Sciences, Mark T. Stauffer (Ed.), *InTechOpen*, 2016.

[31]. Kowada, Y., Ozeki, T., Minami, T., Preparation of silica-gel film with pH indicators by the sol-gel method, *Journal of Sol-Gel Science and Technology*, 33, 2, 2005, pp. 175-185.

[32]. Wencel, D., et al., The development and characterisation of novel hybrid sol-gel-derived films for optical pH sensing, *Journal of Materials Chemistry*, 22, 23, 2012, pp. 11720-11729.

[33]. Krıltz, A., et al., Covalent immobilization of a fluorescent pH-sensitive naphthalimide dye in sol–gel films, *Journal of Sol-Gel Science and Technology*, 63, 1, 2012, pp. 23-29.

[34]. Nordin, M. N. a., et al., Immobilization of Bromocresol Purple in Inorganic-Organic Sol-Gel Thin Film with Presence of Anionic and Non-ionic Surfactants, *Procedia Chemistry*, 19, 2016, pp. 275-282.

[35]. Gupta, S., et al., Fluorescence quenching of rhodamine 6G by 1, 3, 5-Trinitroperhydro-1, 3, 5- triazine entrapped in porous sol–gel silica, *Journal of Luminescence*, 168, 2015, pp. 88-91.

[36]. Afshani, J., et al., A simple nanoporous silica-based dual mode optical sensor for detection of multiple analytes (Fe^{3+}, Al^{3+} and CN^-) in water mimicking XOR logic gate, *RSC Advances*, 6, 7, 2016, pp. 5957-5964.

[37]. Fidalgo, A., Ilharco, L. M., Chemical tailoring of porous silica xerogels: Local structure by vibrational spectroscopy, *Chemistry-a European Journal*, 10, 2, 2004, pp. 392-398.

[38]. Cappeletti, L. B., et al., Determination of the Network Structure of Sensor Materials Prepared by Three Different Sol-Gel Routes Using Fourier Transform Infrared Spectroscopy (FT-IR). *Applied Spectroscopy*, 67, 4, 2013, pp. 441-447.

[39]. Walcarius, A., et al., Exciting new directions in the intersection of functionalized sol-gel materials with electrochemistry, *Journal of Materials Chemistry*, 15, 35-36, 2005, pp. 3663-3689.

[40]. Taei, M., et al., Fast and selective determination of phenazopyridine at a novel multi-walled carbon nanotube modified ZnCrFeO4 magnetic nanoparticle paste electrode, *RSC Advances*, 5, 47, 2015, pp. 37431-37439.

[41]. Etienne, M., et al., Electrochemical approaches for the fabrication and/or characterization of pure and hybrid templated mesoporous oxide thin films: a review, *Analytical and Bioanalytical Chemistry*, 405, 5, 2013, pp. 1497-1512.

[42]. Gholivand, M. B., et al., Synthesis of Fe–Cu/TiO2 nanostructure and its use in construction of a sensitive and selective sensor for metformin determination, *Materials Science and Engineering: C*, 42, 2014, pp. 791-798.

[43]. Ghoreishi, S. M., et al., Preparation of a manganese titanate nanosensor: Application in electrochemical studies of captopril in the presence of para-aminobenzoic acid, *Analytical Biochemistry*, 487, 2015, pp. 49-58.

[44]. Zhang, J., et al., A molecularly imprinted electrochemical sensor based on sol–gel technology and multiwalled carbon nanotubes–Nafion functional layer for determination of 2-nonylphenol in environmental samples, *Sensors and Actuators B: Chemical*, 193, 2014, pp. 844-850.

[45]. Rezaei, B., M. K. Boroujeni, and A. A. Ensafi, A novel electrochemical nanocomposite imprinted sensor for the determination of lorazepam based on modified polypyrrole@sol-gel@gold nanoparticles/pencil graphite electrode, *Electrochimica Acta*, 123, 2014, pp. 332-339.

[46]. Walcarius, A. and M. M. Collinson, Analytical Chemistry with Silica Sol-Gels: Traditional Routes to New Materials for Chemical Analysis, in Annual Review of Analytical Chemistry, *Annual Reviews,* Palo Alto, 2009, p. 121-143.

[47]. Qian, N., et al., 4, 4′-Diaminodiphenyl Sulfone Functionalized SBA-15: Toluene Sensing Properties and Improved Proton Conductivity, *The Journal of Physical Chemistry C*, 118, 4, 2014, pp. 1879-1886.

[48]. Dudas, Z., et al., Hybrid silica-porphyrin materials with tailored pore sizes, *Materials Research Bulletin*, 45, 9, 2010, pp. 1150-1156.

[49]. Yadavi, M., A. Badiei, and G. M. Ziarani, A novel Fe^{3+} ions chemosensor by covalent coupling fluorene onto the mono, di- and tri-ammonium functionalized nanoporous silica type SBA-15, *Applied Surface Science*, 279, 2013, pp. 121-128.

[50]. Ungureanu, S., et al., Syntheses and characterization of new organically grafted silica foams, *Colloids and Surfaces A: Physicochemical and Engineering Aspects*, 360, 1–3, 2010, pp. 85-93.

[51]. Beaucage, G., Approximations leading to a unified exponential power-law approach to small-angle scattering, *Journal of Applied Crystallography*, 1995, 28, pp. 717-728.

[52]. Beaucage, G., Small-angle scattering from polymeric mass fractals of arbitrary mass-fractal dimension, *Journal of Applied Crystallography*, 29, 1996, pp. 134-146.

[53]. Capeletti, L. B., et al., The effect of the sol-gel route on the characteristics of acid-base sensors, *Sens Actuators B*, 151, 1, 2010, pp. 169-176.

[54]. Islam, S., et al., Correlation between structural and optical properties of surfactant assisted sol–gel based mesoporous SiO2–TiO2 hybrid nanoparticles for pH sensing/optochemical sensor, *Sensors and Actuators B: Chemical*, 225, 2016, pp. 66-73.

[55]. Hussain, M., et al., Oxygen sensing and transport properties of nanofibers of silica, bismuth doped silica and bismuth silicate prepared via electrospinning, *Sensors and Actuators B: Chemical*, 192, 2014, pp. 429-438.

[56]. Xue, Z., et al., A novel electrochemical sensor for capsaicin based on mesoporous cellular foams, *Analytical Methods*, 7, 3, 2015, pp. 1167-1174.

Chapter 24
Induction Electrostatic Sprays in Agriculture Industry

Murtadha Al-Mamury, Nadarajah Manivannan, Hamed Al-Raweshidy and Wamadeva Balachandran

24.1. Introduction

In this book chapter, the work in induction electrostatic sprays in agriculture industry agriculture for the last thirty-five years is reviewed. The research and advancements in electro-static spray for agriculture have been divided into three periods for this review purpose; 1978-88, 1989-2000 and 2001-2015. Furthermore, the literature related to the factors those affecting the electro-static spraying are also reviewed and various published research in this topic are compared and contrasted in this review. Though electrostatic-spray has started to become popular in developed countries like USA and Europe for more than a decade, it faces many challenges in terms of it efficiency and environmentally friendliness. Therefore, electrostatic spray has not successfully used in the large agriculture in developing world like Iraq and India. The main purpose of this review is to identify the challenges those exists; hence potentially open up more research and development to address these challenges.

In Section 24.2, background to electrostatic spray is presented and in Section 24.3 is dedicated to a through literature review of the research work in electro-static spray based on three major periods. Conclusions are also given for each period in the Section 24.3. The major factors those have an effect on electrostatic spraying efficiency are reviewed in Section 24.4. In Section 24.5, the atomization properties of various electro-static spray are investigated. Finally existing challenges in electro-static spray in agriculture industry are identified (Section 24.6). Section 24.7 concludes this book chapter.

---

[1] Nadarajah Manivannan
 Centre for Electronic Systems Research, College of Engineering Design and Physical Science,
 Brunel University, United Kingdom

24.2. Background

In the 1600s, when the phenomenon of electrostatic force was first discovered, the simple explanation assigned to it was the attraction forces between two objects having different values of electric potential. At that time, there was no notion how to apply it in practical applications, so nobody understood its usefulness. In fact, the first attempts at industrial applications did not happen for another three hundred years, and these were to do with the paint spraying of grounded objects in the 1950s.

The term electro-spray, when the phenomenon of electrostatic force was discovered, the simple explanation assigned to it was still holds for doing so. The first is when the liquid becomes atomized by a mechanical or aerodynamic means, and then the droplets are charged, a process called Post Atomization Charging (PAC). It is widely used in industrial and commercial applications of which crop spraying is one example. The second method is when the charged droplets are produced by Electro Hydrodynamics Atomization (EHDA). This is caused by (EHD) related instabilities being produced at the surface of the liquid by an electric field. EHDA depends on liquid conductivity, which has resulted in the division of liquids into three categories: conductive, semi-conductive and insulating according to the charge relaxation time (τ) [1]. Atomization will occur based on electrical conductivity (σ) and electrical permittivity (ε) of the liquid. The basic definition/attributes of the three types of liquids as illustrated in Table 24.1 below.

Table 24.1. Classification of liquids [2].

Physical properties	Insulating l.iquid	Semi-conducting liquid	Conducting
Conductivity (σ) in comparison to metals	Extremely low	Very low	Low
Relaxation time (τ) and suitability for electro-spray	Long relaxation time Charge carriers take a long time to reach the surface when an electric field is applied. Cannot be sprayed by induction, due to lack of charge carriers. Charge injection necessary to atomize these liquids by EHDA	Moderate relaxation time. Relaxation time and droplet formation times are comparable. Stable jet obtained as a result of tangential stress and surface charge induced by the (radial) electrostatic field.	Short relaxation time Charge relaxation time must be shorter than the droplet formation time. Surface becomes equipotential when subjected to an electric field. •Harder to attain EHDA. Ideal for charging by induction.
Example liquids	Corn Oil $\sigma = 5 \times 10^{-11}$ Sm^{-1} $\tau = 4.5$ s $\varepsilon = 2.7$	Ethylene glycol $\sigma = 1 \times 10^{-6}$ Sm^{-1} $\tau = 0.4$ ms $\varepsilon = 41$	Tap water $\sigma = 1 \times 10^{-2}$ Sm^{-1} $\tau = 21$ ns $\varepsilon = 80$

Electro-spray is defined as the process of simultaneous droplet generation and charging, which happens when the electric force is stronger than the surface tension force. It consists of charged droplet generation, conveyance of particles to the target object and target deposition [3]. Liquid coming out from the nozzle at high pressure is subjected to an electric field, with this pressure causing elongation of the meniscus so as to form a jet spindle. This process allows for the generation of fine droplets of charge of magnitude close to one-half of the Rayleigh limit, which is the magnitude of charge on droplets that overcomes the surface tension force, thereby resulting in their fission [4]. The interactions that happen between charges on the surface of the liquid and the applied electric field give out two results: the first being the acceleration of the liquid and subsequent disruption into droplets, whilst the second is the build-up of the charges on the droplets [1].

Agricultural applications began after the success in the industrial field, with the first pertaining to the development of pesticide materials in powder form. Subsequently, electrostatic techniques were applied to liquid agricultural pesticides and herbicides, because there are a number of benefits in using the electrostatic principle in spraying technology, which include: good distribution of droplets/particles, more uniform deposits on the surfaces of the intended targets as well as low or even no drift. Furthermore, pollination and biomaterial in agriculture have delivered satisfactory results through electrostatic application. However, the direct application of electrostatic paint spraying equipment in orchards is not feasible for three reasons, as explained below [5].

1. The object in electrostatic painting to be coated is located relatively close to the paint gun (usually < 50 cm), which results in a significant increase in deposition efficiency on the object surface due to the high voltage applied to the gun. However, in orchard spraying the object to be coated is often much further away from the sprayer and hence, the technique is not effective.

2. In electrostatic painting, the process takes place in controlled environment with the conditions characterized as having minimum outside influences. But in the case of agricultural spraying, this takes place under a range of environmental conditions, such as temperature, relative humidity, wind velocity, and vibration, which can hinder spraying.

3. There are different skills requirements between operators working in the two different fields. Regarding electrostatic painting, the equipment is usually maintained by a full-time technician and the spraying is carried out by a skilled operator. While in the context of agriculture, pesticide spraying is only one of a number of operations that a farmer must carry out and thus, the equipment must be simple to operate, extremely robust, and reliable.

In general, electrostatic applications have been applied in several domains regardless of the percentage of the efficiency of this technique.

24.3. Classification of the Literature Review into Three-time Periods

In order to understand the developments that happened in the last quarter of the twentieth century regarding electrostatic spray application, this is classified into three stages, because each of those identified focuses on distinct aspects of how to advance this technology.

24.3.1. Literature Between 1978 and 1988

The first study during this period, conducted by Law, focused on designing an induction nozzle for producing fine droplets using a lower voltage when compared to corona charging. The latter method was used at that time in agricultural applications and needed voltages of more than 20 kV. The induction nozzle design has good properties and can achieve high values in the applications. It provides a compact, cheaply fabricated droplet charging method and dependability has been greatly increased by the complete embedding of its electrode in order to prevent mechanical damage and personal hazard. The results of this early experiment were good as they fitted well with the design constraints imposed. The close spacing of the electrode to the droplet formation zone within the charging nozzle now permitted greatly reduced design requirements regarding size, output voltage and current drain of the power supplies necessary for operating an electrostatic – deposition system. It produced sprays of the desired size (such, as 30 to 50, μm volume median diameter) and created a 1.6×10^6 V/m charge-inducing field at the droplet formation zone with a corresponding -14 μC/m^2 surface charge density on the grounded liquid jet (i.e. a free electron surface density of approximately 108/mm^2). In addition, it could be used with low levels of high voltage supplies ranging from 0.85 to 1.0 kV [6].

Field experiments were conducted by Law on four types of plant: cabbage, broccoli, cotton and corn. Spherical shaped metal targets were fixed inside the rows of plants in order to determine the amount of the deposition on them. Two types of nozzles were used: an induction nozzle operated with and without electrostatic charging, and a conventional nozzle. The purpose of this study was to determine the increases in the amount of spray deposition attributable to electrostatics application and compare these figures with those for the conventional one. The results obtained indicated that the charged versus uncharged deposition increase ranged from 1.8-fold to 7-fold. However, deposition-limiting conditions were encountered with certain plants at charge/mass levels as low as -1.6 mC/kg and charged versus conventional deposition benefits for these diverse plants ranged from 1.9 fold to 4.4 fold [7].

The applied levels of voltage and air pressure are factors influencing pesticide spraying efficiency, which were studied by Frost and Law, who wanted to determine their influence on nozzle performance. Four voltages were used, 0.5, 1.0, 2.0, and 3.0 kV, and six air pressure values ranging between 69 and 414 kN/m^2. The results showed that the charge to mass ratio increased with both increasing electrode voltage and increasing atomizing air pressure. By contrast, the charge to mass ratio decreased with both increasing the liquid flow rate and nozzle orifice diameter [8]. The relationship between the charge to mass ratio and liquid flow rate with nozzle orifice diameter was found to be inversely proportional in a study conducted by Inculetet *et al*. In this research, electrostatic and mechanical sprayers were used in an apple orchard, 8 hectares in area, containing 2,000 trees. 1000 were allocated for experimenting using electrostatic assisted spraying, whilst the rest were assigned for testing with mechanical spraying, and the purpose was to compare the efficiency of the two in depositing pesticides. The results obtained illustrated that the deposition was markedly improved on the upper canopy of apple trees and more uniform distribution within the whole plant when compared with conventional (mechanical) application. In general, the results were classified into different tree parts. Regarding the upper canopies of the trees, this region received 85 % more deposition of pesticide and approximately the same deposition was found on the lower canopies. Moreover, electrostatic assisted spraying yielded better uniform distribution of the pesticides within the whole tree. In fact, the total variation in deposition over the whole tree was within 3 % for electrostatic spraying as compared with 49 % for mechanical [9].

A comparative study between electrostatic induction and conventional spraying was conducted by Law and Lane. The first nozzle produced fine droplets of 25 μm size VMD, and those for the second ranged in size from 150 to 200 μm. The experimental works was carried out in a laboratory, with the purpose being to investigate the effects of point presence on the target and the intensity that spray-droplet charging has upon the mass transfer of spray liquid by droplet deposition onto the target and the charge transfer to the target system as partitioned into point discharge and droplet deposition current. The results obtained illustrate that 1) The number of charged droplets onto smooth metal increased seven-fold for spherical, and planar targets compared with uncharged spray; 2) Electrostatic spraying deposition onto a vertical planar target increased 24 fold compared with a conventional hydraulic-atomizing nozzle; and 3) Spray deposition onto the sides of a vertically oriented planar target was typically at least 80 percent more than onto the planar target in its horizontal orientation. Fig. 24.1 shows the trend in the surface charge to mass ratio magnitude versus spray cloud current for spherical targets [10].

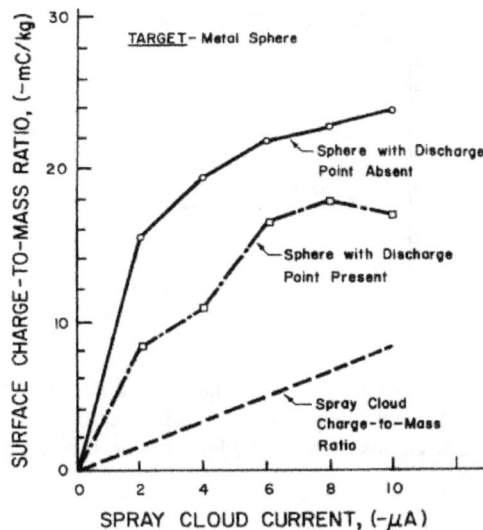

Fig. 24.1. The relationship between the charge to mass ratio values versus spray cloud current for spherical targets Law and Lane [10].

To overcome the major problems with induction charging of spray from a hydraulic nozzle and so be able to design a workable system, a study was conducted by Merchant and Green. A small flat fan nozzle tip was used and connected to a high voltage supply with a range of 0-10 kV. The purpose of this experiment was to investigate how to overcome the major obstacles in induction charging spraying using a hydraulic nozzle and consequently, understanding how a workable system could be designed. The results obtained can be summarized as follows: The maximum charge to mass ratio achieved was 2 mC/kg at a nozzle voltage of 9 kV. Charging increased the uniformity of deposit on cylindrical artificial targets by providing underside deposits and increasing those on the sides of the targets. In addition, charging increased the total deposit on the artificial targets by a factor of 1.6 to 2.8 at a nozzle speed of 2m/s and an operating voltage of 6 kV. Also, the droplet sizes exhibited differences between the charged and uncharged systems. Regarding which, it can be seen that the volume contained in drops above 250 μm diameter is less in the charged case (an average of 1.5 % of the total) than in the uncharged case (an average of 7.2 % of the total). Fig. 24.2 shows the distribution of spray volume by drop diameter [11].

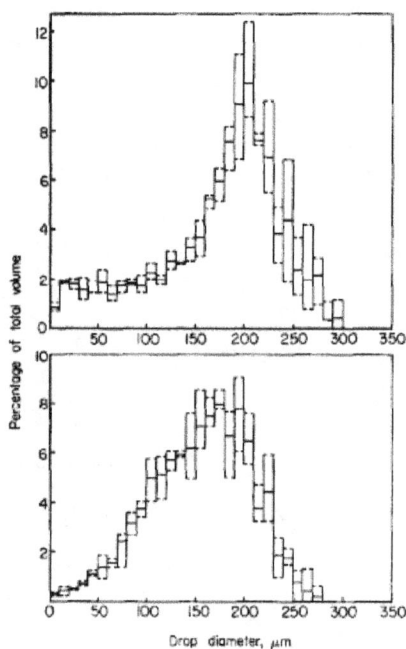

Fig. 24.2. Distribution of spray volume by drop diameter: uncharged spray (top); and charged spray (bottom). Solid lines represent mean values, whilst broken lines show the maximum and minimum values measured Merchant and Green [11].

Image and space charge forces play a very important role in enhancing the deposition of liquid on the target. The former is limited to when the droplet is a negligible distance from the target while the latter relates to driving the charged droplets over greater distances within the plant canopy. These two forces become evident when the electrostatic forces predominate over mechanical forces like drag and gravity. Several experiments were conducted in the field and laboratories in order to measure both of these values. One of these is a study carried out on orchard trees conducted by Castle and Inculet using an electrostatic spray system developed at the University of Western Ontario. It uses a number of air shear nozzles which, when exposed to air velocities in the order of

300 km/h, atomize the liquid pesticides into a very dense fine spray (MMD of 80 μm). The electrode is embedded flush with the surface and is energised from an 18-kV high-voltage power supply connected to a tractor's battery. The main purpose of this study was to present a way of applying electric space charge theory to liquid droplets and to show some experimental results that demonstrated improved deposition. The results obtained are summarized below [5].

1) Image forces have a negligible effect on pesticide deposition;

2) Since the drag force predominates in high-velocity regions, the space charge forces will have a negligible effect and hence, the mechanical depositions will predominate;

3) When the space charge forces are larger than both the gravitational and drag forces, enhanced deposition should be expected due to electrostatic forces;

4) There is no significant effect between the electrostatic and mechanical forces on the deposition on the lower parts of the tree. Electrostatic forces provide more than 85 % of pesticides deposition on the upper part of the plant compared to the condition when no electrostatic forces are applied;

5) Space charge improves the uniformity of deposition on the tree. For example, the ratio of residue in the upper and lower part of the plant is 0.95 with charging while the same ratio without charging is 0.51;

6). Space charge law effectiveness was observed in several field testing's and, in general, there is agreement between the analytical results and experimental deposition. Table 24.2 shown below provides the results of the above experiments.

Table 24.2. Force ratios for particle diameters 50 several field testing's and, in general, there is agreement between the analytical Castle and Inculet [5].

Electrostatic Forces	Droplet Diameter (μm)	Position A		Position B	
		F_e / F_g	F_e / F_d	F_e / F_g	F_e / F_d
Image force	50	0.15	0.0001	0.15	0.01
	100	0.3	0.001	0.3	0.1
Space charge force	50	6	0.006	60	5.5
	100	3	0.01	30	11
Space charge induction force	50	12	0.012	120	11
	100	6	0.01	60	22

The meanings of F_e, F_g and E_d are Electrostatic force, Gravitational force and Drag force, respectively.

The nozzle designed by Law in 1978 was used and tested again by Law in two places. The first was in a field cultivated with crops and vegetation, including, crop rows, planar turf grass type crops and orchard trees. The other one was executed in a laboratory on different metallic shaped targets, including spherical and planar forms, with vertical and horizontal orientations. The aim was to provide an overview of research and development efforts directed towards the incorporation of electrostatic forces for increasing mass transfer efficiency of the basic droplet-deposition process inherent in the application of agricultural pesticides to living plant systems. The results obtained from this study indicated that the deposition efficiency of an induction nozzle increased from two to seven-fold for the various models and biological targets, as compared to similar uncharged sprays and conventionally applied ones. Table 24.3 below indicates the superiority of induction charged spraying over the conventional one in terms of deposition efficiency [3].

Table 24.3. Relative increases in spray deposition onto various agronomic plants and model targets as a result of electrostatic spray application Law [3].

Type Target	Application Comparison	
	Charged Versus Uncharged	Charged Versus Conventional
Cabbage heads	7.0**	2.9
Broccoli plant	1.8**	1.9**
Cotton plants	2.5**	N/A
Corn plants	2.0**	4.4**
Smooth metal sphere	7.0**	7.8**
Metal sphere with point	3.5**	3.5**
Horizontal metal plate	2.5**	1.6**
Vertical metal plate	3.0**	24.0**

**Significant difference ($p < 0.01$)

On a grass strip in a field, the induction charging nozzle type ConJet Boom and a conventional one were used in experiments conducted in a study by Inculet *et al.* This had two aims: 1- finding a sufficiently large charge to mass ratio to develop the necessary electrical attraction forces; 2 - atomization of the liquid into small enough particles for the electrostatic force to be effective, yet large enough for the droplets not to evaporate or be easily entrained by winds. The results demonstrated that the induction charging nozzle improved coverage of the vegetation (up to 40 % increase over conventional mechanical spraying using the same mechanical sprayer without charging the droplets). The development and experimental work carried out in this project showed that with relatively simple modifications to a standard unrefined crop sprayer a substantial increase in the efficiency of deposition can be achieved and hence, reduce environmental drift [12].

In an experimental work on the barley plant, three types of nozzles that formed different droplet sizes were used by Lake and Marchant. The variables investigated were nozzle size, nozzle height, charge and wind speed. In addition to these, vertical and horizontal targets were distributed within the plant rows. The purpose of this study was to measure the effect of charged spray from a hydraulic nozzle on deposition on targets between the rows of the plant. The results obtained indicated that the charging increased the deposition on the vertical targets and tended to decrease it on horizontal ones. In the wind, the upper surface of the horizontal target showed no effect of electrostatic charging on spray deposits. The wind tended to increase the deposits from uncharged spray on the vertical target, whilst decreasing those from charged spray. Additionally, the model developed was used to explain some of the unexplained effects, for example, those related to nozzle orifice size and wind. Fig. 24.3 shows the relationship between the deposition on the target and nozzle size, nozzle target and nozzle type [13].

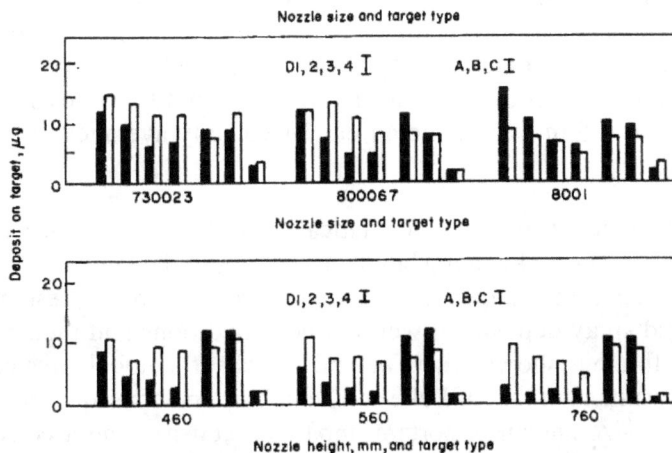

Fig. 24.3. The relationship between the deposition on the target and nozzle size, nozzle target and nozzle type. The letters A, B, and C represent horizontal targets, whilst D1, 2, 3 and 4 represent the vertical targets and, represent charged and uncharged sprays, respectively Lake and Marchant [13].

In the theoretical field, development of a model for enhancing space charge deposition of airborne particulates within an electrostatically shielded target system has been studied by Law and Bowen. The purpose of this study was to present a theoretical basis of the new concept of electroporation of airborne particulates and the four principles for the charged particles involved are listed below:

1) The electric mobility of an airborne particle charged by the ionised field process is a linear function of the particle radius;

2) The resultant space-charge field that exists within a target region is due to the superposition of the fields generated by all species of charged airborne particles within that region;

3) The charge to mass ratio varies inversely with the radius for particles charged by the ionised field process;

4) The fraction of saturation charge which a particle attains in the ionised field charging process is independent of particle radius. In general, dual particle-specie offers further benefit to the management of the total space–charge external to the target as a means of lessening the present limitation on the overall electrostatic deposition process caused by the induced corona at the earth points and edges on the target boundary [14].

In order to evaluate experimentally the contribution of electrostatic forces in improving the depositional characteristic of fine droplets carried in high-velocity air streams and to document the beneficial effects, a comparative study was conducted by Law and Cooper. Two types of nozzles were used, one with small droplets with a size of 30 μm (VMD) and the other big droplets, sized 370 μm (VMD). The results obtained revealed that the former did not reach the magnitude of that observed during conventional spraying by the latter. Moreover, when managing finely divided droplets in the 20 to 40 μm range, the electrostatic force increased deposition by 1.5 to 2.9 fold when compared to conventional spraying. Additionally, electrostatic wrap around benefits were observed at radial distances as close as 2.5 m, where the air carrier velocity averaged above 15 m/s as far away as 8 m [15].

A comparative study between two mathematical models of the transport of charged spray that were put forward by Lake and Marchant and Dix and Marchant was conducted by Hadfield. Specifically, this paper compared the results from these two models by measuring charged spray deposits in well-defined conditions and their relationship with droplet size. The findings showed that Dix and Marchant's model gave closer agreement to the measured results than that of Lake and Marchant and consequently, the variation in charge and droplet size has an important impact on constructing a correction model of deposition distribution. Fig. 24.4 below illustrates the effect of droplet size on the charge to mass ratio [16].

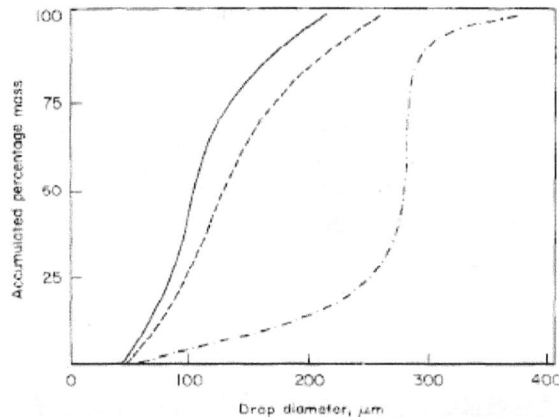

Fig. 24.4. The relationship between the droplet sizes and accumulated percentage mass measured. Drop size spectra: ____, 4000 rev/min; - - -, 3000 rev/min; - . - , 2000 rev/min Hadfield [16].

An embedded electrode electrostatic spray nozzle was used on a field planted with barley by Law *et al* , being operated at 172 kPa (25 psi) to atomize the pesticide liquid to 40 μm (VMD). The purpose of the experiments was to determine the spatial distribution of the droplet mass and charge transfer characterising these various modes of electrostatic cereal-crop spraying. The results obtained from this study can be summarised as follows: The values of droplet charging ranged between 1.5 and 4.5 mC/kg, which resulted in increased spray deposition onto all plant surfaces within the cereal crop tested. Additionally, these increases were from 5.5 fold for the top halves of the barley leaves, to 2.5 fold for their base halves and to 1.5 fold on the lower growing broadleaf weeds. Finally, approximately 10 fold and 34 fold greater charges were collected by the barley plant as compared with a weed and the underlying soil, respectively, as shown in Fig. 24.5 [17].

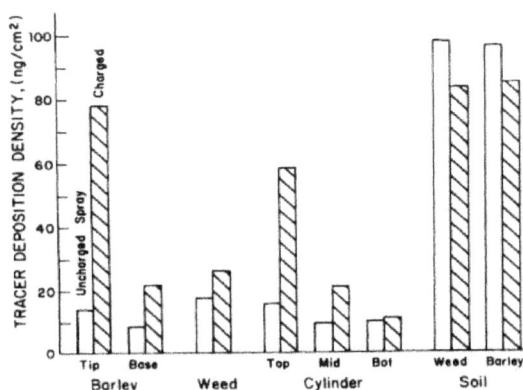

Fig. 24.5. Average charged versus uncharged spray deposition achieved at stated locations throughout a barley-weed-soil target system Law et al. [17].

In another study, experiments using the flat fan hydraulic spray type Teejet 800 were conducted by Marchant and Wilson. The nozzle was operated with two levels of air pressure values, 200 and 600 kPa, to atomize the water flow rate into fine droplets and the power supply that was connected to the nozzle provided a high voltage in the range of 1 to 6 kV. The purpose of this study was to predict the properties of a liquid sheet emerging from a flat fan hydraulic spray nozzle and to produce a formula for the charge to mass ratio of the spray. The results indicated that both the original and the modified equations predicted that charge to mass ratio increases with greater charging voltage and nozzle angle but reduces with an increase in the nozzle orifice area and the electrode spacing. So whilst the original equation predicted that the effect of increasing pressure reduces the charge to mass ratio, the modified equation predicted that the effect depends on the nominal spray angle. Furthermore, the charge to mass ratio was predicted to increase with increasing pressure for brass nozzles having angles of less than 66° and to decrease for larger angles. In addition to these outcomes, the equation predicted that the charge to mass ratio was considered reasonably acceptable up to an operating voltage of 2 kV [18].

In laboratory experiments, an induction nozzle and metal–spherical target with and without a point were used in a study by Law and Cooper. The nozzle was connected to a 700 V electric power source to charge the spray and operated at 138 kPa to give 30 μm droplets a study by considered reasonably was to document space charge field losses caused by induced target coronas and to establish any polarity–dependent advantage that could be exploited for improving electrostatic crop spraying. The results indicated that the masses of the charged deposition for an ionized target were 1.5 and 1.8 fold for the positive and negative sprays, respectively and four-fold for a non-ionized target regardless of spray polarity. In general, target deposition values with negative spray treatments were 410 ng/cm^2 and 180 ng/cm^2 for target-points-absent and target-points-present, respectively, while the positive spray treatments values were 393 ng/cm^2 and 150 ng/cm^2 [19].

Table 24.4. Peak interior electric fields during transient spray trials onto a parallel-plate target [Law and Cooper 1987].

Point Condition	Spray-Cloud Current (-μA)		
	1	2	3
Absent	54.6	62.1	53.4
Interior	53.4	61.6	58.6
Exterior	53.4	52.5*	46.8*

*Significant difference for point-absent treatment mean of the same spray cloud current ($p < 0.05$).

Three types of nozzles made of brass "Teejet type 800067" (Spraying Systems Co., Wheaton, Illinois) with artificial targets consisting of vertical columns of horizontal discs were used by Lake in his work. These nozzles were operated at air pressure of 300 kPa to produce 212 μm (VMD) droplets and were connected to high power supply of 5 kV. The purpose of the study was to measure the deposits from electrostatically charged hydraulic nozzles on the vertical columns of horizontal discs. A summary of the results is that: 1- disc spacing had only a small effect on the target deposits. 2- With no wind, total target deposits with charged spray increased with greater target spacing, but there was no such effect with uncharged spray. 3- The effect of wind was generally a decrease in target deposits with the charged spray. 4- Deposits on the underside of discs were low with both charged and uncharged spray under all conditions (See Fig. 24.6) [20].

24.3.2. Conclusion to the First Period

This time period can be characterized in terms of two important objectives. The first was increasing the amount and value of deposition on the surfaces of the intended targets, which included both the front and back surfaces. The second objective was comparing the efficiency of the pesticide application methods of novel systems using electrostatic spraying with the conventional or classic system. The reason for this comparative work was to encourage workers in the agricultural field to use electrostatic spraying because it was more efficient. Some mathematical formulae and predictive equations were tested during these projects, such as in studies by Inculet [12] and Hadfield [16]. The aim regarding these was to reduce the amount of pesticide waste by controlling the droplet size distribution and decreasing the level of drift.

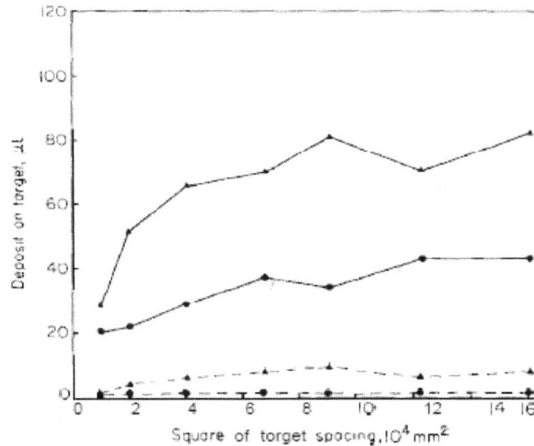

Fig. 24.6. The effect of target spacing in wind: ▲——▲ Charged spray, upper disc surfaces; ●——● uncharged spray, upper disc surfaces; ▲ – – ▲ charged spray, lower disc surfaces; ●– –● ; uncharged spray, lower disc surfaces Lake [20].

24.3.3. Papers Published in the Period between 1989 and 2000

An electrostatic sprayer equipped with a pneumatic atomizing mechanism and induction charging of liquid film was developed by Bologa and Makalsky. The aim of this study was to investigate the influence of the interaction of high-speed air flow with liquid film and to find out how the electric field distribution in the nozzle affects the efficiency of the electrostatic sprayer. The results showed that intensification of the turbulent interaction of the airflow with the liquid film in the initial zone of its formation in the electric field made it possible to change the size of droplets and to increase the spray cloud current. Moreover, the values of the charge to mass ratio exceeded 6 to 8 times the corresponding values for existing electrostatic sprayers and Fig. 24.7 shown below illustrates this interaction [21].

In aerial work, two nozzles were used by Inculet and Fischer when they developed a novel system that consisted of generating two coplanar clouds of insecticides. One formed the port side wing and the other the starboard side. The first type was the spraying system nozzle SS-11004 negatively charged with liquid insecticide while under the strong positive electric field of an induction electrode I E (+). The second type was an SS-110015 charged positively by an induction electrode I E (-), which was connected to a 36 kV supply. The results obtained indicated that the electrically charged spray influences the spray deposition and the effects become more pronounced especially with fine droplets, but it could also enhance deposition (recovery) under arid conditions. Additionally, it emerged that charging coarse spray electrostatically does not influence spray deposition significantly [22].

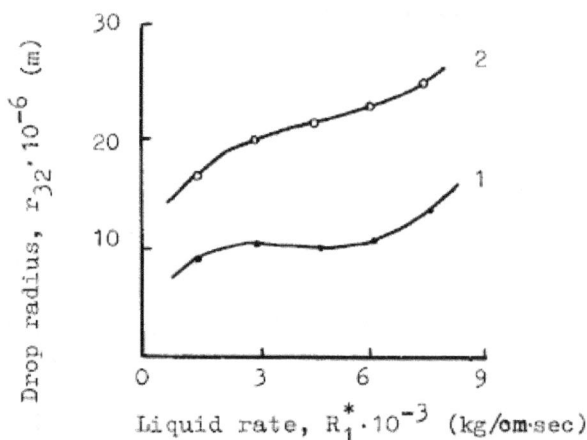

Fig. 24.7. The dependence of the aerosol drop radius on liquid rate from a unit of the nozzle edge length of the sprayer in the presence (1) and in the absence (2) of a turbulent spraying of liquid Bologa and Makalsky [21].

In order to determine theoretically the feasibility and experimentally to confirm the practicality of focused or target–selective deposition of agricultural sprays, an induction nozzle with a copper sphere (row target) of 75 cm diameter was used in laboratory tests conducted by Giles. Polyethylene film with a thickness of 0.1 mm and a dielectric constant of 2.3 was use underneath earthed targets. The results of the study indicate that passive manipulation by increasing the air–gap spacing between the dielectric film and the underlying earthed planar boundaries that ranged from 0 to 4 cm significantly increases lower hemisphere spray deposition, whilst decreasing film deposition. While the active manipulation of the film 4 cm above the earthed plane by recharging it to a surface charge density of -15 $\mu C/m^2$ resulted in greatly increased target and decreased film deposition. Additionally, electrical charging of aerodynamically–delivered spray droplets, in combination with the lifting and recharging of the dielectric film, resulted in an 84 % to 16 % ratio of target sphere to non-target film deposition compared to a corresponding 44 % to 56 % ratio for uncharged spray deposition along a 31 cm section of one sphere and the adjacent film of the test row [23].

An electrostatic induction handgun sprayer with two embedded-electrodes charging and a conventional spray with high volume application were used by Law *et al*. This was tested in laboratory experiments as well as on rows of strawberry plants in a greenhouse. The aim was to evaluate several recent air-assisted electrostatic crop-sprayer machines incorporating this specific type of pesticide spray-charging capability. A second aim was to compare the laboratory and field-test outcomes to establish whether the deposition efficiency and insect control efficacy could be attributed to this reduced–volume aerodynamic-electrostatic application technology. The findings obtained from this study suggested that the air-assisted charged sprayer deposited significantly more insecticide onto plant foliage than the high volume conventional spray method. These values were 1.29 µg/cm^2 and 0.35 µg/ cm^2, giving a 3.7 fold electrode position benefit. Finally, the results of the air–assisted charged sprayer showed that it was appropriate for penetrating charge conductive pesticide to inner plant regions and electrostatically depositing droplets onto leaf undersides through its space charge field [24].

An electrostatic induction twin nozzle especially constructed for generating fine droplets of pesticides with a metal plant model was used by Machowski and Balachandran in a greenhouse. The nozzle was operated at air pressure ranges of 0.3 - 0.8 MPa to atomize 0.5 L / min of solution into fine droplets and the high voltage connected to the nozzle was 2.5 kV. The aim of this research was to design and construct an induction charging electrode arrangement in an electrostatic spraying nozzle for greenhouse application. Laplace's equation, with the finite element software ALGOR, was used to solve the modelling calculation of the field distribution in the vicinity of the nozzle. The results obtained indicated that the reverse field modelling technique can be used successfully to optimise an induction electrode arrangement for twin fluid nozzles with respect to the charging efficiency of spray droplets. Moreover, the level of measured charge to mass ratio increases proportionally to the applied voltage. As a result, significant improvement of deposition efficiency on the underside of leaves was seen when high potential was applied to the charging electrode (see Fig. 24.8) [25].

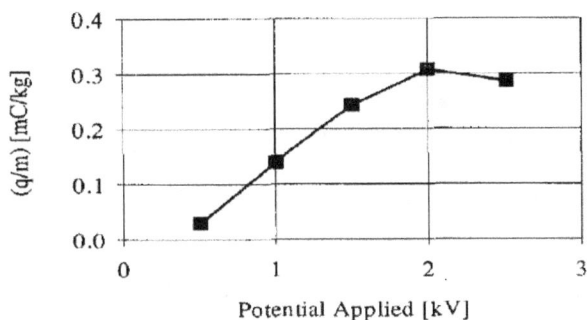

Fig. 24.8. Charge to Mass ratio (q/m) of the spray generated by a 'TL" fog generator [Machowski and Balachandran [25].

In the experimental field, a study that included several operating conditions was conducted by Law *et al*. The nozzle operated at an air pressure of 207 kPa (30 psi) to atomize the water flow rate of 73mL/min into fine charged droplets of 20-50 µm connected to as connected to a high power supply ranging from 1-3 kV and a Faraday cage was used to collect the charged spray from different distances between 0 and 50 cm so that the current of droplets could be measured. The purpose of this study was to present as a function of the degree of bounding of the charge spray–cloud, the experimentally observed effect being hypothesised as space charge suppression of induction spray charging. The results regarding the charge conditions demonstrated that the capability to charge electrically conductive spray by an electrostatic induction nozzle linearly diminishes by 41 % from a

-7.3 mC/kg value of charge to mass one of -4.3 mC/kg. In addition to this, the countering space-charge effect of the dispensed negative spray cloud is attributed to being the cause of this reduced effectiveness in the induction charge-transfer process at the earthed droplet-formation liquid jet within the dielectric nozzle body [26].

Jahannama *et al.* used an air-assisted induction charging nozzle type, namely a MaxCharge, based on the original design of Law from 1978. However, the material of this nozzle is different to that of Law ranging from 1-3 kV and a Faraday cage was used to collect the charge influence of the charging electrode on the primary atomization zone. The theoretical program employed in the study was the VOF model, which was used to simulate two phases, air and water. In addition, the FLUENT CFD package was used to give an account of the mixing and atomizing process inside the nozzle. The aim was to compare the performance of this nozzle with the Law's original. The results obtained indicate that the droplet size distributions of horizontal sprays showed a significant difference between the charged and uncharged cases, whereas there was no remarkable difference between the vertical ones. Moreover, the computational results confirmed that the creation of a vortex and negative pressure field in the middle part of the nozzle plays a significant role during the mixing of the fluids and the liquid atomization processes. Fig. 24.9 shows droplet size distributions for charged and uncharged sprays in the horizontal situation [27].

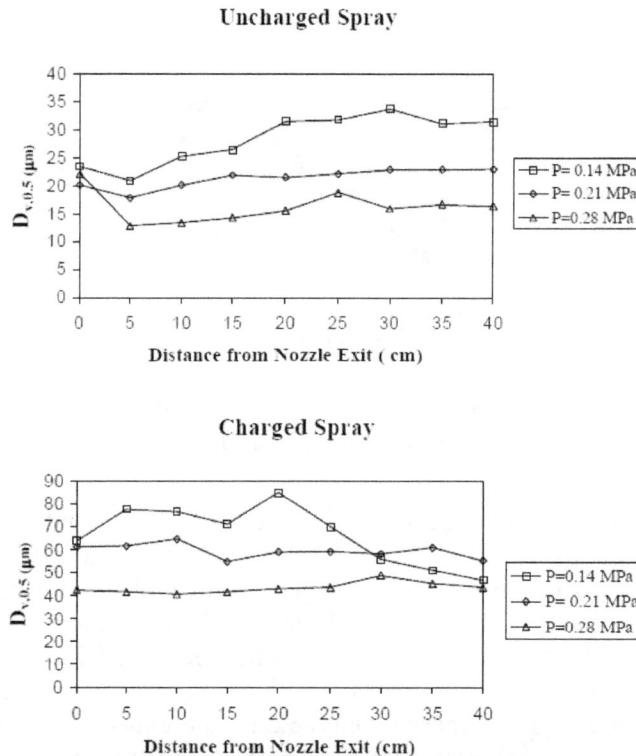

Fig. 24.9. Droplet size distributions for charged and uncharged sprays in a horizontal situation (Jahannama et al. [27]).

24.3.4. Conclusion of the Second Period

In this period, the studies concentrated on new aspects of the electrostatic spray phenomenon. In particular, the focus was no nozzle technique, which led to new nozzle designs that improved deposition and nozzle efficiency. The first concern was regarding how to develop a suitable nozzle for crop spraying, that is, one the increased the amount of deposition on the intended targets and to this end, researchers conducted experiments using different geometry of nozzles. The second concern was to explain the nature of the interaction inside the nozzle between the air pressure and the liquid as well as to study the interaction involved in changing the sizes of charged droplets and the amount of the deposition. Finally, all of these nozzle experiments aimed to increase the amount of deposition on both surfaces of the targets.

24.3.5. Papers Published in the Period between 2001 and 2013

The first study in this period involved experimental and theoretical work conducted by Cook and Law. The induction nozzle designed by Law in 1978 was used in the experimentation, with the purpose being to investigate and explain why the performance of a dielectric embedded-electrode induction spray charging nozzle is affected by the charged cloud of droplets it generates. Poisson's equation was solved using a COMSOL finite element axisymmetric model consisting of 3,704 triangular elements and 1,946 nodes for various placements of nearby earthed surfaces. The results obtained from this study showed that [28]:

1) The potential due to the presence of the negatively charged cloud issuing from an electrostatic induction spray charging nozzle can become substantially larger in magnitude than the positive electrode potential within the dielectric charging nozzle;

2) A grounded Faraday cage placed axially downstream in the cloud effectively suppresses the potential produced by the charged cloud by limiting its size and hence, negatively affects the nozzle performance as well as the location of the grounded boundary.

Fig. 24.10 shows the distribution of potential and voltage obtained with and without space charge.

An electrostatic pressure swirl nozzle with a brass ring electrode was used by Laryea and Young No on orchard trees and the experiments were conducted under charged and uncharged conditions. The aim of this study was to develop a suitable electrostatic nozzle for an orchard sprayer. The results showed that the droplet size exhibits an increase in the VMD as the applied voltages are increased and decreases when the operating pressure is increased (Fig. 24.11 and Fig. 24.12 illustrate these effects). Moreover, the spray width increases as the axial distance is increased, but the applied voltages have no effect on this width and this could affect the spray trajectory. In addition, under all conditions, the charged spray shows an increase in spray deposit and drift reduction as compared to the uncharged spray [29].

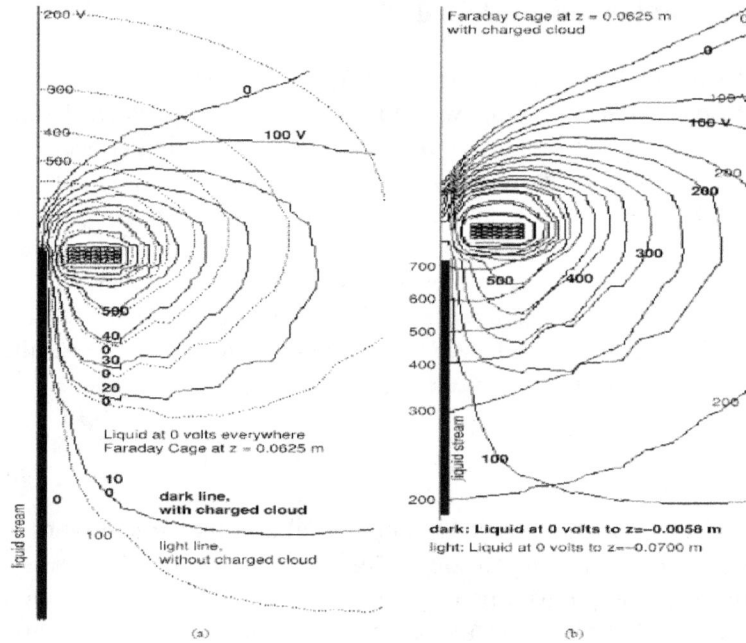

Fig. 24.10. (a) Potential, volts, without and with a space charge with the liquid grounded everywhere. (b) Potential, volts, with a space charge and with the liquid grounded at various locations (0–1000 V shown) Robert and Law [28].

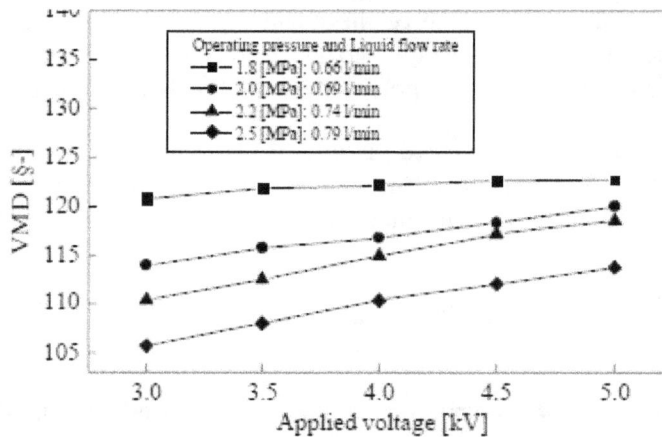

Fig. 24.11. The effect of applied voltage on volume median diameter Laryea and Young No [29].

A comparison study between conduction and induction charging in liquid spraying was carried out by *Zhao et al* [30]. The aim was to clarify that the differences that pertain to the charging mechanism, electric field distribution, energy conversion and the effect of space charge theoretically and experimentally for three-electrode conduction and induction charging management. The results obtained explained that the three-electrode spraying system, conduction and induction charging differ both in the source of charge and electric field distribution. Furthermore, the target current in induction charging depends only upon the space charge field, which causes a larger electrode current and surface discharge at a lower voltage. For the three-electrode geometry, conduction charging produces larger current and smaller electrode leakage current when compared to induction charging. In addition, as the nozzle-to-target distance increases, the space charge effect increases and the target current decreases (see Fig. 24.13) [30].

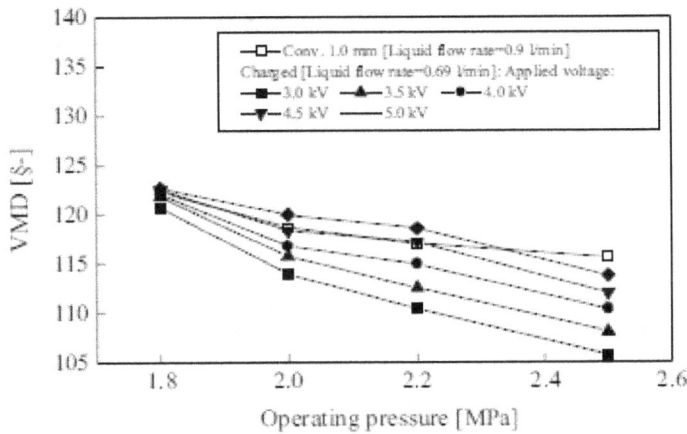

Fig. 24.12. The effect of operating parameters on volume median diameter [Laryea and Young No 2003].

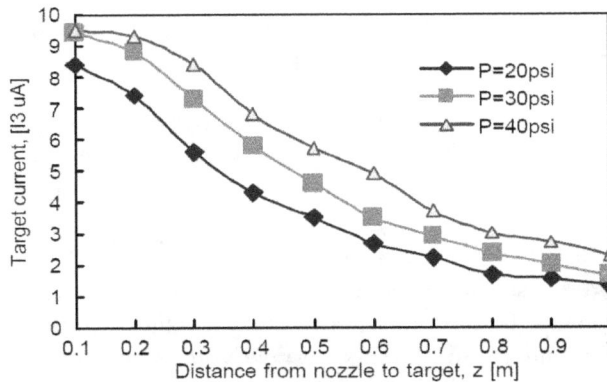

Fig. 24.13. Relationship between target current and distance to target when U ism, electric field distribution Zhao et al [30].

In a theoretical study, Zhao used FLUENT software to model the charged droplet trajectories towards a spherical target for different droplet sizes, charge to mass ratios and nozzle-to-target distances. The purpose of this simulation was to determine the conditions for maximum deposition, while reducing the drift of charged droplets to nearby ground in the electrostatic pesticide spraying process for uniform sized droplet sprays. The results obtained indicate that an increase in the charge to mass ratio increases the deposition significantly, while the radial drift may also increase. Moreover, the droplet size, charge to mass ratio and nozzle-to-target distance need to be carefully selected in order to achieve the best results [31].

Two types of spray nozzles, electrostatic and conventional, were implemented to control the sweet potato whitefly (SWF) pests of cotton in the UK in experimentation work undertaken by Latheef *et al* [32]. The first spray system assembled was the electrostatic spray boom, which included a large number (82) of spray charging nozzles that were necessary to accomplish a spray rate of 4.68 l/ha. These were operated at an air pressure of 482.7 kPa so as to atomize a flow rate 225 ml/min of pesticide into fine droplets and they were connected to a high-voltage supply with an output that ranged from 1 to 13 kV. The second type of nozzle in a boom set-up was the conventional spray applicator, which was attached to provide an application rate of 46.8 Lha^{-1} at 193.1 kPa. Four treatment applications were carried out, three with electrostatic spray, two charged and the third without, whilst the last involved conventional spraying. There were two main aims, with the first being to determine whether electrostatic charged sprays could be effectively used to achieve season-long control of whiteflies using pest control materials of diverse chemistries. The second aim was to ascertain whether the low spray rate (4.68 L/ha) required by the current prototype electrostatic spray charging nozzles would favor the development of resistance in whiteflies to mixtures of insecticides compared with conventional spray applications at 46.8 /ha. The results obtained can be summarized as follows: 1) the potential exists for obtaining increased efficacy against whiteflies using an electrostatic spray charging system. 2) Electrostatic treatments are as good as conventional ones and the amount of water used is 10 times less. 3) Counts of whitefly adults in electrostatic charged spray treatments are comparable with those regarding conventional sprays. However, counts of dead adults in electrostatic treatments are greater than those under conventional means after spraying with insecticides [32].

Maynaghet *et al.* conducted a series of laboratory experiments using an electrostatic sprayer with an ultrasonic nozzle. The operational condition for these experiments included the following factors: two nozzle charging electrodes of radii 10 and 15 millimeters, which were connected to four levels of high voltage supply: 1.5, 3.0, 5.0 and 7 kV, in sequence. Furthermore, six levels of air flow speeds of 14, 14.9, 17, 20.2, 21.6 and 22.0 m/s were used to atomize three levels of water flow rates, which were 5, 12 and 25 ms^{-1}. Finally, four levels of horizontal distances between the electrode and nozzle tip of 1.5, 6, 10 and 15 millimeters, respectively, were deployed. The aim of this study was to evaluate and quantify the charging of the droplets created by an electrostatic sprayer with an ultrasonic nozzle. The results obtained show that while increasing the liquid flow rate, the charged spray current increased at first then started dropping (Fig. 24.14) and the reason for this relates to ionisation. In addition, increasing the electrode distances caused increases in the amount of the spray current (see Fig. 24.15). Finally, the optimum combination of independent parameters was found to be as follows [33]:

Flow rate = 5 ml/min, high voltage = 3 kV, 12100 (rpm) 20.2 (ms^{-1}) radius of electrode = 15 mm and horizontal distance = 10 mm.

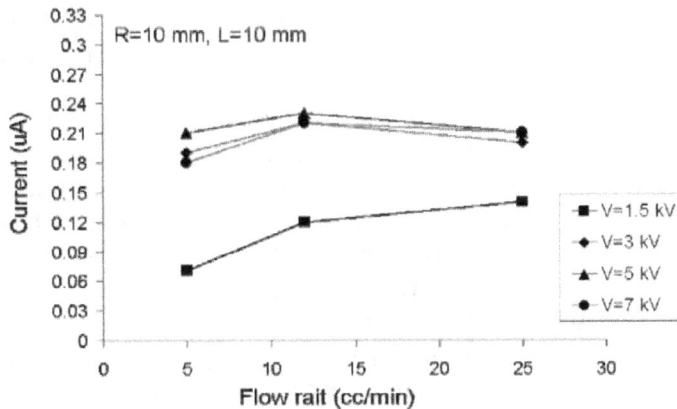

Fig. 24.14. The effect of liquid flow rate on the target current Maynagh et al. [33].

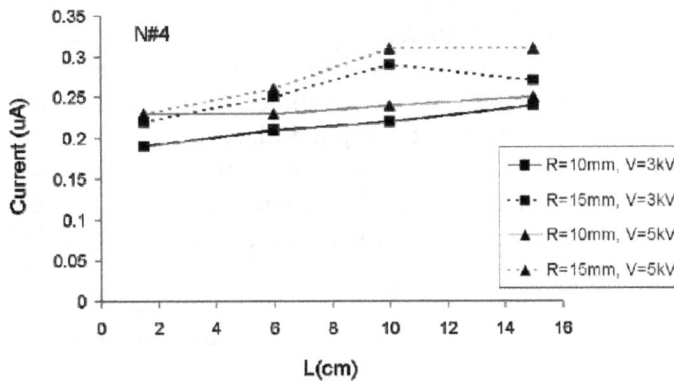

Fig. 24.15. The effect of electrode distance on the current value Maynagh et al [33].

Nader *et al.* investigated the dynamics of induction charging for spherical particles assuming finite volume and surface conductivities by using mathematical and simulation models. The purpose of this study was to illustrate a simulation model for investigating the induction charging of spherical particles with finite surface conductivity. In addition, the aim was to investigate the effect of volume conductivity on the surface layer, surface layer thickness and particle permittivity. The results showed that an increase in the contact area between the particle and the electrode significantly shortens the charging time. Charge accumulation on the particle's surface was fastest for a hemisphere having the largest contact area and slowest for a particle with a point contact. It was demonstrated that even for a short relaxation time constant, the actual charging time constant can be significantly larger, if the contact area is large, due to high contact resistance. Assuming a small finite contact area with the grounded electrode, fixed permittivity and bulk conductivity, the conductivity for the surface area was varied. The results of these simulations show that when conductivity of the surface layer increases, charge accumulation on the particle's surface is faster. For fixed values of the conductivities,

481

increasing the particle's permittivity results in slightly faster charging. However, the effect of increasing the permittivity is negligible when compared to the effect conductivity [34].

In lab experiments, an air-assisted electrostatic induction charging spray- (AEISC) system was investigated by Maski and Durairaj. It operated with higher and lower flow rates depending on nozzle condition. When the jet diameter was 0.35 mm, it provided 30, 45, and 60 ml/min, which represented the lower water flow rate. The high water flow rates used were 250, 450 and 600 ml/min when the nozzle jet diameter was 1.0 mm. The nozzle electrode was connected to a high-voltage supply with an output that ranged from 3 kV to 4 kV for charging the spray liquid. A Faraday pail was used to collect the droplets generated and to measure the chargeability in order to calculate the charge to mass ratio. The objective of this study was to investigate the spray chargeability of an air-assisted electrostatic induction charging spray with due consideration to the effects of electrode voltage, spray liquid flow rate and liquid properties. The results indicated that increasing the electrode voltage between 3 and 4 kV and decreasing liquid flow rates to 30 ml/min maximizes spray chargeability to about 18.5 mC/kg for ground water and 7.5 mC/kg for tank water sprays. The rate of increasing spray chargeability with decreasing liquid flow rate was higher in the lower flow rates than at the higher ones of 250, 450, and 600 ml/ min. The reason for this was attributed to the larger surface area of finely atomized spray-droplets exposed to the applied electric field strength. In general, this means that the charge to mass ratio increases as spray flow rate decreases and its response is non-linear with flow rates in the two distinct ranges mentioned above. Fig. 24.16 shows the relationship between the charge to mass ratio and liquid flow rate [35].

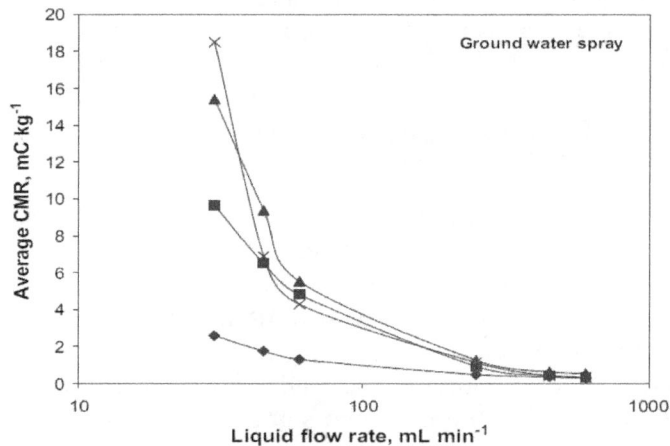

Fig. 24.16. The relationship between the charge mass ratio and liquid flow rate Maski and Durairaj [35].

In the same year, Maski and Durairaj studied the values of depositions on the upper leaves (adaxial) and bottom leaves (abaxial) of an artificial cotton plant with other parameters, which were: spray charging, application speed, height and the orientation of the target. The artificial plant leaf orientations were 0°, 30°, 45° and 60° to the horizontal (below a sheet of aluminum). The high voltage power supply connected to the spraying system was 7.5 kV, while the application speed moves were 0.278, 0.417, and 0.555 m/s. The fluorescent tracer technique was used to determine the spray deposition. The objective of this study was to investigate the effect of these parameters on the values of deposition on abaxial (underside) and adaxial (upper) leaf surfaces. The results can be summarised as: 1) the uncharged case (0 kV) was nearly zero value on the abaxial surface (under) and increased for all target heights, whereas the deposition value of the adaxial surface (upper) yielded the maximum 2) the abaxial deposition of charged spray gradually increased with the target orientation from 30° above to 30° below the leaves, whereas the adaxial deposition was maximum at 0° (horizontal) orientation. 3) The lower height always received a higher adaxial deposition (2.0 mg/cm^2) than the higher (0.95 m) and 4) an increase in abaxial deposition (1.5 mg /cm^2) with the spray charge being distinct at all target heights [36].

Toljic *et al.* [37] employed COMSOL, finite element analysis software package, in order to investigate the relationship between the value of charge on a conductive particle and the particle radius in the process of induction charging. The aim was to determine the value of the radius exponent in the charge to radius dependency for the conductive particles atomized in an electric field. The results obtained explained that the single most predominant parameter affecting the particle charge to radius dependency is the ligament length. Moreover, the exponent in the particle charge to radius dependency is equal to two when the particle is in direct contact with the bulk material. In addition, the maximum charge to mass ratio in liquid spraying will be achieved by maximizing the ligament length [37].

Turning to applications in the field, two types of nozzles were compared on wheat plants by Esehaghbeygi *et al.* The first, air-assisted electrostatic induction charging, required a high-velocity air flow (30 m s^{-1}) within the spraying head assembly to keep the charge in the electrode 8 kV. The second, a spinning charging. The aim was to determine the value of this with a disc speed of 2000 rpm. The aim of this study was to investigate the effectiveness of different herbicide application methods of electrostatic charge and spinning discs under natural weed flora in an irrigated wheat field. The obtained results indicated that electrostatic forces on small droplets are more prominent than gravitational ones and spray droplets can provide an improved deposition with reduced drift. Despite the spinning disc nozzle having more droplet uniformity, it does not significantly improve herbicide efficacy in dense canopy compared with the electrostatic charging sprayer. Furthermore, the spinning disc sprayer decreased water use and so was cheaper to operate, but did not improve herbicide efficacy [38].

Bayat and Sarıhan conducted experiments on maize plants used six spraying methods. The types of nozzles used were (1) conventional boom manufactured with domestic cone (DCN; Toyman Company Đzmir, Turkey), (2) boom with tail boom plus domestic cone (TBDCN), (3) air induction (AI; Spraying System Co. USA), (4) twin jet (TJ; Spraying System Co. USA), (5) air assisted spraying with domestic cone (AADCN) and (6) air assisted TX cone (AATX; Spraying System Co. USA. These nozzles were used at two stages of the plants' growth, the first when the height ranged between 50 and 60 cm and the second when it was > 210 cm. Two application rates of pesticide were used during the growth of the plants, 150 and 300 l/ha. Water sensitive paper was used to determine the coverage rate achieved. The purpose of this study was to determine the efficiency of different types of new nozzles and air-assisted spraying in second-crop maize by measuring spray deposits, coverage, and deposits as losses on the ground. The results revealed that the air-assisted domestic cone nozzles with an application volume of 300 Lha⁻1 achieved the highest deposits for both plant stages, with the coverage rates being 21.3 % and 27.6 %, respectively. This increase over conventional spraying was remarkable especially in the second application period on the middle and bottom parts of the plants. In addition, the ground deposits as losses in both application volumes in the second period were lower than those in the first period [39].

In order to protect tobacco plants, a hollow cone nozzle was used in a study conducted by Chao *et al*. The nozzle was operated at different spray heights, (0.2 m, 0.3 m, 0.4 m) and different spray angles (90°, 45°, and 0°). To atomize the flow rate into fine droplets, the air pressure ranged from 0.4 to 0.6 MPa and water sensitive paper was used to evaluate spray coverage, whereas the spray deposition was measured by a quantitative method (Allura-Red). The aim was to investigate the potential effects of spray height and spray angle on the spraying of the upper side of the leaves, the penetration of the spray liquid and the uniformity of the liquid distribution on the dense crop. The results obtained indicated that the spray coverage on tobacco plants decreased from the upper to lower layers. The deposition value of the upper layer increased as the height grew and the maximum value was located at 0.4 m. Moreover, the utilisation of liquid rate was lower than 45 % at a spray angle of 0° and above 50 % when it was 90°. Furthermore, the coverage of liquid was 52.9 - 86.0 %, 29.5 - 52.7 % and 18.7 - 39.6 % for the upper, middle and lower layers, respectively [38].

Five different spray delivery systems were used by Roten *et al* on a land plot planted with potato plants. The nozzles were the conventional boom, a canopy submerged drop sprayer combination, a pneumatic electrostatic spraying system, an air-assisted rotary atomizer and a high-volume air-assisted boom. The purpose of this study was to assess deposition from conventional and novel spray delivery systems in a potato canopy and investigate a digitised method for the analysis of this deposition. The results indicated that the deposition on the upper side of a leaf increased by 82 to 97 % for all treatments when compared to conventional treatments. Secondly, all those that consisted of one or more novel technology consistently gave higher coverage to the underside of the potato leaves than the conventional boom. Thirdly, the drop-sprayer and both electrostatic spraying system (charged and uncharged) treatments were similar in both the lower and middle canopy of the plant. Fourthly, the air–assisted rotary and air-assisted sustained gave the best coverage in the middle strata with 4.8 % and 7.7 % margin of increase in the coverage area, respectively, compared with conventional spraying. Fifthly, the hardest to reach area, i.e. the underside of the leaf at the lower canopy level, was covered best by the air-assisted rotary nozzle using the conventional method by 6 %, electrostatic spraying charging (ESS) by 5.64 %, and the air–assisted treatment by 3.71 %. Sixth and last, the conventional boom consistently achieved the least amount of deposition to the underside of the leaf only ranging from 0.1 to 0.41 % at best [40].

An electrostatic spraying nozzle, type MBP 4.0, with a 15 L capacity that uses the pneumatic principle for the formation and fractionation of droplets was experimented on in work carried out by Sasaki *et al*. A Faraday pail was used to collect the charged spray from five different distances, 0, 1, 2, 3, 4 and 5m in order to determine the intensity of charge on the droplets. Artificial targets made of wood were constructed and fixed with two metal plates to evaluate deposition, one in transverse and the other in longitudinal alignment to the spraying jet. The aim of this study was to evaluate the factors that affect electrostatic spraying, especially the effect of the distance between the spray tip and the target, the charge to mass ratio (q/m) and the liquid deposition under different positions relative to the target. The results obtained demonstrated that the q/m ratio is inversely proportional to the distance between the sprayer and the target. Additionally, the system was more efficient regarding to droplet deposition when the target was longitudinal to the spray jet. Furthermore, the minimum q/m ratio on which the liquid deposition was increased by the electrostatic system was 0.6 mC kg^{-1} [41].

Mamidi *et al* designed a hand-held electrostatic induction pressure swirl nozzle and used it in laboratory experiments with different artificial targets, including a glass beaker, plastic ball and aluminum. This nozzle had two swirl holes of diameter 1.0 mm and the orifice disc a hole of diameter 0.8 mm. The position of the nozzle electrode ranged from 1.0 mm to 7.0 mm and was connected to a high voltage supply ranging between 0 and 10 kV. All the experiments were conducted in ambient conditions. The nozzle operated at air pressure ranging from 0 - 35 psi in order to atomize a water flow rate of 350 ml/min into fine droplets. The purpose of this study was to establish the optimisation of certain parameters, including the electrode position, applied pressure, electrical conductivity of the spray liquid, and applied voltage so as to enhance the efficiency of a system, thus making it attractive to small and medium scale farmers. A remarkable phenomenon, the "wraparound effect", was utilised, which provided underside deposition efficiently with increased uniformity. In fact, the deposition of liquid quantity was enhanced 2-3 fold when this effect was included. Moreover, the charge to mass ratio increased with the applied voltage and the maximum achieved was 0.419 mC/kg at 3.25 kV (see Fig. 24.17). In addition, the charge to mass ratio increased with increases in both the air pressure and the electrode distance (see Fig. 24.18). In fact, even a 1.0mm difference in the electrode position could change the system from being fully efficient to completely inefficient [42].

To explain the relationship between nozzle orifices and droplet size under different air speeds, Martin and Carlton used an electrostatic induction charging nozzle (Spectrum Electrostatic Sprayers, Houston, Tex.) in laboratory experiments. The nozzle orifice diameters were 1.19 and 1.70 mm. The operational wind and air speeds ranged between 24 to 346 km/h and 177 to 306 km/h, respectively. In all the spray testing, the air pressure used to atomize the liquid was 517 kPa and the nozzle electrode ring was connected to a positive voltage of 6 kV to induce a negative charge on the spray. The study had two objectives: first, to quantify the effects of typical fixed-wing airspeeds and nozzle orifice sizes on the atomization of charged spray. The second objective was to quantify the electrostatic performance characteristics (q/m, charge to mass ratio) of the nozzle for each of the test orifices and at every airspeed tested. The results obtained showed that an increase in nozzle orifice size increased the coarseness of the spray droplet spectra at all airspeeds tested. In addition to this, increases in airspeed produced smaller spray droplets for all the nozzle orifices tested. Fig. 24.19 illustrates the findings [43].

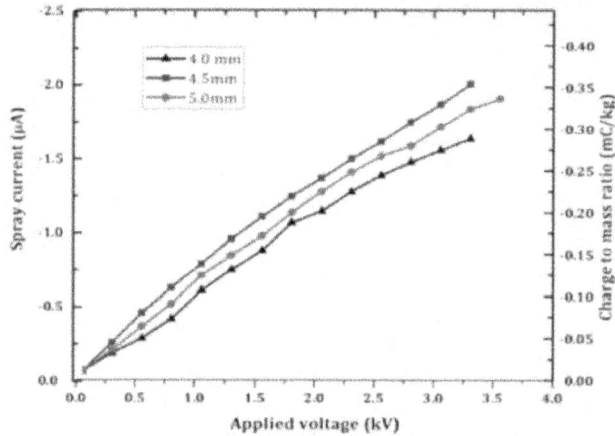

Fig. 24.17. The effects of applied voltage on spray current and CMR at various electrode positions Mamidi et al [42].

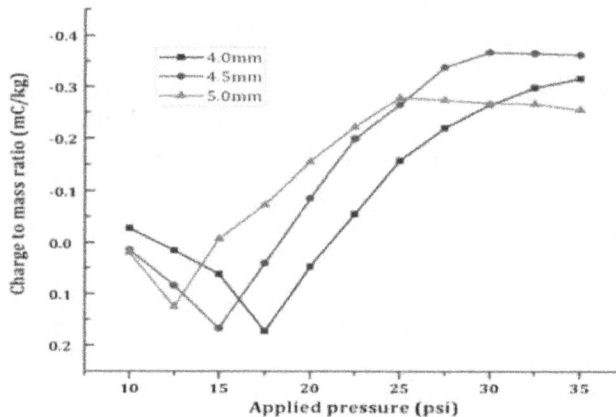

Fig. 24.18. The relationship between the charge mass ratio and applied pressure Mamidi et al. [42].

An air-assisted induction charged nozzle spray was used by Patel *et al* in air atmosphere at ambient conditions (Temperature 20 degrees °C, Relative humidity = 46 ± 3 %). The diameter of the nozzle orifices was 1.60 mm and the air pressure entered the nozzle through five holes into the atomization zone. A nozzle embedded electrode was placed inside the nozzle at 4.0 mm from the nozzle tip and the air pressure used was 145 psi to atomize the water at a flow rate of 90 ml/min into fine droplets. In addition, the values of voltage used ranged from 0-3 kV. The purpose of this study was to perform materialistic investigations regarding electrode material in accordance with the theory behind the use of specific materials in electrostatic spraying for high voltage applications. The materials tested were: nickel, copper, stainless steel, brass, and aluminum. The findings showed that the experimental results had good agreement with the relevant theoretical explanations. Specifically, nickel proved to be a good alternative as an electrode material for high voltage application in electrostatic spray nozzles since the performance of its electrode was better than the others in terms of charge to mass ratio and the mechanical parameters mentioned in subsequent sections (see Fig. 24.20) [44]:

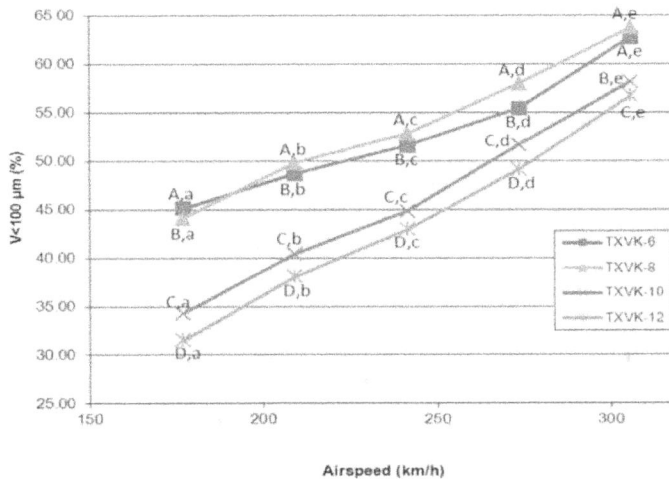

Fig. 24.19. Effects of airspeed and nozzle orifice size on the percent of the spray volume that is contained in spray droplets of 100 μm or less from the Brazilian aerial electrostatic nozzle Martin and Carlton [43].

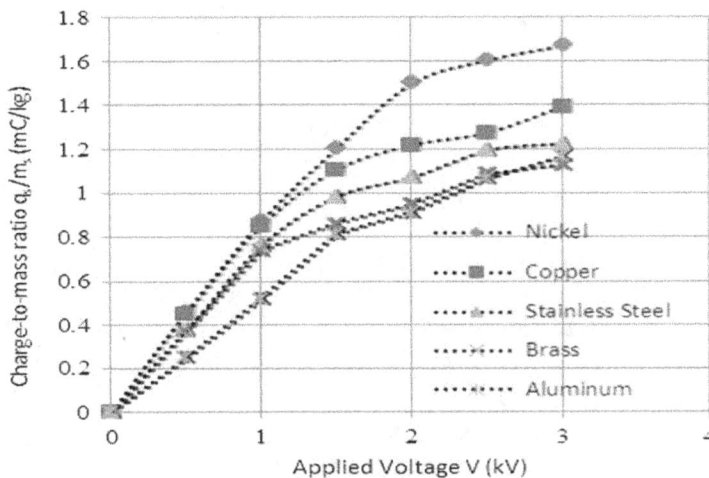

Fig. 24.20. Variation of charge to mass qs/ms (mC/kg) with applied electrode voltage kV Patel et al. [44].

487

Another study conducted by Laryea and Young No used the same electrostatic pressure–swirl nozzles with two types of orchard apple trees, namely M9 and M26. Their aim was to evaluate the contribution of electrostatic forces on spray deposition by using a traditional orchard sprayer and the effect of the sprayer fan speed on spray deposition. The three levels of fan speed were 1000, 1500, and 2000 rpm. The results showed that a significant difference could be observed at the top level for one pass and at the middle level for two passes, at 2000 rpm. In addition, it emerged that the charged spray proved superior to the uncharged one, at a fan speed of 2000 rpm with both passes for both types of trees [45]

24.3.6. Conclusion to the Third Period between 2001 and 2015

This period represents a big advance compared to the two previous ones. It is characterised by the use of advanced technology with excellent software, such as, MATLAB, FLUENT and COMSOL, which have been providing much greater understanding regarding the factors affecting nozzle behavior and deposition value than before. In addition how to improve the nozzle efficiency by selecting the types of nozzle materials and identifying other important factors has been at the forefront during this period. For instance, understanding has been acquired regarding the interactions between the inside and outside of the nozzle in relation to how these affect the efficiency of the application. Finally, during this period, some formulae aimed at predictions the drift value and atomization behaviors have been constructed. The computer simulation models developed have facilitated better understanding of induction charging nozzles for agricultural spraying.

24.4. Factors those Have an Effect on Electrostatic Spraying Efficiency

Fig. 24.21 illustrates four factors that can impact on the efficiency of the induction nozzle and on the electrostatic spraying agricultural application, these being: nozzle geometry, nozzle air pressure, nozzle induction voltage and the properties of the sprayed liquid used. The nozzle geometry factor is determined by three things: the nozzle electrode material, the nozzle electrode radius and the distance between the nozzle electrode and the nozzle liquid film. That is, the value of surface charge on the spread liquid and the chargeability value on the droplet surface depend on these three aspects. For example, if the nozzle distance is the right distance from the nozzle liquid film, the value of the surface charge on the latter will be satisfactory. Furthermore, when the radius of the nozzle and the electrode are the correct size along with the conductivity of the material being high, this will give a value of surface charge to the droplets in the range of the Rayleigh limit, i.e. they will have sufficient chargeability. The second factor is the nozzle air pressure, which controls the volume of the spray and the type of droplet size distribution. For instance, increasing the level of nozzle air pressure means there will be a greater volume of nozzle spray.

On the other hand, the droplet size distribution depends on the pressure levels and can be split into three categories, two of which are beyond the scope of the electrostatic forces zone. The first pertains to small droplets that evaporate through the atmosphere due to their tiny mass and drag forces, while the second refers to big droplets that drop onto the soil due to gravitational force. These two types constitute the total liquid spray drift. Finally, the third category coming within the limits of the electrostatic zone provides the values of the target deposition and coverage area. The liquid spread properties, such as viscosity and surface tension, have effects on the droplet size distribution. They play an inverse role to the electrostatic forces and try to avoid the atomization process or weaken it. Specifically, high values of surface tension and viscosity will make the process insufficient by producing big sized droplets as well as increasing the amount of drift due to gravitational force. Hence, to produce droplet size within the electrostatic forces zone, suitable air pressure is needed in order to destroy the forces of surface tension and viscosity. The fourth factor in nozzle performance and efficiency, as shown in Fig. 24.21, is the induction voltage level, which is important for determining the value of the surface charge on the liquid film and then on the surface of the droplets. If the value of the chargeability on the surface of the droplet is within the Rayleigh limit and the droplet size falls within the electrostatic forces zone, the numbers of charged droplets on the intended target is going to increase. Furthermore, the numbers of the charged droplets reaching the back surfaces of the target due to the role of electric field lines also increases. That is, the probability of increasing the deposition and coverage area on the front and back surfaces of the target will rise, whereas the drift and soil contamination values will drop owing to the dominant electrostatic forces controlling the distribution of the droplets onto the targets. As a result, the amount of chemicals used and the money spent for spraying are reduced.

24.5. Factors Affecting Atomization Properties

The spraying operation involves changing a liquid sheet into smaller droplets by using different devices called atomizers and just as with different nozzles being used different liquids types can be applied. For example, water, oil, emulsion, alcohol and other liquids can be mixed to make new solutions. Many studies have focused on liquid atomization inside and outside the atomizer in order to identify the behavior of interaction that occurs. Specifically, the purpose of these studies was to understand the atomization mechanisms completely in order to use them well in the applications and they have led to the discovery of several factors. These can be divided into two groups, which should be well known by researchers and anyone who intends to work in this field. The first group consists of external forces which are relevant to devices (equipment) and weather conditions, while the second pertains to the internal forces which relate to the liquid properties [46].

Deposition quantity and distribution on the target, risk of spray drift and mode of action and uptake of the chemical particles at the target surface are three functions influenced by external forces. While the droplet size velocity and volume distribution, distribution pattern, entrained air characteristic, spray structure and the structure of individual droplets are influenced by internal factors [47, 48]. Results reported in [49] indicate that a change in liquid properties due to the use of adjuvant materials causes significant change in the spray quality for a flat fan nozzle. Three features of the spraying process have been changed according to nozzle type: atomization trajectory, droplet size and droplet distribution. Results obtained from three types of hydraulic nozzle demonstrated that the reduction in the surface tension of liquid tends to reduce droplet size [50], while this increases with nozzle size increase and decreases with nozzle angle as well as with rising pressure [51, 18].

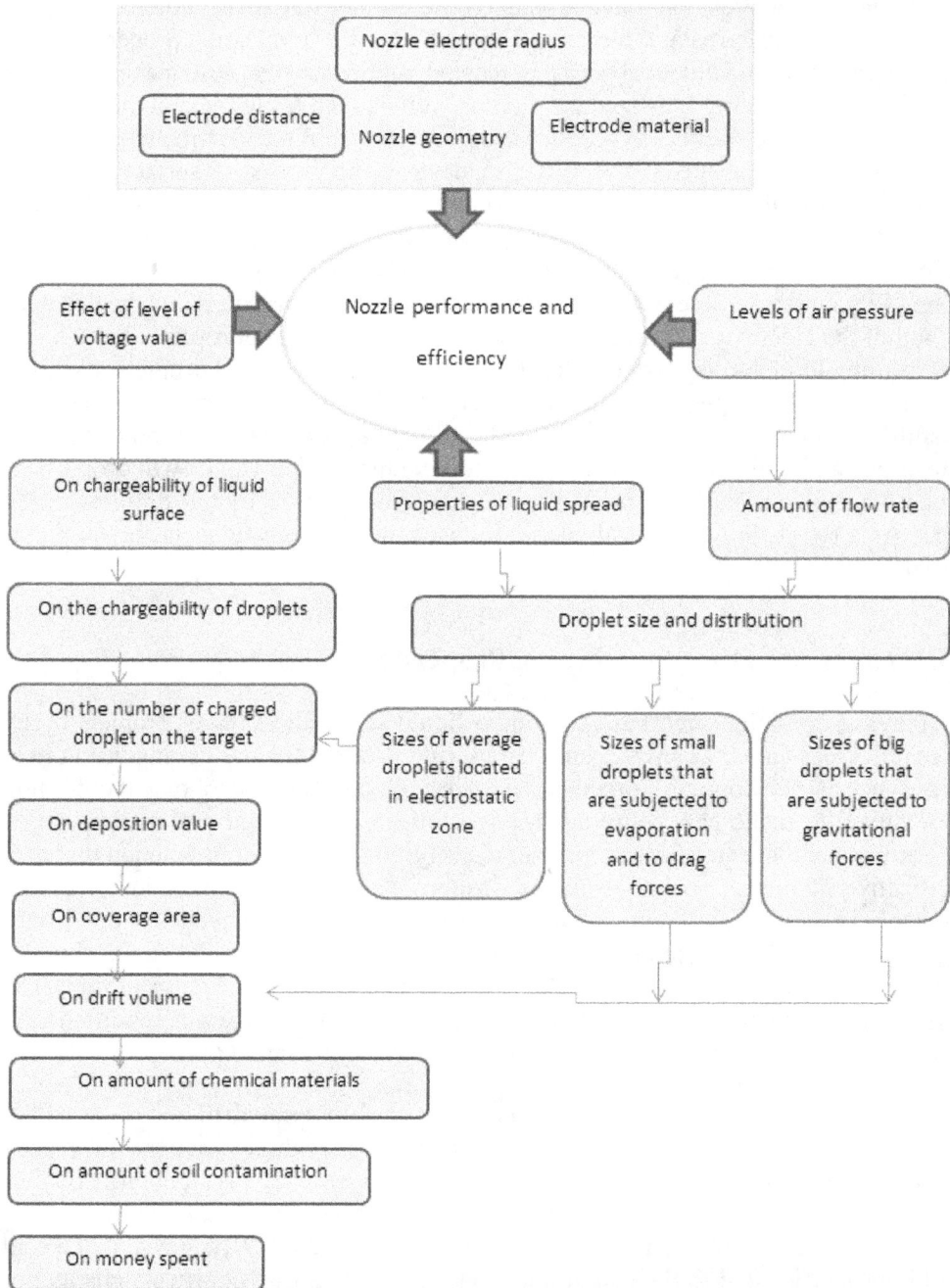

Fig. 24.21. Explained the factors that affect the nozzle performance and efficiency.

24.5.1. Charge to Mass Ratio (CMR)

This refers to the amount of spray current (measured in Coulombs) divided by the mass flow rate (measured by units of mass) also known as chargeability, which is the total amount of electrostatic charge that is carried by the spray droplets and is a measure of the performance of the nozzle system [44]. This is very important for the optimization of the system operational parameters and effectiveness of charged spray application [35]. For, from this value the optimum peak of the nozzle design under experimental conditions can be shown in addition to evaluating and estimating the change of the value of deposition and drift [45]. Several electrostatic processes depend on this value which governs the behavior of particle trajectories in a charged cloud of particles [52, 53, 49 ,50]. Factors impacted by it are electrode voltage, liquid flow rate, the type of liquid, nozzle angle, nozzle orifice area, pressure and the velocity of droplet distribution [54, 56, 57, 33, 20, 16] Inverse proportionality has been found between the charge to mass ratio and nozzle target distance [58]. The reasons for this are a long path, which leads to loss of electrical charge and the droplets being subject to air pressure resistance during their trajectory. On the other hand, the value of the charge mass ratio becomes non-linear with voltage increases above 2 kV, which can be attributed to ionisation of the air in the charging region [51]. The final parameters that need to be identified are: the high voltage magnitude, nozzle to target distance and droplet size, which are chosen so as to yield the best possible results [31].

24.5.2. Droplet Size and Distribution

Droplet size distribution is one of the three important features that determine the efficiency of electrostatic spraying technique (see the Fig. 24.21), with the other two being the ability of the nozzle to generate the desired droplet size efficiently and the construction design having the capacity to generate high voltage [57]. Factors that must be considered when selecting the droplet size are: environmental conditions (temperature, humidity, relative humidity and wind velocity), charge to mass ratio, the velocity of the charged droplets and the type of target [29]. The characteristics of the droplets govern the trajectory of liquid sheet and hence, can increase in efficiency of the pesticide application [59]. That is, the trajectory of the liquid sheet is a function of some of its characteristics, such as: surface density, liquid density and sheet thickness. In addition, droplet diameter and the droplet size distributed can be affected by changing the voltage [56] and/or the water flow feed rate [60, 61].

24.5.3. Back Ionisation

This phenomenon occurs during electrostatic processing and has profound consequences. It is due to droplet deposition between the earthed target and the nearby charged spray cloud as a result of electric field intensification [62]. Given the importance of deposition, it is crucial to understand which factors influence or reduce its efficiency. Two reasons account for the electrical migration back towards the electrified cloud where droplet coalescence is likely to occur. The first is the ionic discharge from target points that can neutralise part of the oppositely charged spray cloud, which can cancel beneficial Columbic forces leading to the probable collapse of the droplet-driving space-charge field. Secondly, all momentum fluxes and charges associated with gas discharge can be shown to repel droplets from the target [63].

24.5.4. The Drift

This has a negative impact on pesticide application in terms of the amount of fluid lost through the spraying process. Some of this is lost as a vapor or small droplets in the atmosphere due to their very light mass, known as exo-drift [64].While conversely, some is lost on the ground due to gravitational forces owing to the droplets having a large mass, termed endo-drift [5] (See flow chart in Fig. 24.3). The third way through which droplets are lost is when they are off target owing to drag forces. This can be attributed to several reasons, including: characteristics of the pesticide solution, weather conditions, droplet size, travel speed, nozzle type, boom height, spray pressure, nozzle spacing as well as the level of attention and skill of the operator [65]. The size of the spray angle plays an important role in determining the drift amount, regarding which it has been found that a fan angle of 110° increases airborne drift by 29 percent more than one of 80° [66]. In addition to this, wind speed, the absorption coefficient and surface roughness lead to increases in the drift amount [67, 68].

The results of drift include damage to neighboring crops that are not resistant to the pesticide, spoiled ecosystems, contaminated waterways and threats to human health. The last is particularly of concern in areas known to have intensive agriculture and relatively small fields that are scattered and close to living spaces and hence, the need for a good understanding of drift from field sprayers is of high priority [69, 70]. Drift not only has a negative environmental impact, for it also results in wasted pesticide and consequently, unnecessary expenditure. In agriculture, most of the pesticides types used are aqueous based liquids and more than 90 percent of them are applied by the spraying method, which means that the amounts of exo- and endo- drift are seriously on the increase.

24.6. Existing Challenges

Despite researchers through their great efforts having achieved satisfactory results in agricultural application of electrostatic sprayers, too many chemicals are still being used, substantial amounts of money continue to be wasted owing to inefficiency and the problems of drift and soil contamination remain unresolved. In the first ten years covered by this literature review, researchers focused on two things: increasing the amount of charge mass ratio on the living target by comparing the values obtained from the methods that used induction voltage with those that did not. The second focus during this period pertained to comparing the efficiency of the induction nozzle with the conventional one. Two factors remained unaddressed at this time: the amount of overdosing of the chemical material and huge quantities of wasted money. While during the second stage, the interest lay in the geometry of nozzle and which of the different types could produce the highest deposition values on the target. At this time, there was little concern about drift loss and the amount of pesticides being used and instead, drift was considered only a secondary result caused by several primary factors. During the third period of the literature review, it became apparent that the work was concentrated on using advanced software simulation nozzle models. These studies sought to comprehend the role of internal and external conditions in nozzle efficiency so as to be able to increase the amounts of the deposition on the targets. A little research has been carried out investigating drift behavior through mathematical formulae so as to reduce the losses from the agricultural application. From the projects and studies that have been reviewed, it is clear that the drift amount and overdosing with chemicals still lack sound remedies. Issues that still remain unresolved are:

1. Conventional inefficient nozzles are still used in many applications;

2. The whole plant is still being treated even when there are parts that do not need it and hence, local infection detection needs to be pursued;

3. No advanced technology has been developed to control the duration of the operation of nozzles.

24.7. Conclusion and Contributions

It has become clear that electrostatic spraying yields more coverage area and deeper penetration into the spatial canopy of plants when compared to conventional systems. Even though researchers have used this technology for some time now to improve nozzle efficiency, performance and to provide uniform distribution of deposition on the target, as explained above, large quantities of money and huge amount of pesticides are wasted because leaves that do not need remediation are being treated farmers and workers. This situation can be addressed by targeting only the infested area, which will not only greatly reduce the amounts of wasted chemical materials and hence, save money, for it would also result in less environmental pollution.

References

[1]. J. Grace and J. Marijnissen, A review of liquid atomization by electrical means, *J. Aerosol Sci.,* Vol. 25, 1994, pp. 1005-1019.

[2]. Z. A. Huneiti, Electrohydrodynamic atomisation of conducting liquid using an AC field superimposed on a DC field, 2000.

[3]. S. E. Law, Electrostatic pesticide spraying: concepts and practice, *IEEE Transactions on Industry Applications,* 1983, pp. 160-168.

[4]. A. Jaworek, Micro-and nanoparticle production by electrospraying, *Powder Technol,* Vol. 176, 2007, pp. 18-35.

[5]. G. P. Castle and I. I. Inculet, Space charge effects in orchard spraying, *IEEE Transactions on Industry Applications,* 1983, pp. 476-480.

[6]. S. E. Law, Embedded-electrode electrostatic-induction spray-charging nozzle: theoretical and engineering design [in the overall process of electrostatic deposition of liquid pesticide

droplets onto agricultural plants], *Transactions of the ASAE [American Society of Agricultural Engineers]*, Vol. 21, 1978.

[7]. S. Law, Electrostatic deposition of pesticide spray onto foliar targets of varying morphology, *ASAE Paper,* 1980.

[8]. A. Frost and S. Law, Extended flow characteristics of the embedded-electrode spray-charging nozzle, *J. Agric. Eng. Res.,* Vol. 26, 1981, pp. 79-86.

[9]. I. Inculet, G. Castle, D. Menzies and R. Frank, Deposition studies with a novel form of electrostatic crop sprayer, *J. Electrostatics,* Vol. 10, 1981, pp. 65-72.

[10]. S. E. Law and M. D. Lane, Electrostatic deposition of pesticide sprays onto ionizing targets: charge-and mass-transfer analysis, *IEEE Transactions on Industry Applications*, 1982, pp. 673-679.

[11]. J. Marchant and R. Green, An electrostatic charging system for hydraulic spray nozzles, *J. Agric. Eng. Res.,* Vol. 27, 1982, pp. 309-319.

[12]. Inculet, G. S. Castel and R. S. Vermey, Electrostatic Spraying of Row Field Crops, Vol. IAS84:37A, 1984, pp. 1058-1060.

[13]. J. Lake and J. Marchant, Wind tunnel experiments and a mathematical model of electrostatic spray deposition in barley, *J. Agric. Eng. Res.,* Vol. 30, 1984, pp. 185-195.

[14]. S. E. Law and H. D. Bowen, Dual particle-specie concept for improved electrostatic deposition through space-charge field enhancement, *IEEE Transactions on Industry Applications, ,* pp. 694-698, 1985.

[15]. S. Law and S. Cooper, Depositional characteristics of charged droplets applied by an orchard air-blast sprayer, *American Society of Agricultural Engineers,* 1985.

[16]. D. Hadfield, The modelling of charged spray deposition on artificial targets, *J. Agric. Eng. Res.,* Vol. 36, 1987, pp. 45-56.

[17]. S. E. Law, J. A. Marchant and A. G. Bailey, Charged-spray deposition characteristics within cereal crops, *IEEE Transactions on Industry Applications*, 1985, pp. 685-693.

[18]. J. Marchant, A. Dix and J. Wilson, The electrostatic charging of spray produced by hydraulic nozzles: Part II. Measurements, *J. Agric. Eng. Res.,* Vol. 31, 1985, pp. 345-360.

[19]. S. C. Cooper, S. E. Law, Transient Characteristics of Charge Spray Deposition Occurring Under Action of Induction Taget Coronas: Space charge Polarity Effect, The Institute of Physics, Static Electrification Group, in *Proceedings of the Oxford Univ. Conf. Electrostatics,* Vol. 85, 1987, pp. 21-26.

[20]. J. R. Lake, The deposition of electrostatically charged sprays on parts of targets shaded from the spray, *J. Agric. Eng. Res.,* Vol. 39, 1988, pp. 9-18.

[21]. A. Bologa and L. Makalsky, Electrostatic pneumatic sprayer of water solutions, *J. Electrostatics,* Vol. 23, 1989, pp. 227-233.

[22]. I. Inculet and J. Fischer, Electrostatic aerial spraying, *IEEE Transactions on Industry Applications,* Vol. 25, 1989, pp. 558-562.

[23]. D. Giles, Y. Dai and S. Law, Enhancement of spray electrodeposition by active precharging of a dielectric boundary, in *Institute of Physics Conference Series,* 1991, pp. 33-38.

[24]. S. E. Law, S. Cooper and R. Oetting, Advances in air-assisted electrostatic crop spraying of conductive pesticides, in *Proceedings of the American Society of Agricultural Engineers Meeting,* 1992.

[25]. W. Machowski and W. Balachandran, Design of electrostatic fog generator using a reverse field modelling technique, in *Proceedings of the IEEE Industry Applications Conference, Thirty-Second IAS Annual Meeting (IAS'97),* 1997, pp. 1784-1789.

[26]. S. Edward Law, J. Robert Cooke and S. C. Cooper, Space charge suppression of electrostatic-induction spray charging, *J. Electrostatics,* Vol. 40, 1997, pp. 603-608.

[27]. M. Jahannama, A. Watkins and A. Yule, Examination of electrostatically charged sprays for agricultural spraying applications, in *Proceedings of the ILASS- Europe,* 1999, pp. 120-127.

[28]. J. R. Cooke and S. E. Law, Finite-element analysis of space-charge suppression of electrostatic-induction spray charging, *IEEE Transactions on Industry Applications,* Vol. 37, 2001, pp. 751-758.

[29]. G. N. Laryea and S. No, Development of electrostatic pressure-swirl nozzle for agricultural applications, *J. Electrostatics,* Vol. 57, 2003, pp. 129-142.

[30]. S. Zhao, G. Castle and K. Adamiak, Comparison of conduction and induction charging in liquid spraying, *J. Electrostatics,* Vol. 63, 2005, pp. 871-876.

[31]. S. Zhao, G. Castle and K. Adamiak, Factors affecting deposition in electrostatic pesticide spraying, *J. Electrostatics,* Vol. 66, 2008, pp. 594-601.

[32]. M. A. Latheef, J. B. Carlton, I. W. Kirk and W. C. Hoffmann, Aerial electrostatic-charged sprays for deposition and efficacy against sweet potato whitefly (Bemisia tabaci) on cotton, *Pest Manag. Sci.,* Vol. 65, 2009, pp. 744-752.

[33]. Maynagh, B. M., Ghobadian, M. J., and T. T. Hashjin, Effect of Electrostatic Induction Parameters on Droplets Charging for Agriculture Application, *J. Agric. Sci.,* 11, 3, 2009, pp. 249-257.

[34]. B. F. Nader, G. P. Castle and K. Adamiak, Effect of surface conduction on the dynamics of induction charging of particles, *J. Electrostatics,* Vol. 67, 2009, pp. 394-399.

[35]. D. Maski and D. Durairaj, Effects of electrode voltage, liquid flow rate, and liquid properties on spray chargeability of an air-assisted electrostatic-induction spray-charging system, *J. Electrostatics,* Vol. 68, 2010, pp. 152-158.

[36]. D. Maski and D. Durairaj, Effects of charging voltage, application speed, target height, and orientation upon charged spray deposition on leaf abaxial and adaxial surfaces, *Crop Protection,* Vol. 29, 2010, pp. 134-141.

[37]. N. Toljic, G. Castle and K. Adamiak, Charge to radius dependency for conductive particles charged by induction, *J. Electrostatics,* Vol. 68, 2010, pp. 57-63.

[38]. A. Esehaghbeygi, A. Tadayyon and S. Besharati, Comparison of Electrostatic and Spinning-discs Spray Nozzles on Wheat Weeds Control, *Journal of American Science,* Vol. 12, 2010, p. 6.

[39]. A. Bayat, A. Bolat, A. Soysal, M. Güllü and H. Sarıhan, Efficiency of different spray application methods in second crop maize, in *Proceedings of the 23rd Annual Conference on Liquid Atomization and Spray Systems ILASS*, Czech Republic, September 2010.

[40]. R. Roten, A. Hewitt, M. Ledebuhr, H. Thistle, R. Connell, T. Wolf, S. Sankar, S. Woodward and S. Zydenbos, Evaluation of spray deposition in potatoes using various spray delivery systems., *New Zealand Plant Protection,* Vol. 66, 2013, pp. 317-323.

[41]. R. S. Sasaki, M. M. Teixeira, H. C. Fernandes, Monteiro, Paulo Marcos de Barros, D. E. Rodrigues and C. B. d. Alvarenga, Parameters of electrostatic spraying and its influence on the application efficiency, *Revista Ceres,* Vol. 60, 2013, pp. 474-479.

[42]. V. R. Mamidi, C. Ghanshyam, P. Manoj Kumar and P. Kapur, Electrostatic hand pressure knapsack spray system with enhanced performance for small scale farms, *J. Electrostatics,* Vol. 71, 2013, pp. 785-790.

[43]. D. Martin and J. Carlton, Airspeed and orifice size affect spray droplet spectrum from an aerial electrostatic nozzle for fixed-wing applications, *Appl. Eng. Agric.,* Vol. 29, 2013, pp. 5-10.

[44]. M. K. Patel, C. Ghanshyam and P. Kapur, Characterization of electrode material for electrostatic spray charging: Theoretical and engineering practices, *J. Electrostatics,* 2012.

[45]. G. Laryea, S. Kim and S. No, Depositional studies of a charged spray application in an orchard, https://www.researchgate.net/publication/266450505

[46]. M. Mokeba, D. Salt, B. Lee and M. Ford, Simulating the dynamics of spray droplets in the atmosphere using ballistic and random-walk models combined, *J. Wind Eng. Ind. Aerodyn.,* Vol. 67, 1997, pp. 923-933.

[47]. P. Miller and B. Ellis, Effects of formulation on spray nozzle performance for applications from ground-based boom sprayers, *Crop Protection,* Vol. 19, 2000, pp. 609-615.

[48]. D. Giles and E. Ben-Salem, Spray droplet velocity and energy in intermittent flow from hydraulic nozzles, *J. Agric. Eng. Res.,* Vol. 51, 1992, pp. 101-112.

[49]. M. Ellis, C. Tuck and P. Miller, The effect of some adjuvants on sprays produced by agricultural flat fan nozzles, *Crop Protection,* Vol. 16, 1997, pp. 41-50.

[50]. B. Ellis, C. Tuck and P. Miller, How surface tension of surfactant solutions influences the characteristics of sprays produced by hydraulic nozzles used for pesticide application, *Colloids Surf. Physicochem. Eng. Aspects,* Vol. 180, 2001, pp. 267-276.

[51]. J. Marchant, A. Dix and J. Wilson, The electrostatic charging of spray produced by hydraulic nozzles: Part I. Theoretical Analysis, *J. Agric. Eng. Res.,* Vol. 31, 1985, pp. 329-344.

[52]. N. Toljic, K. Adamiak, G. Castle, H. H. Kuo and H. C. Fan, Three-dimensional numerical studies on the effect of the particle charge to mass ratio distribution in the electrostatic coating process, *J. Electrostatics,* Vol. 69, 2011, pp. 189-194.

[53]. N. Toljic, K. Adamiak and G. Castle, Determination of particle charge to mass ratio distribution in electrostatic applications: A brief review, in *Proceedings of the ESA Annual Meeting on Electrostatics,* 2008.

[54]. Y. Ru, H. Zhou and J. Zheng, Design and Experiments on Droplet Charging Device for High-Range Electrostatic Sprayer, in Pesticides in the Modern World - Pesticides Use and Management, Margarita Stoytcheva (Ed.), *InTech*, 2011.

[55]. K. D. Kihm, B. Kim and A. McFarland, Atomization, charge, and deposition characteristics of bipolarly charged aircraft sprays, *Atomization and Sprays,* Vol. 2, 1992.

[56]. J. Wilson, A linear source of electrostatically charged spray, *J. Agric. Eng. Res.,* Vol. 27, 1982, pp. 355-362.

[57]. G. Laryea and S. No, Spray characteristics of charge injected electrostatic pressure-swirl nozzle, *Zaragoza,* Vol. 9, 2002, pp. 11.

[58]. R. S. Sasaki, M. M. Teixeira, H. C. Fernandes, P. M. Monteiro, D. E. Rodrigues and C. B. Alvarenga, Effect of space on droplets electrical charge during electrostatic spraying, in *Proceedings of the Power and Machinery. International Conference of Agricultural Engineering: Agriculture and Engineering for a Healthier Life (CIGR-AgEng'12)*, Valencia, Spain, 8-12 July 2012, pp. P-1864, ref. 9.

[59]. D. Nuyttens, K. Baetens, M. De Schampheleire and B. Sonck, Effect of nozzle type, size and pressure on spray droplet characteristics, *Biosystems Engineering,* Vol. 97, 2007, pp. 333-345.

[60]. K. Asano, Electrostatic spraying of liquid pesticide, *J. Electrostatics,* Vol. 18, 1986, pp. 63-81.

[61]. S. C. Cooper and S. E. Law, Bipolar spray charging for leaf-tip corona reduction by space-charge control, *IEEE Transactions on Industry Applications*, 1987, pp. 217-223.

[62]. S. E. Law and A. G. Bailey, Perturbations of charged-droplet trajectories caused by induced target corona: LDA analysis, *IEEE Transactions on Industry Applications*, 1984, pp. 1613-1622.

[63]. S. E. Law, Electrical interactions occurring at electrostatic spraying targets, *J. Electrostatics,* Vol. 23, 1989, pp. 145-156.

[64]. S. Carlsen, N. H. Spliid and B. Svensmark, Drift of 10 herbicides after tractor spray application. 1. Secondary drift (evaporation), *Chemosphere,* Vol. 64, 2006, pp. 787-794.

[65]. S. Edward Law, Agricultural electrostatic spray application: a review of significant research and development during the 20th century, *J. Electrostatics,* Vol. 51, 2001, pp. 25-42.

[66]. T. M. Wolf, R. Grover, K. Wallace, S. R. Shewchuk and J. Maybank, Effect of protective shields on drift and deposition characteristics of field sprayers, *Canadian Journal of Plant Science,* Vol. 73, 1993, pp. 1261-1273.

[67]. N. Thompson and A. Ley, Estimating spray drift using a random-walk model of evaporating drops, *J. Agric. Eng. Res.,* Vol. 28, 1983, pp. 419-435.

[68]. A. Hewitt, Spray drift: impact of requirements to protect the environment, *Crop Protection,* Vol. 19, 2000, pp. 623-627.

[69]. K. Baetens, D. Nuyttens, P. Verboven, M. De Schampheleire, B. Nicolaï and H. Ramon, Predicting drift from field spraying by means of a 3D computational fluid dynamics model, *Comput. Electron. Agric.,* Vol. 56, 2007, pp. 161-173.

[70]. K. Baetens, Q. Ho, D. Nuyttens, M. De Schampheleire, A. Melese Endalew, M. Hertog, B. Nicolaï, H. Ramon and P. Verboven, A validated 2-D diffusion–advection model for prediction of drift from ground boom sprayers, *Atmos. Environ.,* Vol. 43, 2009, pp. 1674-1682.

Index

www.ingramcontent.com/pod-product-compliance
Lightning Source LLC
Chambersburg PA
CBHW080120220326

41598CB00032B/4900